Conference on Fundamental and Applied Science for Advanced Technology (ConFAST 2019)

AA007023

Journal of Physics: Conference Series Volume 1373

Yogyakarta, Indonesia
21 - 22 January 2019

Editors:

Damar Yoga Kusuma
Afroz Ahmad Shah
Abhishek Kumar
Darsono Bahrun
Hanik Humaida
Qonitatul Hidaya
Nursyiva Irsalinda
Iwan Tri Riyadi Yanto
Joko Purwadi

ISBN: 978-1-7138-1130-5
ISSN: 1742-6588

Printed from e-media with permission by:

Curran Associates, Inc.
57 Morehouse Lane
Red Hook, NY 12571

Some format issues inherent in the e-media version may also appear in this print version.

Institute of Physics
Dirac House, Temple Back
Bristol BS1 6BE UK

Phone: 44 1 17 929 7481
Fax: 44 1 17 920 0979

techtracking@iop.org

Additional copies of this publication are available from:

Curran Associates, Inc.
57 Morehouse Lane
Red Hook, NY 12571 USA
Phone: 845-758-0400
Fax: 845-758-2633
Email: curran@proceedings.com
Web: www.proceedings.com

TABLE OF CONTENTS

MARITIME, OCEANIC, AND ATMOSPHERIC SCIENCE

SUSTAINABLE ENERGY AND SMART MATERIALS

ENVIRONMENTAL SCIENCE AND DISASTER MANAGEMENT

INTELLIGENT SYSTEMS IN INFORMATICS AND COMPUTING

Journal of Physics: Conference Series **1373** (2019) 011001 doi:10.1088/1742-6596/1373/1/011001

Committee

Editor in Chief

Asst. Prof. Damar Yoga Kusuma, Ph.D. (Universitas Ahlamd Dahlan)

Editors

Asst. Prof. Afroz Ahmad Shah, Ph.D. (Universiti Brunei Darussalam)
Dr. Abhishek Kumar (Solar Energy Research Institute of Singapore)
Prof. Darsono Bahrun (Center for Science and Accelerator Technology)
Dr. Hanik Humaida (Center for Volcanology and Geological Hazard Mitigation)
Qonitatul Hidayah, M.Sc. (Universitas Ahlamd Dahlan)
Nursyiva Irsalinda, M.Sc. (Universitas Ahlamd Dahlan)
Iwan Tri Riyadi Yanto, M.I.T (Universitas Ahlamd Dahlan)
Joko Purwadi, M.Sc. (Universitas Ahlamd Dahlan)

Scientific Committee

Prof. Dahelmi – Universitas Andalas (Scopus ID: 56152808500)
Prof. Darsono Bahrun – National Agency for Nuclear Power (Scopus ID: 55367047200)
Prof. Dyah Aryani Perwitasari – Universitas Ahmad Dahlan (Scopus ID: 36114858900)
Prof. Eti Yerizel – Universitas Andalas (Scopus ID: 57190031424)
Prof. Hariyadi – Universitas Ahmad Dahlan (Scopus ID: 8833969200)
Prof. John A. Carver – Australian National University (Scopus ID: 7102839510)
Prof. Khrishna P. Candra – Universitas Mulawarman (Scopus ID: 57196435267)
Prof. Lee Pooi See – Nanyang Technological University (Scopus ID: 7406117427)
Prof. Maizatul Akmar binti Ismail – University Malaya (Scopus ID: 25825070200
Prof. Setyawan P. Sakti – Universitas Brawijaya (Scopus ID: 6507450797)
Assoc. Prof. Anton Yudhana – Universitas Ahmad Dahlan (Scopus ID: 36660093000)
Assoc. Prof. Budi Purnama – Universitas Sebelas Maret (Scopus ID: 16029748100)
Assoc. Prof. Fajar Adi Kusumo – Universitas Gadjah Mada (Scopus ID: 24480624100)
Assoc. Prof. Jason J. Jung – Chung-Ang University, South Korea (Scopus ID: 7402896966)
Assoc. Prof. Kitichate Sridith – Prince of Songkla University (Scopus ID: 6506521909)
Assoc. Prof. Kuwat Triyana – Universitas Gadjah Mada (Scopus ID: 8597558700)
Assoc. Prof. Lydia Helena Wong – Nanyang Technological University (Scopus ID: 12041027200)
Assoc. Prof. Maelita R. Moeis – Institut Teknologi Bandung (Scopus ID: 35776885000)
Assoc. Prof. Moh Edi Wibowo – Universitas Gadjah Mada (Scopus ID: 54975678100)
Assoc. Prof. Nurkhasanah – Universitas Ahmad Dahlan (Scopus ID: 56610264900)
Assoc. Prof. Raju K. Gupta – Indian Institute of Technology Kanpur (Scopus ID: 37009709600)
Assoc. Prof. Rino A. Nugroho – Universitas Sebelas Maret (Scopus ID: 57190759643)
Assoc. Prof. Risa Suryana – Universitas Sebelas Maret (Scopus ID: 14523651300)
Assoc. Prof. Suparman – Universitas Ahmad Dahlan (Scopus ID: 6507107541)
Assoc. Prof. Thomas Reindl, Solar Energy Research Institute of Singapore (Scopus ID: 6603154919)
Assoc. Prof. Yan Jian – Hefei University of Technology (Scopus ID: 55262372600)

Journal of Physics: Conference Series **1373** (2019) 011002 doi:10.1088/1742-6596/1373/1/011002

Foreword: The 2019 Conference on Fundamental and Applied Science for Advanced Technology (ConFAST 2019)

We are pleased to present you the Proceeding of The 2019 Conference on Fundamental and Applied Science for Advanced Technology (ConFAST 2019). Held in Yogyakarta on January 25th 2019, the conference is organized by Universitas Ahmad Dahlan (UAD) and supported by local and international partners such as Center for Science and Accelerator Technology (PSTA-BATAN), Center for Volcanology and Geological Hazard Mitigation (BPPTKG), Prince of Songkla University (PSU), Universiti Brunei Darussalam (UBD), and Solar Energy Research Institute of Singapore (SERIS).

The 2019 ConFAST themed "Bridging the Gap: from Academic Research to Industrial Applications", aimed to strengthen the society with advanced science and technology in the wake of industrial revolution 4.0 in two folds: (1) Empowering the sci-tech-based enterprize by adopting advanced science and technology to compete globally and, (2) Imparting meaningful purpose on academic research through research to solve a real problems encountered by societies and industries.

The 421 contributors from academics, government agencies and industries, who contribute 189 abstracts on the field of maritime, energy and advanced materials, disaster mitigation to intelligent systems, truly make the conference a unique platform for a productive, interdisciplinary exchange of ideas. Out of 189 paper presented, 54 manuscripts passed the academic peer review process and are included in this proceeding volume.

The success of this conference cannot be achieved without the involvement from numbers of individuals. We would therefore, express our sincere gratitude to all of the organizing committees, scientific committees, keynote speakers, presenters, participants, and all contributors who have put their finest effort to make this conference a big success. Finally, I hope all participants have gained a rewarding experience from the event.

Sincerely,
Damar Yoga Kusuma, Ph.D.
Editor in Chief
website: http://confast.org
email: confast@uad.ac.

Table of Content

The 2019 Conference on Fundamental and Applied Science for Advanced Technology (ConFAST 2019)

	Nasori	Comparative Study of p-type Photocathode CuBi2O4 Films and CuBi2O4 Nanopillars Photocathode for High Performance Photoelectrochemical Water Splitting
	Rakino	A Passive Cooling System for Increasing Efficiency of Solar Panel Output
	Rafiqoh	Effect of Temperature in Ionic Liquids Pretreatment on Structure of Lignocellulose from Corncob
	Sanjaya	Effect of Premium-Butanol Blends on Fuel Consumption and Emissions on Gasoline Engine with Cold EGR Systems
	Sinaga	Experimental and Computational Study on Heat Transfer of a 150 kW Air Cooled Eddy Current Dynamometer
	Sinaga	Numerical Simulation of the Width and Angle of the Rotor Blade on the Air Flow Rate of a 350 kW Air-Cooled Eddy Current Dynamometer
	Surya	Enhancements in Cure Rate Index and Mechanical Properties of Silica-Filled Natural Rubber Using Octadecanol-Fatty Alcohol
	Surya	The cure and tensile properties of montmorillonite-natural rubber composites in the incorporation of alkanolamide
	Syahbardia	Study of Desiccant Rice Husk Ash for Air Conditioning System
	Tarigan	The Effect of Dust on Solar PV System Energy Output Under Urban Climate of Surabaya, Indonesia
	Tresnawati	Sensor Fabrication for Determination of Cadmium Using Nafion/Ionic Liquid/Graphene Composite with Thick Film Technology
	Triwikantoro	The Crystallization Process of Zr-Based Metallic Glass by the Influence of Heating Rate and Alloy Composition
	Utomo	The Real Time Vital Sign Monitor for Heart Rate and SPO2 Parameter Using Internet of Things Technology
	Wustha	Thickness and Resistivities of Cu/Ni Film Resulted by Electroplating on the Various Electrolyte Temperature
D. Environmental Science and Disaster Management	Danardono	Potential of Carbon Stocks and Its Economic Values in Tropical Karst Landscape (Case Study in Biduk-Biduk Karst, East Kalimantan, Indonesia)
	Lestari	Earthquake Risk Reduction Study with Mapping an Active Fault at the Southern of East Java
	Muriyatmoko	Landslide Prone Maps Using Weighting Parameters in Pulung, Ponorogo East Java
E. Intelligent Systems in Informatics and Computing	Afnia	The Valuation of Deposit Insurance with Risk Using Fourier Transform
	Albab	Forecasting of Vannamei Shrimp Production Based on Weather Factors Using Radial Basis Function Neural Network Approach (Case Study: Lamongan District)
	Ardiansyah	Determination Of Location And Numbers Of Monorail Stops In Surabaya With Max Covering Problem Model
	Barus	Web-Based Application Applying Newton Gregory Forward Method for Prediction of Closing Stock Prices

	Cahyaningrum	Indonesian community welfare levels clustering using the fuzzy subtractive clustering (FCM) method
	Fadlil	Mushroom Images Identification Using Orde 1 Statistics Feature Extraction with Artificial Neural Network Classification Technique
	Fadlil	Mobile Forensics in Android Smartwatch
	Harahap	Implementation of Cloudflare Hosting for Speeds and Protection on the website
	Hidayati	Optimal Control of Dynamic IS-LM Business Cycle Model with Two Time Delays
	Irfan	Development of Question Answering System Based on Ontology in Indonesian Traditional Medicinal Plants
	Ma'arif	Application of Intelligent Search Algorithms in Proportional-Integral-Derivative Control of Direct-Current Motor System
	Meliana	Identification of Cyber Bullying by Using Clustering Methods On Social Media Twitter
	Nafi'Iyah	Fuzzy Features for Facial Shape Classification on Panoramic Dental Image
	Purwanto	Digital Forensic Challenges Integrity data Point-of- Sale (POS) Cloud-based Software as a Service (SaaS)
	Rahmawati	An Application of Stochastic Dominance with Truncated Normal Distribution on LQ45 Index
	Rosnafian	Stability Analysis on the Spread of The Ebola Virus Considering Changes in Human Behaviour and Climate (Case Study in the Democratic Republic of the Congo)
	Setiady	Design of Automatic Braking System Using Interval Type-2 Fuzzy Logic System
	Sugiyanto	Indonesian Financial Crisis Prediction Using Combined Volatility and Markov Regime Switching Model Based on Indicators of Real Interest Rate on Deposits and Lending Interest Rate/Deposit Interest Rate
	Suherman	Link quality decrement and spectrum crowded impacts to wireless local area network performances
	Sukmarini	Minimizing Total Cost in Integrated Scheduling of Production and Distribution Using Ant Colony Optimization
	Sunardi	Design and Development of Calculation System for Normalized Difference in Vegetation Index (NDVI) Using Landsat 8 Satellite Image
	Surya	Identification of Pekalongan Batik Images Using Backpropagation Method
	Triana	Exponential Distribution Parameter Estimation with Bayesian SELF Method in Survival Analysis
	Triwiyanto	Investigation of The Number of Features and Muscles for an Effective Hand Motion Classifier Using Electromyography Signal
	Uddin	Adaptive Trajectory Tracking Control System of Two-Wheeled Robot

	Yudhana	Development of Door Safety Fingerprint Verification Using Neural Network
	Yudhana	The Monitoring of Corn Sprouts Growth Using The Region Growing Methods

Journal of Physics: Conference Series **1373** (2019) 011004 doi:10.1088/1742-6596/1373/1/011004

Welcome Message from the Dean, Faculty of Applied Science and Technology (FAST)

On behalf of the faculty member of FAST, Universitas Ahmad Dahlan, we are delighted to welcome you to The 2019 Conference on Fundamental and Applied Science for Advanced Technology (ConFAST 2019). The Conference in this series has been successfully organized in the previous events starting from 2012 (ICGWBT), 2014 (ICGWBT) and 2016 (ConFAST). Since then, ConFAST has become a "trademark" scientific forum for our faculty to interact with local, regional and international partners.

I would like to express my sincere gratitude to the distinguished keynote speakers for sharing their expertise in our forum, especially for Prof. Setiawan from Universitas Brawijaya, Dr. Warsito P. Taruno from CTECH Laboratory, Prof. Afroz Ahmad Shah from Universiti Brunei Darussalam, Dr. Abhishek Kumar from Solar Energy Research Institute of Singapore, and Prof. Kitichate Sridith from Prince of Songkla University. I hope this conference can serve as initial step stone for a productive collaboration in the near future. Gratitude also goes to all presenters and participant for their active contribution to the conference, we hope to seeing you again in the next series of ConFAST 2021. Also, I would like to congratulate the organizing committee and editorial boards for their rigorous efforts in preparation, execution, and post publication of the conference.

Last but not least, I sincerely hope that all the presenters and participants will benefit from the technical contents, as well as genuine strategic networking forged through this conference.

Sincerely,
Imam Azhari, M.CS.
Dean, Faculty of Applied Science and Technology
Universitas Ahmad Dahlan

Peer review statement

All papers published in this volume of *Journal of Physics: Conference Series* have been peer reviewed through processes administered by the proceedings Editors. Reviews were conducted by expert referees to the professional and scientific standards expected of a proceedings journal published by IOP Publishing.

Journal of Physics: Conference Series **1373** (2019) 012001 doi:10.1088/1742-6596/1373/1/012001

Prediction system of carbon monoxide toxic gas monitoring using double moving average method

Eriansyah Hasibuan[1], H Harahap[1], S Hardi[1], S Suherman[1], F Fahmi[1,*]

[1]Department of Electrical Engineering, Universitas Sumatera Utara, Medan 20155, Indonesia

[*]Email: fahmimn@usu.ac.id

Abstract. Health information at this time is very essential to support the continuity improvement of human life. One of the causes of problems to health in humans is the air pollution. It is necessary to have a monitoring system to provide information on air pollution. In this research, a monitoring system for air pollutant gas, specifically Carbon Monoxide (CO) was developed and calibrated with a prediction system using the double moving average method. The prototype for monitoring CO gas used MQ7 sensor, connected with an Arduino microcontroller and GPRS module. Data readings and prediction results were then displayed in real time connected to the web server. The result of developed system were carried out in the KIM area Port Belawan I, by measuring using CO gas prototypes and sending data in real time to the web server in the internal time of 1 hour 13 minutes. We collected 44 log data from the measurement results of data received from CO gas prototypes. The time range of predictive data with real-time gas CO data is 4 minutes on average. The prediction results using the double moving average method have a maximum mean average percentage error MAPE of 3.39% and a minimum MAPE of 0.74%. While the maximum mean square error MSE is 5.84 and minimum MSE of 0.27.

1. Introduction

The problem of air pollution is a significant issue for our health. Air pollution can contain ozone (O_3^+), dust particles (PM10), nitrogen dioxide (NO_2), and sulfur dioxide (SO_2), and carbon monoxide (CO). Of these pollutants, carbon monoxide poisoning is considered as the most severe and deadly poisoning components in some countries. The effect of toxic CO on the body is mainly caused by a reaction between CO and hemoglobin (Hb). Hb forms molecules that cannot carry oxygen to various parts of the body, causing headaches, dizziness, nausea, vomiting, and so on. [1].

Air pollution is one of the essential steps to be monitored in remote monitoring system for hydroponic planting media [2] integrated pollution monitoring system for smart city [3] supporting the survival of living things. Therefore, the right method is needed to control and monitor air pollution. One method presented in this study is a combination of monitoring methods and prediction methods. The system help predicts CO gas in the future so that it can take the right step if the CO gas concentration is above the average threshold.

There are two main types of predictions, namely models (time series) and regression models (causal). The time series model is the sequence of data points, usually measured at consecutive times with uniform time spacing. Time series analysis is a method for analyzing time series data statistically to determine the characteristics of other data. Regression analysis is one method for determining causal relationships between variables [4]. There are different types of time series models, one of which is the Average model and its derivative of double moving average prediction model.

Previous research done by C.H. Cheng and S.F. Huang (2009) discussed the prediction of air quality monitoring systems, especially Ozone, by predicting moving average displacement methods. In this study the prediction accuracy was significant with the parameter $\alpha = 0.7$ resulting in Mean Absolute Percentage Error (MAPE) = 18.48% and Root Mean Square Error (RMSE) = 5.47 [5].

In this research, a monitoring system for air pollutant gas, specifically Carbon Monoxide (CO) was developed and calibrated with a prediction system using the double moving average method.

2. Material and methods

2.1. Carbon monoxide sensor (CO)
MQ 7 is a gas sensor used to detect carbon monoxide (CO) gas. The MQ-7 sensor is a sensor that has a high sensitivity to CO gas, and its calibration results are stable and durable. This sensor uses a 5VDC power supply and can measure CO 20 - 2000 ppm gas [1].

(a) (b) (c)

Figure 1. The physical form of (a) MQ-7 gas sensor, (b) Arduino Mega 2560 board, and (c) SIM800 module board

2.2. Arduino mega 2560
Arduino Mega 2560 is an Arduino-based microcontroller development board using the ATmega2560 chip. This Arduino Mega2560 board has a lot of I / O pins, some 54 digital I / O pins (15 of them are PWM), 16 analog input pins, 4 UART pins (serial hardware port).

2.3. SIM800 module
SIM800 is a quad-band GSM / GPRS module that works on GSM 850MHz, EGSM 900MHz, DCS 1800MHz and PCS 1900MHz frequencies .SIM800 has a GPRS class 12 / class10 multi-slot feature (optional) and supports GPRS CS-1, CS-2, CS-3 and CS-4 coding schemes. With a small configuration of 24 * 24 * 3mm, SIM800 can meet almost all requirements. Spaces used by applications, such as M2M, smart phones, PDAs, and other mobile devices [6]. The typical form of MQ-7, Arduino Mega 2560 board and SIM800 Module Board are as shown in Figure 1.

2.4. Cloud server
The cloud server is a paradigm of the existence of information, that is, information is permanently on the internet server and information is on the client temporarily. There are terms of cloud computing (cloud computing) in the cloud server. Client computers do not need to do heavy computing with the existence of a cloud computing system to run applications that are needed as well as computer clients do not need to

install the applications needed on each computer. Is available the two main components in the cloud server, the web server and database [7].

2.5. Double moving average model

The Moving Average method consists of a single and a double top. Everything has the same nature, used to estimate future conditions by calculating the average value of a time series. According to the characteristics of the moving average method, the longer the time interval, the better the stability of the trend if the shorter the time interval. It is also implied that the worse the stability of the trend that can be reflected (data that changes very up and down)[8]. To estimate future data using double-moving average by assuming that X = Variable value, t = period and has a past value of N, formula equation S '= single moving average as follows:

$$S'_t = \frac{X_t + X_{t-1} + X_{t-2} + \cdots + X_{t-N+1}}{N} \quad (1)$$

Assume that all moving averages (S') have been calculated. Double moving average (S") written as follows:

$$S''_t = \frac{S'_t + S'_{t-1} + S'_{t-2} + \cdots + S'_{t-N+1}}{N} \quad (2)$$

Furthermore, referring to the adjustment of a single moving average, with differences in looking for differences in the single and multiple moving averages equation a can be written as follows :

$$a_t = S'_t + (S'_t - S''_t) = 2S'_t - S''_t \quad (3)$$

Determine estimated trends from one time period to the next

$$b_t = \frac{2}{N-1}(S'_t - S''_t) \quad (4)$$

Regarding accuracy is an element to show goodness according to which in the end shows how far the prediction model is capable of producing data that has been known [9]. There are several methods to find the value of accuracy in predictions, with the error rate method looking for the percentage value of prediction errors with the equation:

$$PE_t = \frac{X_t - F_t}{X_t} \quad (5)$$

$$MAPE = \frac{\sum_{i=1}^{n} |PE_i|}{n} \quad (6)$$

To further test the accuracy of the predictions of the relative size of predictive values can be calculated with:

$$ME = \frac{\sum_{i=1}^{n} e_i}{n} \quad (7)$$

$$MSE = \frac{\sum_{i=1}^{n} e_i^2}{n} \quad (8)$$

3. System design

In the design of hardware we considered about how to assemble all the hardware that was used in simple and compact form. While software design we dealt with how to get data that has been sent from the connected hardware and as a display to monitor the data on the system then analyzed using a double moving average method to determine predictive value data for future carbon monoxide gas real time. The design of this system can be seen in Figure 2.

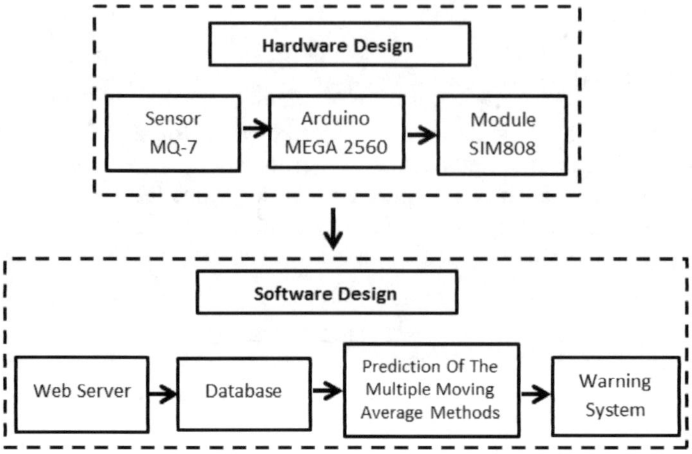

Figure 2. Design of carbon monoxide toxic gas forecasting system

3.1. CO gas sensor

MQ-7 As a CO gas sensor connected to Arduino Mega2560 which functions as a microcontroller and the SIM800 Module functions as the sender of data to the web server using a GPRS network. CO gas prototype design can be seen in Figure 3.

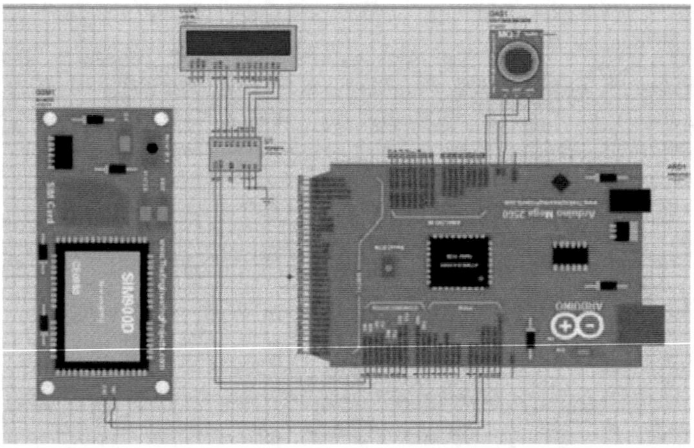

Figure 3. CO gas prototype

3.2. Web server design

In this study the address used to access the web server is https://sisteminformasico.000webhostapp.com/. The program for building a web server page uses HTML format, with data storage using the MySQL database.

4. Result

4.1. CO Gas Prediction Results

The testing procedure was carried out within 1 hour 13 minutes. Data starts to be analyzed after there are 17 measurement data that are displayed based on the latest data. The test group as a prediction sample is data for periods 8 to 17. CO Gas prediction Web server page can be seen in Figure 4.

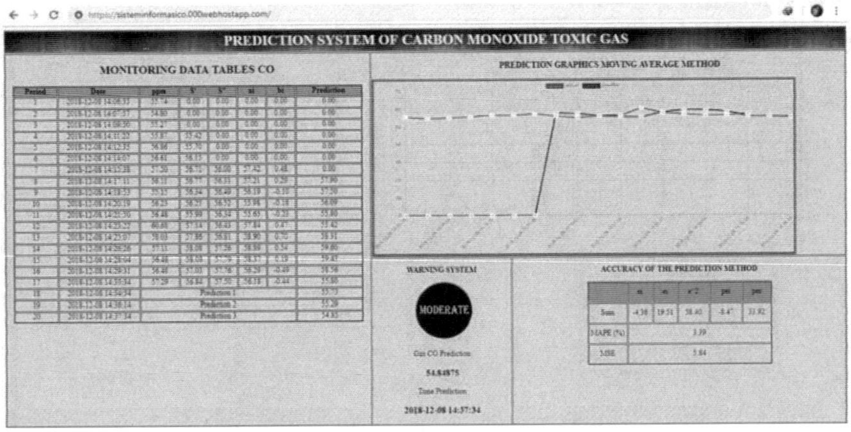

Figure 4. CO gas prediction web server page

As a sample calculation carried out in Table 1, the following below shows predictive systemic calculations in period 17 with values N = 4 and M = 1 :

$$S'_t = \frac{56{,}48 + 56{,}23 + 55{,}74 + 55{,}50}{4} = \frac{223{,}95}{4} = 55{,}99$$

$$S''_t = \frac{56{,}33 + 56{,}24 + 56{,}20 + 55{,}99}{4} = \frac{224{,}76}{4} = 56{,}19$$

$$a_t = (2 * 55{,}99) - 56{,}19 = 55{,}79$$

$$b_t = \frac{2}{4-1}(55{,}99 - 56{,}19) = 0{,}666 * -0{,}2 = -0{,}13$$

$$Prediction\ 1 = 56{,}12 + (-0{,}06)(1) = 55{,}06$$

$$Prediction\ I\ (Periode\ 18) = 55{,}79 + (-0{,}13)(1) = 55{,}65$$

$$Prediction\ 2\ (Periode\ 19) = 55{,}79 + (-0{,}13)(2) = 55{,}52$$

$$Prediction\ 3\ (Periode\ 20) = 55{,}79 + (-0{,}13)(3) = 55{,}38$$

The following below shows the systematic calculation of prediction errors in period 17 (Table 1) :

$$ei = X_i - F_i = 57{,}11 - 57{,}58 = -0{,}47$$

$$e_i^2 = -0{,}47^{\,2} = 0{,}22$$

$$PE_i = \left(\frac{X_i - F_i}{X_i}\right) x\ 100 = \left(\frac{57{,}11 - 57{,}58}{57{,}11}\right) x\ 100 = -0{,}82$$

$$|PE_i| = Absolute\ PE_i = 0{,}82$$

$$MAPE = \frac{\sum_{i=1}^{n} |PE_i|}{n} = \frac{8{,}54}{10} = 0{,}85\ \%$$

$$MSE = \frac{\sum_{i=1}^{n} e_i^2}{n} = \frac{3{,}39}{10} = 0{,}39$$

The test results were carried out on Saturday, December 8, 2018, at KIM Belawan, precisely at the intersection of Jl. The port starts from 13:20 until 14:33 as many as 44 CO gas data. The average delivery delay is 1 minute 30 seconds and the prediction time span 3 is from the actual measurement of 4 minutes. Figure 5 shows the results of the difference in predictions of 1 to 3 in each recent data measurement.

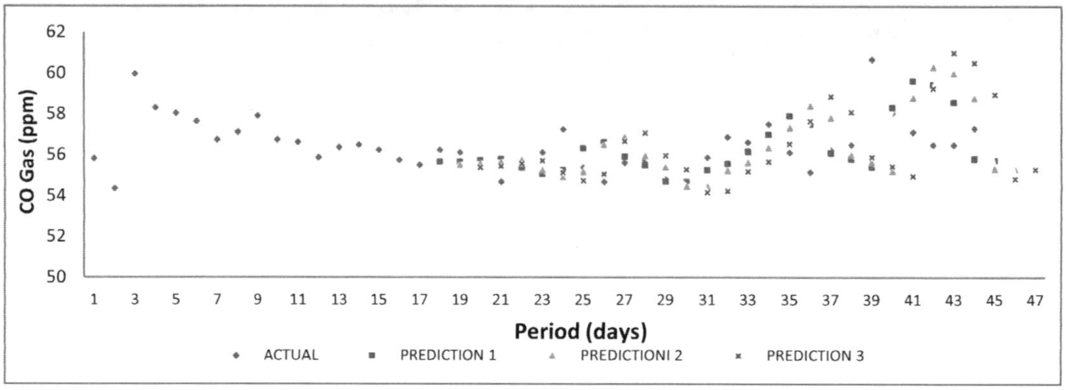

Figure 5. Graph of actual CO gas values and predictions

4.2. The accuracy of prediction results

(a)

(b)

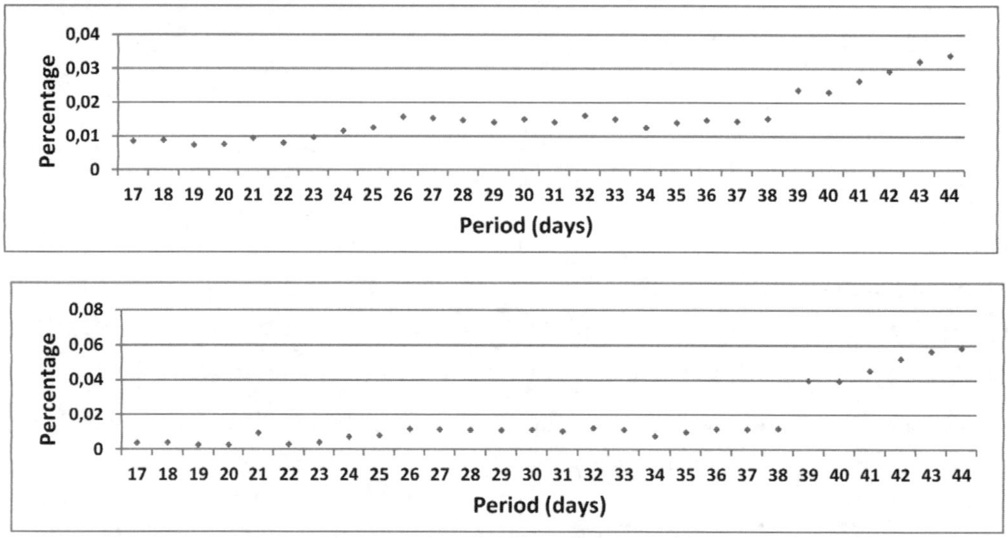

Figure 6. (a) Graph of mean absolute percentage error (MAPE), (b) Graph of mean squared error (MSE)

In this test, the error rate of predictive data was the result of monitoring of Carbon Monoxide Gas analysis with the equations of Mean Percentage Error (MAPE) and Mean Squared Error (MSE). In Figure 6.(a, the MAPE error rate with the results of testing of 44 data using the moving average method has a maximum error rate of 3.39% and a minimum error rate of 0.74%.

In Figure 6.(b), the MSE error rate with the results of testing of 44 data using the moving average method has a maximum error rate of 5.84 and a minimum error rate of 0.27.

5. Conclusion

Test results carried out in the KIM area Port Belawan I, by measuring using CO gas prototypes and sending data in real time to the web server in a 1 hour 13 minute evaluation time. From the results of measurement of data derived from CO gas prototypes as much as 44 data. The time span of the data predicts with real-time gas CO data averaging 4 minutes. The results of the prediction use the moving average method which has a MAAP maximum of 3.39% and a minimum MAPE of 0.74%. While the maximum MSE is from 5.84 and MSE minimum of 0.27

6. References

[1] Karamchandani S, Gonsalves A, and Gupta D. *Pervasive monitoring of carbon monoxide and methane using air quality prediction.* In: Proceeding of the International Conference on Computing for Sustainable Global Development; 2016 Mar 16; New Delhi, India. New York: IEEE; 2016. p. 2498–2502.

[2] Siregar B, Efendi S, Pranoto H, Ginting R, Andayani U, and Fahmi F. *Remote monitoring system for hydroponic planting media.* In: Proceeding of the International Conference on ICT for Smart Society; 2017 Sep 18-19; Tangerang, Indonesia. New York: IEEE; 2017. p. 1–6.

[3] Siregar B, Nasution A B A, and Fahmi F.2016 *Integrated pollution monitoring system for smart city.* Proceeding of the International Conference on ICT for Smart Society; 2016 Jul 20-21; Surabaya, Indonesia. New York: IEEE 2016. p. 49–52.

[4] Bhardwaj R, and Pruthi D 2017 *Time series and Predictability analysis of Air Pollutants in Delhi.* Proceeding of the International Conference on Next Generation Computing Technologies; 2016 Oct 14-16; Dehradun, India. (New York: IEEE) p. 553–560.

[5] Cheng C H, and Huang S F. *Forecasting of ozone concentration using frequency MA-OWA Model.* In: Proceeding of the International Conference on Systems Man and Cybernetics; 2009 Oct 11-14;San Antonio, USA. New York: IEEE 2009. p. 3461–3466.

[6] *Datasheet SIM800 Hardware Design V1.09in* SIMCOM

[7] Amli M S, Yuliarto B, and Nugraha. *Jurnal Otomasi Kontrol dan Instrumentasi* **7** 1-8.

[8] Zhang Z, Jiang Z, Meng X, Cheng S, and Sun W. *Research on Prediction Method of API Based on The Enhanced Moving Average Method.* In Proceeding of the International Conference on Systems and Informatics; 2012 May 19-20; Yantai, China. New York: IEEE 2012. p. 2388–2392.

[9] Makridakis S, Wheelwright S C, and McGee V E 1983 *Forecasting: Methods and Applications 2nd ed.* (Hoboken NJ: Wiley) p 72-79.

Journal of Physics: Conference Series **1373** (2019) 012002 doi:10.1088/1742-6596/1373/1/012002

Characteristics of the Extreme Rainfall over Indonesian Equatorial Region based on the Madden-Julian Oscillation Index Data Analysis

Eddy Hermawan [1,a)], Siti Hairunnisa Norfahmi [2,h)], Arief Suryantoro[1,b)], Teguh Harjana[1,c)], Trismidianto[1,d)], Anis Purwaningsih[1,e)], Wiwiek Setyawati[1,f)], and Siti Azizah[1,g)],

[1]*Atmospheric Science and Technology Center of LAPAN, Indonesia*
[2]*Institute Teknologi Bandung, Indonesia*

Email: *a)eddy_lapan@yahoo.com b)ariefsurya61@gmail.com c)teguh.harjana@lapan.go.id d)trismidianto@lapan.go.id e)anispurwaningsih49@gmail.com f)wiwieksetyawati21@gmail.com g)visibilika@gmail.com*
h)hairunnisa_norfahmi@yahoo.com

Abstract. Indonesia Maritime Continent (IMC) is one the most important region at the equator which affected by many atmospheric phenomena, especially Madden-Julian Oscillation and Monsoon. Asian Winter Monsoon causes the increasing of rainfall in several places in Indonesia in December-January-February (DJF), and since MJO related to a propagation of convective cloud, somehow it affects the rainfall variability in equator, especially over Indonesia. Interaction of both phenomena simultaneously bring rainfall is increasing in Indonesia, especially in Western Part of Indonesia. Therefore, a research about the impact of MJO in rainfall especially in Bandung and other cities over Indonesia is important. This study aims to investigate impacts of MJO events on the rainfall variability in Bandung and other cities focusing when Asian Winter Monsoon (December-January-February/DJF) for period of DJF 2002/03 – DJF 2012/13 passing over Indonesia. The results show that from 10 MJO events in period DJF 2002/03 – DJF 2012/13, the event increased the rainfall continuously in phase 3, phase 4, and phase 5. In phase 3, the increasing happened in every station except Supadio in Pontianak with the total precipitation > 30 mm/day. The condition of rainfall in phase 4 fluctuates in every station, but with the dominant increasing in Equator. In phase 5, the decreasing of rainfall happened in North of the Equator and Equator, while in South of Equator, the rainfall in Cengkareng Station in Jakarta increased about 20% but decreased about -42% in Bandung Station.

1. Introduction

The geographical condition of the Indonesian Maritime Continent has resulted in Indonesia's territory having a very strong potential for convective activity [1]. Therefore, various atmospheric phenomena occur in Indonesia [2]. Atmospheric phenomena, including the MJO and Monsoon phenomena, affect the variability of Indonesian rainfall. Monsoons are changes in annual trade flows associated with variability in rainfall in each monsoon region [3]. The Asian-Australian Monsoon classification by [4] states that increased rainfall in Indonesia generally occurs during the Asian Winter Monsoon (DJF).

Whereas MJO is associated with the formation of convective clouds at the equator and has an intraseason pattern [5-7], thus it affecting the variability of rainfall around tropical regions including

Indonesia[8,9]. The western region of Indonesia is bordered by the Indian Ocean where connectivity is very intensive, so it has a direct response to the MJO phenomenon. Studies on MJO in the Indonesian Maritime Continent have been carried out [1,10,11] to the effect of MJO on cases of extreme rainfall in Jakarta [12,13]. The results of these studies show varied results regarding the impact of MJO on rainfall, including in the western region of Indonesia. [8] and [14] agree that the "complex" Indonesian territory has resulted in the absence of a specific pattern regarding the state of rainfall in the Indonesian region during active MJO. Therefore, this study is intended as a further study to understand the impact of active MJO on rainfall in the western part of Indonesia during the Asian Winter Monsoon.

This study aims to analyze rainfall patterns in the western part of Indonesia using the rainfall data from BMKG station which is represented by rainfall in Cut Nyak Dien and Polonia Stations for the northen region, rainfall for Kototabang and Supadio Stations for the Equator region, and rainfall at Cengkareng and Bandung stations for the southern region; to identify MJO events in the DJF period 2002/03 to DJF 2012/13 in phases 3, 4 and 5; and analyze the influence of MJO in each phase on rainfall in the western region of Indonesia.

2. Data and Method

The data used includes rainfall data from TRMM satellite precipitation and BMKG observation rainfall represented by 6 stations: Cut Nyak Dien and Polonia Stations representing the North, Kototabang and Supadio Stations representing the Equator region, and Cengkareng and Bandung Stations representing the Southern region. It is estimated that rainfall patterns in the Equator region have an equatorial pattern, therefore zonal wind data from Equatorial Atmospheric Radar / EAR (Kototabang, 0.2°S, 100.32°E), RMM index data, Outgoing Longwave Radiation (OLR) data, and rainfall data are used.

While the used methodology of this study is analysing of regional rainfall characteristic which divided into the Northern part of Equatorial region, near the Equatorial region, and the Southern part of the Equatorial region, respectively. Then, analysing of Monsoon in the Equator region, and identification of MJO events using the RMM index and analysis of OLR anomaly using the Hovmoller diagram, before finally analysing of the influence of MJO on rainfall data (based on observation data and spatial data). The lower threshold used for spatial results is 5 mm/day, which is the lower limit of light rainfall according to BMKG (2014).

3. Results and Discussion

3.1. Precipitation Pattern Analysis

As shown in Figure 1, the rainfall in the North region shows that some areas have a monsoonal pattern, as shown in the results of the FFT of the Polonia Station rainfall. Others tend to have an equatorial pattern, as shown by composite and FFT results at Cut Nyak Dien Station. However, the average monthly rainfall in the Cut Nyak Dien region ranges from 250-350 mm, which indicates rainfall in the DJF period in the region. Moreover, the 12-month oscillation peak (monsoonal) seen in the rainfall at Cut Nyak Dien station, which shows the Cut Nyak Dien region is also affected by the Monsoon phenomenon.

Generally, based on composite results, the rainfall of the Equator region tends to have rainfall characteristics in the equatorial pattern. Aside the results of the FFT showing a 12-month (monsoonal) oscillation pattern, the monsoon effect in the Equator region can also be identified through the results of zonal wind wavelets in Kototabang as shown in Figure 2. Whereas the South region shows a clear monsoonal pattern so that the southern region can be said to be influenced by an annual oscillation pattern such as the monsoon as shown in Figure 3.

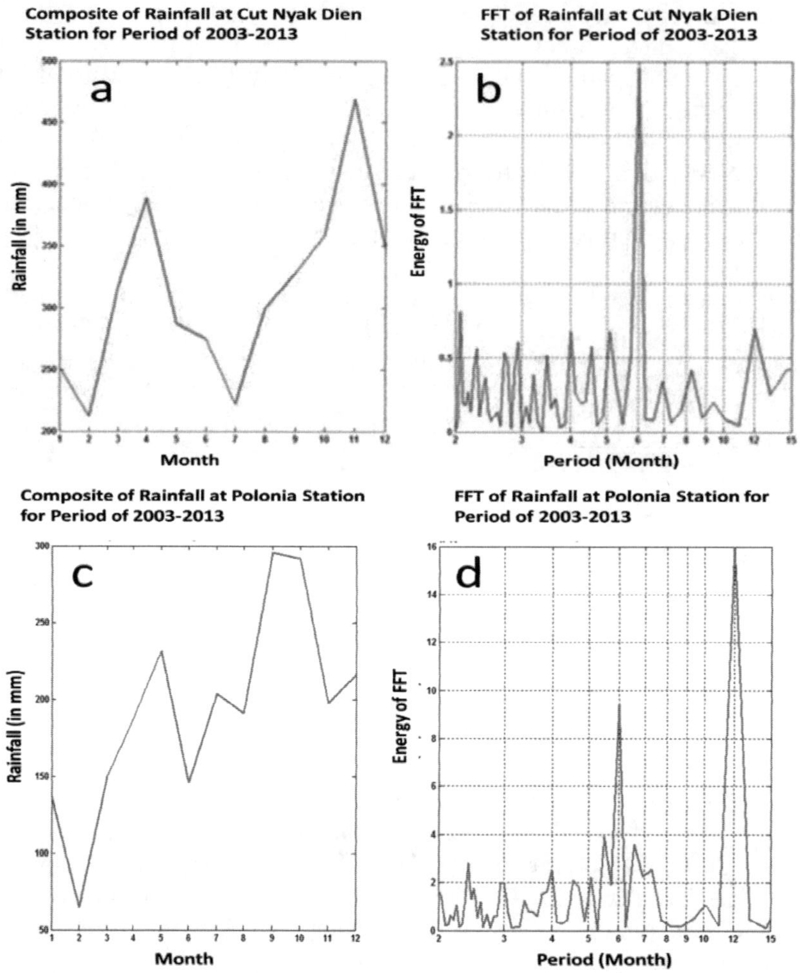

Figure 1. The Composite and FFT of Rainfall over the Northern part of the Equatorial Region of Indonesia for Cut Nyak Dien (represented as (a) and (b)), and Polonia (represented as (c), and (d), respectively)

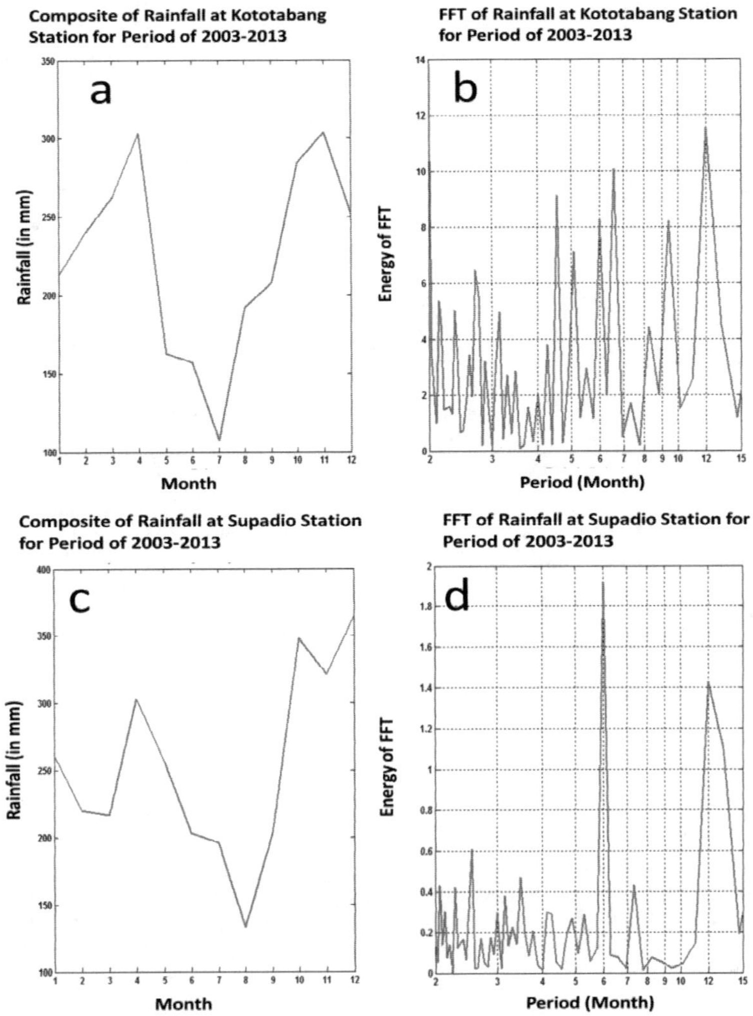

Figure 2. As the same as Fig. 1, but for near Equatorial line for Kototabang (represented as (a) and (b)), and Supadio (represented as (c), and (d), respectively)

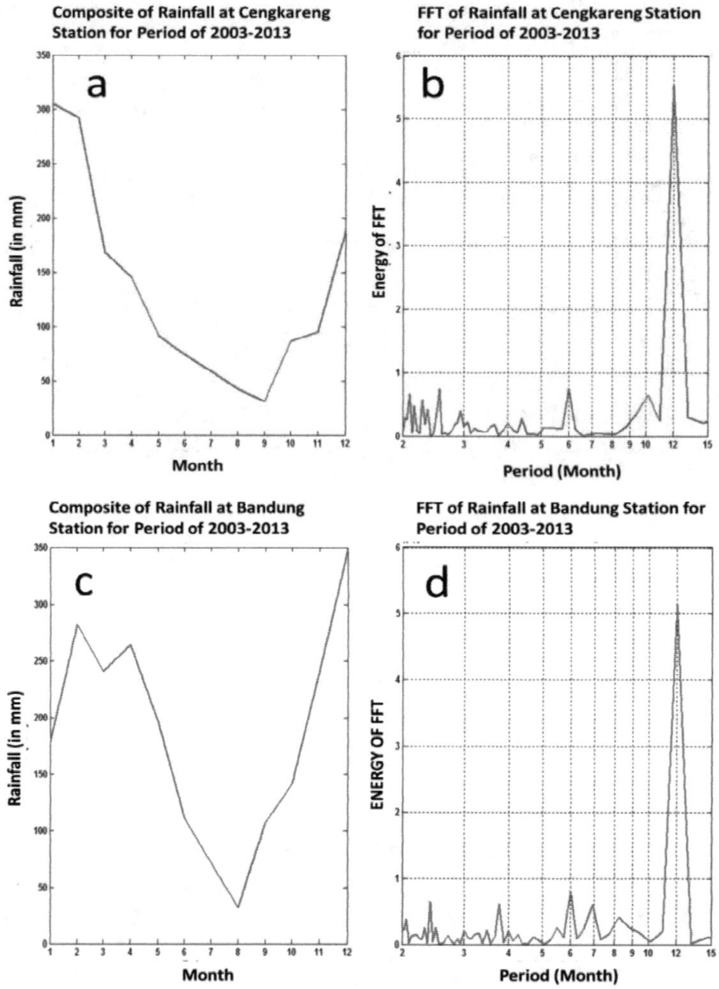

Figure 3. As the same as Fig. 1, but for Southern of Equatorial line for Cengkareng (represented as (a) and (b)), and Bandung (represented as (c), and (d), respectively)

3.2. Analysis of Monsoonal Pattern using Wind Data from EAR over Western part of Indonesia

The monsoonal pattern in the Western Region of Indonesia is shown by the results of wavelet analysis on both wind zonal and meridional wind data in the Lower Level Troposphere (~ 850 hPa) and in the Upper Troposphere (~200 hPa). The result, (Figure 4) shows that there is a 12 monthly dominant oscillation pattern in each wavelet result.

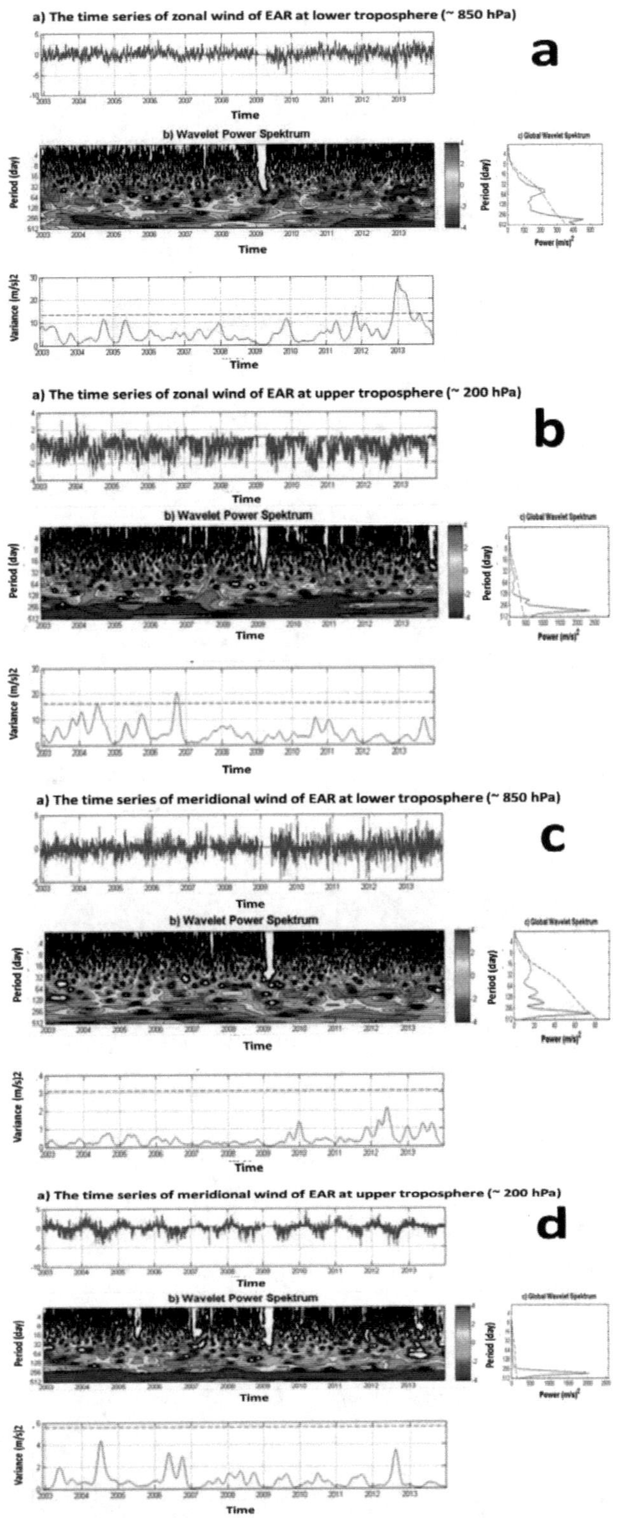

Figure 4. Wavelet plot of wind data taken from Equatorial Atmospheric Radar (EAR) at Kototabang, West Sumatera that showing the zonal and meriodional wind of EAR for lower and upper atmosphere (represented as 850 and 200 hPa, respectively)

MJO Identification

An identification of overall MJO events using the RMM index time series for the period December 2002 - December 2013 is shown in Figure 5. There were 10 MJO events in the DJF 2002/03 - DJF 2012/13 period (Table 1).

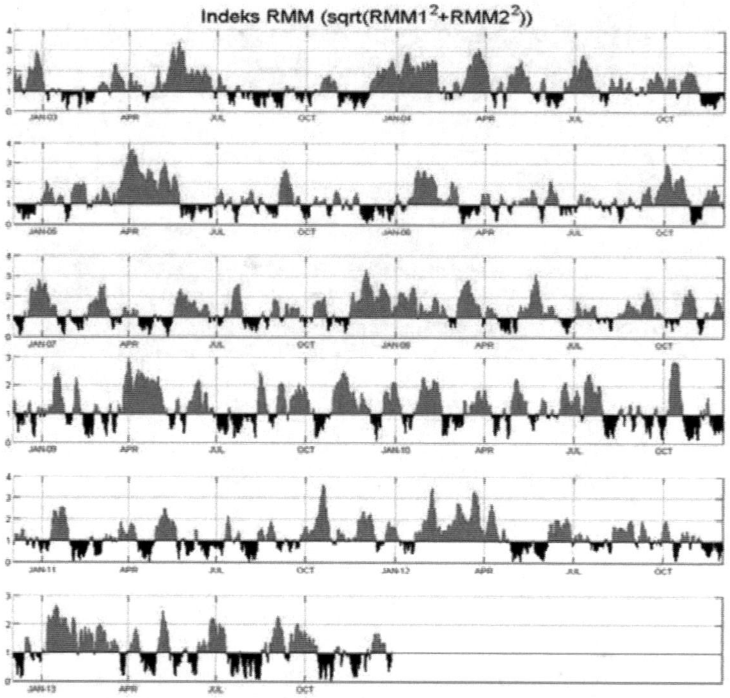

Figure 5. Time Series of RMM Index started from 2003 to 2013

Table 1. Active MJO events during Asian Winter Monsoon in DJF 2002/03 – DJF 2012/13

MJO	Periods	Phase 3		Phase 4		Phase 5	
		Date	Duration	Date	Duration	Date	Duration
1	DJF 2002/03	23 Dec - 27 Dec	5 days	28 Dec - 31 Dec	4 days	-	-
2	DJF 2003/04	7 Dec - 13 Dec	7 days	14 Dec - 16 Dec	3 days	17 Dec - 22 Dec	6 days
3	DJF 2003/04	27 Jan - 31 Jan	5 days	1 Feb - 7 Feb	7 days	8 Feb - 11 Feb	4 days
4	DJF 2005/06	11 Jan - 14 Jan	5 days	15 Jan - 17 Jan	3 days	18 Jan - 23 Jan	6 days
5	DJF 2006/07	17 Dec - 29 Dec	13 days	30 Dec - 1 Jan	3 days	2 Jan - 8 Jan	7 days
6	DJF 2007/08	18 Dec - 21 Dec	4 days	22 Dec - 26 Dec	5 days	27 Dec - 4 Jan	9 days
7	DJF 2007/08	1 Feb - 3 Feb	3 days	4 Feb - 8 Feb	5 days	9 Feb - 12 Feb	4 days
8	DJF 2011/12	-	-	-	-	20 Jan - 26 Jan	7 days
9	DJF 2012/13	31 Dec - 1 Jan	2 days	2 Jan - 6 Jan	5 days	7 jan - 10 jan	4 days
10	DJF 2012/13	15 Feb - 16 Feb	2 days	17 Feb - 22 Feb	6 days	23 Feb - 24 Feb	2 days

Contribution of Active MJO during Asian Winter Monsoon to Precipitation

Table 2. Rainfall during active MJO and Asian Winter Monsoon in DJF 2002/03 – DJF 2012/13

Stations	Precitpitation during MJO		Phase 3	Phase 4	Phase 5
Cut Nyak Dien	Average Percentage (%)		77	70	-38
	Frequency	Above normal	6	6	1
		Normal	0	0	0
		Below normal	3	3	8
Polonia	Average Percentage (%)		26	-19	-84
	Frequency	Above normal	4	3	0
		Normal	0	0	0
		Below normal	4	5	7
Kototabang	Average Percentage (%)		49	22	-54
	Frequency	Above normal	4	4	1
		Normal	0	0	0
		Below normal	5	5	8
Supadio	Average Percentage (%)		2	47	-50
	Frequency	Above normal	3	5	1
		Normal	2	1	1
		Below normal	4	3	7
Cengkareng	Average Percentage (%)		63	16	20
	Frequency	Above normal	3	5	5
		Normal	0	0	1
		Below normal	6	4	3
Bandung	Average Percentage (%)		47	2	-42
	Frequency	Above normal	4	1	0
		Normal	3	1	4
		Below normal	2	7	5

The Badan Meteorologi, Klimatologi dan Geofisika (BMKG) observation data showed an increase in rainfall in nearly all of the western part of Indonesia when MJO in phase 3. The rainfall then declined in phase 4 and phase 5.

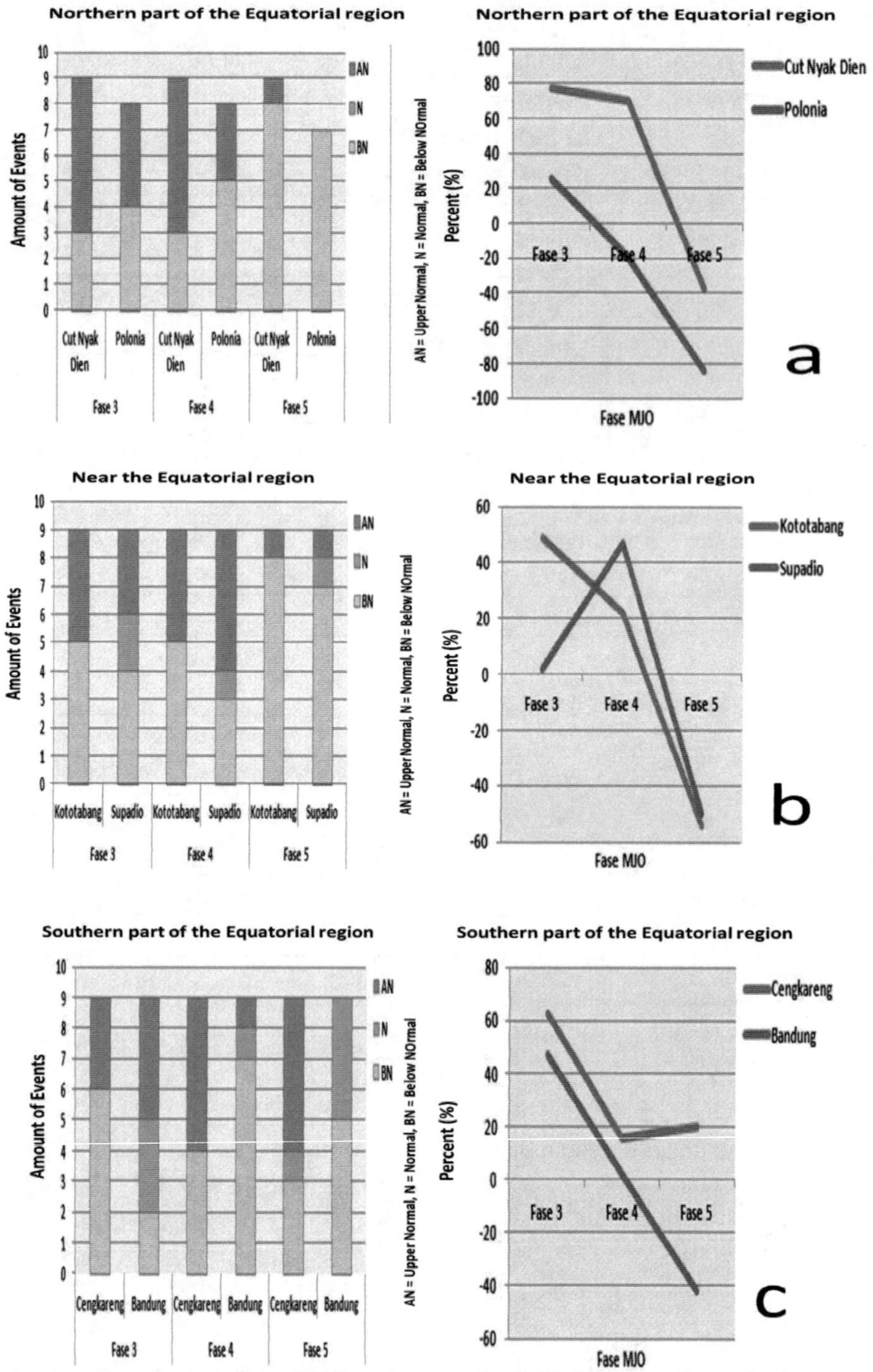

Figure 6. Rainfall during active MJO and Asian Winter Monsoon in DJF 2002/03 – DJF 2012/13 in Northern part of the Equatorial region (a), Near the Equatorial region (b), and Southern part of the Equatorial region (c), respectively

In the Northern area, consistently, an increase in rainfall is seen in the Cut Nyak Dien region reaching 77% when MJO phase 3 with the frequent number of above normal rainfall events. It means that the probability of increasing rainfall in the Cut Nyak Dien region when MJO in phase 3 is greater is around 6/9 or 66%. Likewise with the Polonia region with an increase in rainfall of 26% but with a probability of increasing and decreasing rainfall ≈ 50%. In the Equatorial Area, phase 3 of MJO increases rainfall in the Kototabang area by 49% but with a probability of increasing and decreasing rainfall ≈ 50%, in the other hand rainfall in the Supadio region shows normal condition. In the Southern Region, increased rainfall also occurs in both the Cengkareng and Bandung regions, but with the different probability of increasing and decreasing rainfall. In Cengkareng, the probability of decreasing rainfall is 66%, whereas in Bandung probability of increasing rainfall is ≈ 50%.

During phase 4 of MJO, the increasing of rainfall in the Northern Region occurred in the Cut Nyak Dien area reaching 70%, however in the Polonia region it decreased by -19%. In the Equatorial Area, an increase in rainfall occurred at 22% in the Kototabang region and 47% in the Supadio region. However, the probability of increasing rainfall in the Equatorial Area is ≈ 50% in both the Kototabang and Supadio regions. In the Supadio region, the increase in rainfall in phase 4 was significant compared to phase 3 and phase 5. When MJO phase 4, in the Southern Region, the rainfall conditions tend to be normal (16% in the Cengkareng area and 2% in the Bandung area) with a probability of decreasing rainfall more often occurs at 77% in the Bandung area.

In phase 5, the consistency of decreasing rainfall tends to occur in the North and Equatorial Areas. In the South, an increase in rainfall of 20% is seen in Cengkareng with a probability of increasing rainfall ≈ 50%. Whereas in the Bandung region, there was a decrease in rainfall of -42% with a probability of a decrease in rainfall of ≈ 50%.

Overall, the spatial data of average precipitation (Figure 6) shows a similar pattern with the results of observations, namely an increase in rainfall in almost the entire West Indonesia when MJO in phase 3 then decreases in phase 4 and phase 5. Consistently, precipitation occurs in almost all regions in the western part of Indonesia during MJO phase 3. The ocean area appears to have greater precipitation than the land area. In the coastal region of Sumatra island bordering the Indian Ocean, precipitation ranges from 15 - 30 mm / day. Likewise, with the ocean area around the islands of Java and Kalimantan. Mainland in Java seems to have considerable precipitation around 10-30 mm / day compared to Sumatra and Kalimantan (5 - 25 mm / day). In phase 4, several regions in the western part of Indonesia began to show precipitation conditions <5 mm / day. In the Northern Region and the Equatorial Area shows precipitation ranges from 5-15 mm / day. Whereas in the South, precipitation reaches> 30 mm / day. while the Equator and North regions, ranging from 5 - 20 mm / day. In phase 5, the South Region experiences precipitation ranging from 5-20 mm / day, while the Northern Region and the Equatorial Region of the western region of Indonesia show precipitation <5 mm / day.

Active MJO tends not to occur during the Australian Summer Monsoon Season (JJA). Figure 7 shows that the active MJO event in the JJA 2003 - JJA 2013 period starts in the transition season I (MAM) and ends around the beginning of June.

Journal of Physics: Conference Series **1373** (2019) 012002 doi:10.1088/1742-6596/1373/1/012002

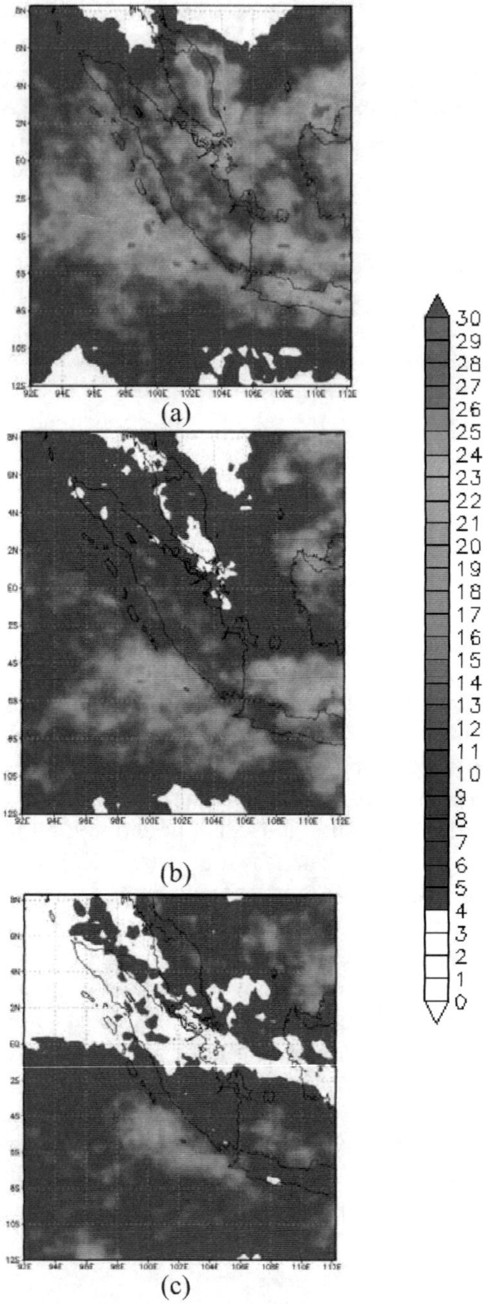

Figure 7. Average rainfall during phase 3 (a), 4 (b), and 5 (c) during active MJO and Asian Winter Monsoon in DJF 2002/03 – DJF 2012/13

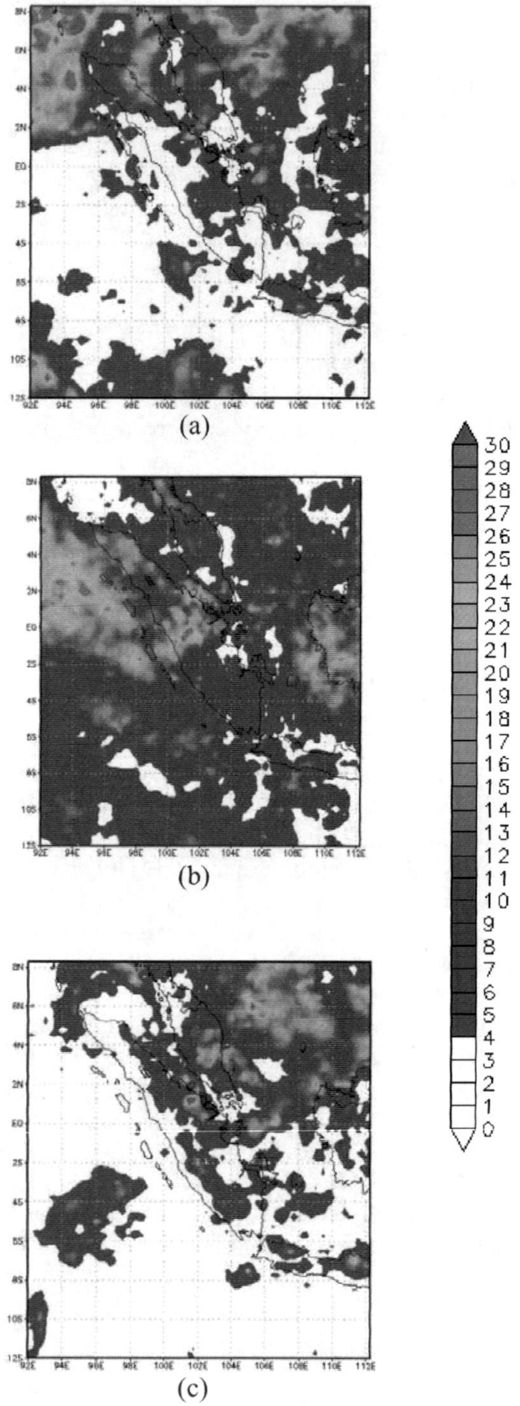

Figure 8. Average rainfall during phase 3 (a), 4 (b), and 5 (c) during active MJO and Australian Summer Monsoon in DJF 2002/03 – DJF 2012/13. *MJO inactive in early June

4. Conclusion

Based on the results of the study, it can be concluded that of the 10 MJO events in the DJF 2002/03 - DJF 2012/13 period, rainfall increased when MJO phase 3 , then it continued to decline in phase 4 and phase 5. The influence of Monsoon was seen through an increase in rainfall in the Western Region of Indonesia (it was seen during MJO phase 3, before entering the Western Region of Indonesia). The condition of rainfall varies when MJO phase 4 (currently in the Western Region of Indonesia), and tends to decrease when MJO phase 5 (after passing through the Western Region of Indonesia). In phase 3 there was an increase in rainfall (except Supadio area); in Cut Nyak Dien region by 78%, at Polonia Station at 25%, at Kototabang Station at 49%, at Cengkareng Station 63%, and at Bandung Station 47% with precipitation reaching> 30 mm / day. Rinfall in phase 4 shows quite fluctuating results. In the Northern Territory rainfall increased 70% in the Cut Nyak Dien area but decreased by -19% in the Polonia region. In the Equatorial Area, an increase in rainfall occurred at 22% in the Kototabang region and 47% in the Supadio region. In the Southern Region, the condition of rainfall tends to be normal around 16% in the Cengkareng area and 2% in the Bandung area. In phase 5, the decrease in rainfall is seen in the Northern Region (38% Cut Nyak Dien Station, Polonia Station -84%) and the Equatorial Area (Kototabang Station -54%, Supadio Station -50%). In the Southern Region, rainfall in Cengkareng Station increased by 20%, and at Bandung Station decreased by -42%. For further research, more BMKG data is needed to see the effect of MJO on rainfall. It is also needed the use of other variables (humidity and wind) to analyze air mass transport.

References

[1] Hidayat R, Kizu S. 2010 *International Journal of Climatology* **30** 1816-1825.
[2] Hermawan E. 2011. "Peran MJO dalam membangkitkan Intensitas Monsun di Indonesia". Seminar Nasional Sains Atmosfer dan Antariksa (SNSAA) 2011
[3] Ramage C. 1971 Monsoon Meteorology. International Geophysics Series, Vol. 15, San Diego: Academic Press.
[4] Prawirowardoyo S. 1996. Meteorologi. Bandung: Penerbit ITB, pp. 85-86
[5] Madden R A and Julian P R. 1971. *Journal of the Atmospheric Sciences* **28** 702-708.
[6] Madden R A and Julian P R. 1972. *Journal of the Atmospheric Sciences* **29** 1109-1123.
[7] Lau K M and Chan P H. 1985. *Monthly Weather Review* **113** 1889–1909.
[8] Zhang C. 2005 *Rev. Geophys.* **43** RG2003.
[9] Zhang C. 2013 *Bull. Am. Meteorol. Soc.* **94** 1849-1870.
[10] Kamimera H, Shuici M, Yamanaka M D and Syamsudin F. 2012. *SOLA* **8** 111-114.
[11] Fujita M, Yoneyama K, Suichi M, Nasuno T. 2011. *Journal of the Meteorological Society of Japan* **89A** 317-330.
[12] Trilaksono N J, Otsuka S and Yoden S. 2011. *SOLA* **7** 193-196.
[13] Wu P, Arbain A A, Suichi M, Hamada J I, Hattori M, Syamsudin F and Yamanaka M D. 2013. *SOLA* **9** 79-83.
[14] Peatman S C, Mathews A J and Stevens D P. 2014 *Q. J. R. Meteorol. Soc.* **140** 814-825.

Journal of Physics: Conference Series **1373** (2019) 012003 doi:10.1088/1742-6596/1373/1/012003

Analysis of Radiosonde Temperature Profile Measurement in Pamengpeuk

S A Rahayu [1,*], E Maryadi[1], R Sunarya[1], G A Nugroho[1], L R Triani[1]

[1] Center for Atmospheric Science and Technology LAPAN, Jl. Dr. Djunjunan No.133 Bandung, Indonesia

[*]Email : soni.aulia@yahoo.com; soni.aulia@lapan.go.id

Abstract. Indonesia is a tropical maritime country that has high complexity in terms of atmospheric dynamics. Understanding of atmospheric conditions is very important to obtain valid knowledge and atmospheric models. This understanding can be obtained through adequate data support. Vertical atmospheric data is very limited so it is necessary to develop equipment to be able to measure and provide valid data. one of the equipment used to measure the atmosphere vertically is called radiosonde. The development of this radiosonde has been carried out by various parties, one of them by PSTA LAPAN. This development is focused on the antenna system and the accuracy of the temperature data received. The antenna that is designed is an Omni 5/8λ antenna integrated with LNA. Receiving radiosonde data using antenna up to a height of 36 km. The temperature data that has been obtained is validated using radiosonde Vaisala and Meisei data. The temperature analysis is done by using Mean Bias Error (MBE) and Root Mean Square Error (RMSE). After being compared with the reference data, the RMSE values of 1.2 and MBE 2.3 were obtained with a correlation of 0.9. From these results it can be seen that the temperature data obtained is in accordance with atmospheric conditions, so this temperature sensor can be used for vertical atmospheric measurements.

1. Introduction

Understanding of atmospheric conditions is very important to obtain valid knowledge and atmospheric models. This understanding can be obtained through adequate data support. Vertical atmospheric data is very limited so it is necessary to develop equipment to be able to measure and provide valid data. one of the equipment used to measure the atmosphere vertically is called radiosonde.

Radiosonde is one of the equipment that is flown using a balloon vehicle with a payload that can measure atmospheric parameters. There are atmospheric parameters that can be measured including pressure, humidity, temperature, carbon dioxide and others [1]. Vertical atmospheric observation is very much needed in the military, weather forecast and aviation world both at home and abroad [2]. Vertical atmospheric information is used to determine atmospheric conditions at each height level so as to minimize events caused by weather changes.

The radiosonde system can measure atmospheric parameters from the surface to a height of ± 35 km [3]. Along with the development of the times, meteorological changes are closely related to human life, and directly affect people involved in industry, agriculture, military, aerospace, etc. [4-7]. Radiosonde development can be done from various aspects, from the transmitter system, receiver system, radio communication system or sensor. This research discuss further about the development of temperature sensors focused on the accuracy of temperature data received by the radiosonde receiving system, so that this data can later be utilized in the field of weather observation and is used as a reference for measuring atmospheric parameter data going forward.

2. Data and Research Methods

This research uses field testing data on December, 1st and 2nd 2018 at the Center for Testing Technology and Space and Atmosphere Pameungpeuk, Garut West Java with coordinates 007^{0}38'46" S, 107^0 41'15" E at an altitude of ± 7 masl. This test aims to test the performance of home-made radiosonde against two other commercial radiosonde. The radiosonde that has been made while given the R-sonde symbol, this is to distinguish from the others.

Development of this R-sonde is carried out on the antenna system and the accuracy of the temperature data received. The antenna designed is an Omni 5/8λ antenna integrated with LNA. This paper will focus on the performance of the temperature sensors used compared to the reference temperature data. In the previous research, the development of temperature sensors has been carried out using the DS 1820 brand temperature sensor [8]. The results obtained are not good, it can be seen from the pattern of recorded temperature data not following the pattern of the reference data used. After the data has been further processed, a large RMSE value of 11.01 and MBE 7.43 [9]. Therefore this research is focused on the improvement of accuracy of the temperature data received by the radiosonde receiving system, so that this data can later be utilized in the field of weather observation and used as a reference for measuring atmospheric parameter data going forward.

The temperature sensor used is the Honeywell brand with specifications that can measure from -90^0 -s/d + 60^{0}C, accuracy of 0.5^{0}C [10]. The selection of these sensors is in accordance with the specifications of the measured minimum temperature at a certain height which is around -85^{0}C. R-sonde specification can be seen in Table 1 below.

Table 1. R-sonde specifications

Spesification	R-sonde
Frequency	433 MHz
Dimension	11x 8 x 9 (mm)
Output power	100 mW
Modulation	LORA
Boudrate	1200 bps
Temperature range	+60°C to -90 °C
Humidity range	0 to 100 %RH
Pressure range	1030.0 hPa to 5 hPa

The R-sonde test was carried out by comparing the measurement results from two other comparable radiosonde, namely the Japanese Meisei brand and Finland Vaisala brand. The radiosonde specifications of the two types of radiosonde can be seen in Table 2 and the images of the two brands can be seen in Figure 1.

Table 2. Specifications of Meisei and Vaisala [11,12]

Spesification	Meisei	Vaisala
Type	iMs-100	RS41 SG
Frequency	400 – 406 MHz	400.15 – 405.99 MHz
Dimension	53 x 55 x 131 (mm)	155 x 63 x 46 mm
Output power	< 100 mW	60 mW minimum
Modulation	Digital PCM	GFSK
Boudrate	1200 bps	4800 bps
Transmitter type	FM	Crystal stabilized
Temperature range	+60°C to -95 °C	+60°C to -90 °C
Humidity range	0 to 100 %RH	0 to 100 %RH
Pressure range	1050.0 hPa to 3 hPa	From surface pressure to 3 hPa

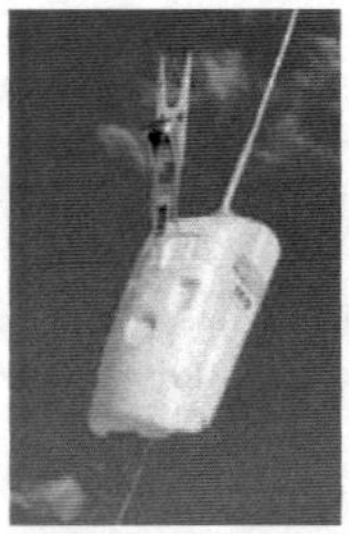

(a)

(b)

Figure 1. Payload Meisei (a) and Payload Vaisala (b) [11,12]

Illustrations from this test can be seen in Figure 2. This test uses a balloon vehicle where the balloon will carry 3 payloads consisting of Vaisala payload in Figure 2 no 1, Meisei in Figure 2 no 2 and R-sonde in Figure 2 no 3 which is placed in one cross section. The load of the three payloads that will be flown has almost the same weight so that the distance between the payload is set to ± 50 cm. This aims to prevent collisions between payloads when flown due to wind.

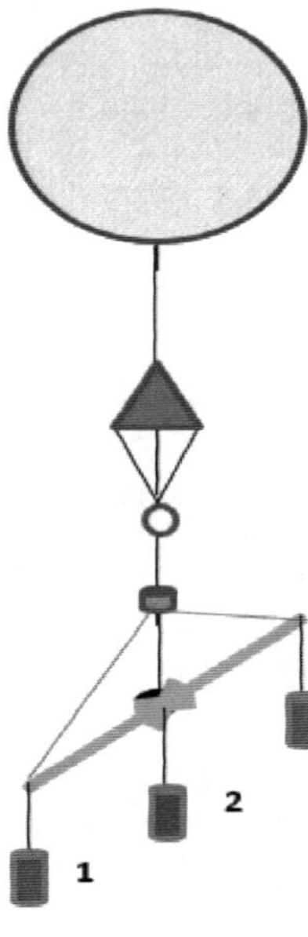

Figure 2. Illustration of the launch of the R-sonde with reference radiosonde (1) Vaisala, (2) Meisei, (3) R-sonde

After the balloon is flown, at each ground station each payload will receive data through the GUI from each receiving system. Data obtained in each GUI, are then processed to see the performance of the temperature sensor carried by the R-sonde. To see the performance of the temperature sensor carried by the R-sonde, the Root Mean Square Error (RMSE) method and Mean Bias Error (MBE) are used. RMSE shows the standard deviation value of the difference in data samples predicted by observation data [13]. This RMSE has been used as a standard statistical measure in testing the performance of models in meteorological, air quality and climatology studies [14]. Mathematical calculation of RMSE can be seen in Formula 1 [13], [15]:

$$RMSE = \sqrt{\frac{\sum_{t=1}^{n}(\hat{y}_t - y_t)^2}{n}} \quad (1)$$

MBE describes the direction of bias errors from the data being evaluated. MBE can be negative and can be positive [16]. The MBE calculation is mathematically like Formula 2 below [13],[15]:

$$MBE = \frac{1}{n}\sum_{t=1}^{n}(\hat{y}_t - y_t)^2 \qquad (2)$$

Where :

$\hat{y}_t$: Predicted data
y_t : Observation data ke -t
N : Amount of data

The three radiosonde data received by the receiving system will be processed using the RMSE and MBE methods. So that the results of the comparison of R-sonde values are obtained with reference data. It also aims to test the reliability of sensors that have been developed.

3. Results and Discussion

After the payload is flown by carrying all three payloads, namely Vaisala, Meisei and R-sonde data from all three can be received by each recipient system. This balloon flight is carried out 3 times, namely 2 times on December, 1st 2018 at 06.00 and 18.00 WIB then once on December, 2nd 2018 at 06.00 WIB. In each launch temperature data obtained with different heights depends on the stability of data communication between the transmitter and receiver and the eruption of the balloon.

At the first launch on December, 1st 2018 at 6:00 a.m, the altitude of 25.3 km was obtained, can be seen in Figure 3 (left). The second launch on December, 1st 2018 at 6:00 p.m. arrived at a height of 17.3 km in Figure 3 (center). While the third launch is on December, 2nd 2018 at 06.00 the data can be received as far as 36 km, in Figure 3 (right).

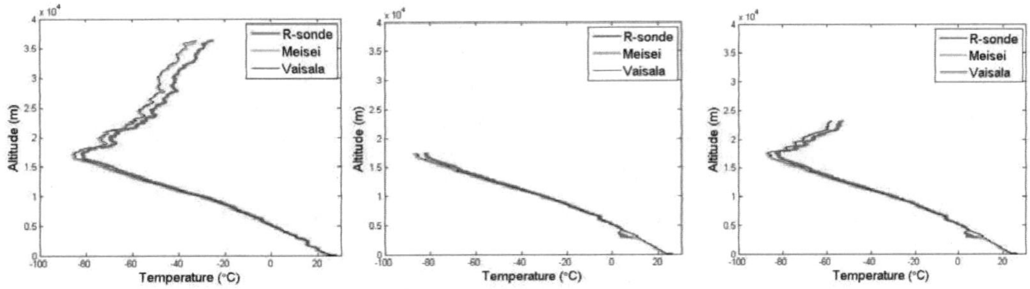

Figure 3. First (left), second (middle) and third (right) launch

Seen from the three launches, different heights were obtained due to the resistance of the balloon. Basically, balloon resistance is very dependent on the upper atmosphere. Where the more upward, the pressure will decrease, causing the balloon to grow bigger and faster to erupt.

In Figure 3 above, purple line is R-sonda data, blue line is Meisei data and black line is Vaisala data. Where the vertical axis is the height in meters (m) to a height of 40 km, while the horizontal axis is the unit temperature data ^{0}C with a range of -100^0 to 30^{0}C.

From the three launches, the vertical temperature pattern is the same, where the vertical temperature pattern of the R-sonda is close to the Vaisala and Meisei patterns. However, after reaching the lowest temperature, the R-sonda temperature data is higher than the two reference data. The lowest temperature value is at an altitude of 17.3 km with varying values, R-sonda data -81.3 ^{0}C, Meisei data -85.1 ^{0}C, and Vaisala data -84.3 ^{0}C.

Furthermore, the temperature data obtained is calculated mathematically by the RMSE value per altitude level using Formula (1). This aims to see how much the same temperature data obtained by the R-sonde with Vaisala data or Meisei data. In addition to calculating the RMSE value, the MBE value is calculated by using Formula (2). This MBE indicates bias from two observational data. This MBE can be positive and has a negative value depending on the observation data.

This research, the RMSE and MBE values will be analyzed from the R-sonde data with Vaisala data and R-sonde with Meisei data, as well as correlations obtained from the two data compared at each launch.

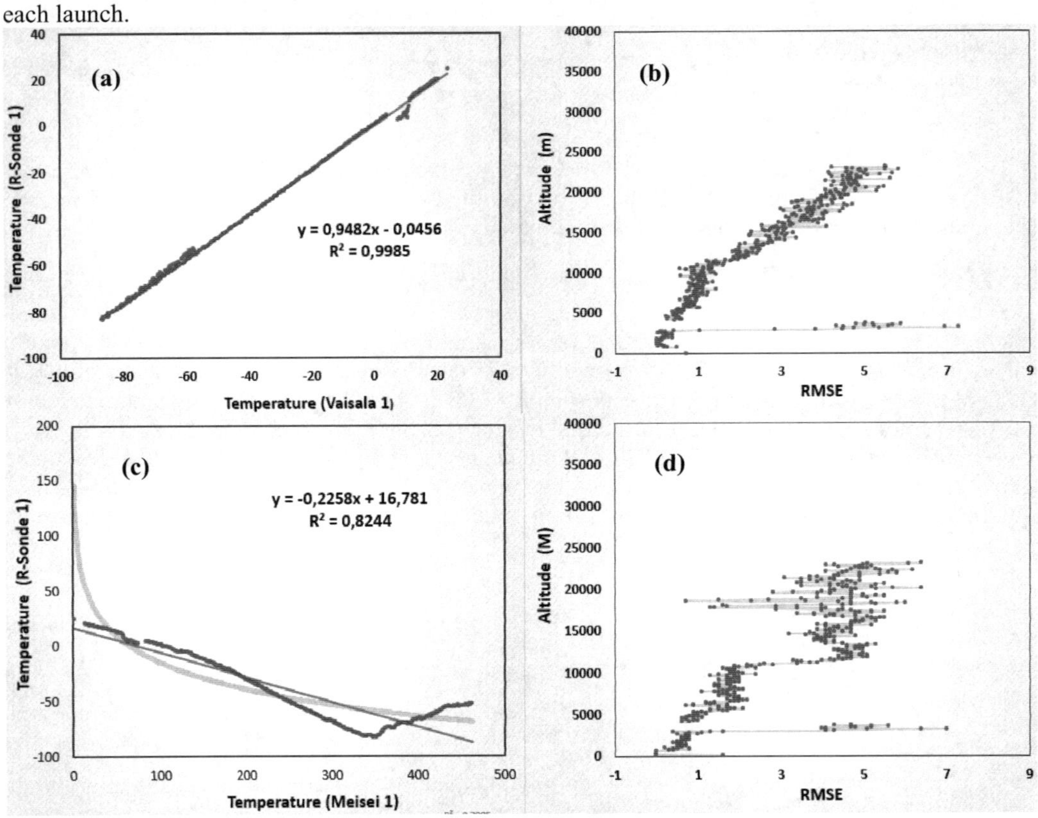

Figure 4. Data for the launch of December, 1st 2018 at 06.00, R-sonde with Vaisala (a and b) and R-sonde with Meisei (c and d)

In Figure 4a, 5a, 6a is the R-sonde launch data with Vaisala where the vertical axis is the R-sonde temperature and the horizontal axis of the Vaisala temperature. Figure 4b, 5b, 6b is the RMSE value per altitude level for R-sonde with Vaisala. Figure 4c, 5c, 6c are Meisei (horizontal) R-sonde (vertical) data. Figure 4d, 5d, 6d data results in the RMSE value per height level. However, the calculation results of each RMSE, MBE and correlation values can be seen in Table 3.

Table 3. Comparison of RMSE, MBE and correlation values

Date and Time	R-sonde - Vaisala			R-sonde - Meisei		
	RMSE	MBE	Correlation	RMSE	MBE	Correlation
December, 1st 2018-06.00	4.6	1.2	0.99	4.9	2.0	0.90
December, 1st 2018-18.00	1.7	-0.5	0.99	1.7	0.008	0.99
December, 2nd 2018-06.00	1.2	0.7	0.99	2.0	1.2	0.99

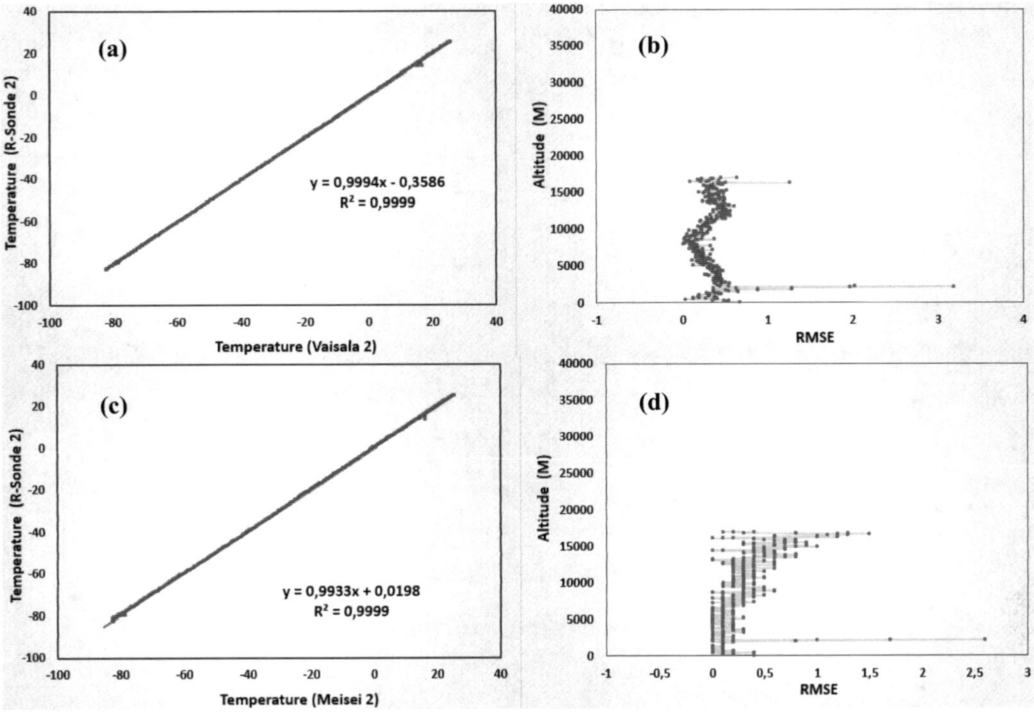

Figure 5. Data launch date December, 1st 2018 at 18.00, R-sonde with Vaisala (a and b) and R-sonde with Meisei (c and d)

From Table 3 it can be seen that the largest RMSE value on December, 1st 2018 at 06.00 both the ratio of R-sonde with Vaisala and Meisei is worth 4.6 and 4.9. The smallest value on December, 2nd 2018 at 06.00 with a value of 1.2. RMSE value from three experiments with six RMSE values obtained, from the calculation of the RMSE value is smaller. It means that the temperature value of the R-sonda when compared to Vaisala and Meisei has data that has no significant difference. So the standard deviation can be said to be small.

For the largest MBE value at the launch of December, 1st 2018 at 6:00 a.m., the R-sonde with the Meisei is worth 2.0. This MBE can be positive or negative. If it is positive it means that the launch value of the R-sonde is greater than the measured temperature value by Meisei. This can be proven by looking at Figure 7 (left and right). As for the launch of December, 2nd 2018, a negative MBE value was obtained which means that the temperature value of the R-sonda is worth less than the value of the Vaisala.

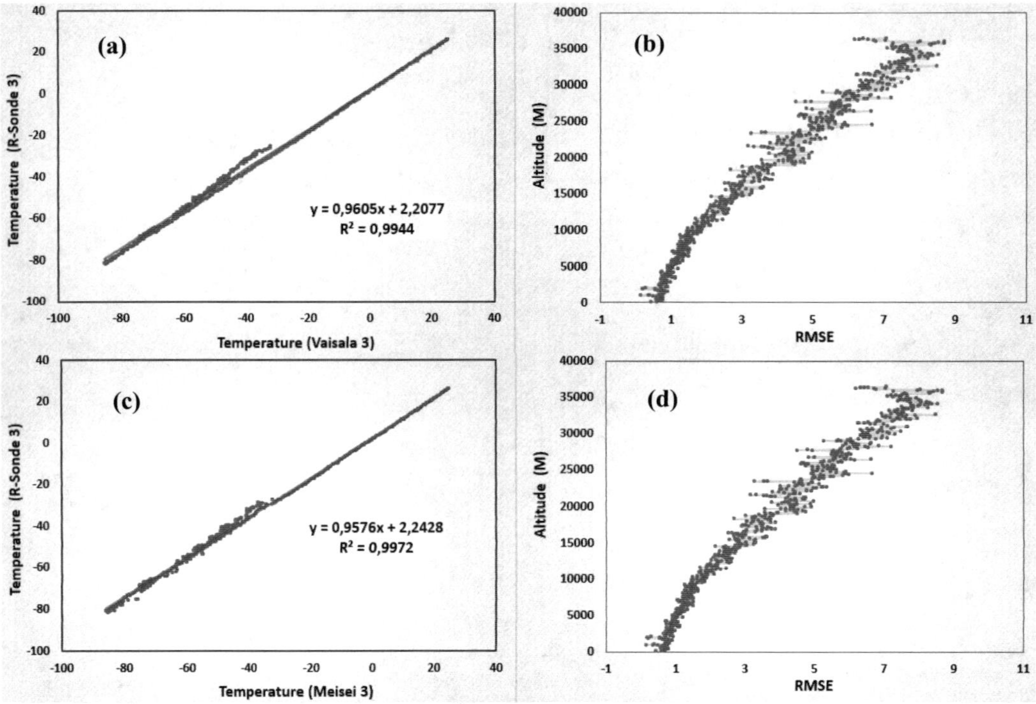

Figure 6. Data of launching December, 2nd 2018 at 06.00, R-sonde with Vaisala (a and b) and R-sonde with Meisei (c and d)

In general, the RMSE value obtained from the three launches has a good result which is marked by a small RMSE value. However, to see in more detail the difference in R-sonde data with reference data, it will be analyzed based on the temperature value when heading to the lowest point and the temperature value after the lowest point. The data analysis division is divided at an altitude of 0-17.3 km (Figure 7) and 17.3 to the maximum height (Figure 8). Where at an altitude below 17.3 km, the value of the R-sonde temperature data is almost the same as the reference temperature data value, whereas for above 17.3 km, the R-sonde value is greater than the reference data.

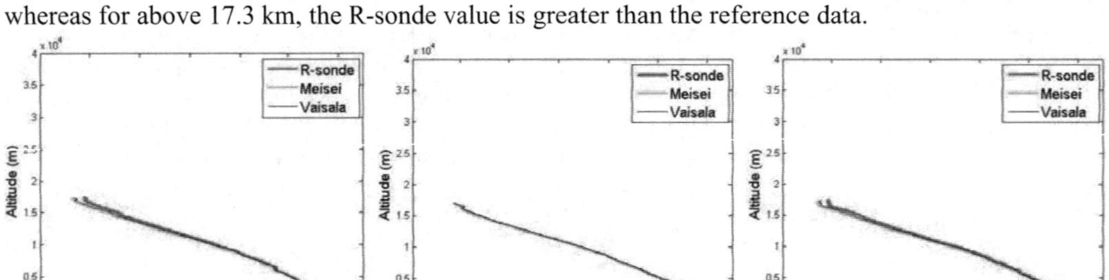

Figure 7. Temperature profile data to a height of 17.3 km

RMSE values for altitudes from 0-17.3 km can be seen in Table 4. From Table 4, the RMSE value is smaller than the previous RMSE. This means the standard deviation and the MBE value are also smaller.

Table 4. RMSE and MBE values at an altitude of 0-17.3 km

Date and Time	R-sonde - Vaisala		R-sonde - Meisei	
	RMSE	MBE	RMSE	MBE
December, 1st 2018 06.00	2.3	0.99	3.1	2.2
December, 1st 2018 18.00	0.4	-0.3	1.7	0.008
December, 2nd 2018 06.00	1.9	1.7	2.9	2.6

After knowing the RMSE value at an altitude of 0 - 17.3 km, it will now be analyzed at an altitude above 17.3 km. From the temperature profile on Figure 7 it can be seen that there are differences in the measured temperature values, it appears that the R-sonde temperature value is greater than the Vaisala and Meisei values. To see the diversity of data, the values of RMSE and MBE will also be analyzed.

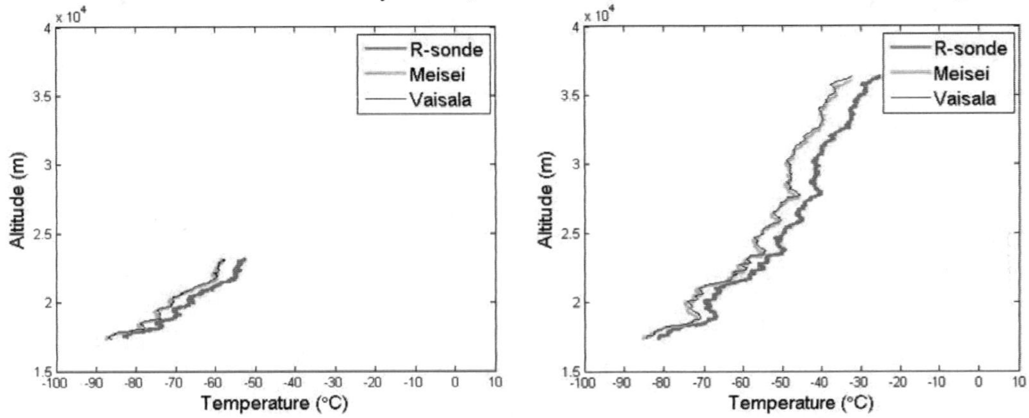

Figure 8. Temperature profile data at altitudes above 17.3 km

After calculating the value of RMSE and MBE can be seen in Table 5. The RMSE value at an altitude above 17.3 km looks larger than the RMSE value at an altitude below 17.3 km. This difference in value is very significant according to the profile graph in Figure 8. Likewise for the MBE value, it appears that the MBE value is positive, which means that the R-sonde value is greater than both the Vaisala and Meisei reference values.

Table 5. RMSE and MBE values at altitudes above 17.3 km

Date and Time	R-sonde - Vaisala		R-sonde - Meisei	
	RMSE	MBE	RMSE	MBE
December, 1st 2018 - 06.00	4.9	5.4	4.9	5.3
December, 1st 2018 - 18.00	-	-	-	-
December, 1st 2018 - 06.00	5.9	5.7	14.1	0.3

After analyzing measured temperature data from the R-sonde payload, as a whole it can be used as a reference for temperature measurement vertically. This is based on the results of the temperature sensor analysis used which has good RMSE and MBE values when compared to using Vaisala and Meisei. And the correlation that occurs between the temperature sensor in the R-sonde with Vaisala and Meisei has a correlation value of 0.9 This shows a strong correlation.

4. Conclusions

The development and performance comparison of radiosonde has been carried out from various aspects. From this research, the development of temperature sensors aimed at accurately obtaining temperature data, so that it can be used for measuring atmospheric parameters. Based on the results of data analysis obtained from the launch of R-sonde, Vaisala and Meisei radiosonde, it was found that the values of RMSE and MBE at altitudes below 17.3 km were smaller than those at the altitude above 17.3 km. This is in accordance with the visual analysis of the graph obtained. That the temperature value of the R-sonde at an altitude below 17.3 km has a pattern and a value that is confined to a pattern from Vaisala and Meisei. However, at an altitude above 17.3 km the measured temperature value is greater than the measured reference data (Vaisala and Meisei) value, after being analyzed, the RMSE and MBE values obtained are greater than the under-altitude of 17.3 km.

ACKNOWLEDGEMENTS

The authors are gratefull for the generous funding from the Ministry of Research, Technology and Higher Education, Republic of Indonesia. All authors also acknowledge the assistance from research team of PT. INTI, BMKG, and LAPAN consortium.

REFERENCES

[1] Ratnam M V, Pravallika N, Babu S R, Basha G, Pramitha M and Murthy B V K 2014 *Atmospheric Measurement Technique* **7** 1011-1025.
[2] Zhao S J, Su X Y, Gao T C 2012 *Meteorol. Sci. Technol.* **2** 31–33.
[3] Shi Y, Luo Y, Zhao W, Shang C, Wang Y and Chen Y 2013 *Sensors* **13** 8977–8996.
[4] Fuke M V, Adhyapak P V, Mulik U P, Amalnerkar D P, Aiyer R C 2009 *Talanta* **78** 590–595.
[5] Lv X, Li Y, Hong L, Luo D, and Yang M, 2007 *Sensors and. Actuators B: Chemical* **124** 347–351.
[6] Dessler A E, Sherwood S C 2009 *Science* **323** 1020–1021.
[7] Traversa E 1995 *Sensors and Actuators B: Chemical* **23**, 135–156.
[8] Dallas DS1820, http://pdf1.alldatasheet.com/datasheet-pdf/view/58548/DALLAS/DS1820.html accessed on date 9 Januari 2017
[9] Rahayu S A, Nugroho G A and Maryadi E 2017 Analisis Kinerja Sensor Suhu dan Tekanan dengan Menggunakan Wahana Balon, Dinamika dan Teknologi Atmosfer Benua Maritim Indonesia, ed S K Prabotosari et al (Bandung: Media Akselerasi) p. 59-71.
[10] Integrated RadioSonde Module SPI Interface Specification Document, Version: 1.0, Honeywell Confidential & Proprietary, available online from https://sensing.honeywell.com/index.php?ci_id=159835&la_id=1&No=0&Nrpp=50&Ntt=HTSL, accessed on February 10, 2018.
[11] Meisei GPS Radiosonde iMS100 product catalogue, accessed online from http://www.meisei.co.jp/english/products/ims-100-e.pdf , accessed on 9 January 2017.
[12] Vaisala Radiosonde RS41-D datasheet, accessed online from https://www.vaisala.com/sites/default/files/documents/RS41-D-Datasheet-B211609EN.pdf, accessed on 9 January 2017.
[13] Risyanto, Sipayung S B and Maryadi E 2014 Analisis Profil Suhu Vertikal Atmosfer dari Data MODIS Terra/Aqua di Wilayah Indonesia (*Prosiding Seminar Sains Atmosfer 2014*) LAPAN Bandung
[14] Chai T and Draxler R R 2014 *Geoscientific Model Development* **7** 1247–1250.
[15] Willmott C and Matsuura 2005 *Clim. Res.* **30** 79–82.
[16] Falayi E O and Rabiu A B 2011 *Archives of Physics Research*, **2** 99-109.

Journal of Physics: Conference Series **1373** (2019) 012004 doi:10.1088/1742-6596/1373/1/012004

Analysis of Drought Potential in Sumba Island until 2040 Caused by Climate Change

Sinta Berliana Sipayung[1,*], Indah Susanti[1], Edy Maryadi[1], Amalia Nurlatifah[1], Bambang Siswanto[1], Muhammad Nafayest[1], Fanny Aditya Putri[1], Eddy Hermawan[1]

[1] Center of Atmospheric Science and Technology, Indonesia National Institute of Aeronautics and Space (LAPAN)

*Email: s_berlianasipayung@yahoo.com

Abstract. Climate Change is one of the events that can change climate components due to global warming events. One of the changes in climate components can cause changes in rainfall in an area. While changes in rainfall conditions in an area can cause increased vulnerability of a region to potential of drought. In this research, the effects of climate change on the drought potential in Sumbawa Island region was investigated, where this region is one of the areas prone to drought in Indonesia. Climate projection until 2040 was carried out using the Conformal Cubic Atmospheric Model (CCAM) model with 14 km spatial resolution and monthly temporal resolution. While the drought index is calculated by the SPI (Standardize Precipitation Index) method where positive numbers tend to mean wet, lift between -1 to 1 is neutral, and increasingly negative numbers tend to mean dry. It is known that in 2040 January was known that there would be no drought on Sumbawa Island with values ranging from -1to 2. While in August of 2040, the potential for drought disaster should be watched out. This is because in this month the drought index value ranges from -2.5 to - 2. The potential for drought itself is known to increase in August 5 years until 2040 when compared with 5-year data, 2025 and 2030 in August in the Bima region. Meanwhile, based on time series analysis, it is estimated that there will be some extreme dry conditions, in August 2022, 2024, 2028, 2030, and 2033. Knowing these extreme conditions, it is hoped that this will become the basis for regional planning systems, starting from planning infrastructure and social systems to climate change mitigation and adaptation.

1. Introduction

As a unique Maritime Continent country that located at the equatorial region, Indonesia is surrounded by two great continents (Asia and Australia) and two great oceans (Indian and Pacific). Therefore, that Indonesia has two seasons namely wet and dry season. Based on these two seasons it will affect through meteorological parameters especially on rainfall, so that there are rainfall patterns in Indonesia, namely monsoon, equatorial, and local depends on its location and normally the dynamic of monsoon is similar. But with the issue of global warming packaged in IPCC [1] all the information has changed in terms of climate variability and has been reviewed from various models how the climate condition at the past, present and in the future. Based on a number of model results from various scenarios, the average global climate change will rise at rate of 0.03°C per year. Some models analyze that rainfall will increase continuously at the point between 2020 and 2060. Even world institutions including scientist collaborate to fight climate change, as concrete step is to overcome rising temperatures and reduce the increasing of CO_2 emissions.

Global warming in the last 100 years has shown that global average temperatures have increased by 0.748C (1906-2005) [2]. The increasing amount of rainfall and air temperature especially in Indonesia has been proven by previous researchers [3]. Rainfall and temperature are the most influential variables in the field of climate science and hydrology which are often used to determine the magnitude of climate change and variability [4][5]. Furthermore, dry season rainfall is significantly strongly correlated with South El Nino Oscillation (ENSO), while rainy season rainfall is weakly correlated with ENSO [6]. Water is scarce source in Indonesia which is seasonally affected by droughts and floods. Related to climate change, the importance of wetland to improve irrigation, household water, agriculture, farm and fishery as well as industrial water. Therefore, hydrological cycle is key factor in the atmospheric dynamics for long term.

The phenomenon of MJO (Madden Julian Oscillation) greatly affects weather and climate globally. One of global phenomenon that affects Indonesia's weather and climate is MJO, according to Madden & Julian [7]. Weather and climate with rainfall parameters in Indonesia generally affected by atmospheric circulation phenomenon both on a global, region and local scale. But for West Nusa Tenggara islands rainfall affected by orographic [8]. Projection result by using scenario model output data are known that there are has been rainfall increasing in Indonesia except in southern Indonesia [9][10][11]. Further projection results also mentioned that increasing annual rainfall on most islands, except in southern Indonesia, which decreased to 15%. Seasonal changes rainfall in parts of Sumatera and Kalimantan are around 10-30% wetter in 2080 during December-February period, while Jakarta is projected to be drier around 5 to 15% during June-August [12]. Rainfall increasing occur 0,2%-0,3% in tropic (10^oN, 10° S) and rain decreases in sub tropic (10° – 30° N) [5]. Based on region climate projections in four seasons there are changes in rainfall and rising temperatures in the West Nusa Tenggara region [13]. With that existence of climate change system will impact on other sectors especially in the hydrological system. Variability in rainfall on a year-to-year time scale provides useful for the management of water and agriculture resources while on a time scale of decades to century allows more informative strategic decision making for investment in infrastructure such as irrigation and water storage.

Climate change is a process of changing the pattern of world climate so that it can lead to erratic weather phenomenon, climate change can be characterized by changing climate variables such as temperature or rainfall that occurs in long periods between 50-100 years [12]. Climate condition itself is one thing that can affect many sectors. Therefore, climate change can affect many sectors in many places. One of them is climate change which resulted in the changing pattern of rainfall and also the water availability. The impact of climate change on the condition of water availability in an area such as the occurrence of extreme wet weather such as floods, or vice versa that is dry extremes that cause drought. Historical records of precipitation, streamflow and drought indices all show increased over many land areas [14]. Previous studies [15] show that changes in sea surface temperatures have large influences on land precipitation.

The impact of climate change on water sources in Indonesia can be significant due to increasing frequency of floods and droughts. In general, the impact of climate change shows that with increasing temperatures it has the potential to increase demand for water consumption and can causing decrease in soil water content so that is vulnerable to water resources. Variability in rainfall on a year-to-year time scale provides useful guidance for the management of water and agricultural resources while on a time scale of a decade to century allows more informative strategic decision making for investment in infrastructure such as irrigation and water storage. SPI is designed as meteorological drought index on region and temporal time scale.

The regional climate of the province of West Nusa Tenggara, eastern Indonesia was simulated using downscaled climate data [13] starting from historical to scenario data (1994-2005 and 2006-2060) 14 km resolution for islands of Lombok and Sumbawa. Generally, a resolution simulate 14 km CCAM produces rainfall, maximum and minimum temperature which similar to observation.

However, 14 km simulate has rain bias around 5 mm/day in December-February rainy season and lower bias in others season. Climate projections are examined for two future time intervals centered in 2006 and 2060. Rainfall change simulation by 2060 suggested both increase and decrease by 5% in December-February, with more acute decline of 10% in some areas and decreasing to 10% in March-May. For other season, generally small changes are simulated. Region temperature projected to increase around 10C in 2030 and 1.6–2 C by 2060. High resolution model output allows detailed differentiation between location across the islands. Result shows that because orographic effect there is abrupt climate gradient, producing significant local differences in climate projections [8]. Drought is natural phenomenon which is mainly caused by under normal rainfall over a long period. Sumbawa is the biggest island in West Nusa Tenggara with broad reach 15.448 km2. The aim of this research is to analyze the potential of drought occurrence in Sumbawa Island, Nusa Tenggara Barat due to climate change.

2. Data and Method

2. 1. Data

The data used in this research include:

2.1.1. GFS (Global Forecast System) Data

GFS data is used as input for CCAM model. This data has $1^{o}x1^{o}$ resolution containing various atmospheric parameters which later can be used as an initial condition for the CCAM model. Later this data can be downscaling with a more detailed resolution than $1^{o}x1^{o}$. In this study, the data is downscaling to a resolution of 0.125^{o} or about 14 km.

2.1.2. BMKG (Meteorology, Climatology, and Geophysics Agency) in-situ Data

In-situ data used to validate CCAM output. In-situ data used in this study is rainfall data at Bima Station.

2. 2. Method

Generally, the process of this research as follows:

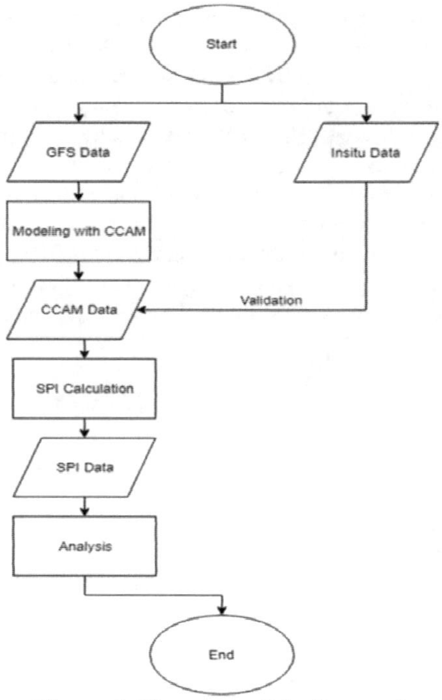

Figure 1. Flowchart of the Research

2. 2. 1. CCAM Modelling

CCAM (Conformal Cubic Atmospheric Model) is global atmospheric model developed for climate modelling and region climate downscaling [16][17]. This model applies quasi-uniform grids, or the principle of projecting the earth's surface into a cube panel [18]. Because of that, this model does not require lateral boundaries in the process of running it. This model uses an advection approach to lagrangian systems with horizontal bicubic interpolation [13]. The CCAM model is downscaling until the data has a spatial resolution of 0.125° and temporal resolution per month. In this research, the CCAM output of 14 km resolution used is rainfall data which will be processed and calculated the SPI value.

2. 2. 2. SPI Calculation

According to McKee (1993)[19], we can calculation of SPI by using the gamma distribution. The calculations are used as follows :

$$G(x) = \frac{\int_0^x x^{a-1} e^{(-x/\beta)}\, dx}{\beta^{\alpha} \Gamma(\alpha)}$$

By using the gamma function, SPI is obtained like the equation below

$$Z = SPI = -\left(t - \frac{c_0 + c_1 t + c_2 t^2}{1 + d_1 t + d_2 t^2 + d_3 t^3}\right); \quad 0 < H(x) \leq 0.5$$

$$Z = SPI = +\left(t - \frac{c_0 + c_1 t + c_2 t^2}{1 + d_1 t + d_2 t^2 + d_3 t^3}\right); \quad 0.5 < H(x) \leq 1.0$$

Where

$$t = \sqrt{\ln\left(\frac{1}{H(x)^2}\right)};\quad 0 < H(x) \le 0.5$$

$$t = \sqrt{\ln\left(\frac{1}{(1.0 - H(x))^2}\right)};\quad 0.5 < H(x) \le 1.0$$

$$c_0 = 2.515517$$
$$c_1 = 0.802583$$
$$c_2 = 0.010328$$
$$d_1 = 1.432788$$
$$d_2 = 0.189269$$
$$d_3 = 0.001308$$

Where the meaning of the SPI value is presented in table 1

Table 1. SPI Category [19]

SPI Value	Drought Category
$SPI \le -2$	Extremely dry
$-2 \le SPI \le -1.5$	Very dry
$-1.5 \le SPI \le -1$	Dry
$-1 \le SPI \le 1$	Normal
$1 \le SPI \le 1.5$	Wet
$1.5 \le SPI \le 2$	Really wet
$SPI \ge 2$	Extremely wetness

2. 2. 3. Pearson Correlation Calculation

In CCAM data validation with observation data used pearson correlation and RMSE. Pearson correlation is a measure of correlation or linear relationship between two variables. The values range from -1 to +1. Correlation values are calculated by the formula as follows :

$$r = \frac{n(\sum xy) - (\sum x)(\sum y)}{\sqrt{[n\sum x^2 - (\sum x)^2][n\sum y^2 - (\sum y^2)]}}$$

Where x is the observation data and y is the data model.

2. 2. 4. Bias calculation with RMSE (Root Mean Square Error)

RMSE is the average value of the number of squared errors, it can also state the size of the error produced by a forecast model. A low RMSE value indicates that the variation in value produced by a forecast model is close to the variation in the value of its observation [20]. The RMSE value is formulated as follows [21] where fi is the forecast produced by the model and Oi is the observation value. :

$$RMSE = [\frac{\sum_{i=1}^{N}(f_i - O_i)^2}{N}$$

3. Results and Discussion

Rainfall data is the main input of SPI calculation *(Standardized Precipitation Index)*. Meteorologically, rainfall is also the main parameter in determining whether drought will occur or not in an area. This is because rainfall is one of the main sources of water availability in an area.

The output of the CCAM model needs to be validated before it used to project the value of future rainfall. In case of Sumbawa area, the CCAM rainfall value is validated by the value of in situ rainfall in Bima, NTB. Validation uses the Pearson correlation calculation method to determine the similarity of trends and calculate the bias or error as the distance between the CCAM data and in situ data with RMSE (Root Mean Square Error).

The correlation value between CCAM rainfall data and in situ rainfall data is around 0.42. When viewed from the plot results, for some trends CCAM appears and the insitu data is quite a trend. However, there are several points where CCAM appears to overestimate the insitu data (Figure 2). Generally CCAM is quite good at projecting the value of rainfall in the dry season around June, July and August. Where as in the wet season like December, January and February CCAM tends to overestimate. To determine how much the overestimate value is, an error calculation is performed with the RMSE search method. The RMSE value of the CCAM data and in situ data ranges from 139.64. This value is quite large when compared with the average rainfall value in Bima data which ranges from 148.18. This means that it is biased between CCAM data and internal data almost close to the monthly average rainfall value of the data. However, with medium correlation values (close to 0.5) and RMSE below the average, the CCAM data from this simulation can still be used in terms of research on climate change.

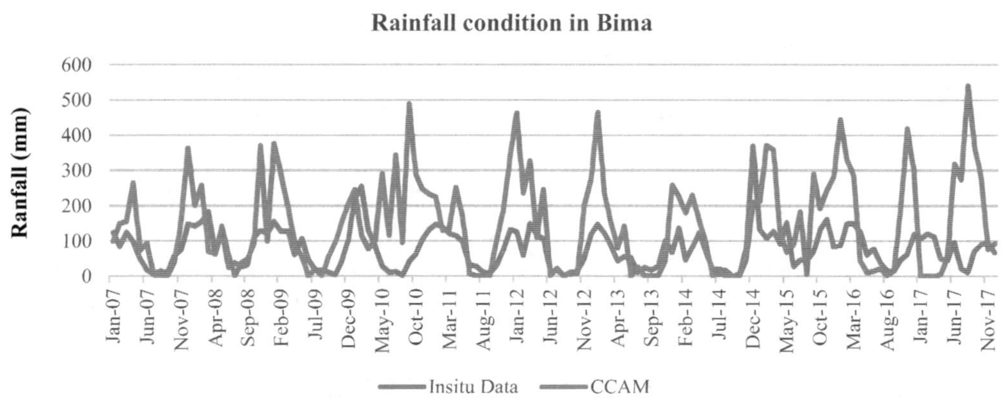

Figure 2. Rainfall condition in Bima

After being validated, CCAM data is used to be able project rainfall conditions in Sumbawa until 2055. In January 2040, it is estimated that rainfall in Sumbawa is around 50-300 mm on land (Figure 3a). This value is quite good and is estimated to be quite large for the value of rainfall in an area. Furthermore, the calculation of the value of SPI in the Sumbawa region was carried out in 2040 based on the results of CCAM input. It is known that the value of SPI in the Sumbawa region generally shows a positive value which means that in January 2040 there will be no expected drought in the Sumbawa region. The SPI values in this region range from -0.5-1.5 (Figure 3c). A relatively small SPI value generally occurs in the area around Mount Tambora and Bima region (Figure 3c). However, the value of -0.5 is not indicated as a drought but a neutral condition according to McKee's classification and calculation. NTB region itself has a rainfall pattern that tends to be monsoonal which means that in January rainfall will tend to be large.

From the description above, it can be concluded that in August 2040 there was a potential danger of a meteorological drought in Sumbawa Island region. Whereas in January 2040 drought is not expected to occur.

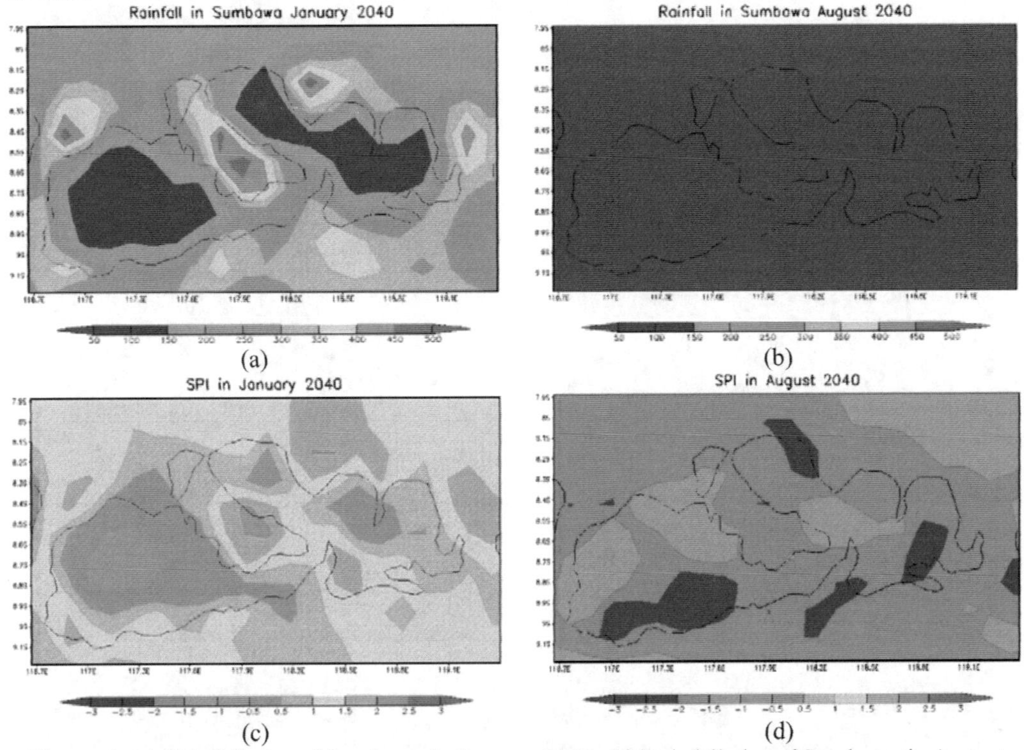

Figure 3. (a) Rainfall plot of Sumbawa in January 2040, (b) Rainfall plot of Sumbawa in August 2040, (c) SPI plot of Sumbawa in January 2040, and (d) SPI plot of Sumbawa in August 2040

To be able to provide further understanding of evolution SPI condition on Sumbawa Island in 2040, an SPI plot was carried out in 2040 for February, March, April, May, June, July, September, October, November and December. In February, the SPI value on Sumbawa Island is estimated to be quite normal and predicted in neutral conditions with SPI values in the range of -0.5-1.5 (Figure 4a). This is not surprisingly enough remember even in February in Sumbawa island the rainfall tends to be high. In March 2040, the SPI value tends to be normal to wet with a range of 0-1.5 (Figure 4b). In April, May, June, and July where in these months is the transition month to the dry season, the SPI value on Sumbawa Island decreases around -1-0.5 (figure 4c-f). In September where rainfall tends to be small, the value of SPI in the Sumbawa Island area generally decreases to -1-0.5 (Figure 4g), while in October the value decreases again in the range -1.5 - (- 1) (Figure 4h). Dry conditions with SPI reaching -1 extend in November including in the East Sumbawa and Central Sumbawa regions (Figure 4i). In December the drought was more severe in some areas marked by SPI values which even reached -2.5 for example in the area around Mount Tambora (Figure 4j).

Journal of Physics: Conference Series **1373** (2019) 012004 doi:10.1088/1742-6596/1373/1/012004

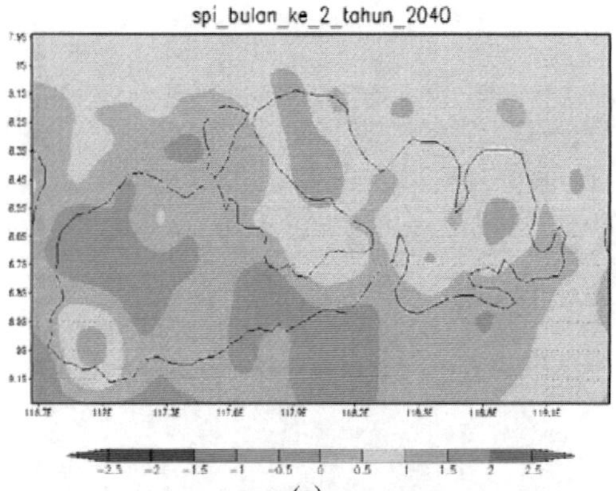

(a)

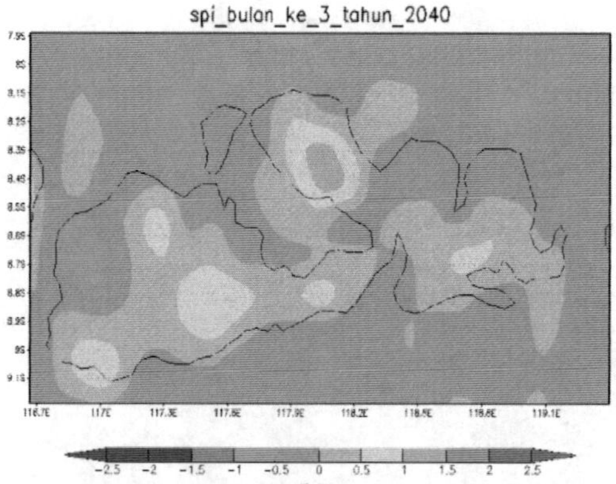

(b)

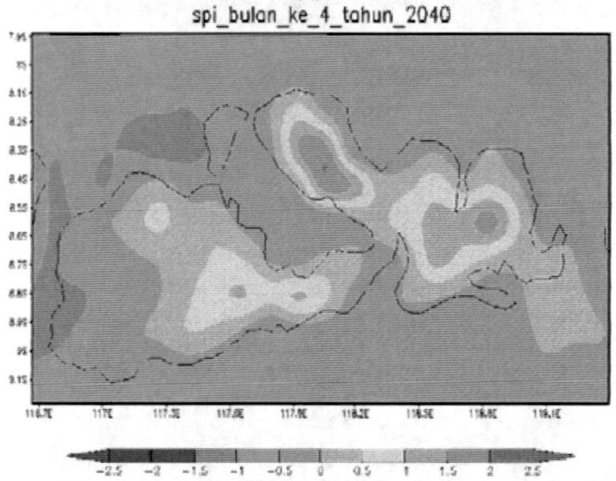

(c)

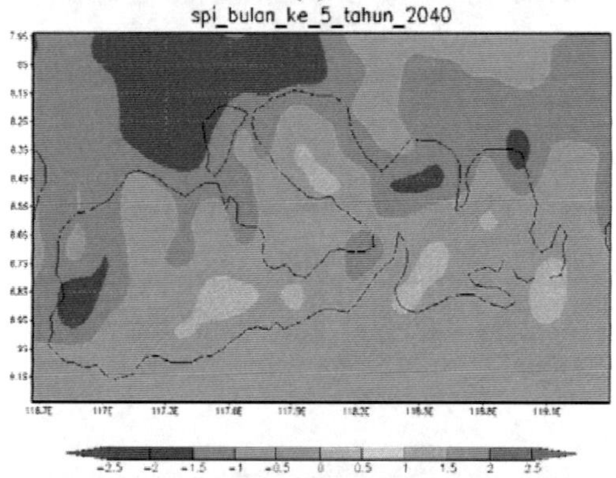

(d)

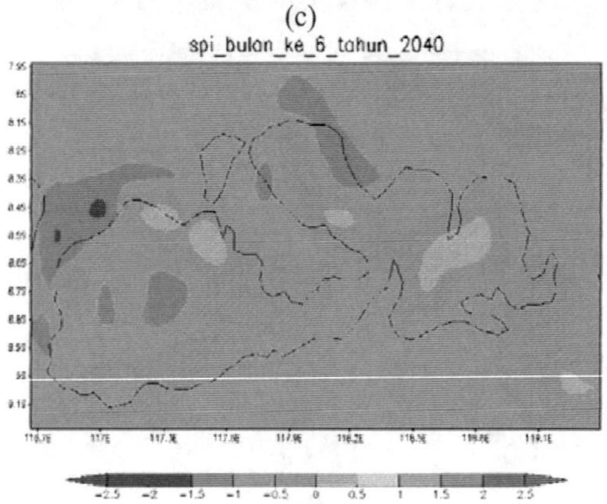

(e)

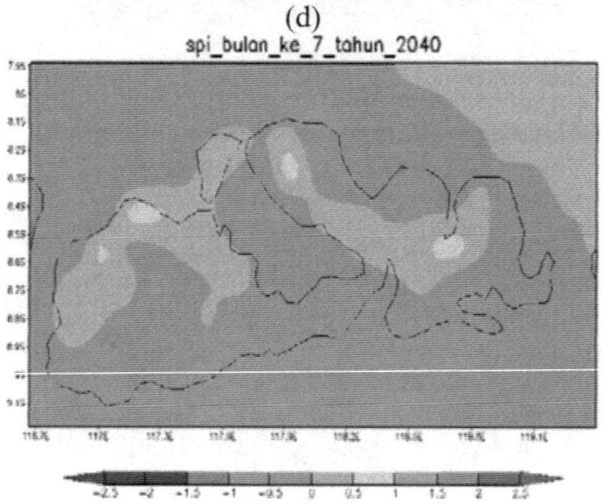

(f)

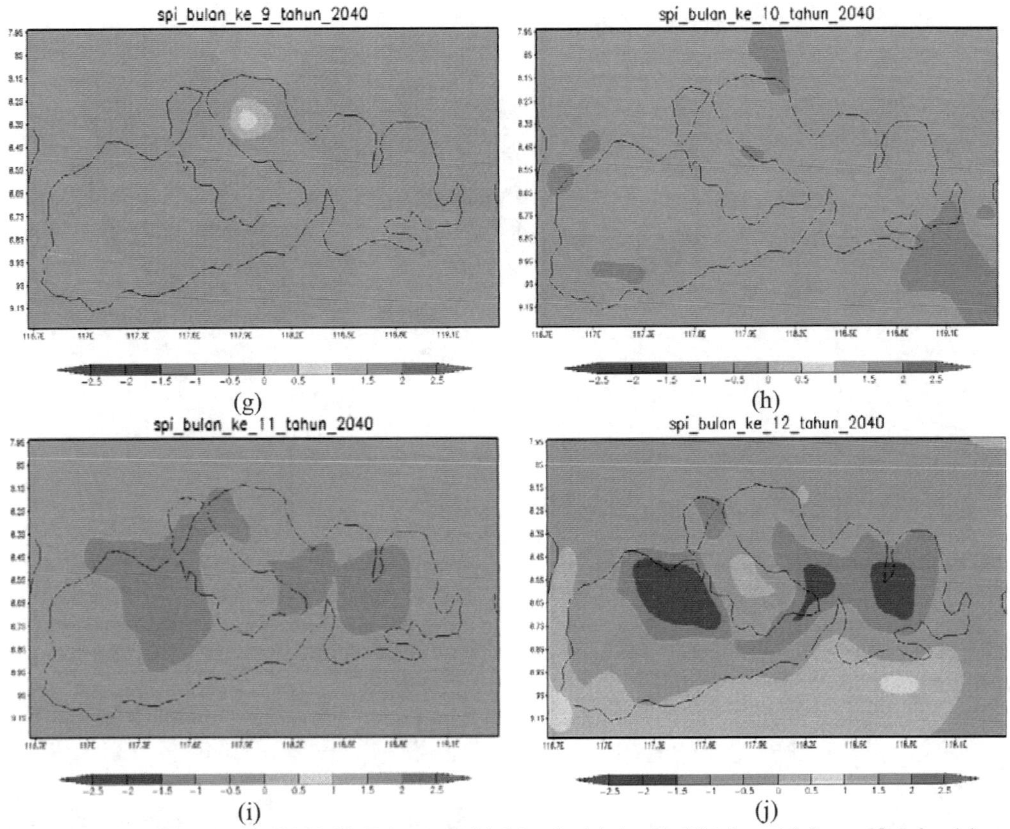

Figure 4. SPI value in 2040 (a) February, (b) March, (c) April, (d) May, (e) June, (f) July, (g) September, (h) October, (i) November, (j) December

August itself is the month where rainfall in Sumbawa island has decreased (the value tends to be small). Therefore, the potential danger of drought in this month is quite large compared to other months. To be able to find out the 5-year evolution of drought conditions in August on Sumbawa Island, analyzed the average 5-year value of SPI until 2050. 2020 here means August average data in 2016-2020, 2025 means August average data in 2021-2025, etc. The temporal value of SPI plot is done at Bima point.

Based on the results of the 5-year evolutionary average, data are obtained that in 2016-2020 the average value of SPI is in the range of 0.27, which means that this condition is normal and tends to be wet (Figure 4). In the period of 2021-2025 changes in conditions are expected to occur, causing conditions to be normally dry, with SPI values ranging from -0.25 (Figure 5). The thing to watch out for is in the period 2026-2030 where droughts will occur with SPI values reaching below -0.5 (Figure 5). Inverted conditions occur in the period 2031-2035 where wet conditions are expected to occur with an SPI value close to 1 (Figure 5).

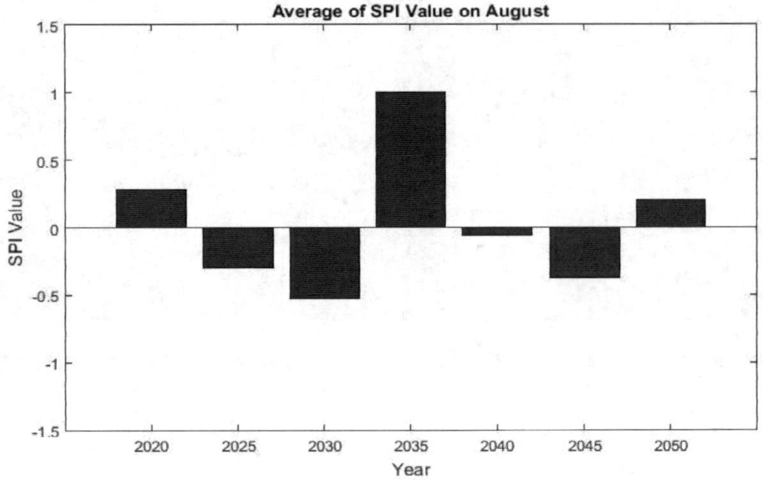

Figure 5. Five years-average of SPI value on August

To be able to provide further understanding of when droughts will occur in August in the 2011-2040 period, a plot of SPI values will be carried out in August 2011-2040. Referring to the classification carried out by McKee, an area is said to be dry if it has an SPI value below -1. In this condition, drought in Bima is expected to occur in August in 2022, 2024, 2028, 2030 and 2033 (Figure 6).

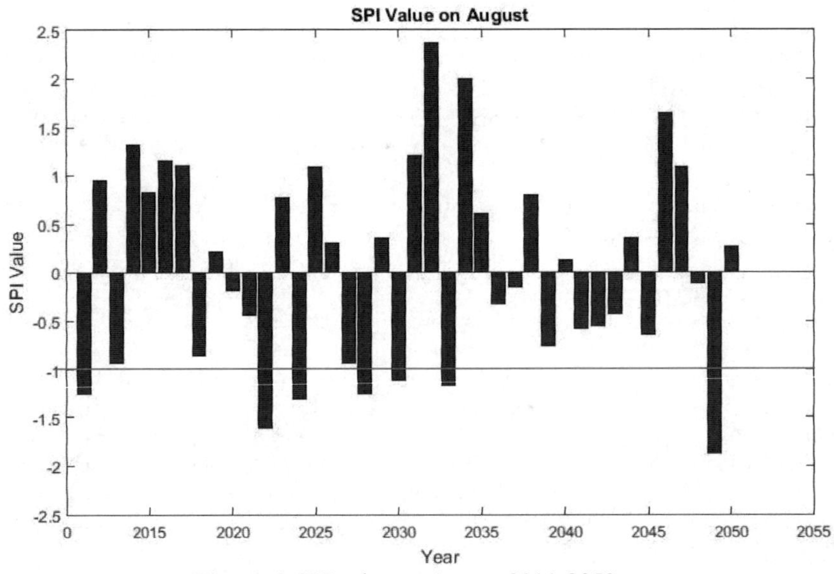

Figure 6. SPI value on August 2011-2050

4. Conclusions

In 2040 a drought is expected to occur on Sumbawa Island in August 2040 with an SPI value in the range of -2.5 - (- 2). While in January where rainfall on Sumbawa Island was quite high, meteorological droughts were not expected to occur. In 2040, apart from January wet conditions (with

SPI values generally above 1) and expected to occur in February and March. While neutral conditions will occur in April, May and June. While in July, August, September, October, November and December, Sumbawa Island is expected to tend to dry out. Drought in August is estimated not only to occur in 2040 but also in 2022, 2024, 2028, 2030 and 2033.

5. Conclusions

In 2040 a drought is expected to occur on Sumbawa Island in August 2040 with an SPI value in the range of -2.5 - (- 2). While in January where rainfall on Sumbawa Island was quite high, meteorological droughts were not expected to occur. In 2040, apart from January wet conditions (with SPI values generally above 1) and expected to occur in February and March. While neutral conditions will occur in April, May and June. While in July, August, September, October, November and December, Sumbawa Island is expected to tend to dry out. Drought in August is estimated not only to occur in 2040 but also in 2022, 2024, 2028, 2030 and 2033.

6. Acknowledgement

Author would say thank you to Centre of Atmospheric Science and Technology for Material and Moral Supporting so that this research can be done.

7. References

[1] Intergovernmental Panel on Climate Change (IPCC) 2001 in McCarthy J J, Canziani O F, Leary N A, Dokken D J, and White K S (eds.); *Climate Change 2001 Impacts, Adaptation, and Vulnerability*, Cambridge: Cambridge University Press. p. 1032.

[2] Solomon S, Qin D, Manning M, Chen Z, Marquis M, Averyt K B, Tignor M, and Miller H L (eds.) 2007 *Contribution of Working Group I to the Fourth Assessment Report of the Intergovernmental Panel on Climate Change.*

[3] Hidayati R, 1990 *Kajian Iklim kota Jakarta, Perubahan dan Perbedaan dengan daerah sekitarnya. Thesis Program Studi Agroklimatologi.* Institut Pertanian Bogor

[4] Cheung W H, Senay G B, and Sing A. 2008 Trends and spatial distribution of annual and seasonal rainfall in Ethiopia. *International Jurnal Climatology*, 28 (2008), pp. 1723-1734, 10.1002/joc.1623

[5] Intergovernmental Panel on Climate Change IPCC 2007 IPCC-Task Group on Data and Scenario Support for Impacts and Climate Analysis (TGCIA), 2007 *General Guidelines on The Use of Scenario Data for Climate Impact and Adaptation Assessement*, Version 2, pp. 66.

[6] Saji N H, Goswami B N, Vinayachandran P N, and Yamagata T. 1999. *Nature* **401** 360-363.

[7] Madden R A, and Julian P R 1971 *J. Atmos. Sci.* **28** 702-708.

[8] Kirono G D, Butler J R A, McGregor J, Ripaldi A, Katzfey J, Nguyen K 2016 *Climate Risk Management* **12** 45–58.

[9] Hulme M, Mitchell J F B, Ingram W, Johns T C, Lowe J A, New M G, and Viner D, 1999 *Global Environmental Change* ***9**, S3-S19.*

[10] Boer R and Faqih A 2004 *Current and Future Rainfall Variability in Indonesia. In An Integrated Assessment of Climate Change Impacts, Adaptation and Vulnerability in Watershed Areas and Communities in Southeast Asia, Report from AIACC Project No. AS21 (Annex C, 95-126). International START Secretariat*, Washington, District of Columbia. http://sedac. ciesin. org/aiacc/ progress/ Final Rept_AIACC_AS21. Pdf.

[11] Naylor R L, Battisti D S, Vimont D J, Falcon W P, and Burke M S 2007 Assessing risks of climate variability and climate change for Indonesian rice agriculture. *Proceedings of the National Academy of Sciences of the United States of America* 104(19) 7752-7757.

[12] Kementerian Lingkungan Hidup (KLH) 2012 *Perubahan iklim global.* accesed online on 27 July 2014, from: http:/climatechange.menlh.go.id.

[13] McGregor J L, Kim R, Nguyen C, Kirono D G C, Katzfey J J, 2016 High-resolution climate

Journal of Physics: Conference Series **1373** (2019) 012004 doi:10.1088/1742-6596/1373/1/012004

projections for the islands of Lombok. *Climate Risk Management* **12** 32–44.
[14] Dai A, 2011 *WIREs Climatic Change* **2**, 45–65.
[15] Giannini A, Saravanan R, and Chang P, 2003 *Science* **302** 1027– 1030.
[16] McGregor J L, Gordon H B, Watterson I G, Dix M R, Rotstayn L D, 1993 The CSIRO 9-level atmospheric general circulation model. *CSIRO Division of Atmospheric Research Technical* Paper No. 26.
[17] McGregor J L, 2005 C-CAM: Geometric aspects and dynamical formulation [electronic publication]. *CSIRO Atmospheric Research Technical* Paper No. 70.
[18] McGregor J L, 2015. *Climatic Change* **119** 369-380
[19] McKee T B, Doesken N J, and Kliest J, *The relationship of drought frequency and duration to time scales*. In Proceeding of the Eighth Conference of Applied Climatology, 1993 Jan 17-22; Anaheim, CA. Massachusetts: American Meteorology Society. 1993. p. 179–184.
[20] Makridakis S, Andersen A, Carbone R, Fildes R, Hibon M, Lewandowski R, Newton J, Parzen E, Winkler R, 1982 *Journal of Forecasting* **1** 111-153.
[21] Barnston A G, 1992 *Weather Forecasting* **7** 699-709

Analytical Hierarchy Process (AHP) in selecting suitable Marine Protected Area (MPA) site in Pulo Breuh (Breuh Island), Indonesia

F Syahputra *[1,2], A M Muslim [2], W I A W Talaat [2], N Irsalinda[3]

[1] Fakultas Perikanan Universitas Abulyatama, Aceh Besar, Indonesia
[2]Institut Oseanografi dan Sekitaran, Universiti Malaysia Terengganu, Terengganu, Malaysia
[3]Department of Mathematics, Faculty of Applied Science and Technology, Univeristas Ahmad Dahlan, Yogyakarta, Indonesia

*Email: faisalsyahputra@yahoo.com

Abstract. Marine protected area (MPA) is one of marine management tools that effective in regulating human activities within certain area. Therefore, choosing the location for the MPA is a critical step due to the nature of this tool that is prone to cause conflict and has broad impact to local and larger community. The selection processes require many criteria to consider thus this research use Analytical Hierarchy Process (AHP) method in Geographical Information System (GIS) to calculate MPA location. AHP has ability to simplify multidimensional criteria into single dimensional scale and GIS make is easier to analyze wide area. The research is conducted in Pulo Breuh (Breuh Island) Aceh Besar district of Aceh province, Indonesia. There are 6 components to calculate in determining MPA location for this island, the Settlement; Tourist Attraction; River; Aquaculture site; Ferry port and Fishing port, and each of this criterion is depicted into single layer of map based on the distance. The MPA needs to be close to the settlement and tourist attraction to ensure the advantages as well as far enough from the river, aquaculture site and ports. in the weighing criteria the priorities are, from highest to lowest, the settlement, aquaculture site, river, ferry port, fishing port, and tourist attraction. The result will be 3 levels of color-coded overlay map ranges from suitable, less suitable and not suitable place for MPA.

1. Introduction

Marine and coastal management becomes more and more important in the recent years, environmental degradation and human activities in land and in the sea put pressure in marine ecosystem that affect marine ecosystem health and productivity. Marine Protected area (MPA) is one of management tools to help maintaining the pressure by regulated human activities within certain area. Selecting the area for the MPA location is a complex process since it will affect environment and local community that close to the MPA; moreover, MPA also prone to conflict of interest due to its nature that regulated (limited) human activities within the respective area. Therefore, deciding which area is suitable for MPA is multi-criteria problem that requires multi-criteria analysis to get the decision.

One of the multi-criteria analysis tools is Analytical Hierarchy Process (AHP). AHP is a robust tool in solving multi-criteria problem. It works by dividing the problem into several alternatives or criteria hierarchically in regards to solving the problem. Furthermore, the criteria or alternatives is assigned weight by comparing one another through Pair-wise Comparison Method. The decision is made by cumulation of the weight of criteria for each alternative solution. AHP has ability to simplify complex problem and make the decision easy to understand. Combining AHP with Geographical Information System (GIS) will make the decision processes for MPA site can be easily followed by stakeholders. Stakeholders understanding of how the site for the MPA selected is important because stakeholder's involvement in MPA development will determine if the MPA success or not, especially the local community, local government and people that their activities or livelihood is influenced by the MPA.

This study is conducted in Pulo Breuh (Breuh island), a small island situated off the northern coast of Sumatera Island; the precise location is in 05^{0}38’49’’ N- 05^{0}45’19’’ N and 095^{0}0’6,23’’- 095^{0}8’39’’ E. It consists of 12 (twelve) villages. The island is part of Pulo Aceh islands and sub-district of Aceh Besar regency. The island is closer to Province Capital, Banda Aceh than to its own regency capital, Jantho, that make the resident more often doing their economic activities in Banda Aceh. The island can only be accessed by boat from Banda Aceh. Pulo Aceh unique location, bordering Andaman Sea and Indian Ocean, create a good ecosystem for coral reef and pelagic fish to thrive in the water around these islands and also make this island in the migration path of big pelagic fish such as tuna and shark [1]. Therefore, this island is famous fishing ground for fishermen of Aceh and recreational angler; furthermore, the natural condition of this island and the sea attracts tourist to come to this island [2].

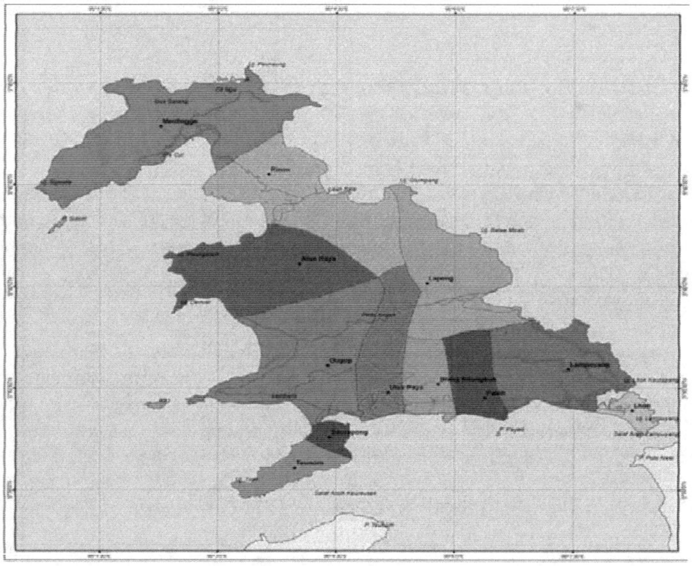

Figure 1. The Map of Pulo Breuh (Breuh Island)

Unfortunately, several areas in Pulo Breuh coral reef is badly damage because of destructive fishing method especially from fish bombing and octopus' extraction. The tsunami disaster in 2004 also adds damage to Pulo Breuh coastal area where this area has the highest number of broken and upturned coral reef of all Aceh coastal area [3] and the rapid development during rehabilitation and reconstruction program after tsunami also adds more pressure to the island environment. The damaged coral reef habitat affects the island community livelihood and, in some extent, all ecosystem of the island. Therefore, some conservation program like MPA is needed in this island.

Effective MPA requires effective management in maintaining the regulated area from human activities intervention. Study shows that small area MPA, close to living area and managed by local community gives better result [4]. Thus, this paper will try to emulate this approach in selecting suitable MPA location.

2. Theoretical Framework

2.1. Analytical Hierarchy Process (AHP)
Analytical hierarchy process (AHP) is structured process that help user to make a decision based on their goal and their understanding of the problem. It bases on comparison of relative preference of the user. AHP has ability to measure tangible and intangible object with similar measurement [5].

AHP is developed during arms tradeoff in Arms Control and Disarmament Agency (ACDA) where they need to gather recommendation from many experts with different background and reference of measurement; therefore, they need method which capable to combine all input into one framework and one final recommendation. The idea is combining the measurable and unmeasurable with some ratio or scales that can be calculated in normal manner. AHP combines multidimensional scales into single dimensional priorities scale which determine the outcome that makes it capable to put individual judgment as input in the function. In addition, this method can process judgment from several people or group in the decision process [6]. It is important in selecting the criteria not to enter the criterion which will definitely determine the solution [7]. In the MPA site selection case, the location of coral reef or other important habitat and species is better to be avoided because this criterion will overshadow other criteria.

2.1.1. How AHP Work.
AHP work as thought guidance process in decision making process. First step is to determine the objective of this analysis, the criteria and the alternative solution and put it in hierarchy order. These criteria determine the choice of alternative that could be used to reach the decision. Each of criteria is assigned weight as measure of importance of each criterion in developing of alternative solution. It is performed using Pair-wise Comparison Method (PCM) also developed by Saaty [8], [9].

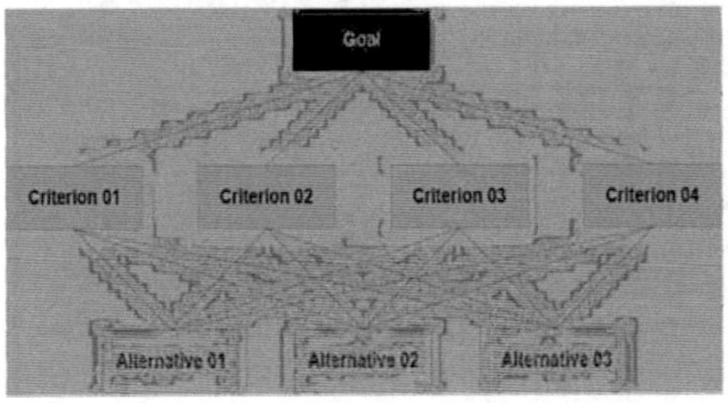

(a)

	Alternative 1	*Alternative 2*	*Alternative n*
Alternative 1	1	A12	A1n
Alternative 2	1/A12	1	A2n
Alternative n	1/A1n	1/A2n	1

(b)

Figure 2. How AHP process work (a). Hierarchical order of criteria and alternatives (b) Pair-wise matrix comparison to get the weight for each criterion.

The value of the criteria must be consistent throughout the process or the calculation will be incorrect; hence, the AHP process includes consistency test, if the consistency ratio (CR) value is above 0,1% the calculation must not be accepted.

2.1.2. AHP and GIS.
Combination of AHP analysis and GIS is an effective tool for spatial base decision making due to ability of AHP to simplify the decision making process and GIS ability to analyze, manage and display data or result in spatial manner [10]. Incorporated AHP in GIS is a matter to define the criteria and

alternative solution in spatial data and perform overlay analysis. Each criterion is assigned a thematic map and the overlay analysis is performed follow the weighing process of AHP.

3. Methodology

In this study, the AHP analysis in GIS is conducted simultaneously in Quantum GIS software using Easy AHP plugin by Mehmet Selim Bilgin [11]. The AHP objective for this study is to find the suitable location for Pulo Breuh MPA. There are 6 (six) criteria for this AHP calculation:

1. Settlement
   Location of the MPA needs to be as close as possible to the inhabitant settlement which makes monitoring of violation of MPA regulation easier. Therefore, the value of the settlement thematic map gets lower the farther the point from the settlement.
2. Aquaculture site
   Aquaculture site should be far from the MPA due to the high human activity and the possibility of influence from fish inside the pent to the protected area and vice versa. Therefore, the value of the aquaculture site thematic map gets higher the farther the point from the Aquaculture site.
3. Ferry port
   The MPA site must be far from ferry port to avoid high human activities and pollution from the ship. To this day, only small ship that ferry the passenger from and to the island; however, the possibility of pollution from this ship still high. The map value will increase the farther the point from the Ferry port.
4. Fishing port
   The fishing port in Pulo Breuh even has more activities than ferry port. The value on the map also get higher the farther the point from fishing port.
5. River
   There is no big river in this island; however, in rain season the river will bring mud to the sea as the effect of deforestation in the island. Thus, the MPA site needs to be not to close to the river and the value of point in the river thematic map will be higher the farther it is from the river.
6. Tourist attraction
   Most tourist attractions in this island are sea-related object or activities and preserving these attractions is important to the community livelihood. Consequently, the MPA is better located close to the tourist attraction. The value of the tourist attraction thematic map will be lower the farther the point from the tourist site.

Basic setting of GIS software in assigning value of thematic map is the closer to the point the lower the value so for the Settlement and Tourist Attraction the value needs to be reversed.

All the map of the criteria then converted into layers in uniform spatial format and the calculation of AHP analysis of this map layers will be conducted in the EasyAHP plugins. The result will be the map with value that can be classified into suitable area (high value) and non-suitable are (low value).

4. Result and Discussion

The result of AHP analysis shows the most importance criteria is the Settlement with 37.7 % and the lowest criteria is the Tourist Attractions with 5.4%. the complete table is as follows

Table 1. The result of the weight criteria based on pair-wise comparison

Category		Priority	Rank
1.	Settlement	37.7 %	1
2.	Aquaculture	24.3 %	2
3.	River	15.0 %	3
4.	Ferry Port	8.8 %	4
5.	Fisheries Port	8.8 %	4
6.	Tourist Attraction	5.4 %	6

The complete pairwise comparison is as follow:

Table 2. Complete pair-wise comparison of all criteria

	1	2	3	4	5	6
1	1	2.00	3.00	4.00	4.00	5.00
2	0.50	1	2.00	3.00	3.00	4.00
3	0.33	0.50	1	2.00	2.00	3.00
4	0.25	0.33	0.50	1	1.00	2.00
5	0.25	0.33	0.50	1.00	1	2.00
6	0.25	0.25	0.33	0.50	0.50	1

There 15 (fifteen) comparison with the consistency ratio CR= 0,011 (value < 0.10) indicates the judgment of the criteria is consistent. The suitable location is generated by Quantum GIS software based- on this result can be seen in Fig. 3.

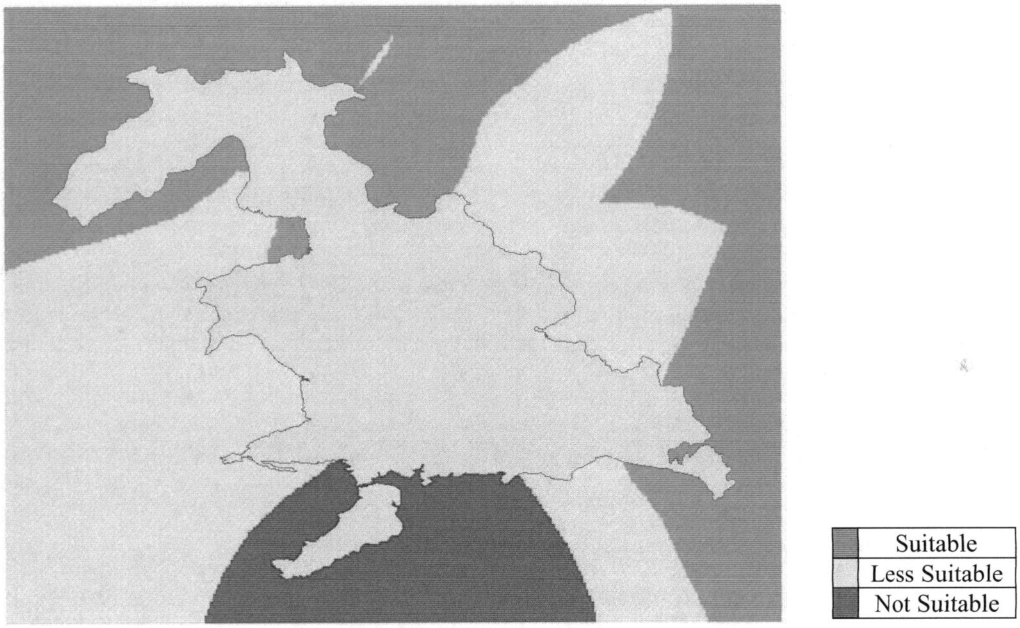

Figure 3. The result of suitable MPA site analysis using AHP.

The suitable area mostly located in the northern area and small part of eastern area of the island. It is mostly due to the distribution of settlement, as main criteria, in this island. Most of village in southern part of the island situated in coastal area and leave the middle section of the island settlement free. The northern area which is smaller has two cluster of settlement but situated strategically close to the bay area; meanwhile in eastern part of area only has one cluster of settlement (Fig.4.). with the aquaculture site and several ports are located in southern part of the island; the condition is clearly not suitable for the MPA site which means the result of this AHP analysis already gives a general picture where is the MPA can be established. Determining the exact location for the MPA requires more complex study regarding the habitat and organism in the area.

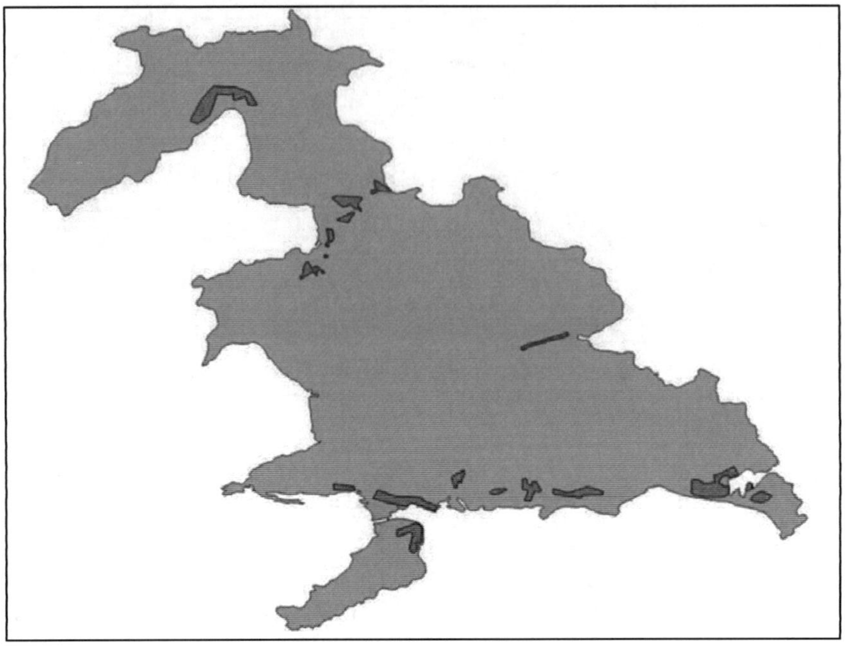

Figure 4. Location of settlement in Pulo Breuh

5. Acknowledgements

The Authors wish to express and thank you their gratitude to the 2019 Conference on Fundamental and Applied Science for Advanced Technology (2019 ConFAST) Committee and also for the help from Aceh Provincial Government, INOS of Universiti Malaysia Terengganu, Univeristas Abulyatama and Universitas Ahmad Dahlan.

References

[1] Cochard R 2016 *Sustainable Natural Resources Management in Dynamic Asia* **1** 231-271.
[2] Fadillah A, Dewi T G, and Hardjanto A 2012 *International Journal of Social Science and Humanity* **2** 206-212..
[3] Campbell S J *et al.* 2007 *Atoll Res. Bull.* **544** 55–78.
[4] McClanahan T R, Marnane M J, Cinner J E, and Kiene W E 2006 *Curr. Biol.* **16** 1408–1413.
[5] Saaty R W 1987 *Math. Model* **9** 161–176.
[6] Covas M T, Silva C A, Dias L C 2013 *International Transactions in Operational Research* **20** 269–299.
[7] Russo R F S M and Camanho R 2015 *Procedia Computer Science* **55** 1123-1132.
[8] Siddayao G P, Valdez S E, and Fernandez P L 2014 *Int. J. Mach. Learn. Comput.* **4** 450-457.
[9] Chabuk A, Al-Ansary N, Hussain H M, Knuttson S, Pusch R 2016 *Waste Management Research* **34** 427-437.
[10] Şener Ş, Şener E, Nas B, and Karagüzel R 2010 *Waste Management* **30** 2037-2046.
[11] A. C. Aydinoglu and Bilgin M S 2015 *Nat. Hazards Earth Syst. Sci.* **15**, 335–347.

Kinetics of Reducing Sugar Formation from Coconut Husk by Subcritical Water Hydrolysis

Alfain Noerdin Alimny, Maktum Muharja, Arief Widjaja*

Department of Chemical Engineering Sepuluh November Institute of Technology, Surabaya 60111, Indonesia

*Email: arief_w@chem-eng.its.ac.id

Abstract. Coconut husk hydrolysis using subcritical water (SCW) with N_2 as pressurizing gas was carried out in a batch reactor at temperature of 150 °C-190 °C, at pressure of 60-100 bar and the reaction time from 20-60 min. The effect of SCW pretreatment could reduce lignin and produced reducing sugars. The highest TRS (total reducing sugar) yield was 0,4904 at 170 °C, 80 bar, and the reaction time of 20 min. Reducing sugars concentration increased when temperatures rised from 150 to 170 °C. However, reducing sugar concentration decreased at temperature above 170 °C. The decrease of reducing sugar was due to degradation of reducing sugars into its degradation products. A first-order reaction model was determined to describe the reaction rate constant of the formation of reducing sugar and its degradation products. Kinetic parameters were obtained by fitting the proposed equation with the experimental data. The reaction rate constant increased when temperature increased from 150 to 170 °C, but then decreased when the temperature increased above 170 °C. Activation energy was also determined based on the Arrhenius equation.

1. Introduction

The development of modern transportation and increasing vehicle usage requires large-scale energy sources. Currently, the petroleum and gas sector is the main factor for energy source. The ever increasing energy requirements necessitate alternative energy sources which might complementing the current energy usage. From some alternative energy sources developed, biomass is a top priority because biomass feedstock can be converted into renewable energy. In the economic sector, biomass provide a high value on product like carbon waste as a substrate of fermentation for the production of biofuel known as the second generation of energy production.

Coconut husk is a biomass produced with large quantity in the plantation. According to the Directorate General of Plantations in Indonesia, coconut husk generates more than one million tons in 2016. As plantation waste, coconut husk production became a serious problem for the environment since it is mostly dumped in the ground and burned that produces greenhouse gases responsible for the global warming. The coconut husk contained 26,72 % cellulose and 17,73 % hemicellulose that can be converted into reducing sugars as the substrate for biofuels [1].

The hydrolysis of lignocellulose biomass is mostly conducted through thermochemical, chemical and biochemical methods. Gasification, pyrolysis, and liquefaction are some thermochemical methods to produce variety of biofuels such as biochar and bio-oil. Biochemical pathway includes enzymatic hydrolysis and fermentation of waste biomass to obtain hydrogen and alcohol production [2]. Chemical pathway includes alkali and acid methods. Unfortunately, some methods have weaknesses. Acid and alkali hydrolysis are corrosive, using toxic solvents, and require process of neutralization before being dumped into the waste [3].

Subcritical water (SCW) pretreatment attracted attention because it is eco green without using organic solvents. Another interesting property of SCW is increasing H^+ and OH^- as increasing temperature so that water behaves as acid or base catalyst [4]. In addition, the SCW process is fast, non-corrosive, and produce little waste (Prado *et al*, 2014). SCW pretreatment also has unique properties of density, solvent ability, and dielectric constant. At subcritical condition, the density of water decreases, increasing the rate of diffusion, and decreasing viscosity and surface tension [5]. Although SCW pretreatment have some weaknesses such as the requirement of higher safety due to extreme condition, high investment cost and the formation of degradation product, however the overall economic and technical viewpoint encourage the implementation of this process [4].

In the previous research, the kinetics of SCW pretreatment have been carried out with several biomass, such as rice straw, bagasse, and peanut waste [6-8]. The results of the kinetics from previous studies provide information about understanding the hydrolysis of biomass as a source of reducing sugars. However, coconut husk hydrolysis kinetics studies have not been reported. This study aimed to study the kinetics of production of reducing sugars (TRS) by subcritical water pretreatment. Kinetic parameters was obtained by fitting experimental data with the proposed equation The effect of temperature, reaction time and pressure on subcritical was investigated in this study.

2. Material and Methods

2.1. Materials

Coconut husk was obtained from Manado, North Sulawesi, Indonesia. The biomass was first dried in the sun for 2 days, then it was milled and screened by using screener (Retsch GmBh Rheinische Strade 36 4278, Haan, Germany) to obtain a particle size of 100-120 mesh. N_2 gas was from Samator Indonesia. All chemicals and reagents were purchased from Sigma Aldrich (Germany).

2.2. Subcritical Water Pretreatment

Subcritical water equipment consisted of heaters, control devices (PID), and subcritical reactors made of stainless steel. The reactor connected with PID, pressure indicator, and heater as shown in Figure 1. The process was carried out under batch conditions and flowed with N_2 gas.

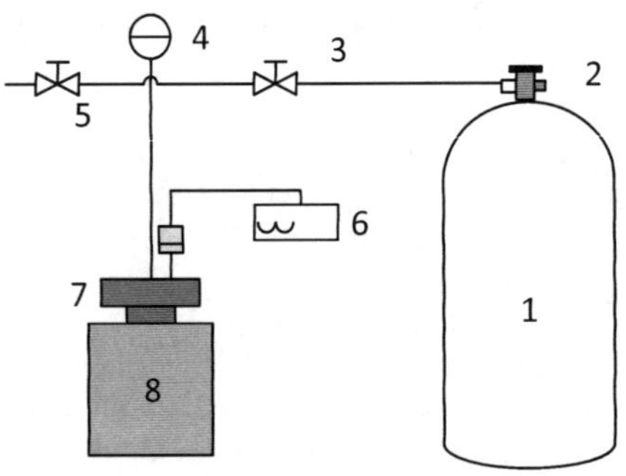

1 = Nitrogen gas tube
2,3,5 = Valve
4 = *Pressure Indicator*
6 = PID
7 = Reactor
8 = Heater

Figure 1. Design of SCW Pretreatment

6 g of dried coconut husk and 120 ml of deionized water were added to the reactor. N_2 gas was flowed to the reactor with pressure of 60, 80 and 100 bar. Then, heating was carried out with temperature of

150 °C, 170 °C, and 190 °C and the reaction was conducted for 20, 40 and 60 min. In the final process, the temperature of reactor was cooled to room temperature and the liquid fraction was separated from solids using a vacuum filter.

2.3. Analytical Method

The content of cellulose, hemicellulose and lignin in coconut fiber was determined by gravimetric method [9]. The results of reducing sugars after SCW on the liquid were analyzed by the DNS method [10]. The furfural content was analyzed by gas Chromatography Flame Ionization detection (GC-FID) (GC-2010 Plus Shimadzu, Kyoto, Japan) using a modified method from Gaspar and Lopes [11]. The separation was performed using RTX-Wax column (Restek, Pennsylvania, USA), 30 m x 0,25 mm ID, film thickness 0,25 μm. The temperature of the injection and detector was set as 250 and 275 °C, respectively.

2.4. The Determination of SCW Kinetics

Hydrolysis of the lignocellulose in SCW is very complex which requires a simplified reaction model to determine the kinetics that is based on irreversible series of reactions [6, 12].

$$\text{Coconut husk (SH)} \xrightarrow{k_1} \text{Reducing Sugars (R)} \xrightarrow{k_2} \text{Degraded Product (P)} \quad (1)$$

where k_1 and k_2 are the rate constants of the production of reducing sugar and degraded products, respectively. The reaction rate of the first step of the reaction is shown in equation 2 and 3.

$$-\frac{dCSH}{dt} = k_1.C_{SH} \quad (2)$$

$$C_{SH} = C_{SHO}.e^{-k1.t} \quad (3)$$

The reaction rate of sugar production from lignocellulose followed by its degradation can be represented by Eq. (4).

$$\frac{dCR}{dt} = k_1.C_{SH} - k_2.C_R \quad (4)$$

Next, substitution of Eq (3) to Eq (4) results in the following first order differential equation:

$$\frac{dCR}{dt} + k_2.C_R = k_1.C_{SHO}.\,e^{-k1t} \quad (5)$$

where C_{SH} is coconut husk concentration (g/L), C_R is reducing sugar concentration (g/L), and C_{SHO} is the initial concentration of coconut husk (g/L).

By integration of Eq (5) and solved at a boundary condition of t = 0, Cs = 0, the following equation can be obtained.

$$C_R = \frac{k1.CSHO}{k2-k1}.(\exp(-k_1.t) - \exp(-k_2.t)) \quad (6)$$

Reducing sugar concentration in Eq. (6) can be represented by sugar yield as follows:

$$R_S = \frac{CR}{CSHO} = \frac{k1}{k2-k1}\,(\exp(-k_1.t)\text{-}\exp(-k_2.t)) \quad (7)$$

The reaction rate constants k_1 and k_2 in different condition were obtained by fitting the experimental data using Solver provided in Excel 2017. Activation energy was determined by correlation between rate constant with temperature described by Arrhenius equation.

$$\ln k = \frac{Ea}{RT} - \ln A \quad (8)$$

where Ea is activation energy ($KJ.mol^{-1}$), R = 8,314 J mol^{-1} K^{-1}, A is pre-exponential factor (min^{-1}). Plot -ln (k) vs 1/T leads to a straight line with slope indicating the activation energy and intercept the pre-exponential factor.

3. Results and Discussion

3.1. The Chemical Composition of Coconut Husk

Chemical composition of coconut husk before and after pretreatment was analyzed by gravimetric method [9] and shown in Tabel 1.

Table 1. Chemical Composition of Coconut Husk After SCW Pretreatment

P	T	t	Component						
bar	°C	min	Residual solid	Hemicellulose		Cellulose		Lignin	
				(%)	g/L	(%)	g/L	(%)	g/L
Control	-	-	6	12,877	6.441	20,641	10.322	62,272	31,142
		20	4,228	9,010	3,174	17,901	6,307	59,155	20,841
60	150	40	4,165	8,316	2,886	17,607	6,111	57,247	19,869
		60	3,302	7.124	1.960	16.017	4.407	56,635	15,584
		20	3,977	10.193	3.378	18.934	6.275	57,607	19,093
80	150	40	3,799	10.061	3.185	18.777	5.945	56,739	17,963
		60	3,421	9.387	2.676	16,696	4,760	56.464	16,097
		20	4,119	10.976	3.768	14.698	5.046	56.446	19,384
100	150	40	3,792	10.597	3.349	14.792	4.674	54.495	17,219
		60	3,611	10.226	3.077	13.230	3.171	53.138	15,990
		20	3,998	10.388	3.461	18.980	6.130	57.918	18.706
60	170	40	3,802	9.671	3.064	18.663	5.913	56.148	17.788
		60	3,626	9.062	2.738	17.030	5.146	55.207	16.682
		20	3,889	7,714	2,5	18,055	5,851	55.121	17,864
80	170	40	3,785	6,175	1,948	16,637	5,248	52.098	16,434
		60	3,349	5,787	1,615	13,149	3,670	51.270	14,309
		20	3,849	9.464	3.036	20.181	3.849	56.623	18,163
100	170	40	3,920	8.926	2.916	18.879	3.920	55.630	18.173
		60	3,582	8.559	1.962	18.842	3.582	54.872	16.381
		20	3,628	8,676	2,623	19,529	5,904	57.907	17,505
60	190	40	3,289	8,454	2,556	17,806	4,881	53.608	14,695
		60	3,296	7,376	2,026	16,882	4,637	53.390	14,661
		20	3,589	9,029	2,403	18,951	5,668	53.725	16,068
80	190	40	2,982	8,202	2,038	18,222	4,527	53.325	13,249
		60	2,977	7,314	1,815	17,907	4,443	52.905	13,126
		20	3,813	10,193	3,239	20,097	6,386	55.368	17,679
100	190	40	3,702	7,087	2,186	20,553	6,341	52.075	16,065
		60	3,411	7,643	2,172	22,012	6,257	51.567	14,658

It can be seen from Table 1 that under all condition of temperature and pressure, concentration of lignin decreased compared to its initial concentration. The lowest lignin concentration of 13,126 g/L was obtained at 190 °C and 80 bar. The SCW process that is a physical process performed under high temperature and pressure effectively degraded lignin in coconut husk as also reported previously [4]. Lowering lignin concentration is very important to open access for enzymes to degrade cellulose and hemicellulose to produce reducing sugar.

3.2. Time Course of Sugar Yield on SCW Pretreatment

Table 1 also shows that the concentration of cellulose and hemicellulose also decreased indicating a degradation of both substances to sugar. The time course of sugar yield produced during the SCW process are shown in Figure 2. The yield of reducing sugar based on the weight ratio of the reducing sugar concentration to the initial cellulose and hemicellulose concentration was used instead of sugar concentration. This was in order to have better description of the degree of conversion of total cellulose and hemicellulose contained in coconut husk to reducing sugar. If cellulose and hemicellulose comprised of mainly monomeric sugars, the maximum yield obtained from SCW would therefore be 1.0.

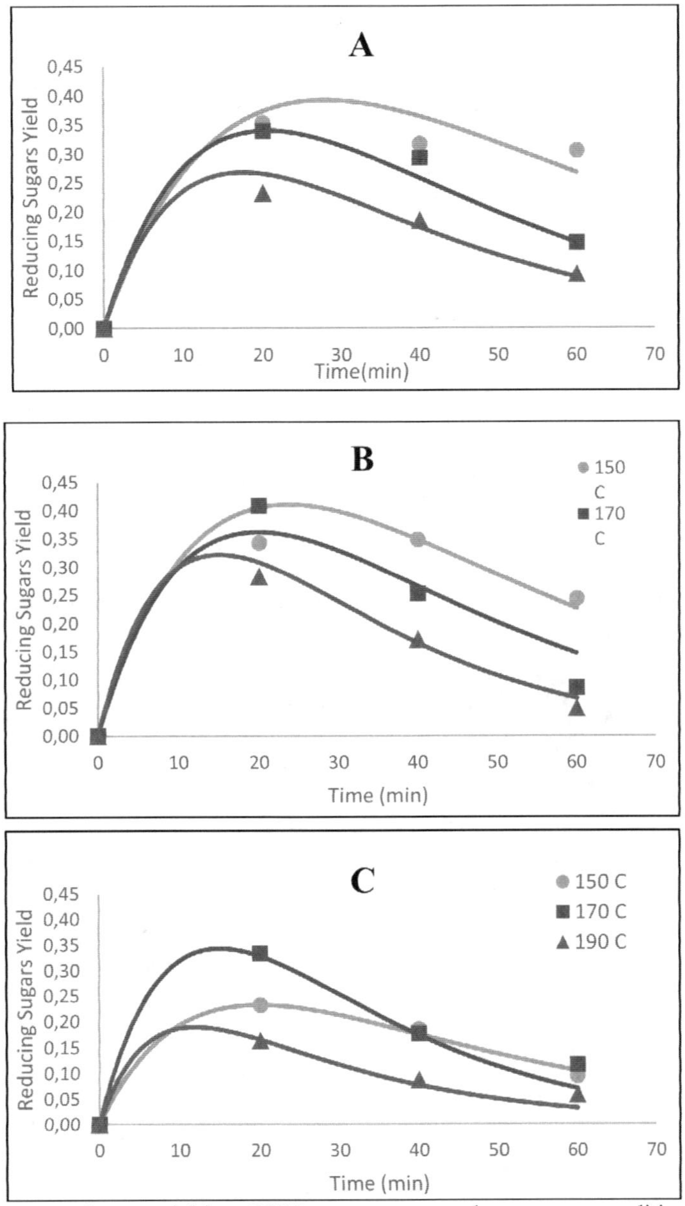

Figure 2. Time course of sugar yield on SCW pretreatment under pressure conditions: (A) 60 bar, (B) 80 bar, and (C) 100 bar.

SCW hydrolysis in coconut husk was studied at reaction time ranging from 20, 40 and 60 min. From Figure 2, TRS yield increased with increasing time from 0 to 20 minutes and then decreased with further

Journal of Physics: Conference Series **1373** (2019) 012006 doi:10.1088/1742-6596/1373/1/012006

increase of reaction time from 20 to 60 min. The increase of reducing sugar was resulted by the degradation of cellulose and hemicellulose to monomeric sugars such as glucose, xylose, arabinose. This coincided with the data of degradation of cellulose and hemicellulose in Table 1. The decrease in TRS yield was caused by decomposition of reducing sugars into the degraded product such as furfural. From the results of analysis by GC, furfural content after SCW was 21,7 mmol/ml under 170 °C and 60 bar. Degraded products such as furfural, 5-HMF (5-hydroxyl-methyl-furfural), dihydroxyacetone, glycol was produced by other studies [7, 13]. However, TRS yield at 80 bar and 150 °C increased until 40 min. This indicated that degradation of sugar under this SCW condition was a little bit slower. The highest TRS yield was 0,4094 obtained at 170 °C, 80 bar and the reaction time of 20 min.

3.3. Effect of Temperature on Reducing Sugar Concentration during SCW Pretreatment

The effect of temperature on reducing sugar concentration was studied as shown in Figure 3. Variation temperature on SCW pretreatment was 150°C, 170 °C and 190 °C which considered the range of low to medium severity factor as reported by Muharja et al. [1]. Higher temperature or longer time will be of higher severity condition that will result in higher degradation product. It is expected that the data chosen may cover the condition of good operating condition useful for real application of the system. Figure 3 shows that under this temperature condition reducing sugar concentration tended to decrease with an increased in temperature. This data indicated that the SCW condition employed in this experiment has high severity condition under which the sugars produced by the degradation of cellulose and hemicellulose was degraded into its degradation product such as furfural and HMF [1]. Under low to mild severity condition, the reducing sugar concentration should increase when temperature of SCW is increased. This was associated with increasing ionization constant (also called dissociation constant) at high temperatures by which the concentration of H^+ and OH^- increased. It caused broken down of the glycosidic bond to glucose with attack by electrophilic hydrogen atoms [14]. Some data in Figure 3 also shows an increase of sugar concentration when temperature was raised. This mostly occurred during the starting condition of 20 min which indicated that the severity condition did not give the degradation effect yet of the sugar produced.

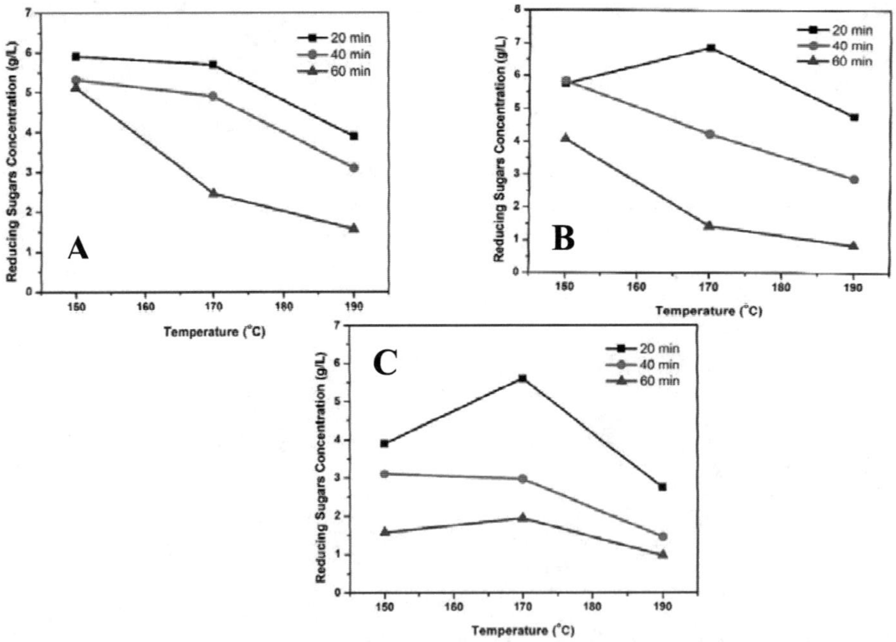

Figure 3. Effect of temperature on SCW pretreatment under pressure condition : (A) 60 bar, (B) 80 bar, and (C) 100 bar.

Journal of Physics: Conference Series **1373** (2019) 012006 doi:10.1088/1742-6596/1373/1/012006

3.4. Clarification of The Kinetics of The Coconut Husk Hydrolysis

The kinetics of the hydrolysis of coconut husk was investigated under different temperature and pressure. Temperature variables were 150, 170, and 190 °C while pressure were 60 bar, 80 bar and 100 bar. In this study, the reaction rate constants were obtained by fitting the experimental data using non-linear regression using solver in Excel 2017 based on Equation (7).

Table 2 shows the rate constants for the reducing sugars and degraded product at different pressure. k_1 is the rate constant values for the formation of reducing sugars and k_2 is the rate constant for the formation of degradation product. As can be seen, k_1 increased with the increasing temperature from 150 to 170 °C. The increase in k_1 was associated with the increase of H^+ concentration that the glycosidic bond in cellulose can be broken down by attacking electrophilic hydrogen from H^+ concentration [6, 14]. However, k_1 at 60 and 100 bar decreases. This result was in accordance with the experimental data that TRS yield increased when temperature was raised to 170 °C and then decreased above 170 °C. The decrease of rate constant was attributed to the degradation of reducing sugars to degraded product such as furfural. The formation of degradation products can be seen from the increase of k_2 when temperature was raised.

The prediction curve based on the obtained kinetic parameters k_1 and k_2 were plotted in Figure 2. It can be seen that the prediction curve fitted the experimental data satisfactorily with $R^2 > 0.9$.

Table 2. The reaction rate constants of the formation of reducing sugars and degradation product at different temperature and pressure

P = 60 bar			
T (°C)	k_1 (min^{-1})	k_2 (min^{-1})	R^2
150	0,03828	0,03356	0,94893
170	0,04483	0,05232	0,98289
190	0,04208	0,07519	0,96138
P = 80 bar			
T (°C)	k_1 (min^{-1})	k_2 (min^{-1})	R^2
150	0,0473	0,03756	0,94870
170	0,04931	0,05089	0,92862
190	0,05818	0,07519	0,96291
P = 100 bar			
T (°C)	k_1 (min^{-1})	k_2 (min^{-1})	R^2
150	0,03267	0,07323	0,99366
170	0,06223	0,07110	0,96188
190	0,04568	0,14001	0,94076

Table 3. Activation energy (Ea) and pre-exponential factor (A) of coconut husk hydrolysis in SCW

	Reducing Sugar		**Degraded Product**	
P (bar)	Ea (kJ.mol^{-1})	A (min^{-1})	Ea (kJ.mol^{-1})	A (min^{-1})
60	15,95	1,981	51,24	0,702
80	11,92	2,117	40,73	0,041
100	11,59	2,21	40,02	0,431

Table 3 shows the activation energy (Ea) and pre-exponential factor (A) for the formation of reducing sugars and degradation product in SCW. The parameters were derived from the plot of ln k vs 1/T based on Eq (8) in which the slope of the straight line indicated activation energy and the intercept indicated pre-exponential factor. It can be seen from Table 3 that Ea decreased as the pressure increased whether for the sugar and also for the degraded product. Furthermore, Ea for the formation of reducing sugar was always lower than that of the degraded product. Basically, lower Ea will result in a higher speed of catalytic process or reaction leading to higher product yield. Reactions with higher Ea are very temperature sensitive than those with lower Ea. Liang et al. [15] reported a different result in which the Ea for sugar formation was higher than that for the degraded product. For this case, it will be necessary to conduct the process at higher temperature to increase the performance of the process. The lower Ea of the sugar formation obtained in the present study is therefore beneficial in which it is not necessary to perform the process at higher temperature. Besides lower temperature of the process will prevent the degradation of sugar, this will also give merit in the viewpoint of energy efficiency.

4. Conclusion

The formation of reducing sugars and the degradation products during SCW pretreatment was investigated. The optimal condition with the highest TRS yield of 0,4904 was obtained at 170°C, 80 bar and reaction time 20 min. The decrease of reducing sugars above 170°C caused the decomposition of reducing sugars to degraded product such as furfural. Furfural content was obtained 21,7 mmol/ml. The rate constants of the formation of reducing sugar and degradation products was determined by fitting the experimental data with the proposed equation by a first order reaction model. The prediction curve fitted the experimental data satisfactorily.

Acknowledgments

The authors express sincere thanks for the funding of this research from The Ministry of Research and Technology Directorate of Higher Education Indonesia.

References

[1] Muharja M, Junianti F, Ranggina D, Nurtono T, Widjaja A 2018 *Bioresource Technology* **249** 268-275.

[2] Nanda S, Mohanty P, Pant K K, Naik S, Kozinski J A, and Dalai A K 2013 *Bioenergy Research* **6** 663–677.

[3] Lachoz-Perez D, Martinez-Jimenez F, Rezende C A, Tompsett G, Timko M, Forster-Carneiro T 2016. *Journal of Supercritical Fluids* **108** 69–78.

[4] Sangian H F, Ranggina D, Ginting G M, and Widjaja A 2015 *Scientific Study and Research: Chemistry and Chemical Engineering* **16** 13–27.

[5] Ravber M., Knez Ž., Škerget M. 2015 *J. Supercrit. Fluids* **104** 145–152.

[6] Lin R., Cheng J, Ding L, Song W, Qi F, Zhou J, and Cen K. 2015 *Bioresource Technology* **186** 8-14

[7] Prado J, Forster-Carneiro T, Rostagno M A, Follegatti-Romero L A, Filho F M 2014. *Journal of Supercritical Fluids* **89** 89–98.

[8] Guangyong Z, Zhu X, Fan Q, and Wan X 2011 *Journal of Analytical and Applied Pyrolysis* **90** 182-186.

[9] Datta R 1981 *Biotechnology and bioengineering* **23** 2167–2170.

[10] Miller G L 1959 *Analytical Chemistry* **31** 426–28.

[11] Gaspar E M S M, Lopes J F 2009 *J. Chromatogr. A* **1216** 2762–2767.

[12] Zhao Y, Lu W J, and Wang H T 2009 *Chemical Engineering Journal* **150** 411-417.

[13] Mohan M, Banerjee T, and Goud V V 2015 *Bioresource Technology* **191** 244-252.

[14] Zhu G 2013 *Clean Technologies and Environmental Policy* **15** 55–61.

[15] Liang J, Chen X, Wang L, Wei X, Wang H, Lu S, Li Y 2017 *Bioresour. Technol.* **228** 147–155.

Experimental Test of the Effect Wet Ethanol on Decreasing Emissions Exhaust in EFI Gasoline Engine with EGR System

M Arief Alfi Ardian, [1], Syaiful [1] dan Nazarudin Sinaga [1]

[1] Deprtement of Mechanical Engineering, Diponegoro University, Semarang, JL. Prof. Soedarto, Tembalang, Semarang, Indonesia

Email: ariefalfi.ardian@gmail.com

Abstract. The increase in the number of vehicles results in increased fuel consumption and air pollution. This also affects the more depletion of fossil fuel availability. An increasing number of these vehicles contribute to air pollution in urban areas. Hazardous substances such as NOx (Nitrogen Oxide), CO (Carbon Monoxide), and HC (Hydrocarbon) are harmful to health. In order to reduce the emissions of CO and HC, wet ethanol is one of the solutions to overcome these problems. Besides ethanol can be obtained from materials that are renewable. In fact, wet ethanol has a value of higher octane and higher oxygen content than gasoline. Blending both the fuel is expected to reduce air pollution. The purpose of this research is to obtain the best premium and wet ethanol percentage for lowering emissions of CO and HC. The fuel used is gasoline (88 octane number) and the wet ethanol (109 octane number). The percentage of wet ethanol blended into the fuel is in the range of 5% to 15%. Experiments conducted an engine speed of 2500 to 4000 rpm with an interval of 500 rpm. This study investigates the effects of wet ethanol and premium blends and the use of EGR (Exhaust Gas Recirculation) against exhaust emissions on a gasoline engine with EFI systems (Electronic Fuel Injection). From the results of this study is found that the exhaust gas temperature increases in the increase of engine speed. The addition of EGR reduces the exhaust gas temperature. The increase in BSFC (brake specific fuel consumption) without the largest EGR at we15 is 3.36% at 0.38 kg / kw. Hour, while the decrease in BSFC with the largest EGR at we5 is 17.61% at 0.30 kg / kw. Hour. Wet ethanol and gasoline blends cause a decrease in emissions of CO and HC. With a 15% wet ethanol blend, CO and HC emissions can be reduced to 60.23% and 44.30%, respectively. However, adding 15% wet ethanol into gasoline fuel increases 34.66% of CO2.

1. Introduction

The growth in the number of vehicles is increasing. More fuel consumption. The air pollution produced is higher. The fuel consumption of a motorized vehicle is affected by the acceleration process. Low octane value reduces the acceleration of motor vehicles. Fuels containing low octane values increase fuel consumption. The octane number is obtained by comparing the n-heptane and iso-octane content on a fuel. The higher the octane value of the fuel, the better the performance produced by the engine [1[. Wet ethanol is a derivative of alcohol. Wet ethanol has a fairly high octane value that can be used as a mixture of fuel in motorized vehicles. The heating effect on Hot EGR raises the combustion temperature and air-fuel mixture, thereby reducing HC and CO emissions [2].

Wet ethanol has a higher octane value than premium, which is 109. Experiments carried out by blend premium fuel (octane value 88) and wet ethanol with composition WE0, WE5, WE10 and

WE15 on a 4 cylinder gasoline engine 1781 cc EFI showed improved engine performance and increasingly good exhaust emissions [3]. Reduced emissions are affected by oxygen levels on wet ethanol. Adding oxygen to the fuel can improve the combustion process for the better [4]. The addition of methanol and ethanol to fuel can be used on gasoline engines and facilitate the initial ignition of the engine when cold. The resulting CO and HC emissions decreased [5].

In this study, wet ethanol was used as a mixture of premium fuel. The mixing is expected to improve engine performance and reduce exhaust emissions. Emissions include measurement of levels of (Carbon Monoxide) CO, (Hydrocarbon) HC, (Carbon Dioxide) CO2, and (Exhaust Gas Temperature) EGT. This research was conducted in consideration of whether or not the emissions produced using the mixture were better.

2. Research Methods

The fuels used in this study are the gasoline engine, castor oil, and wet ethanol. Premium fuel is produced by PT. Pertamina, Tbk. Wet ethanol was obtained from the Indrasari chemical shop, Semarang. The percentage of castor oil volume is 10% while the percentage of wet ethanol volume tested is 5%, 10% and 15% of premium fuel amount, respectively called P100WE0, P95WE5, P90WE5, P85WE15. Mixing fuel is prepared before the start of the experiment to ensure that the mixture is homogeneous. A stirrer was used when mixing the fuel.

Table 1. Fuel properties

Properties	Wet-Ethanol	Premium
Octane Number	109	88.8
Water Content (%v)	>5.0	0.003
Viscosity	1.623	0.22
Density Temperature15°C(kg/m^3)	0.8974	744
Number Value (MJ/kg)	21.73	42.69
Flash Point (°C)	24	7.22
Laten Heat (KJ/Kg)	1139	328.91

Table 2. Machine specifications

Engine Toyota Kijang 7K	
Engine Type	Gasoline engine
Production	Toyota
Number of Cylinders	4 Straight
Engine Capacity	1781 cc
Valve	(SOHC) 8 valve
Maximum Power	94 hp - 5000 rpm
Maximum Torque	155 Nm – 3200 rpm
Fuel System	EFI

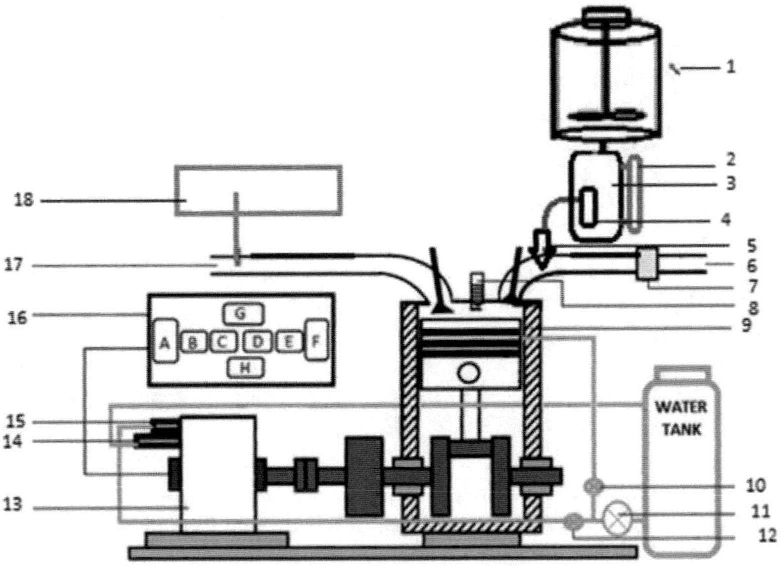

Figure 1. Experimental set-up

The experiment was carried out by using a 4K 7K Toyota Kijang gasoline engine with 4 cylinders with a capacity of 1781 cc and 2500, 3000, 3500 and 4000 rpm engine speed variants. Measuring the flow rate of fuel using a stopwatch by determining the volume of fuel that runs out on each test. Mixing fuel is done using a mixer (1). In the tank, there is a burette (2). Fuels that have been homogeneously mixed are then channeled to the tank (3) and fuel pump (4). A burette is used to determine the volume of fuel used. When the gasoline engine is working, loading is carried out with a constant load measured using a water brake type DYNOmite Land and Sea dynamometer (13). This is done to measure torque and engine power. All measurement data is informed by the display panel (16). At the intake manifold channel and the exhaust manifold are installed thermocouple to measure air temperature. At the exhaust manifold, a Gas Analyzer 898 (18) is installed. to measure exhaust gas HC, CO, CO2, and O2.

3. Result and Discusion

3.1. Test results of Brake Spesific Fuel Consumption (BSFC)

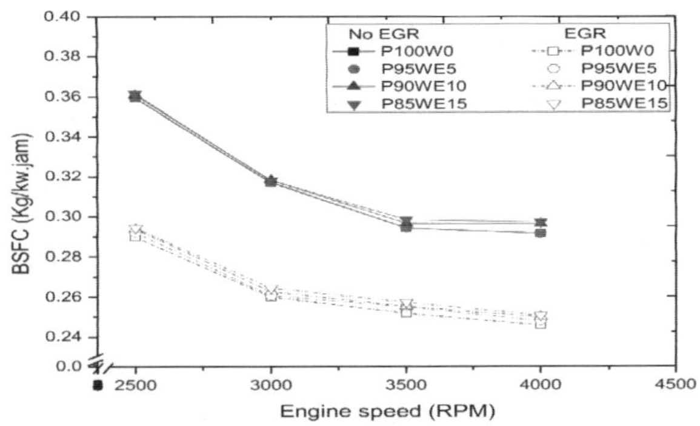

Figure 2. Graph of the effect of wet ethanol on Brake Specific Fuel Consumption (BSFC)

Figure 2 shows the results of BSFC testing with variations in engine speed and fuel blends with EGR and without EGR. Effect of variations in engine speed and fuel mixture with EGR and without EGR on BSFC. The test results show a decrease in fuel consumption along with increasing engine speed and variations in the use of fuel mixtures. The use of EGR shows that fuel consumption is lower than without EGR. BSFC reduction with the use of wet ethanol due to more oxygen content increases the combustion efficiency [6]. BSFC declined as the engine load increased. Wet ethanol mixture BSFC is higher in some engine operating conditions, due to lower heating value than gasoline. Low calorific value on wet ethanol because it contains 50% oxygen in the fuel so it does not contribute to further heat during combustion in the cylinder [4]. BSFC rises as the percentage of wet ethanol rises. The low energy content of the wet ethanol-gasoline mixture causes an increase in BSFC [7]. The value of heat and oxygen atoms in a higher mixture of wet ethanol-gasoline causes an increase in fuel consumption [8]. With energy content and heat value of ethanol/methanol which is lower than gasoline causes an increase in fuel consumption compared to using gasoline [9]. BSFC is lower on low load operating machines with EGR than machines without EGR [10]. BSFC methanol vapor and methanol gas are lower than gasoline, this is due to increased fuel heating value when the fuel is heated with exhaust heat before being injected into the intake manifold so as to increase thermal efficiency, make BMEP rise and lower BSFC compared to gasoline [11]. Whereas according to the use of EGR can reduce BSFC and (nitrogen oxide) NOx emissions [12]. The increase in BSFC without the largest EGR at WE15 is 3.36% at 0.38 kg/kW. Hour, while the decrease in BSFC with the largest EGR at WE5 is 17.61% at 0.30 kg/kW. Hour.

3.2. The test result in hydrocarbon (HC)

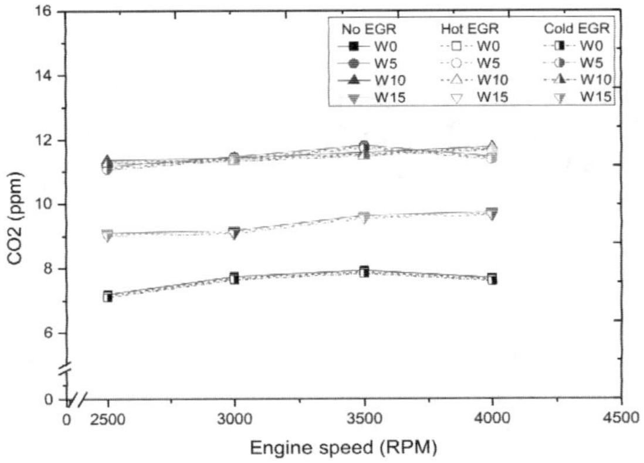

Figure 3. Graph of wet ethanol effect on hydrocarbon (HC)

Figure 3 shows that HC emissions formed from combustion results decrease in each mixture. The use of Cold EGR with P100 HC fuel formed reaches 79 ppm at 4000 rpm engine speed. The most decrease in the P85WE15 mixture reached 34 ppm at 4000 rpm engine speed. The decline reached 56.96%. Decreasing HC occurs due to increased oxygen levels in the combustion reaction. Wet ethanol contains oxygen compounds. When wet ethanol is mixed with the premium, there is an increase in oxygen content in the combustion chamber. Increased oxygen levels trigger the reaction of hydrogen with oxygen more than hydrogen with carbon [13].

3.3. The test result in carbon dioxide (CO2)

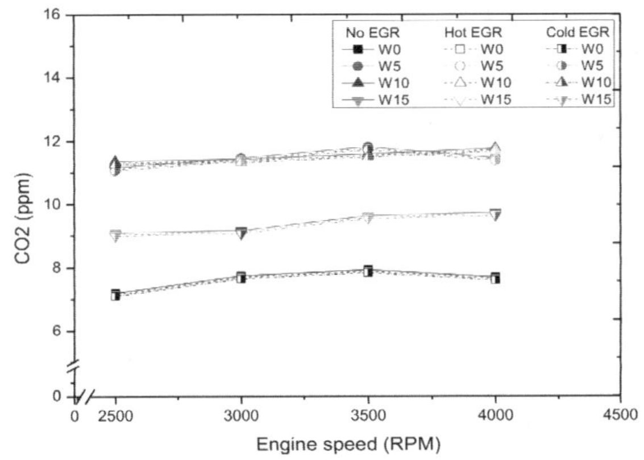

Figure 4. Graph of wet ethanol effect on carbon dioxide (CO2)

Figure 4 CO2 content produced by the P85WE15 mixture reaches 1.1% at 2500 rpm engine speed. At 4000 rpm engine speed, there was a decrease in CO2 levels in the P85WE15 mixture which produced CO2 levels of 1.02%. Increased O2 levels and CO2 levels are influenced by oxygen compounds in alcohol derivatives. The addition of wet ethanol increases combustion efficiency. Combustion efficiency increases Brake Thermal Efficiency (BTE) and CO2 emissions. CO emission levels decrease [14]. Alcohol fuel causes increased CO2 emissions. The ratio of oxygen in alcohol produces combustion efficiency and causes higher CO2 emissions [15]. Wet ethanol has a higher speed of fire spread compared to gasoline, and can increase octane in gasoline and reduce CO and HC emissions [3]. Wet ethanol contains oxygen so that when experiencing the combustion process there is an increase in O2 and CO2 levels in the remaining combustion results [16].

3.4. The test result in carbon monoxide (CO)

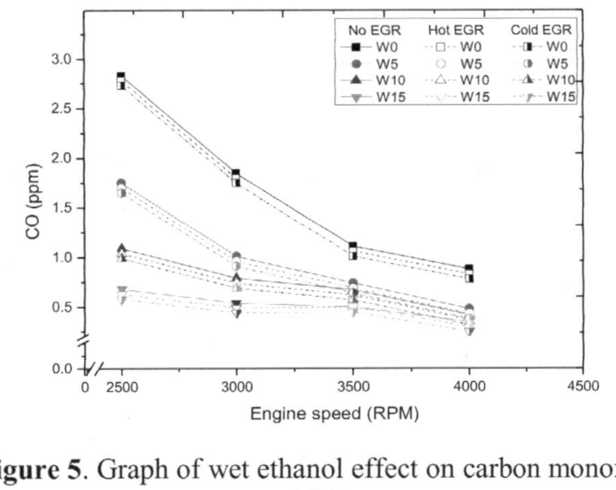

Figure 5. Graph of wet ethanol effect on carbon monoxide (CO)

Figure 5 The highest decrease in CO content occurred in the P85WE15 mixture with Cold EGR. The CO P85WE15 level reaches 10% at 4000 rpm engine speed. The decrease is influenced by the combustion process and the oxygen content on wet ethanol [17]. The addition of wet ethanol to gasoline can reduce CO emissions because the higher oxygen content contributes to the formation of CO2 and NOx. In addition, there is a decrease in CO emissions when using wet ethanol instead of gasoline [18]. Other research also shows that a mixture of ethanol-gasoline (E5 and E10) and methanol-gasoline (M5 and M10) fuels used in Honda Civic 1.4i cars can reduce vehicle exhaust emissions in the form of HC and CO significantly at vehicle speeds of 80 km/hours [4].

3.5. The Test Result in Exhaust Gas Temperature (EGT)

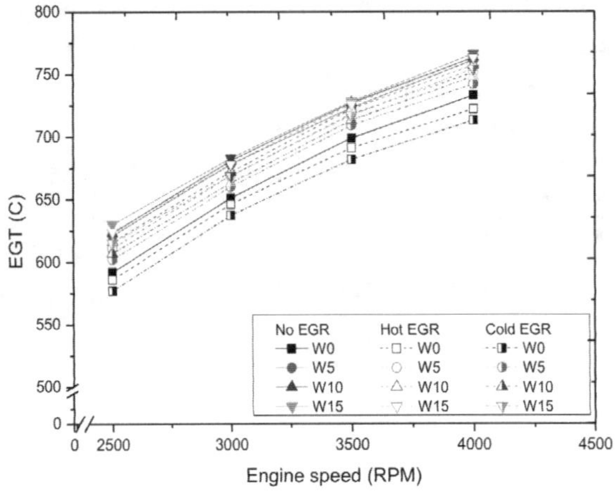

Figure 6. Graph of wet ethanol effect on *Exhaust Gas Temperature (EGT)*

Figure 6 presents a graph of the results of testing the exhaust gas temperature (EGT). In general, the use of gasoline engine, premium, and wet ethanol blends fuels shows an increase in exhaust gas temperature. The highest increase in exhaust gas temperature occurred in the P85WE15 mixture which reached 2.86%. The P85WE15 mixture produces a flue gas temperature of 754 ^{o}C at 4000 rpm engine speed. Premium produces an exhaust gas temperature of 733 ^{o}C at 4000 rpm engine speed. Increasing the exhaust gas temperature which is affected by latent heat on wet ethanol is higher than the premium. Wet ethanol has a higher combustion speed than premium. Oxygen levels on wet ethanol affect the heat generated in the combustion process. This causes the combustion of the fuel in the cylinder to be more perfect [5].

4. Consclusion

The following are the main conclusions obtained from research on engine performance and exhaust emissions from the EFI gasoline engine with premium and wet ethanol mixed fuels:

1. The use of EGR shows that fuel consumption is lower than without EGR.
2. The emission of exhaust gas produced in the testing machine decreases. It can be seen from the decreasing levels of emission of CO and HC contained in the residual gas from combustion.
3. Oxygen levels in wet ethanol reduce the levels of HC and CO formed and increase the levels of O2 and CO2.

Acknowledgment

The author would like to thank the entire Thermofluid Laboratory team, which is one of the laboratories owned by the UNDIP Mechanical Engineering Department. This laboratory examines the phenomena that occur in the existing machining process and machine performance testing, especially in the field of Energy Conversion.

Reference

[1] Anderson JE, DiCicco DM, Ginder JM, Kramer U, Leone TG, Raney-Pablo HE and Wallington TJ 2012 *Fuel* **97** 585-594.
[2] Wei H, Zhu T, Shu G, Tan L and Wang Y 2012 *Applied Energy* **99** 534-544.
[3] Syaiful MN and Bae MW 2016 *ARPN Journal of Engineering and Applied Sciences* **11** 4065-4069.
[4] Elfasakhany A 2015 *Engineering Science and Technology, an International Journal* **18** 713-719.
[5] Liu S, Clemente ERC, Hu T and Wei Y 2007 *Applied Thermal Engineering* **27** 1904-1910.
[6] Elfasakhany, A. 2017 *Renewable and Sustainable Energy Reviews* **71** 404-413.
[7] Siwale L, Kristóf L, Bereczky A, Mbarawa M and Kolesnikov A 2014 *Fuel Processing Technology* **118** 318-326.
[8] Lin SL, Lee WJ, Lee CF and Chen SJ 2010 *Energy & Fuels* **24** 4522-4533.
[9] Iliev S 2015 *Procedia Engineering* **100** 1013-1022.
[10] Elfasakhany A 2016 *Engineering Science and Technology, an International Journal* **19** 1224-1232.
[11] Balki MK, Sayin C and Canakci M 2014 *Fuel* **115** 901-906.
[12] Zhang Z, Zhang H, Wang T and Jia M 2014 *Energy* **65** 18-24.
[13] Canakci M, Ozsezen AN, Alptekin E and Eyidogan M *Renewable Energy* **52** 111-117.
[14] Zhao H, Ge Y, Hao C, Han X, Fu M, Yu L and Shah AN 2010 *Science of the Total Environment* **408** 3607-3613.
[15] Chang YC, Lee WJ, Lin SL and Wang LC 2013 *Applied Energy* **109** 182-191.
[16] Elfasakhany A 2016 *Journal of Cleaner Production* **139** 1057–1067.
[17] Rahman KM, Kawahara N, Tsuboi K and Tomita E 2016 *Fuel* **165** 331–340.
[18] Fagundez JLS, Sari RL, Mayer FD, Martins MES and Salau NPG 2016 *Applied Thermal Engineering* **112** 317-325.

Maximizing Power Tracking of an Offshore Wind Turbine Using Perturb and Observe Method

R Ardiles[1], S Hasan, Z Pane[2]*, S Suherman[3]

[1,2,3]Electrical Engineering Department, Universitas Sumatera Utara, Medan-Indonesia
*Email: zulkarnaen.pane@usu.ac.id

Abstract. As fossil energy sources have been limited, renewable energy sources are now being explored. The wind energy is potential in the coastal area of Indonesia such as Nias island. There is only 0.5% of the potential energy sources in this area has been used. This article studies the wind energy prospect and simulates the implementation of maximum power point tracking based on perturb and observe method for permanent magnet synchronous generator of wind energy converter. Study shows that the method is able to enhance output power of wind energy generator about 7.22 % in average with shorter tracking time of 0.67 s.

1. Introduction

Wind energy can be used as the alternative energy source for electricity. The potential wind sources can be found along the southern coastal line of Java island, Sumatra, and islands in the East Indonesia. According the blueprint of national energy conservation 2005-2025 of energy source ministry, the prospect of wind based electrical generation is about 9.29 GW, but only about 0.5% (0.0006 GW) is used [1].

As one of the prospective islands, Nias island located in North Sumatra is still supported by Diesel based electrical generation. This causes the high cost of electricity production so that the amount of the generated energy is limited due to limited budget. Beside efficiency step [2], alternative energy source should be explored. In fact, various energy sources are available in Nias [3]. One of the developed solutions is the offshore wind turbine.

In order to optimize the electrical energy generation, form the wind energy, many ways have been proposed to intelligently forecast the potential sources, maximize the control generation system, and efficiently convert the energy. Maximum power point tracking (MPPT) is aimed at achieving maximum point of the observed system. The wind turbine has a specific output power versus rotor speed characteristic, where at a certain speed point turbine produces a maximum power. The MPPT within the wind turbine is to maintain generator producing the maximum power. Some control systems to realize MPPT are tip speed ratio (TSR), power signal feedback (PSF) and hill-climb search (HCS) [4]. The TSR adjust requires wind speed data through an anemometer. Data is used to vary the actual rotor speed by changing the PWM inverter switching ratio. This method may reduce the MPPT efficiency if wind speed is inaccurate [5]. The PSF on the other hand, uses turbine performance data in order to determine usable wind energy based on pitch angle of the propeller. The pitch angel adjustment needs properly performed to enable maximum output power [6].

The MPPT based on Perturb and Observe (P&O) method is potential to be applied on wind turbine. The method is able to locating the maximum power generation point with shorter processing time. The maximum power can be approximated without knowing the turbine characteristics as well as wind speed. The maximum power is approximated by using duty cycle adjustment of the pulse width modulation (PWM) and by observing the maximum power point curve [7]. The method uses "track-memorize-reuse" where the system learns from the start. Data collection uses the hill-climb search so that the decision is based on the memory [8]. The method depends on the relation of output power variation versus rotor speed as in Figure 1. The system generally consists of the PMSG, DC-DC power converter and controller. The controller uses PWM and duty cycle.

MPPT P&O control rotor speed (ω) to be always in optimal point for any wind speed based on input mechanic torque (Tm) and electromagnetic torque (Te) [5] as shown in Equation 1:

$$P_{m-opt} = \frac{0.5\,\pi\,C_p\,R^5}{\lambda^3{}_{opt}}\,\omega_r{}^3 = K_{opt}\,.\,\omega_r{}^3 \qquad (1)$$

The generated voltage and current are approximated by Equation 2. In order to simplify the controller, MPPT P&O uses a straight-line approach, K_r (Equation 3) to the output characteristics of the generator (Figure 1) instead of using wind speed sensor [5]:

$$E_g = p\,\omega_r\,\lambda_f = \omega_e\,\lambda_f \qquad (2)$$

$$\mathrm{V_{dc}} = \mathrm{K_r}\,.\,\omega\mathrm{e} \qquad (3)$$

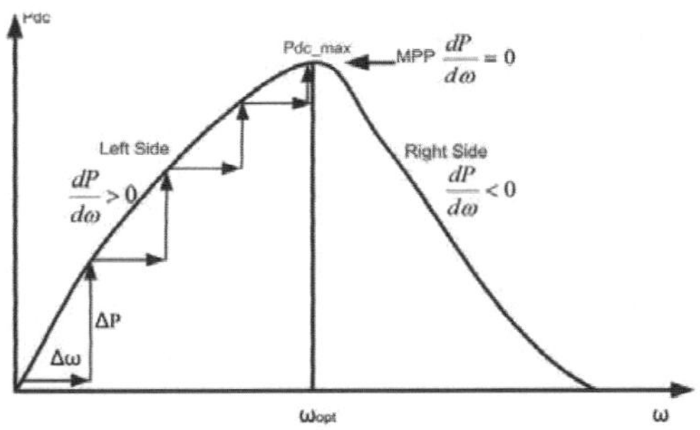

Figure 1. The output power and rotor speed characteristics

This article reports the method implementation on offshore wind turbine in Nias where the wind characteristics in this area are changing overtime. The MATLAB based simulation is performed to simulate the system and to analyze the turbine performances. The implementation is based on the real data where a permanent magnet synchronous generator (PMSG) is employed. The following section describes the research method, the observation results are discussed afterward.

2. Research Method

In order to model system, wind speed information in east Nias coastal from the meteorological station is used as shown in Figure 2. The PMSG used for generating electrical energy has specifications as shown in Table 1.

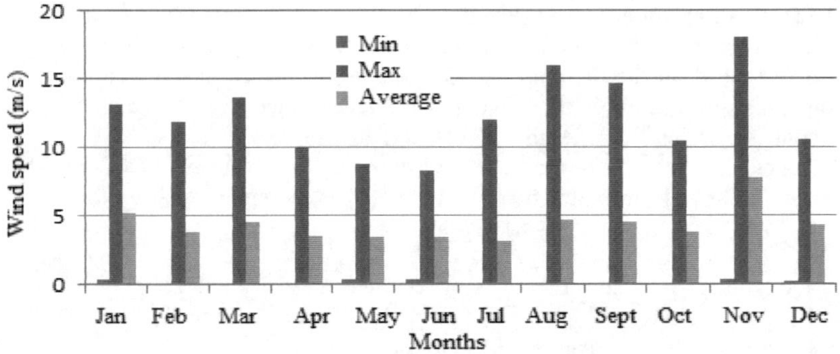

Figure 2. Wind speed in Nias

Table 1. PSMG specifications

Parameter	Symbols	Values
Power (W)	P	6000
Nominal rotor speed (rad/s)	Ω	153
Nominal wind speed (m/s)	V	12
Polar pairs	P	5
Stator resistance (Ω)	R_s	0.425
Inductance of Stator d-axis (mH)	L_d	8.4
Inductance of Stator q-axis (mH)	L_q	8.4
Magnetic flux linkage (Wb)	λ_f	0.433
Torque (Nm)	T	40
Damping coefficient (Nm/s)	B_m	0.001189
Rotor inertia (kg.m^2)	J	0.01197

Figure 3 shows wind turbine model implementation in MATLAB. The assessment is to see the power output, rotor torque of the system for the average wind speed in Table 1 load resistance of 25 to 1000 Ohm.

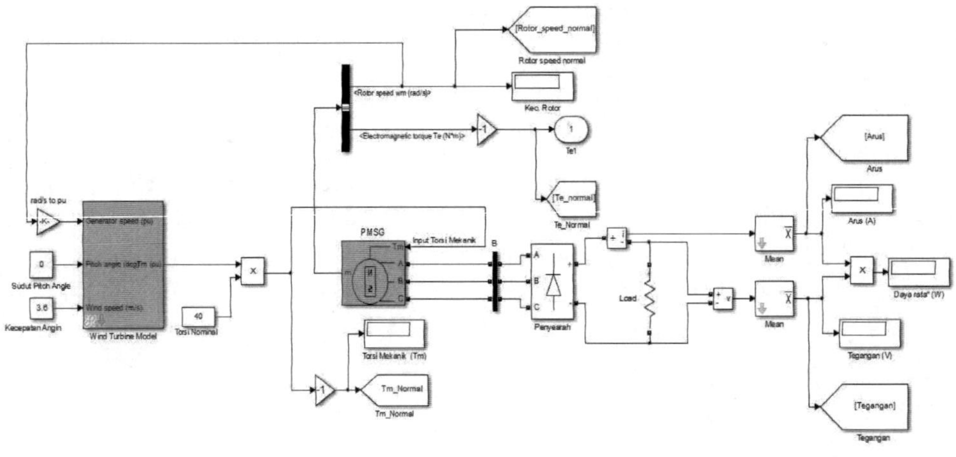

a) Wind turbine model without MPPT P&O

Journal of Physics: Conference Series **1373** (2019) 012008 doi:10.1088/1742-6596/1373/1/012008

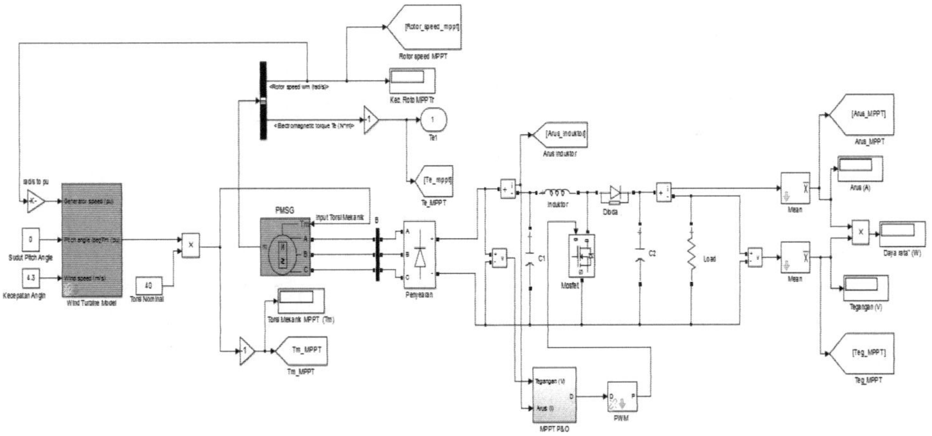

b) Wind turbine model with MPPT P&O

Figure 3. Wind turbine model in MATLAB

3. Results and Discussion

Simulations result show that output power, voltage, torque, inductor current, rotor speed and duty cycle change when MPPT applied on wind turbine. The maximum power for wind speed of 3.2 m/s achieves 95.79 Watt when MPPT applied, it increases about 11.63% from 85.81 Watt at the basic condition. The maximum value is obtained at t=0.7 s (Figure 4), faster than original condition, 0.9 s. The output voltage increases from 131 Volt to 138.4 Volt (not shown).

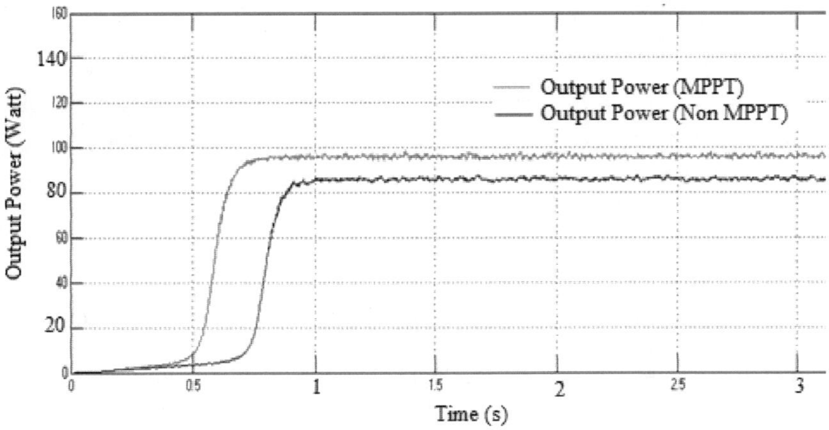

Figure 4. Output power comparisons for windspeed 3.2 m/s

By the addition of MPPT P&O, the maximum power is obtained by electromagnetic torque (T_e) of 2.05 Nm. Much smaller than without MPPT, 2.27 Nm. The average inductor current (I_L) caused by PWM switching at boost converter is about 0.70 Ampere at t=0.7 s. The evaluation results for various wind speed is shown in Tabel 2.

The method consistently increases the output power for all wind speed from 2.56% to 11.6%. Although this improvement is weaker when wind speed increases. The output power is exponentially increasing to wind speed, but the rising time is slower.

Table 2. Output power comparisons

Wind speed (m/s)	R (Ohm)	Non MPPT Power (W)	Non MPPT Rise time t (s)	MPPT P&O Power (W)	MPPT P&O Rise time t (s)	Increment (%)
3.2	200	85.81	0.90 s	95.79	0.70 s	11.63 %
3.4	200	99.90	0.90 s	110.00	0.80 s	10.11 %
3.5	200	106.00	0.80 s	117.00	0.80 s	10.38 %
3.6	150	133.58	1.70 s	144.00	1.00 s	7.80 %
3.8	150	151.76	1.20 s	164.00	0.80 s	8.06 %
3.9	150	162.71	0.95 s	175.00	0.70 s	7.55 %
4.3	150	206.62	0.70 s	222.00	0.60 s	7.44 %
4.6	150	242.32	0.60 s	252.98	0.55 s	4.40 %
4.7	150	253.50	0.60 s	266.00	0.54 s	4.93 %
5.3	150	339.38	0.50 s	355.00	0.50 s	4.60 %
7.8	100	1072.56	0.40 s	1100.00	0.40 s	2.56 %

4. Conclusions

This paper has discussed the maximum power point tracking by using perturb and observe (MPPT P&O) on the offshore wind generator of East Nias, Indonesia. This study was conducted as the matter of fact, wind energy in the area is potential, but is not well explored. The study is carried out by taking the measurement data into simulation model on MATLAB. The simulation shows that the MPPT P & O method is able to cope with various wind speed and consistently increase the maximum output power with shorter rise time. The method is able to enhance output power of wind energy generator about 7.22 % in average with shorter average tracking time of 0.67 s.

5. References

[1] Blueprint Pengelolaan Energi Nasional 2006-2025," Arsip ESDM, vol. Kementeria, 2006.

[2] Bakhtiar B and Suherman 2015 *Jurnal Arus Elektro Indonesia* **1** 23-29.

[3] "Analisis Keekonomian Penerapan Pembangkit Listrik Tenaga Surya Pada Sistem Ketenagalistrikan Nias," EBT dan Konserv. Energi, vol. 14, p. Puslitbang Ketenagalistrikan, 2015.

[4] J. Modelling, "MPPT control in wind energy conversion systems and the application of fractional control (PI α) in pitch wind turbine Sihem Ghoudelbourk Djalel Dib * Amar Omeiri Ahmad Taher Azar *," vol. 26, no. 2, pp. 140–151, 2016.

[5] Putri, R.I., Pujiantara, M & Priyadi, A. 2018. Maximum Power Extraction Improvement using Sensorless Controller Based on Adaptive Perturb and Observe Algorithm for PMSG Wind Turbine Application. *IET Electric Power Applications*, Vol. 12, Issue 4.

[6] Jasmin M. 2007. *Modelling and Control of Wind Turbine*. Master Thesis (London: Imperial College London).

[7] Zammit, D., Staines, C. S., Micallef, A., & Apap, M. (2017, June). MPPT with current control for a PMSG small wind turbine in a grid-connected DC microgrid. *In Research and Innovation on Wind Energy on Exploitation in Urban Environment Colloquium* pp. 205.

[8] Hui J C, Bakhshai A, and Jain P K 2016 *IEEE Transactions on Power Electronics* **31** 4861-4875.

Journal of Physics: Conference Series **1373** (2019) 012009 doi:10.1088/1742-6596/1373/1/012009

Effect of Wet-Ethanol on Fuel Consumption and Soot Emissions on Direct Injection Diesel Engines with Jatropha and Solar Cold EGR System

Ahmad Faoji, [1], Syaiful [1] dan Nazarudin Sinaga [1]

[1] Deprtement of Mechanical Engineering, Diponegoro University, Semarang, JL. Prof. Soedarto, Tembalang, Semarang, Indonesia

Email: faojia9@gmail.com

Abstract. Diesel engines have excellent performance and high efficiency. The increasing use of diesel engines has resulted in increased use of fossil fuels and air pollution. Addition of castor oil to diesel fuel increases smoke emissions due to higher viscosity than diesel fuel. Therefore the addition of wet-methanol in diesel fuel and European oil can solve the problem of diesel engines. This study aims to evaluate the effect of adding wet ethanol to fuel consumption and soot diesel engine emissions. The test uses the Isuzu 4JB1 engine equipped with the Cold EGR system. The percentage mixture of wet ethanol fuel is 5%, 10%, and 15% based on volume. The results showed that the addition of Wet-ethanol to diesel and fuel mixtures in Europe caused a decrease in the value of fuel consumption and soot emissions. The highest reduction in smoke opacity occurred on the DJ10WE15 fuel of 16% while the reduction in fuel consumption occurred in DJ10WE10 fuel by 11%.

1. Introduction

Diesel engines are an essential part of the transportation mode. The use of diesel engines will continue to increase as the volume of motorized vehicles increases. However, smoke emissions continue to be the biggest threat to the diesel engines for health and the environment [1]. Besides, the use of diesel engines in depletion of petroleum reserves [2]. Jatropha is an alternative to fuel to overcome fears of depletion of petroleum reserves. Jatropha was chosen because it was produced from non-food ingredients [3].

The use of jatropha as a mixture of material in diesel fuel produces higher soot emissions than diesel fuel. Besides, the value of power and torque produced by diesel engines is also lower than diesel fuel because of high viscosity and low calorific value of jatropha [4].

According to Sugeng and Syaiful, the addition of liquid alcohol in a mixture of diesel fuel and Europe can reduce the emissions of diesel engines. The higher the alcohol content, the more soot emissions decrease [5]. Wet-ethanol is a high-octane derivative of alcohol so that it can be used as a mixture of fuel. Based on the tests conducted by Abdul K. et al., The addition of ethanol to diesel fuel produces soot emissions and lower fuel consumption than pure diesel fuel. High oxygen content and low viscosity in ethanol are the main factors in the decline. The same tendency also occurs in Cenk S study (2010) [6]. Based on the above, this study observed the effect of using ethanol on fuel consumption and soot emissions from the EGR cold system.

2. Research Methods

The fuels used in this study are diesel, castor oil, and wet ethanol. Diesel fuel is produced by PT. Pertamina, Tbk. jatropha and wet ethanol were obtained from the Indrasari chemical shop, Semarang. The percentage of castor oil volume is 10% while the percentage of wet ethanol volume tested is 5%, 10% and 15% of diesel fuel amount, respectively called DJ10WE5, DJ10WE10, DJ10WE15. Mixing fuel is prepared before the start of the experiment to ensure that the mixture is homogeneous. A stirrer was used when mixing the fuel.

Table 1. Fuel properties

Properties	Diesel	*Jatropha*	Wet-ethanol
Cetane Number	48	41.8	5-8
Water Content (% v)	0.05	3.16	1.8
Viscosity (at 400 c (mPa.s))	2.0-5.0	3.23	2.51
Calorific Value (MJ / kg)	45.21	37.97	26.8
Flash Point (0C)	60	198	38
Oxygen content (%)	-	10.9	34.8

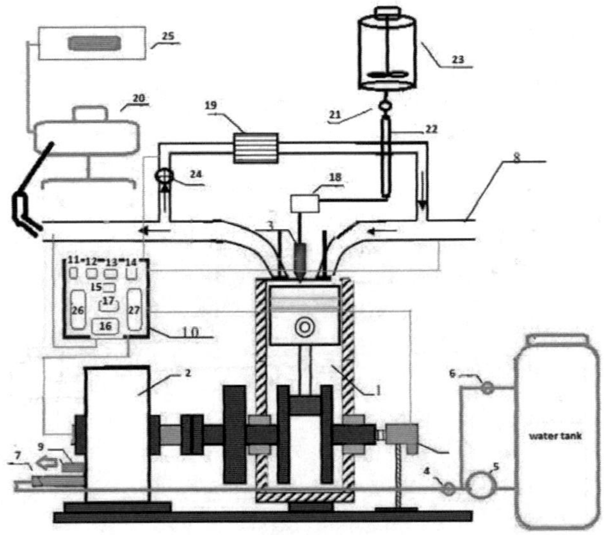

Figure 1. Experimental set-up

Tests carried out with the Isuzu 4JB1 4-cylinder direct injection diesel engine. Mixed diesel, oil distance, and wet ethanol, fuels have arranged a percentage before the experiment begins. Because of the high polarity of wet ethanol, it is difficult to blend with diesel, and the mixer 23 is used to obtain a homogeneous fuel mixture. The location of the mixer is placed higher than the engine so that the fuel mixture flows into the fuel injector (3) via valve (22) and burette (21) based on the principle of gravity and assisted by the fuel pump 18 installed on the engine. A burette (21) is used to measure fuel consumption, where each volume of fuel is 30 ml calculated in their time. After that, the fuel is sprayed into the combustion chamber using fuel injection.

A tachometer (17) was used to monitor the engine speed detected by proximity sensors. A dynamometer (2) was applied to measure the brake torque produced by a diesel engine. This was done by engine shaft coupling with the dynamometer shaft. Engine load varies from 25% to full load capability achieved by the flow rate of water flowing to the dynamometer. Water was pumped by a water pump (5) to the dynamometer. The amount of torque released was read by a tension load sensor mounted on the torque arm on the dynamometer then tells on the load screen (16).

The experiments were carried out at a fixed engine speed of 2500 rpm with variations in the concentration of mixed fuels. Some of the exhaust gas is either directly recirculated (Hot EGR) or cooled by heat exchangers (Cold EGR) (19). The percentage of EGR leads to the intake manifold (8) for mixing with fresh air through the inlet valve openings (24). In each variation, the test was carried out three times, and the uncertainty value in percent was taken from the test results.

The thermocouple K type was installed to measure temperature in the exhaust manifold, EGR inlet, EGR outlet, and the intake manifold and engine block where the measurement results are notified by the thermocouple display (11-15). In order to measure the air mass flow rate and EGR level, the orifice plate applied by U manometer 1 (27) and U manometer 2 (26) was mounted on the intake manifold and the EGR input channel. As for measuring smoke emissions, meter smoke (OTC 495) (20) plus the gas analyzer display (25) was applied.

3. Result and Discusion

3.1. Test results of Brake Spesifik Fuel Consumption (BSFC)

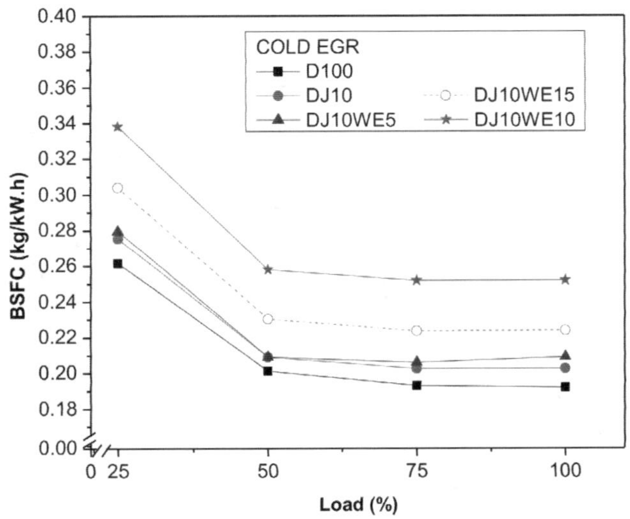

Figure 2. Graph of the effect of wet ethanol on Brake Specific Fuel Consumption (BSFC)

Figure 2 shows the BSFC value of a decrease in each fuel test. The physiochemical properties of fuels, such as viscosity and oxygen content in wet ethanol, affect the combustion of diesel engines [7]. However, the presence of jatropha and wet ethanol in diesel fuel causes the BSFC value of mixed fuels to be higher than pure diesel (D100) because of lower calorific value of jatropha oil [8]. The highest BSFC value is observed in the use of DJ10WE10 fuel by 11% for 25% load.

Journal of Physics: Conference Series **1373** (2019) 012009 doi:10.1088/1742-6596/1373/1/012009

a. The test results in smoke opacity

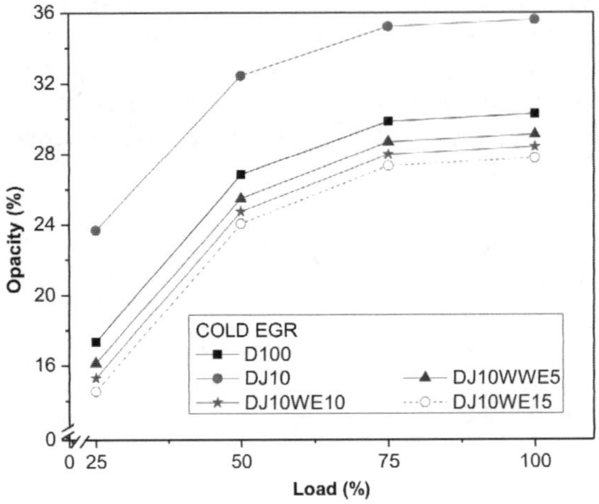

Figure 3. Graph of wet ethanol effect on *Smoke Opacity*

Figure 3 describes the smoke opacity test results. The smoke opacity value increases as the load percentage increases. However, the addition of wet ethanol to a mixture of diesel fuel and jatropha produces a decrease in smoke opacity because the high oxygen content in wet ethanol promotes carbon oxidation [9]. The highest reduction in smoke opacity is for the use of 16%DJ10WE15 fuel at 25% load.

b. The Test result in Exhaust Gas Temperature (EGT)

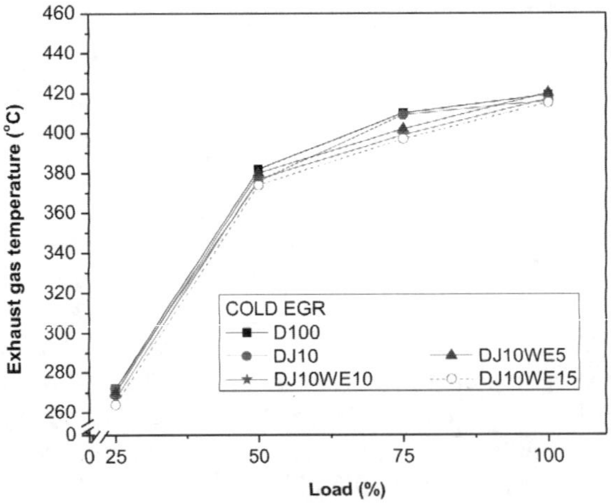

Figure 4. Graph of wet ethanol effect on *Exhaust Gas Temperature (EGT)*

Figure 4 presents a graph of the results of testing the exhaust gas temperature (EGT). In general, the use of diesel, jatropha, and wet ethanol mixed fuels shows an increase in exhaust gas temperature. But the EGT value is still lower than pure diesel. The poor atomization of fuel tends to reduce the quality of combustion [10]. Besides, the low calorific value and high viscosity in jatropha as blended fuels result in lower heat compared to D100 [11]. The highest decrease in EGT value is found for the use of DJ10WE15 fuel by 3% at 100% load.

4. Consclusion

The following are the main conclusions obtained in the study of the effect of low levels of wet ethanol on BSFC and Smoke Opacity on direct injection diesel engines with EGR systems R:

a. The addition of wet ethanol to the use of a mixture of diesel fuel and jatropha produces a higher BSFC value than D100.
b. High oxygen content and low viscosity in wet ethanol cause smoke opacity to be lower than D100.
c. The heating value contained in jatropha and wet ethanol also results in a decrease in the value of exhaust gas temperature.

Acknowledgment

The author would like to thank the entire Thermofluid Laboratory team, which is one of the laboratories owned by the Undip Mechanical Engineering Department. This laboratory examines the phenomena that occur in the existing machining process and machine performance testing, especially in the field of Energy Conversion.

Reference

[1] Altuna S, Bulut H and Oner C 2008 *Renewable Energy* **33** 1791–1795.
[2] Sobri and Syaiful 2014 *Effect of Methanol on Reducing Soot of Diesel Engine Exhaust Gas with Cold EGR Systems Using Biosolar and Jatropha Mixed Fuels* (Semarang: Proceedings of the 4th SNST 2013).
[3] Pandey VC, Singh K, Singh, JS, Kumar A, Singh B, Singh RP 2012 *Renewable and Sustainable Energy Reviews* **16** 2870-2883.
[4] Senthilkumar R, Ramados K and Manimaran R 2013 *Journal of Researches In Engineering Mechanical and Mechanics Engineering* **13** 15-20.
[5] Sugeng and Syaiful 2013 *Effect of Methanol on Low Levels on Thermal Efficiency of Direct Injection Diesel Engines with EGR Systems* (Semarang: Proceedings of the National Seminar on Science and Technology*)* **1** 1-6.
[6] Sayin C 2010 *Fuel* **89** 3410-3415.
[7] Khalik A, Tradevi SK and Dincer I 2011 *Applied Thermal Engineering* **31** 1621-1629.
[8] Tan YH, Abdullah MO, Hipolito CN, Zauzi NSA and Abdullah GW 2017 *Energy Conversion and Management* **132** 54–64.
[9] Praptijanto A, Muharam A and Putrasari Y 2015 *Energy Procedia* **68** 345-354.
[10] Chauhan BS, Kumar N and Cho HM 2012 *Energy* **37** 616-622.
[11] Muralidharan K, Vasudevan D and Sheeba KN 2011 *Energy* **36** 5385-5393.

Effect of Water Content in Methanol on The Performance and Exhaust Emissions of Direct Injection Diesel Engines Fueled by Diesel Fuel and Jatropha Oil Blends with EGR System

Johan Firmansyah[1,*], Syaiful[2] and Eflita Yohana[3]

[1,2,3] Department of Mechanical Engineering, Universitas Diponegoro, Jl. Prof. Sudharto, SH, Tembalang, Semarang 50275, Indonesia

Email: [1,*]johanfirmansyah77@gmail.com, [2,*]syaiful.undip2011@gmail.com, [3,*] efnan2003@gmail.com

Abstract. Diesel engines are usually used because of their durability and high efficiency. Therefore the number of diesel engine uses is increasing. This results in an increase in the amount of diesel fuel consumption and air pollution from diesel engine exhaust gases, especially soot emissions. To reduce this soot emission, methanol which is rich in oxygen is added to diesel. The substitution of alcohol into diesel has an impact on the performance of the diesel engine. Therefore, this study is focused on investigating the effect of water content in methanol on diesel engine performance and soot emissions. The study was conducted using an Isuzu 4JB1 diesel engine equipped with a cold EGR system. The ratio of the diesel-jatropha-methanol mixture used is 55/30/15% on a volume basis. While the water content in methanol used is 0 to 25% at a 5% interval. The test is carried out at a stationary 2500 rpm and given a load of 25% to 100% at a 25% interval. A dynamite Land & Sea dynamometer and stages 898 smoke meter were used to measure performance and soot emissions. The experimental data shows that the higher water content in methanol in the mixture of diesel-jatropha-methanol fuel causes a reduction in performance and an increase in soot emissions of diesel engines

1. Introduction

Diesel engines are usually used in the industrial automotive because they have greater power and higher fuel efficiency than gasoline engines. However, diesel engines produce higher NOx and soot emissions and are harmful to human health. Therefore, the widespread use of diesel engines results in air pollution due to increased exhaust emissions and increased demand for diesel fuel. Jatropha is one of the alternative non-food alternative fuel sources that has been studied both as pure fuel and as a mixture of diesel fuel so that it is expected to reduce dependence on fossil fuels [1]. However, the use of jatropha as a mixture produces higher soot emissions than diesel [2]. In addition, the addition of jatropha as a mixture of fuel produces a decrease in torque and power compared to pure diesel. This is due to the low calorific value in Europe [3].

Methanol is an alcohol group that can be considered to reduce smoke emissions and increase fuel demand when used as additives in diesel fuel [4]. High oxygen content in methanol is a major factor in reducing the emission of smoke produced by diesel engines [5]. Syaiful et al., 2017 [6] in their study observed that adding methanol to a mixture of jatropha and diesel fuels was able to reduce soot emissions. This decrease increases with the increase in the percentage of methanol in the fuel mixture. In addition, the greater the methanol content in diesel mixture fuel results in a decrease in torque and greater power [7]. A decrease. In this study Low Purity Methanol (LPM) is used as a substitute for methanol because in addition to its price is cheaper, LPM has a higher H O level so that it will reduce soot emissions caused by diesel engines.

A diesel engines due to the high temperature in the combustion engine [8]. Exhaust Gas Recirculation (EGR) is an effective pressure and inexpensive method for reducing NOx emissions by recirculating some of the exhaust emissions gas into the combustion engine [9]. A decrease concentration of air in the combustion engine causes a decrease in combustion temperature so that NOx emissions also decrease. The high NOx emissions in diesel engines can be decreased by using EGR system because EGR can reduce the concentration of air in the combustion engine so that the temperature in the combustion chamber decreases. EGR system used in the study is Cold EGR type. Some of the exhaust emissions is circulated into the combustion engine via a heat exchanger while the Hot EGR type is without using heat exchanger media.

2. Methodology
Based on the above, the study observed the effect of methanol with varying moisture content on soot emissions and performance produced by direct injection fuel diesel fueled and European-style system.

Research Methods

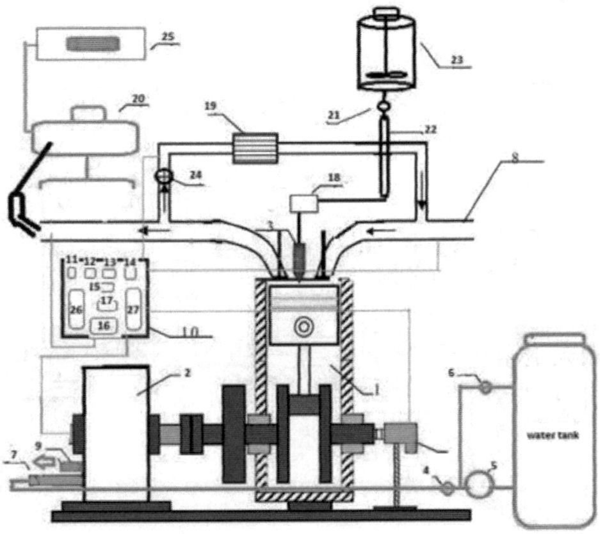

FIGURE 1. Experimental equipment scheme

A fuel used in this study is diesel, jatropha, and methanol with various different water content. Diesel fuel is produced by PT. Pertamina Tbk. The variation of fuel mixture in this study is 100% diesel (D100), 70% diesel and 30% jatropha (DJ30), 55% diesel, 30% jatropha and 15% methanol with a water content of 5%, 10%, 15%, 20 %, and 25% (DJ30M15A5, DJ30M15A10, DJ30M15A15, DJ30M15A20, DJ30M15A25).

3. Result and Discussion
Isuzu 4JB1 diesel engine is connected by a dynamometer. Engine rotation is measured using a digital tachometer with the proximity sensor. A measuring cup and stopwatch are used to measure the fuel flow rate. Exhaust gas temperature was measured using a digital thermometer, while the air flow rate and EGR flow rate were obtained by measuring the difference in pressure on the orifice. The experimental equipment scheme can be seen in Figure 1. Tests are carried out at a constant engine speed of 2500 rpm. In this condition, the EGR input is cooled by giving the cooler to the EGR channel to be inserted into the cylinder (Cold EGR). Variation in valve openings starting at 25%, 50%, 75%,

and 100% is carried out for each variation of the EGR valve opening. The variations above are carried out for each mixture of fuel. The fuel specifications are shown in Table 2.

TABLE 1. Fuel properties comparison of methanol, jatropha and diesel fuel

No	Properties	Diesel	*Jatropha*	Methanol
1	Cetane number	48	41.8	4.8
2	Water Content (%v)	0.05	3.16	0.05
3	Viscosity (mPa.s)	2.0-5.0	3.23	0.6
4	Calorific Value (MJ/kg)	45.21	37.97	22.08
5	Flash Points (°C)	60	198	13
6	Oxygen content (%)	-	10.9	34.8

RESULT AND DISCUSSION

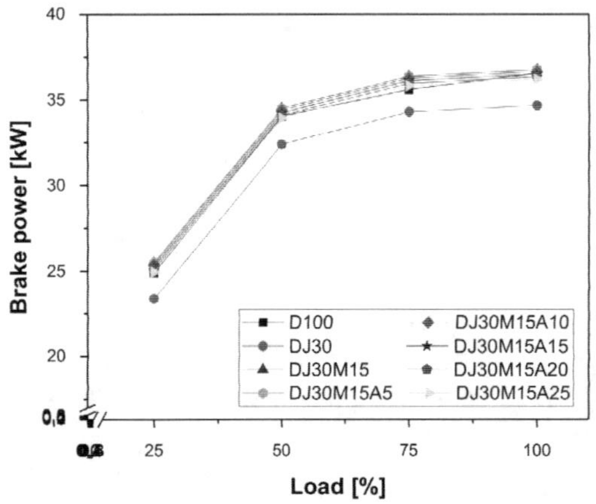

(a)

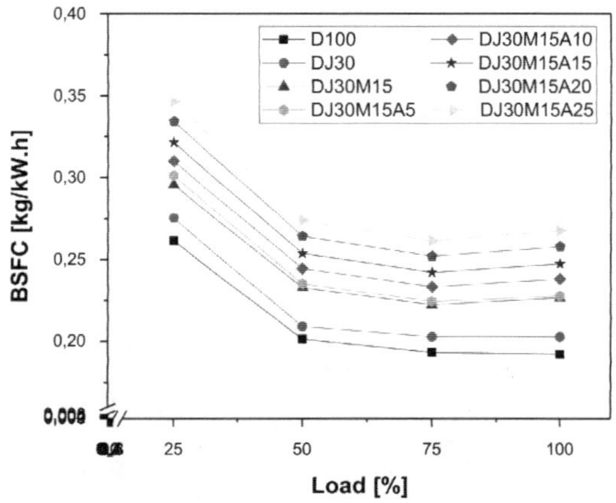

(b)

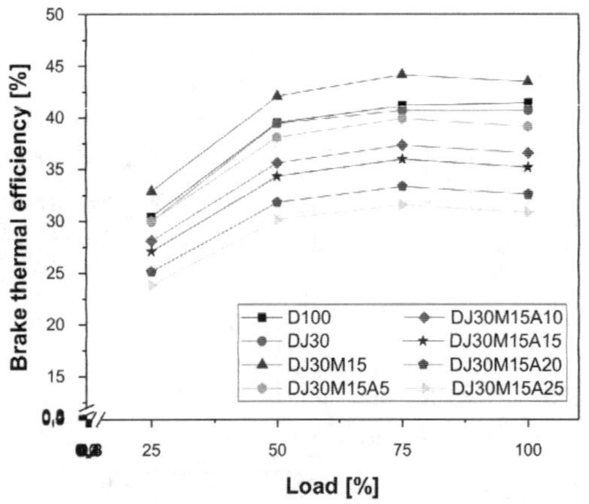

(c)

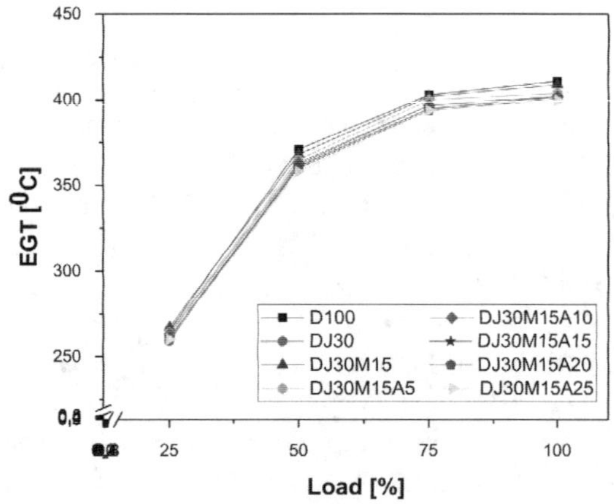

(d)

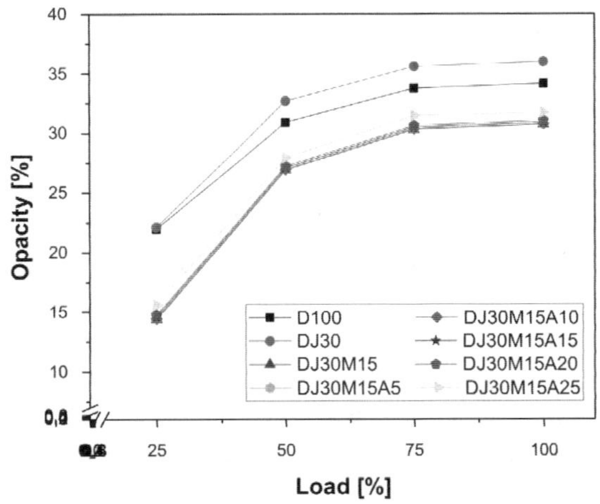

(e)

FIGURE 2. (a) The brake power characteristics, (b) Brake-Specific Fuel Consumption, (c) Brake Thermal Efficiency, (d) Exhaust Gas Temperature, and (e) Exhaust Gas Oppacity Performance of direct injection fuel diesel engines by EGR cold systems using various diesel fuel, jatropha, and methanol compositions.

Figure 2.(a) shows the value of Brake Power increases with increasing load. The use of diesel-jatropha-methanol mixed fuel show results in a slight increase in the value of brake power. The high oxygen content and low viscosity in methanol facilitate the process of injecting fuel [5] thus causing the burning rate to increase [10]. The highest increase in brake power value occurs in DJ30M15 fuel which is 2.53% compared to D100.

Figure 2.(b) shows a reduction in the value of the Fuel Consumption (BSFC) Specific Brake with increasing load. The use of diesel-jatropha-methanol mixture produces a higher BSFC value than D100. This is due to the low calorific value in Europe and methanol in diesel-jatropha-methanol mixed fuels [11]. The highest increase in BSFC value occurs in DJ30M15A25 fuel which is equal to 39.42% compared to D100 at 100% loading.

Figure 2.(c) presents a reduce in Brake Thermal Efficiency (BTE) when using a blended of diesel-jatropha-methanol fuel. This is due to lower calorific value than pure diesel. In addition, high viscosity in Europe also causes BTE to tend to decrease [12], [13]. The highest decrease in BTE value occurred in DJ30M15A25 fuel which was equal to 25.3% at 100% loading.

Figure 2.(d) show the result is a graph of the Exhaust Gas Temperature (EGT) measurement on direct injection diesel engines with diesel-jatropha-methanol mixed fuel. The use of diesel-jatropha-methanol mixed fuels shows results in a reduce in EGT. The poor atomization of fuel tends to reduce combustion characteristics [13]. In addition, the low temperature of the heating value in a blended of diesel-jatropha-methanol fuel causes the heat produced by mixed fuels to decrease. This triggers a reduction in temperature in the combustion engine [14]. The highest decrease in EGT value occurred in DJ30M15A25 fuel at 50% loading.

Figure 2.(e) presents the results of smoke opacity testing on the use of diesel-jatropha-methanol mixed fuel indirect injection diesel engines. The use of diesel-jatropha-methanol mixed fuel results in a decrease in smoke opacity. This decrease is due to the high oxygen content in mixed fuels which promotes carbon oxidation [9]. The highest reduction in smoke opacity occurred in DJ30M15 fuel, which amounted to 34.51% at 25% loading.

4. Conclusion

From the result obtained in this study, it can be concluded that (1) The addition of jatropha in diesel-butanol fuel mixture results reduced the BSFC and BTE performance. This is due to low heating value in Jatropha and methanol compared to D100. The highest reduction in BSFC and BTE occurred in mixed fuels with the highest water content in methanol, namely DJ30M15A25. (2) The use of diesel-jatropha-methanol mixture also lower the exhaust gas oppacity compared to the D100. This is deemed due to the high oxygen content in methanol to facilitate more efficient fuel oxidation. In addition, the low viscosity in methanol also affects the reduction of smoke opacity produced by diesel engines. The highest decrease in smoke opacity occurred in DJ30M15 fuel. (3) The use of diesel fuel mixture, Europe and methanol with the highest water content (25%) is a composition of a mixture of bad fuels for increasing performance and decreasing soot emissions of diesel engines. While the composition of diesel, jatropha and pure methanol (DJ30M15) mixtures are highly recommended to improve the performance of the diesel engine.

5. Acknowledgement

The authors sincerely acknowledge the suport and facilities from Thermofluid Laboratories, Faculty of Engineering, Universitas Diponegoro.

6. Reference

[1] Achten, W.M.J, verchot, Franken, Mathjis, and Sing, V.P. 2008 *Biomass and Bioenergy* **32** 1063-1083.
[2] Sentilkumar R, Ramados K and Manimaran R. 2013 *Global Journal of research in engineering mechanical and mechanic engineering* **13** 1.
[3] Dwivedi G, Jain Siddarth and Sharma M.P. 2011 *Renewable and Sustainable Energy Review* **15** 4633-4641.
[4] Imran A, Varman M, Masjuki and Kalam M.A. 2013 *Renewable and Sustainable Energy Review* **26** 739-751.
[5] Syaiful and Sugeng. 2017 *Nova* **51** 53-94.
[6] Qi D.H, Chen L.M, Geng Y.Z.H and Bian X.C.H. 2010 *Applied Energy* **87** 1679-1686.
[7] Karabates M, Ergen G and Hosoz M. 2011 *Fuel* **04** 36-39.
[8] Syaiful, Stefan M and Kazuyo T. 2013 *The Seventh International Symposium on Mechanics: Japan* **7** 147-151.
[9] Vinod S, Soni S.L and Sharma D. 2012 *International Journal of Hydrogen Energy* **37** 3807-3817.
[10] Pradeep V and Sharma D 2007 *Renewable energy* **32** 1136-1154.
[11] Anand K, Sharma, R.P and Mehta, P.S. 2011 *Biomass and Bioenergy* **35** 533-541.
[12] Chen, L.Y, Chen, Y.H, Hung, Y.S, Chiang, T.H, and Tsai, C.H. 2013 *Journal of Taiwan Institute of Chemical Engine* **44** 214-220.
[13] Chauhan, B.S, Kumar, Naveen and Cho Haeng, M. 2012 *Energy* **37** 616-622.
[14] Muralidharan K, Vasudevan D and Sheeba, K.N. 2011 *Energy* **36** 5385-5393.

Comparison of Hi-Art Tomotherapy Machine Outputs Using AAPM TG-148 and IAEA TRS 483 Codes of Practice

O A Firmansyah[1]*, A F Firmansyah[1], S I Sunaryati[1], W E Wibowo[2]

[1] Center for Technology of Safety and Radiation Metrology, National Nuclear Energy Agency, Jakarta, Indonesia

[2] Department of Radiotherapy, National General Hospital Cipto Mangunkusumo, Jakarta, Indonesia

*Email: okky-agassy@batan.go.id

Abstract. Comparison of Hi-Art Tomotherapy Machine Outputs using AAPM TG-148 and IAEA TRS 483 Codes of Practice. This paper describes a short history and technological aspect from a Hi-Art tomotherapy machine. The comparison between the tomotherapy machine and conventional linear accelerator machine in the technological aspect and therapy features was also described. There are two codes of practice to determinate the absorbed dose to water from the Hi-Art tomotherapy machine, The AAPM TG-148 and IAEA Technical Report Series No. 483. Determination of the absorbed dose to water was carried out at the machine specific reference condition by using the ionization chamber type of PTW TW 30013 SN 6367 volume 0.6 cc, and electrometer PTW Webline T10022/268. The results obtained show that the value of absorbed dose to water using the AAPM TG-148 was 8507 ± 2.4 mGy/minute and while using the IAEA TRS 483 the **value** of absorbed dose to water was 8485 ± 2.4 mGy/minute. The comparison result of the determination of the absorbed dose to water using AAPM TG-148 and IAEA TRS 483 protocol has a deviation of 0.2%.

1. Introduction

Radiotherapy is one option for treating cancer beside surgery and chemotherapy. The data shows that radiotherapy had an important role in the treatment and cure of cancer [1]. The goals of radiotherapy are to provide a measurable dose to a particular target volume to cure the tumor cell with a minimal side effect for normal tissue [2]. Radiotherapy was very helpful for the patient to curing their cancer/tumor.

Teletherapy machine that usually used in Indonesia is Cobalt-60 and linear accelerator machine. The use of cobalt-60 for the medical services is relatively cheaper and simpler but has some lacks and constraints in the operation, which was a low activity and the presence of radioactive waste, making the use of cobalt-60 teletherapy machine began to be abandoned in favor of medical linear accelerator machine. In 2008 the number of radiotherapy services known in Indonesia is as many as 22 service centers with a total of 17 of cobalt-60 and 10 medical linear accelerators [1]. Along with the needs of healthcare facilities, the number of radiotherapy centers in Indonesia in 2014 is rising to 27 service centers with 18 cobalt-60 teletherapy machines, 23 medical linear accelerators and 1 gamma knife. Per May 2018, the number of the radiotherapy service turned into 42 service centers with 25 cobalt-60 teletherapy machine, 40 medical linear accelerators, and two gamma knife.

In the year 2016 National General Hospital Dr. Cipto Mangunkusumo installed a helical Hi-art Tomotherapy machine. Hi-Art tomotherapy machine is a medical linear accelerator with 6 MV photon beam. The therapy modalities with tomotherapy machine are new and the first in Indonesia. The Hi-Art tomotherapy machine can be seen in Figure 1 below.

The tomotherapy machine has a different principle of works with the conventional medical linear machine. The use of helical in this machine is to show that the gantry on fixed position and the patient table is move during the patient treatment. While was on a conventional medical linear accelerator gantry moves and patients table on a fixed position. Hi-Art tomotherapy machine has a treatment mode slice by slice (tomo) and spiral shaped. Some differences from the tomotherapy machine with the conventional medical linear accelerators can be seen in table 1 below.

Table 1. The difference between the tomotherapy machine with the conventional medical linear accelerator

Technology	Conventional Medical Linear Accelerator	Tomotherapy Machine
Flattening filter	Use	Not use
Maximum Field Size	40 cm x 40 cm	5 cm x 40 cm
SSD	100 cm	85 cm
Field Size indicator	Light indicator	None
Detector	None	Xenon Detector Array
Computed Tomography	KVCT	MVCT
Unit	cGy/MU	cGy/minute

Every medical linear accelerator should be measured in accordance with the rules of BAPETEN No. 1 the year 2006. In the year 2016, Assef et.al. have measured the absorbed dose to water of tomotherapy machine using the code of practice AAPM TG-148 that recommended by the IAEA and the AAPM [3]. The measurement was carried out at machine specific reference. AAPM TG-148 described the value of k_Q on the static field and machine specific reference condition.

In November 2017, IAEA issued a protocol IAEA Technical Report Series (TRS) 483: Dosimetry of small Static Fields Used in External Beam Radiotherapy an International - Code of Practice for Reference and Relative Dose Determination. On the IAEA TRS 483 discussed regarding the value of k_Q to determine the absorbed dose to water of the tomotherapy machine at a specific reference condition with SSD 85 cm and field size 5 cm x 10 cm.

This paper describes a short history and a few aspects from Hi-Art tomotherapy machine and also comparison on the determination of absorbed dose to water using the code of practice AAPM TG-148 and IAEA TRS 483. The comparison between the tomotherapy machine and conventional linear accelerator machine in the technological aspect and therapy features was also described.

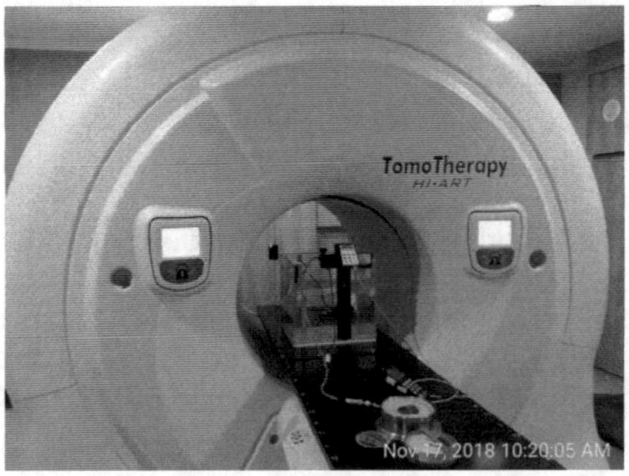

Figure 1. Hi-Art Tomotherapy Machine at RSCM Jakarta

1.1. A Short History of Tomotherapy Machine

Helical tomotherapy machine was developed by Thomas Rockwell Mackie, Ph.D. from the University of Wisconsin, Madison, USA. The development of this tomotherapy machine was expected to be a solution from the conventional linear accelerator machine's problem. These problems are [4] :

- The limitation of target dose that can be delivered due to the presence of neighboring sensitive structures
- Verification of the correct beam shape and position simultaneously while ensuring that the patient is positioned correctly
- Limitation in the safety of dynamic therapy due to the possibility of collision between the patient and the treatment unit.

The use of a continuous motion system of the slip-ring gantry for radiotherapy was introduced by this tomotherapy machine. Tomotherapy has a system that can modulate a beam using a fan beam and also has a temporarily modulated multileaf collimator system (binary MLC).

On 21 August 2002, the University of Wisconsin treated the first patient the first man with bone metastasis cases [5]. Through the first patient treatment, it can be known the MVCT characteristics. The MVCT features still needed further development for optimal performance.

Helical tomotherapy has technological development and it's applications for another disease's treatment. Tomotherapy has a new system called adaptive therapy system. This system is able to display and analyze the dose given to the patient, and also to calculate the treatment dose.

1.2. The Main Component of Tomotherapy Machine

The main components of the Hi-Art tomotherapy machine include the linear accelerator 6 MV, multileaf collimator, megavoltage computed tomography (MVCT), data acquisition system, beam stopper, xenon detector, pulse forming network modulator, and computer control. The main components of the Hi-Art tomotherapy machine can be seen in this figure below.

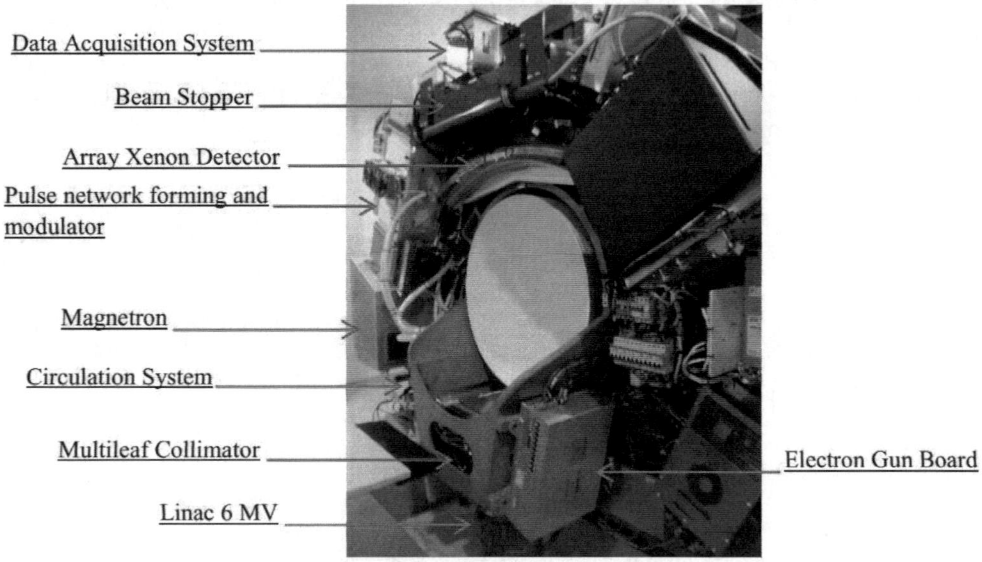

Figure 2. Main component of tomotherapy machine

1.2.1. Radiation Source: Linear Accelerator Medic. The radiation source for tomotherapy is linear accelerator 6 MV. The linear accelerator machine was placed on a ring gantry that can rotate 360° continuously. To produce a homogeneous beam in the conventional medical linear accelerator, the machine should have a flattening filter that placed on the beam path. Flattening filter is a cone which made of a material with a high atomic number. On helical tomotherapy machine, flattening filter is not used, thus the provisions of flatness and symmetry on conventional photon beam are not applied.

Another purpose by not using this flattening filter is the radiation quality will be lower but will increase the dose rate.

1.2.2. Multi-leaf Collimator. The beam's shape from this tomotherapy machine is a fan beam that is produced by a set of collimating jaw with a maximum transverse length of 40 cm and a maximum width of 5 cm longitudinal. The beam passes through 64 binary collimator blades that controlled using pneumatically.

1.2.3. Megavoltage Computed Tomography (MVCT). As described in Table 1, one of the differences between a tomotherapy and a conventional medical linear accelerator machine is the radiation source for imaging process. On the conventional medical linear accelerator machine, imaging process was carried out using an additional radiation source with a kilovoltage x-ray machine (KVCT). While on the tomotherapy machine, the radiation source for the imaging process is photon beam with potential voltage for the linear accelerator is reduced from 6 MV to 3.5 MV. To get a fluoroscopy image, an array xenon detector was placed on the gantry opposite the linear accelerator. Beside for clinical purposes, MVCT can also be used for assistive devices to place the detector accurately at the desired position. The measured dose during this MVCT must be less than 4 cGy.

1.2.4. Beam Stopper. The function of the beam stopper is for the absorber of the main radiation beam emitted by the radiation source (6 MV linear accelerator). The absorbent radiation beam is placed on the gantry opposite with the radiation source. The position of the beam stopper is on the 74 cm from the machine isocenter. The beam stopper will be effective with the dimensions of the tin plate that defines as 89 cm long, 23 cm wide and 13 cm thick [6].

1.2.5. Xenon Detector and Data Acquisition System. The detector used by tomotherapy is the xenon ionization detector. The detector is arranged in such a way as to form an arrangement of xenon gas filling detector systems. This system consists of 738 xenon gas filling detectors and placed opposite the radiation source [7]. The diagnostic results obtained will be continued as input to the tomotherapy software. The main purpose of installing Xenon detector arrays for MVCT is for verification of the delivery and reconstruction dose by the tomotherapy system.

1.3. The Differences of Techniques/Processes in Tomotherapy Machine and Linear Accelerator

Based on the explanation above, there is a significant difference between tomotherapy and conventional medical linear accelerator on the technique or treatment process. The following are some of the differences:

Table 2. The differences of techniques/processes in tomotherapy machine and linear accelerator [8]

Process or Technique	Conventional Linear Medic Accelerator	Tomotherapy Machine
Conformal radiation therapy	3-D treatment planning	Inverse treatment planning
Set-up verification	Laser alignment; Port films or electronic portal imaging	CT projections to obtain detailed anatomical information
Delivery modification	Repositioning of the patient	Adapt delivery according to patient displacement
Beam alignment verification	Port films or electronic portal imaging	Acquire CT scans before, during or after treatment
Delivery verification	Electronic portal imaging	Compute the energy fluence actually delivered to the patient on a pulse by pulse basis
Dose reconstruction	None	Superimpose on a CT representation the dose actually deposited in the patient during treatment and compare to the planned dose

2. Method and Materials

2.1. Determining The Absorbed Dose to Water of Tomotherapy Machine

The use of the AAPM TG-148 protocol is based on the absence of k_Q values for the tomotherapy machine at protocol TRS 398 that usually use for the determination of absorbed dose to water for conventional linear accelerator medic. The conventional linear accelerator medic has K_Q values for reference condition at field size 10 cm x10 cm, SSD 100 cm, and using a flattening filter, while the field size of tomotherapy is 5 cm x 40 cm, SSD 85 cm and using flattening filter free. In the protocol TRS 398, a K_Q value for each detector with various volumes has been provided on a standard field size 10 cm x 10 cm, SSD 100 cm and using a flattening filter, valid for electron beam or photon beam. So based on the machine requirements to determine the K_Q value on the tomotherapy machine, an approach is used in accordance with the AAPM TG-148 protocol.

The absorbed dose to water for the photon beam produced from the tomotherapy machine at the reference depth is given by equation 1 below [9].

$$D_{w,Q_{msr}}^{f_{msr}} = M_{Q_{msr}}^{f_{msr}} . N_{D,w,Q_0} . k_{Q,Q_0} . k_{Q_{msr},Q}^{f_{msr},f_{msr}} \quad (1)$$

Where

Q : is the radiation beam quality [$\%dd(10)_x$] of the conventional reference field size of 10 cm x 10 cm at 100 cm SSD according to TG-148 protocol.

Q_{msr} : is the radiation beam quality [$\%dd(10)_x$] of the machine specific reference field size f_{msr} (5 cm x 10 cm field at 85 cm SSD)

$M_{Q_{msr}}^{f_{msr}}$: is the corrected reading of the dosimeter for the field f_{msr}

N_{D,w,Q_0} : is the absorbed dose to water calibration factor for a reference quality Q_0

k_{Q,Q_0} : is the beam quality correction factor for the quality Q of the conventional reference field f_{msr} (10 cm x 10 cm at 100 cm SSD).

$k_{Q_{msr},Q}^{f_{msr},f_{ref}}$: is the factor to correct for the differences condition of field size, geometry, phantom material, and beam quality of the conventional reference field f_{ref} and the machine specific reference field f_{msr}

Thomas R. Mackie describes a method to determine the correction factor. The function of the correction factor is to correct the difference between the conditions of field size, geometry, phantom material and beam quality of the conventional reference field and the machine specific reference using a third order polynomial expressed in equation 2 [10].

$$\%dd(10)_{x[HTTG-51]} = 1.35805.(\%dd_{(10)x(HTref)})^3 - 244,493.(\%dd_{(10)x(HTref)})^2 + 14672.98.(\%dd_{(10)x(HTref)})^2 - 293479.4 \quad (2)$$

The k_Q value can be obtained using Table 2. The k_Q values are nearly constant for the most commonly cylindrical ionization chambers found in TG-148 and Thomas et al, varying from 0.999 to 0.995.

Table 3. Values of KQ for photon beams as a function of the beam quality %dd(10)x][10]

Ion Chamber	K_Q Beam Quality Specifier [$\%dd(10)_x$] 58	63	66
PTWN3001 0.6 cc Farmer	1	0.996	0.992
PTW N30002 0.6 cc all Graphite	1	0.997	0.994
PTW N30004 0.6 cc Graphite	1	0.998	0.992
PTW N31003 0.3 cc waterproof	1	0.996	0.992

Assef F. F et al. have been done for determining the absorbed dose to water for the machine specific reference field from a Hi-Art helical tomotherapy machine using three different ionization chambers [4]. The measurement of the percentage depth dose has been carried out at the source to the detector distance of 85 cm and a field size of 5 cm x 10 cm. The measurement of the absorbed dose to water has been done using standard detector Farmer 0.6 cc TW 30013.

Table 4. The Result of Determination of Absorbed Dose to water using AAPM TG-148 [3]

Detektor	M_Q (nC)	$N_{D,w}$ (mGy/nC)	k_Q	PDD_{10} (%)	$D_{w,Q(Zmax)}$ (mGy/minute)	$D_{W,Q_{msr}}^{f_{msr}}$ (mGy/minute)
A 1 SL	8.330	603.3	0.9985	5018	58.87	8524 ± 2.5 %
PTW TW30013	92.390	54.27	0.9985	5006	58.87	8503 ± 2.5 %
PTW 31016	2.009	2516	0.9980	5045	58.87	8569 ± 2.5 %

These results indicate that the measurements using the three detectors get insignificant differences. The difference in measurement with a detector is not greater than 0.5% for each detector. Determination of the absorbed dose to water using the AAPM TG-148 protocol recommended by the IAEA and AAPM.

In November 2017, The IAEA issued the IAEA TRS 483: *Dosimetry of small Static Fields Used In External Beam Radiotherapy - an International Code of Practice for Reference and Relative Dose Determination.* IAEA TRS 483 describes k_Q value for the tomotherapy machine at machine specific reference condition. Determination k_Q value for tomotherapy is different from a conventional linear accelerator machine.

Absorbed dose to water for conventional linear accelerator medic were determined using radiation quality (*Q*) can be determined using the ionization chamber which calibrated in absorbed dose to water for Co-60 $N_{D,W}$ (gamma ray). The calculation has been doing by the following equation [11].

$$D_{w,Q} = M_Q . N_{D,w,Q} . k_Q \tag{3}$$

Where

$D_{w,Q}$: Absorbed dose to water with radiation quality Q (mGy)

M_Q : Raw data dosimeters corrected for temperature, air pressure, polarity, k_{pol} and ion recombination, k_s (nC)

$N_{D,w}$: Detector calibration factor for Co-60 gamma rays (mGy/nC)

K_Q : Correction factor for photon beam radiation quality from the detector

In the IAEA TRS 483, the equation to determine the absorbed dose to water is essentially the same with the equation in AAPM TG-148. The k_Q value in IAEA TRS 483 was determined based on calculations using the Monte Carlo program for machine specific reference conditions for the tomotherapy machine. These conditions include parameters:

Table 5. Reference condition for Tomotherapy machine in IAEA TRS 483 [11]

Parameter	Reference value or reference character
Phantom Material	Water
Phantom shape and size	At least 30 cm x 30 cm x 30 cm
Chamber type	Cylindrical
Measurement depth z_{reff}	10 g/cm^2
Position of the reference point of chamber	Measurement at depth z_{reff}
SSD/SDD	85 cm
Field size	5 cm x 10 cm

The k_Q value in the table can be directly used to determining the absorbed dose to water because the condition has been adjusted for the tomotherapy machine. In IAEA TRS 483 for PTW 30006/30013 Farmer ionization detector, k_Q value was obtained at 0.995. The measurement was carried out using PTW TW 30013 SN 6367 volume 0.6 cc and electrometer PTW Webline T10022/268.

3. Result and Discussion

In Table 4 below, the average reading from the electrometer is 92.97 nC for measurements per 60 seconds. The k_Q value is obtained from IAEA TRS 483 for a value equivalent to the specific reference for the tomotherapy machine. The absorbed dose to water at the static conditions of the tomotherapy machine was 8485 ± 2.4 mGy/minute.

Table 6. The Result of Determination of Absorbed Dose to water

Code of Practice	Detector	M_Q (nC)	$N_{D,w}$ (mGy/nC)	K_Q	PDD_{10} (%)	$D_{w,Q}(Z_{max})$ (mGy/minute)
IAEA TRS 483	TW30013	92.97	53.92	0.9950	58.98	8485 ± 2.4 %
AAPM TG-148	TW30013	92.97	53.92	0.9975	58.98	8507 ± 2.4 %

If the determination of the absorbed dose to the water is carried out using the AAPM TG-148 protocol on the same data, then the k_Q value will be obtained at 0.9975 and the value of absorbed dose to the water was 8507 ± 2.4 mGy/minute.

The comparison result of the determination of the absorbed dose to water using AAPM TG-148 and IAEA TRS 483 protocol has a deviation of 0.2%. The deviation obtained does not show a significant difference. The differences between AAPM TG-148 and IAEA TRS 483 were so close, because it was calculated by using the same principal. The calculation k_Q using Monte Carlo program uses that reference condition for the tomotherapy machine. Both protocols can be used to determine the absorbed dose to water in the static conditions of the tomotherapy machine.

4. Conclusion

Tomotherapy machine has different technologies with conventional linear accelerator medic. The unique system of tomotherapy has a multi-bladed collimator (binary MLC), continuous irradiation with a slice per slice irradiation capability, 360° rotating gantry, megavoltage computed tomography (MVCT) and applied adaptive therapy system. The field size of the tomotherapy machine are 5 cm x 10 cm and 5 cm x 40 cm. Because the size of this field size, the cases that can be handled by tomotherapy machine are different from conventional linear accelerator medic.

The protocols that can be used to determine the absorbed dose to water in the static conditions of the Tomotherapy machine is AAPM TG-148 and IAEA TRS 483. The measurement of photon beam uses the PTW 30006/30013 Farmer ionizing and webline electrometer T10022/268. The value of the absorbed dose to water which determined using AAPM TG-148 protocol was 8507 ± 2.4 mGy/minute, while determined using IAEA TRS 483 protocol the absorbed dose to water was 8485 ± 2.4 mGy/minute. The comparison result of the determination of the absorbed dose to water using AAPM TG-148 and IAEA TRS 483 protocol has a deviation of 0.2%. Both protocols can be used to determine the absorbed dose to water in the static conditions of the tomotherapy machine.

5. Acknowledgment

The authors would like to acknowledge to the staff of the Radiotherapy Unit of Dr. Cipto Mangunkusumo National General Hospital for help with the tomotherapy Hi-Art measurements.

6. References

[1] Eri Hiswara 2017 *Prosiding PPI – PDIPTN* (Yogyakarta - PSTA BATAN) p 47–52
[2] Diah FI and Anggraita P 2010 *Prosiding PPI – PDIPTN* (Yogyakarta - PSTA BATAN) p 166–75.
[3] Assef F.F and S I Sunaryati 2017 *Prosiding PPI – PDIPTN* (Yogyakarta - PSTA BATAN) p 413–18.
[4] Mackie TR, Holmes T, Swerdloff S, Recwerdt P Deasy JO, Yang J, Paliwal B and Kinsella T *Med Phys* **20** 1709–19.
[5] Mackie TR 2006 *Phys Med Biol* **51** R427-53.
[6] S. Baechler, F.O. Bochud, D. Verellen, R. Moeckli 2007 Shielding requirements in helical tomotherapy *Phys Med Biol* **52** 5057–67.
[7] Podgorsak EB 2005 *Radiation Oncology Physics : A Handbook for Teachers and Students* (Vienna – International Atomic Energy Agency)
[8] Van Dyk J, Kron T, Bauman G, Battista J, Bauman G and Battista JJ 2002 *Phys Canada* **58** 79–86.
[9] International Atomic Energy Agency 2017 *Dosimetry of Small Static Fields Used in External Beam Radiotherapy : An Code of Practice for reference and Relative Dose Determination* (Vienna – International Atomic Energy Agency)
[10] Langen KM, Papanikolaou N, Balog J, Crilly R, Followill D, Goddu SM, Grant W, Olivera G, Ramsey CR and Shi C. 2010 *Med Phys* **37** 4817–53.
[11] Pedro Andreo, David T Burns, Klaus Hohlfeld, M Saiful Huq, Tatsuaki Kanai, Fedele Laitano, Vere Smyth SV 2006 *Absorbed Dose Determination in External Beam Radiotherapy : An International Code of Practice for Dosimetry based on Standards of Absorbed Dose to Water.* (Vienna – International Atomic Energy Agency)

Performance of dethridge wheel as low head power generator and loss analysis

O Heriyani[1], D Mugisidi[2*], R A Luhung[2], M Y Djeli[2], A Fikri[2]

[1] Electric Engineering, Engineering Faculty, Universitas Muhammadiyah Prof. Dr. HAMKA, Jl. Tanah Merdeka no 6 Kp. Rambutan Ciracas, Jakarta Timur 13830, Indonesia.

[2] Mechanical Engineering, Engineering Faculty, Universitas Muhammadiyah Prof. Dr. HAMKA, Jl. Tanah Merdeka no 6 Kp. Rambutan Ciracas, Jakarta Timur 13830, Indonesia.

*Email: dan.mugisidi@uhamka.ac.id

Abstract. This research to measure the performance of dethridge wheel using a very low head as a driving force and analyze losses in the testing system. The parameters measured in this research are flow rate (*m^3/s*), water velocity (*m/s*), water level (m), torque (*Nm*), and rotation (rpm). Debit is a variabel that is given treatment with variations in the difference between the initial flow rate and the next is 0.002 *m^3/s*. The highest efficiency resulted in the first flow rate of 43.314% with hydro power of 1.862 watts and the power of the water wheel 0.806 watts. The higher increase flow rate causes the losses that occur are also grater, the state of the flow is more turbulent, and the loss due to force and fiction increase water.

1. Introduction

Electricity has become a primary need at this time. Various independent efforts to produce electricity continue to be carried out by researchers. One energy source that is often used is hydropower. The use of hydro energy to produce electrical energy encourages various countries to evaluate available energy sources [1]. The oldest mechanical equipment used to convert the energy of water flow into work is a water wheel. The limited location of high head water, the utilization of water flow with a low head is a concern and Zuppinger wheel is very suitable for producing electricity by utilizing water flow with a head less than 5 *m* [2]. Water flow with head less than 5 *m* can be found in irrigation canals. Water flow in irrigation channels that have a low head can be used to generate electricity [3]. To utilize low head water flow, such as irrigation flow, Dethridge wheel, which was initially used as a flow measurement, is currently used for electricity generation with an efficiency of 60% [4] and its performance increases on the channel with a width of two to three times greater than the wheel width [5]. In this research, Performance and loss of dethridge wheel as an electric generator drive measured.

2. Experimental Set Up

The dethridge wheel in this research was made of steel plate with a thickness of 1.8 *mm*. Runners use a 15 *mm* plywood that coated with waterproof coating. The geometry size of the Dethridge wheel refer to previous reseach [4], as can be seen in figure 1.

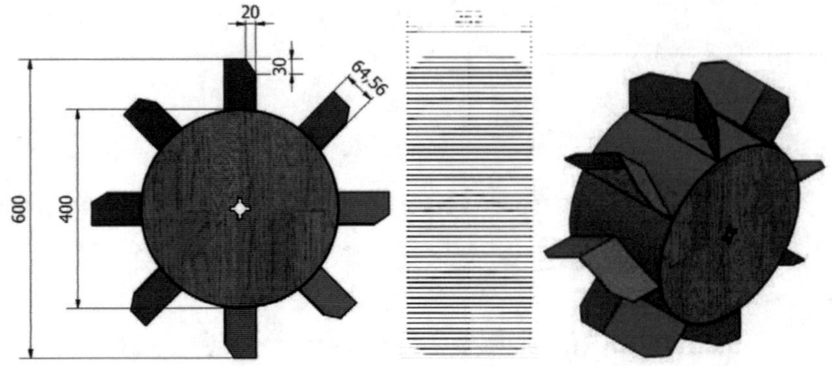

Figure 1 Dethridge Wheel Geometry

Experimental rig material in this research is 15 *mm* plywood that coated with fiber and waterproof coating. Dimension is shown in figure 2.

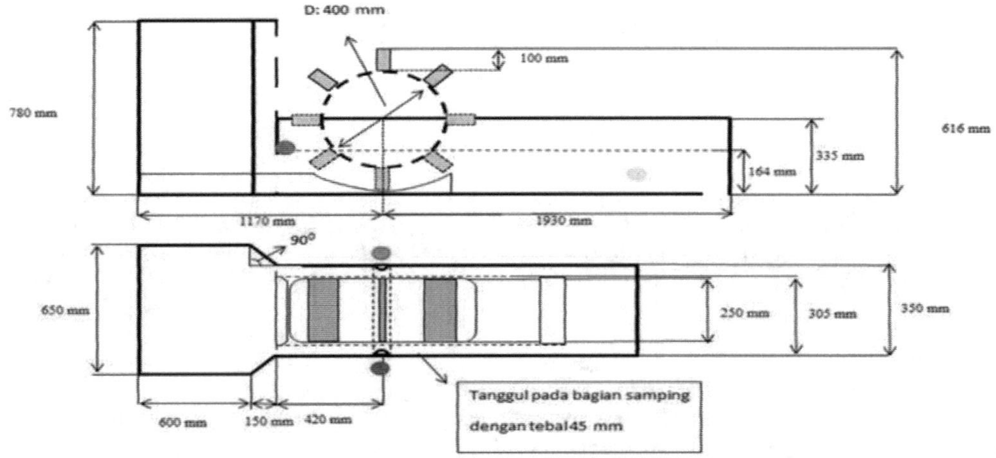

Figure 2 Canal Geometry

The experiment use six variations of flow rate, 0.006 m^3/s, 0.008 m^3/s, 0.010 m^3/s, 0.012 m^3/s, 0.014 m^3/s, and 0.016 m^3/s. Water is pumped from the storage tank to the water column through a flow meter. From the water column, which aims to stabilize the flow, water flows through the canal and then rotates the waterwheel and then flows back to the storage tank.

Water level measured before the wheel by attaching a permanent gauge to the walls of the experimental rig as well as after the wheel. The increase in water height starts from the first flow rate to the last flow rate as in the pattern of figure 3.

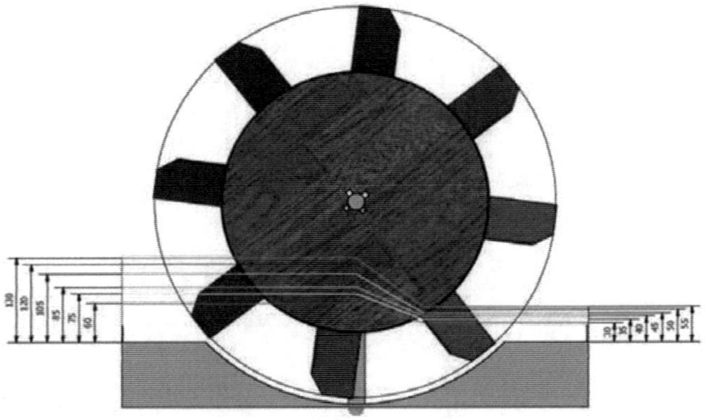

Figure 3 Water Level Pattern

3. Result and Discussion

The results of measurements with six variations of flow rate obtained a comparison of the performance of the wheel for each flow rate as in table 1.

Table 1 Performance Comparison of Debit Variations

Flow Rate (m^3/s)	*Head* (m)	Flow Velocity (m/s)	Rotation (rpm)	Torque (Nm)	Hydro Power (watt)	Wheel Power (watt)	Efficiency (%)
6,360	0,030	0,407	6,908	1,115	1,862	0,806	43,314
8,360	0,040	0,410	10,496	1,144	3,263	1,257	38,516
10,360	0,045	0,452	13,387	1,146	4,549	1,606	35,300
12,360	0,060	0,457	16,361	1,225	7,236	2,099	29,001
14,360	0,070	0,463	18,792	1,189	9,809	2,339	23,845
16,360	0,075	0,484	21,576	1,023	11,973	2,310	19,291

In table 1, it can be seen that there is an increase in flow rate, a decrease in efficiency of the wheel. The highest efficiency is 43.314% with a flow rate variation of 0.006 m^3/s, as shown in figure 4.

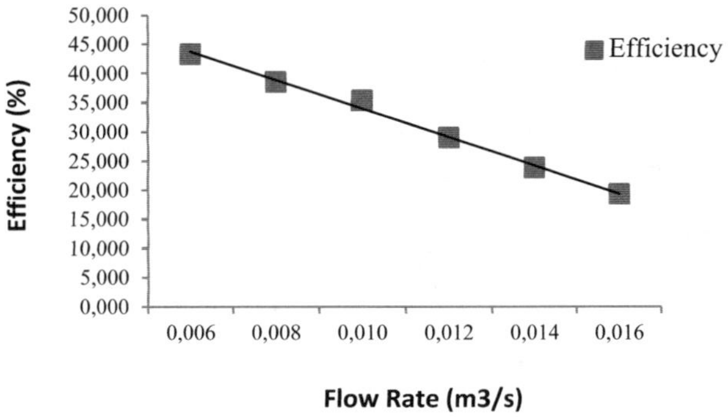

Figure 4 Graph of Flow Rate to Efficiency

The efficiency of the Dethridge wheel decreases when the water discharge is increased. This is because the increase in dethridge wheel rotation is not proportional to the increase in water discharge.

This is due to the water resistance as a reaction when the blade hits the water. In addition, there is friction between the water under the Dethridge wheel with the surface of the water channel.

Because the increase of Dethridge wheel rotation is not proportional to the increase in discharge, the water is retained before the Dethridge wheel so that the water level increases. Increasing the water level causes the potential of water power is also increase. This increase is not proportional as the increase in the power produced by the Dethridge wheel, as shown in figure 5, therefore the efficiency decreases.

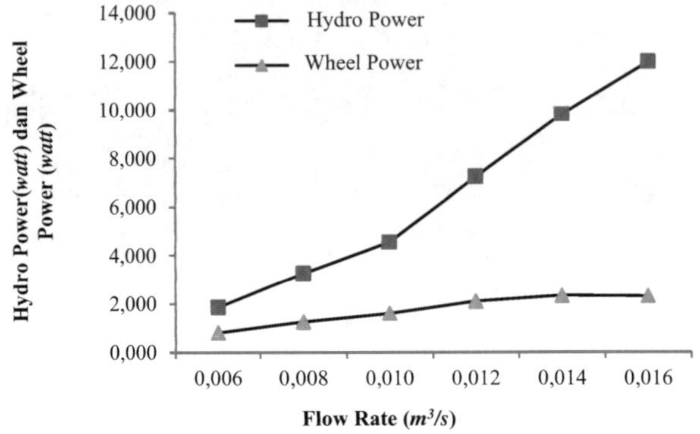

Figure 5 Graph of Flow Rate to Hydro Power and Wheel Power

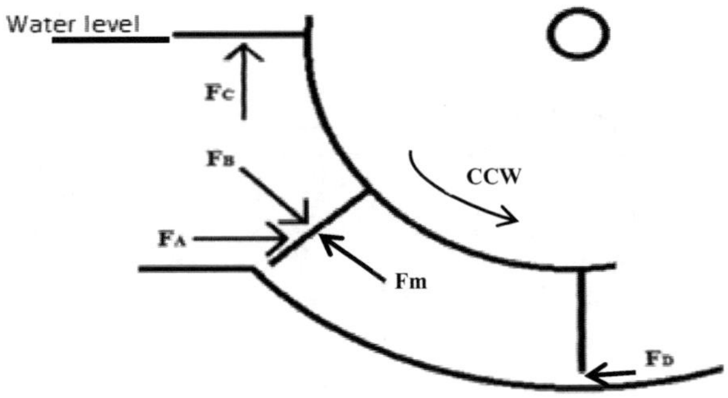

Figure 6. Hydro Force (F_A), Hydro Tangential Force (F_B), Resistance Force (F_c), Hydro Friction Force (F_D) and mass of water (F_m)

Figure 6 show forces when the water pushes the Dethridge to rotate. The force of water that comes from the flow energy pushes the Dethridge wheel to rotate counterclockwise. The thrust of the water, reduced by water reaction when the Dethridge wheel blade hit the water. The water reaction pushes the blade up so that the thrust force decreases. In addition, the friction that occurs between water and the surface of the water channel also reduce the force of water. Furthermore, the water that trapped between the Dethrige wheel blades no longer pushes but becomes an additional loss that also reduces the force of water.

4. Conclusion

Based on the results of testing and data collection, it can be concluded that efficiency the highest is 43.314% with the variation of the first debit amounting to 0.006 m^3/s. The increase in flow rate causes efficiency to decrease, because a lot of hydro power is not maximally used due to hydraulic losses. The cause of efficiency has decreased due to several hydraulic losses. It caused by the drag in the variation of flow rate 0.014 m^3/s and 0.016 m^3/s which causes a decrease in torque at the discharge variation 0.014 m^3/s and 0.016 m^3/s.

References

[1] Laws N D and Epps B P 2016 *Renew. Sustain. Energy Rev.* **57** 1245–1259.
[2] Paudel S, Weber M, Geyer D, Saenger N. Zuppinger water wheel for very low-head hydropower application. In Krope J, Olabi AG, Goričanec D, Božičnik S, editors, 10th International Conference on Sustainable Energy and Environment Protection: Marine and Hydro Power. University of Maribor Press. 2017.
[3] Zema D A, Nicotra A, Tamburino V, and Zimbone S M 2016 *Renewable Energy* **85** 498-506.
[4] Paudel S and Saenger N 2016 *IOP Conf. Ser.: Earth Environ. Sci.* **49** 102007.
[5] Paudel S and Saenger N 2017 *Renew. Energy* **115** 175-182.

Deposition Time Variation on Thickness and Resistivity of Cu/Ni Thin Film Obtained by Magnetic Field-Assisted Electroplating Process

Muthi'ah Lutfia Khansa[1,*], Moh. Toifur[2] Azmi Khusnani[3] , and Yudhiakto Pramudya[4]

[1,2,3,4] Dept. of Physics Education, Ahmad Dahlan University, Jl.Pramuka 42 Sidikan, Yogyakarta, Indonesia

Email: [1,*] mutiakhansa@gmail.com

Abstract. Thin films of Cu/Ni have been synthesized by the electroplating method assisted by a magnetic field on the variation of deposition time. The use of magnetic fields in the deposition process to increases the rate of electroplating and reducing the effect of hydrogen evolution so that the surface morphology of the layers becomes homogeneous. Deposition time affects the thickness of the formed layer. The electroplating is carried out at a DC voltage of 1.5 volts, the magnetic field of 200 gauss, a solution temperature of 60 °C, and an electrode distance of 4 cm with Cu as a cathode and Ni as an anode. The electrolyte solution made from the mixture of H_3BO_3 (30 g), $NiCl_2$ (195 g), $NiSO_4$ (45 g), and H_2O (750 ml) and deposition times varied from 5 s - 45 s with intervals of 10 s. The characterization was done by thickness test using the calculation of the mass, and resistivity of the Ni layer by using the four-point probe. The results indicate that the deposition time is proportional to the thickness of the formed layer and inversely proportional to the resistivity. Ni layer thickness ranges from 1.48×10^{-2} μm to 5.02×10^{-7} μm. while the sheet resistivity is from $(4.87 \pm 0.02) \times 10^{-3}$ Ω/sq to $(1.38 \pm 0.06) \times 10^{-3}$ Ω/sq.

1. Introduction

These Technological developments continue to increase to be more effective and accurate as a basis for conducting research related to innovation in the field of technology. One of the techniques developed in the research is sensor technology. The use of sensors is found in almost all fields [1,2], including in the field of biology related to embryo and gamete preservations, preservation of organ transplants, and preservation of blood products, and preservation of food [3,4].

Preservation usually uses a medium with low temperature such as liquid nitrogen that has a minimum temperature of 200 °C. Studies on materials that have the potential to be used as low-temperature sensors are continuously carried out [5]. Materials that are sensitive to low temperatures can be made from a thin film by utilizing changes in the resistance when the temperature changes. This material is a type of RTD (Resistance Temperature Detector) [6].

Cu and Ni and these combinations in the form of alloy CuNi and thin film Cu/Ni is a material that has the potential to be used as a low-temperature sensor [7-10]. Cu/Ni is a kind of temperature sensor that works based on a Resistance Temperature Detector (RTD). Therefore study about the resistance of the elements in responding to the temperature is an important thing. As sensor material, Cu, Ni or Cu/Ni film are elements that have excellent physical properties that are a linear relationship between resistance and temperature [9]. Plating Ni material on Cu substrate will increase the resistivity of Cu/Ni.

In this research, we synthesize Cu/Ni film by electroplating on the various time depositions. The time deposition in plating will influence the thickness of the film, and also the thickness will influence the resistivity. For metal, resistivity is inversely proportional to the thickness. The more thickness getting less resistivity. It is whished that the high resistivity of Cu/Ni will have a correlation to the increasing sensitivity [5,11]. The problem is the surface morphology of Ni film that is not uniform, and still contains vacancies. By that condition, the resistance of sample in different location become not homogen.

To improve the microstructure on the surface of the sample in this study used electroplating assisted by magnetic fields with the direction perpendicular to the electric field between anode-cathode (E). The utilization of magnetic field (B) in that direction will generate the Lorentz force on the direction perpendicular to E and B. Interaction between the Lorentz force and Ni ions in the electrolyte will generate the magnetohydrodynamic (MHD) effect [12,13]. MHD will impact to arising the mass transport of deposits [14,15]. This is can be seen from the increasing the current along the deposition process [16,17]. Beside that MHD will make the film morphology, in this case, Ni film be homogeneous [12,18], the effect of hydrogen evolution reactions be reduced [13,19].

2. Research Methods

2.1. Material research

The The material used in this research is copper (Cu) plate as a substrate with the size of 2.5 cm × 1.0 cm × 0.02 cm and nickel (Ni) with the size of 3 cm × 1.0 cm × 0.4 cm as a coated material. The electrolyte makes from a mixture of nickel sulfate ($NiSO_4$), nickel chloride ($NiCl_2$), boric acid (H_3BO_3), and distilled water of 195g, 45g, 30g, 750ml, respectively. Alcohol 96% is used to rinse the substrate. Toothpaste is used to clean the substrate of impurities and grease sticking, and metal polisher is used to polish the substrate.

The equipment used for preparing the sample consisted of 1000 ml beaker glass used as a plating bath; Magnetic stirrer is used to stir the solution; the electric stove is used to heat the electrolyte solution. Vernier thermocouple is used to control the temperature of the solution stable at 60 °C, the DC voltage set on 1.5 volts as a power supply. PA-214 Ohaus scale with accuracy to 10-4 g was used to measure the mass of Cu substrate before plating and Cu/Ni film after plating. An ultrasonic cleaner is used to clean the sample and two multimasters that set as voltmeter and amperemeter were used to measure the voltage and current in for point probe with accuracy to 10^{-3} mV and 10^{-2} mA respectively. The voltage and current data will be used to determine the sheet resistivity of the Cu/Ni film.

2.2. Method of experiment

In At the preparation stage, the cut copper substrate was polished with autosol then cleaned with toothpaste and rinsed with distilled water at 60 °C gradually until three times. The last stage substrate was rinsed using 96% alcohol in an ultrasonic cleaner for 3 minutes. The sample then hairdryed. After drying the sample was wrapped in tissue, and then inserted in a plastic clip and strored in drybox. Electrolyte solutions are stirred in a measuring cup using a magnetic stirrer for 4 hours.

The next stage is the electroplating process. An electrolyte solution is heated to 60 °C. Setting the electrolyte temperature 60 °C will produce a more flexible layer [20]. Then poured in plat bath — the tool design as in Fig. 1 the magnetic field to be set at 200 gauss. Nickel anodes and copper cathodes were connected to a 1.5 volt DC voltage. Data was acquised in the form of deposition time, current, and voltage with smpling rate 20 data/s. Deposition time was varied from 5s-45s with intervals of 5 s.

Journal of Physics: Conference Series **1373** (2019) 012013 doi:10.1088/1742-6596/1373/1/012013

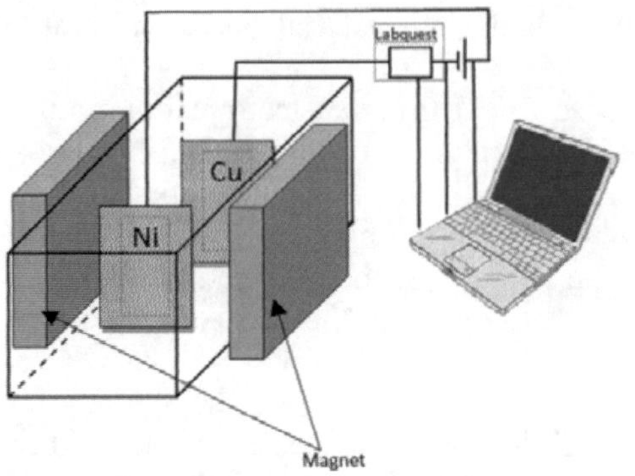

Figure 1. Schematic diagram of the experimental setup

2.3. Method of experiment

The Ni layer thickness was obtained through measuring the mass difference in Cu substrate and after deposition with Ni by the Lowenheim equation,

$$\delta = \frac{m_2 - m_1}{\rho A} \tag{1}$$

And

$$S_\delta = \sqrt{\left(\frac{\partial\delta}{\partial m} S_m\right)^2 + \left(\frac{\partial\delta}{\partial p} S_p\right)^2 + \left(\frac{\partial\delta}{\partial A} S_A\right)^2} \tag{2}$$

Where δ is sheet thickness, m_1 and m_2 is mass of Cu substrate and Cu/Ni sample respectively, ρ is density, A is cross-section area, S_δ is the error of thickness. The sheet resistivity was determined based on a formula,

$$R_s = \frac{\pi}{\ln 2}\frac{V}{I} \tag{3}$$

If the measurement is done by varying I and recording the value V, then by linearization of data (V_i,I_i) we found a as a slope of the curve, and the resistivity becomes,

$$R_S = \frac{\pi}{\ln 2} a \tag{4}$$

The error of R_s is,

$$S_{R_S} = \frac{\pi}{\ln 2} S_a \tag{5}$$

where,

$$S_a = S_{\hat{y}} \sqrt{\frac{N}{N\Sigma x_i^2 - (\Sigma x_i)^2}} \tag{6}$$

and,

$$S_{\hat{y}} = \sqrt{\frac{\Sigma(y_i - \hat{y})^2}{N-2}} \quad (7)$$

with N is the amount of data.

3. Result and Discussion

3.1. Current in the deposition process

In figure 2 a deposition current is displayed during the electroplating. From this picture, it appears that all of the current used in a deposition is relatively stable at around 0.45 A. Unfortunately the time for start stable is different. When the plating process starts the current from zero then rises. For deposition at 15s, 25s, and 35s, currents fast to reach stability in contrast to deposition at 5s and 45s that appear to be rather slow to achieve a stable current of about 0.3s. With different initial stable conditions, this will affect the deposit product attached to the Cu substrate at the cathode. The presence of the ripple, especially in a deposition for 35s, describes the number of electrons passing through the amperemeter. This also illustrates the amount of Ni ions leading to the cathode.

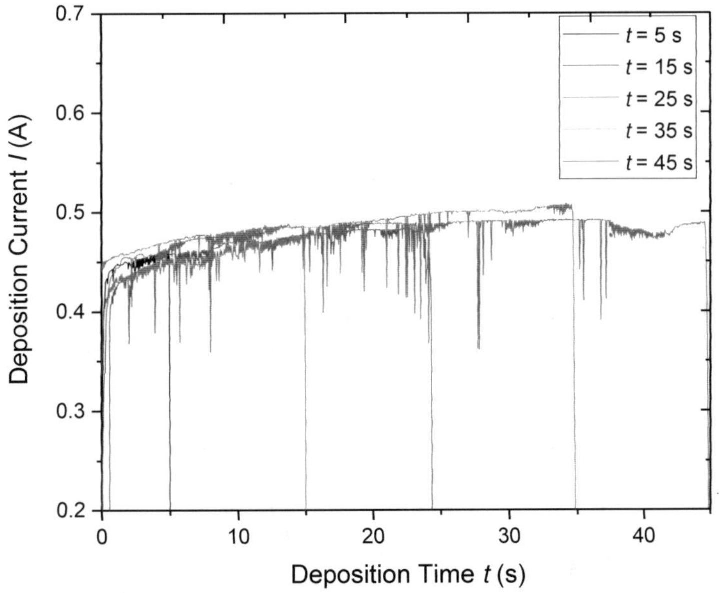

Figure 2. Deposition current at various deposition time

3.2. Film Thickness

The results of measuring the thickness of the film from various variations t with a magnetic field of 200 gauss are shown in Figure 2. It appears that the thickness of the layer is directly proportional to the time of deposition according to the equation.

$$\delta = 0.01195t - 0.05384 \quad (8)$$

where δ is the thickness and t is deposition time. Determination index of this graph is 99%, that shows the presence of a strong relation between time and the thickness. The highest thickness values obtained at 45 s deposition time are 50,2 μm, while at the deposition time 5 s the smallest film is 1,48 μm.

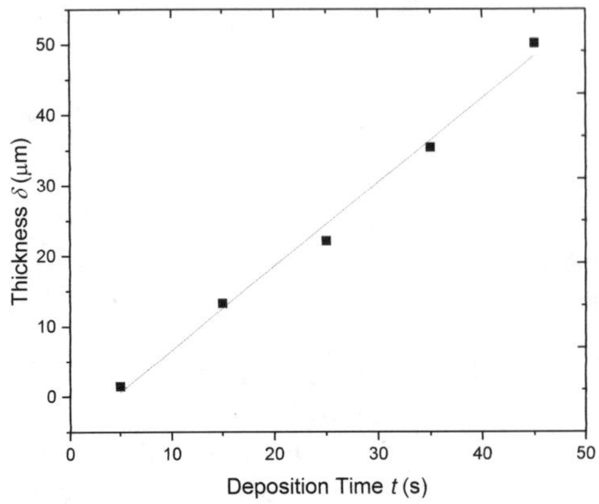

Figure 3. Layer thickness of the Ni layer at various deposition time

In accordance with Faraday's Law that the mass formed at the electrode is proportional to the equivalent weight, current strength, and deposition time.

$$M = \frac{e.i.t}{F} \tag{9}$$

where M the mass of the Ni deposit, and e is the equal weight, i is the current in the deposition process, F is the Faraday constant. Other researchers [20] also obtained the same results that the longer the deposition time, the thicker the layer formed.

As a comparison, we also deposited Ni on Cu substrate during 5s without applying the magnetic field. The thickness of Ni is 2.9 μm. This result is contrary to the purpose of using a magnetic field which is to increase the mass transport of Ni ions to the cathode. The fact that deposit at the cathode is less than the deposit obtained without magnetic fields. The thickness of the deposit only exceeds the deposit without B when the deposition time is 35s. It is assumed that the use of a magnetic field of 200 gauss generates a Lorentz force which causes some Ni ions to deviate from the target cathode, by taking into account that the electrode distance of 4 cm is a significant distance to deviate the Ni ions. Another possibility is that the activation of the magnetic field precedes the activation of the electric field so that Ni atoms are attracted to the cathode but not strong. When field B is removed the Ni does not stick. When the electric field is activated, the flow of Ni ions close to the substrate is blocked with nickel particles that have been attached to the cathode. Therefore, only a little Ni deposit is formed. To avoid that, activation of the magnetic field and electroplating process needs to be carried out together.

3.3. Formatting author affiliations

In Figure 2 is shown the sheet resistivity of Cu and Cu/Ni. From the figure, it appears that the sheet resistivity of Cu increases after being deposited with Ni. In line with the increase in deposition time, the sheet resistivity of Cu/Ni become decrease. The curvature of the curves for R_s-Cu and R_s-Cu/Ni shows that the five Cu substrate used has different R_s. This affects the value of R_s-Cu/Ni. But from the increase in Cu/Ni resistivity from Cu resistivity has a linear relationship to the deposition time according to the equation.

$$\Delta R_s = -4.8\text{E-}07t + 0.0013 \ (\Omega/\text{sq}) \tag{10}$$

Where the determination index $R^2 = 0.99$. The smallest ΔR_s is 1.25×10^{-3} Ω /sq that corresponds to 5s deposition time and the highest ΔR_s is 1.23×10^{-3} Ω/sq related to deposition time 45s.

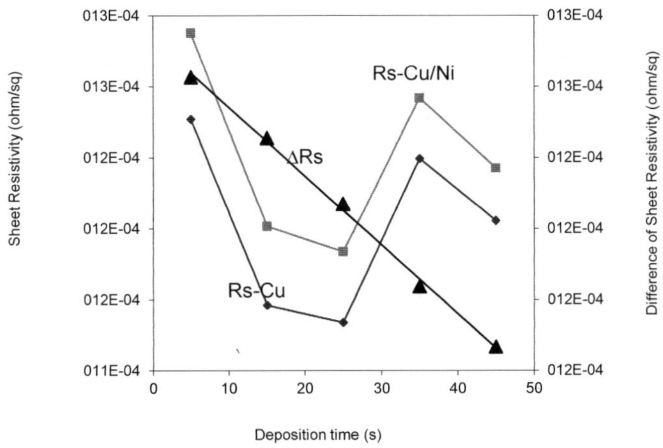

Figure 4. The sheet resistivity of Cu and Cu/Ni

4. Conclusion

Based on the results of the study, it was found that the Ni layer on the Cu substrate had succeeded to be synthesized in the variation of the electrolyte solution temperature. In this study, it is also known that the thickness of the Ni layer is proportional to the temperature of the electrolyte solution and inversely proportional to its resistivity.

Acknowledgments

This is the preliminary research from the main research concerning with Cu/Ni temperature sensor. The researcher expresses his deepest gratitude to the Ministry of Research, Technology and Higher Education of the Republic of Indonesia which has provided research funding through the 2019 Higher Education Advanced Applied Research (PTUPT) scheme.

References

[1] Strite S and Morkoc H 1992 *J. Vac. Sci. Technol.* B **10** 1237
[2] Nakamura S, Senoh M, Nagahama S, Iwase N, Yamada T, Matsushita T, Kiyoku H and Sugimoto Y 1996 *Japan. J. Appl. Phys.* **35** L74
[1] Hsueh C T, Wen C Y and Ouyang Y C 2015 *IEEE Sensors journal*, **15** 3590-3602.
[2] Kulkarni M, Sundaram S M, and Diwakar V 2015. *In 2015 IEEE International Transportation Electrification Conference* (*ITEC*) (pp. 1-3). IEEE.
[3] Khadatkar R M, Kumar S, and Pattanayak S C 2004. *Cryogenics*, **44**. 661–678.
[4] Goswami T K 2017, 1–4
[5] Toifur M, Yuningsih Y, and Khusnani A 2018 *J. Phys.: Conf. Ser.* **997** 012053
[6] Lebioda M and Rymaszewski J 2015 *Prz. Eletrotechnczny* **91** 225.
[7] Santos EJP and Vasconcelos IB 2008 *Proceedings in 2008 26th International Conference on Microelectronics*, Vol. 1 and 2. p. 333-336.
[8] Nusse D, Noetzel R, and Vdi 2008 *Sensoren Und Messsysteme* p. 891-898.
[9] Fraden J 2003 New York: Springer.
[10] Afsarimanesh N and Ahmed PZ 2011 *International Journal on Smart Sensing And Intelligent Systems*, **4** 726-739.
[11] Toifur M, Nuramalia A, Okimustava, Sukarelawan MI 2017 *Jurnal Material dan Energi Indonesia*. **7** 33-43
[12] Yu Y D, Song Z L, Ge H L, and Wei G Y 2014 *Surface Engineering* **30** 83–86.

[13] Zielinsky M, Miękoś E, Szczukocki D, Dałkowski R, Leniart A, Krawczyk B, and Juszczak R 2015 *Int. J. Electrochem. Sci.* **10** 4146-4154.
[14] Ganesh V, Vijayaraghavan D, and Lakshminarayanan V 2005 *Applied Surface Science* **240** 286–295.
[15] Ebadi M, Basirun W J, and Alias Y 2010 *Journal of Chemical Sciences* **122** 279–285.
[16] Aaboubi O and Msellak K 2017 *Journal Applied Surface Science* **396** 375-383.
[17] Shetty S and Hegde A C 2017 *Electrocatalysis* 184-185.
[18] Monzon L M and Coey J M 2014 *Electrochemistry Communications* 42-43.
[19] Kołodziejczyk K, Miękoś E, Zieliński M, Jaksender M, Szczukocki D, Czarny K, and Krawczyk B 2018 *Journal of Solid State Electrochemistry* **22**1629–1647.
[20] Kumar S, Pande S, and Verma P 2015 *International Journal of Current Engineering and Technology* **5**.

The Effect of Water Surface Level in sensible heat material on Yield of Single Basin Solar Still: Experimental Study

Dan Mugisidi[1,*], Berkah Fajar [2], Tony Utomo [2], Syaiful [2]

[1]Doctoral Programe, Department of Mechanical Engineering, Faculty of Engineering, Universitas Diponegoro, Indonesia
[2]Department of Mechanical Engineering, Faculty of Engineering, Universitas Diponegoro, Indonesia

[*]Email: dan.mugisidi@uhamka.ac.id

Abstract. The purpose of this experiment is to study the yield of the single basin solar still as an effect of water surface level in sensible heat material in a single basin solar still. The study is conducted on an open floor roof top at the Faculty of Engineering, Muhammadiyah University Prof. DR HAMKA in Jakarta, Indonesia. The study is conducted within two month and use 3 solar stills made of 3 mm Stainless Steel with dimensions 420 mm x 305 mm. The cover is made of 3 mm glass with a slope of 30 degrees. Iron sand that used as sensible heat material is taken from Glagah beach, Kulonprogo in Yogyakarta, Indonesia. 3 solar stills containing 20 mm height of iron sand. The height of the water in the three solar stills is 15mm (V1), 20mm (V2) and 25mm (V3) so that sequentially the surface of the water will be below the surface of the iron sand, the same as the surface of iron sand and above the surface of iron sand. The results obtained show that the daily yield accumulation of V2 is on average 49% higher than V1 while V3 is 57% higher above V1.

1. Introduction

Solar still is a simple device and only has 1 chamber to evaporate and condense basin water. The solar still is heated by the sun and the heat absorbed by the heat sink, then the heat is transferred to basin water. The hot water evaporates and then condense on the cover glass, expected to flow in the collecting gutter. The water that enters the collecting gutter will be channelled to the reservoir and already turned into fresh water.

Various methods and materials are used to improve the performance of solar still including Efforts to increase productivity are carried out using heat storage from sensible heat material. An experimental study use carbon powder performed charcoal particles is used as absorber medium and increase productivity as much as 15% [1]. El-Sebaii use sand as sensible storage material under basin liner [2]. Sand compare with servotherm medium oil is used as sensible heat storage material and placed beneath basin liner [3]. Carbon powder as a porous solar absorber inside solar still improve yield average 35% [4]. Sensible heat storage materials like quartzite rock, red brick pieces, cement concrete pieces, washed stones and iron scraps placed inside solar still [5]. The black jute clothes act as porous absorbers in basin water inside the solar still [6]. Extended porous fins made up of blackened old cotton rags are dipped partially inside the basin of the still [7]. The use of Portland cement black-blackened, as heat storage, increases freshwater production by 39% [8]. still. A modified solar using silica sand and layered with black coal powder at its surface compared with conventional solar still to study its heat and mass transfer

[9]. To our knowledge, no experimental research has been conducted using iron sand to enhance productivity of solar.

The present work is using iron sand as a porous absorber inside solar still. Since some study already prove that water level inside basin have significant influence [10] then it is motivate to vary water surface to become below, same level and above of iron sand surface to see its effect.

2. Experimental set up

Three single basins solar still are fabricated with stainless steel plate 2mm SUS304 and the cover use 3mm glass. Effective dimensions of the solar still 403mm x 305mm. At figure 1, all solar stills are filled with iron sand up to 20mm and basin water varied in height 15mm, 20mm, and 25mm, which are called V1, V2 and V3 respectively. Data was taken on August 19, 2018, August 29, 2018 and September 1 2018 starting at 08.00 to 17.00 without put solar radiation condition into account. Temperature was measured using the calibrated thermocouple type K, which has an accuracy of 0.1%, in 15 minutes interval. Digital anemometer 0 – 30 with interval 0,1 m/s was used to measure wind speed. The water level inside all solar still maintain at fix level using buffer tank outside the still. The iron content in iron sand varies depending on the area of mining, it can be 59.8% [11] to 76.35% [12]. In this experiment, the iron sand used content 70.3% of iron and was taken from Glagah beach, Kulonprogo in Yogyakarta, Indonesia. Pororsity of the iron sand is calculated using equation:

$$\phi = \frac{V_v}{V_t} \tag{1}$$

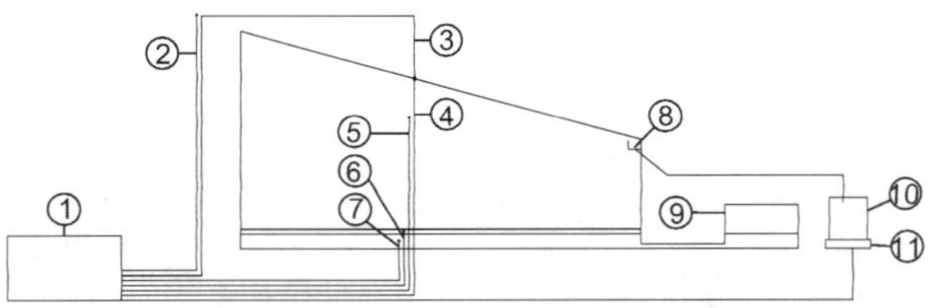

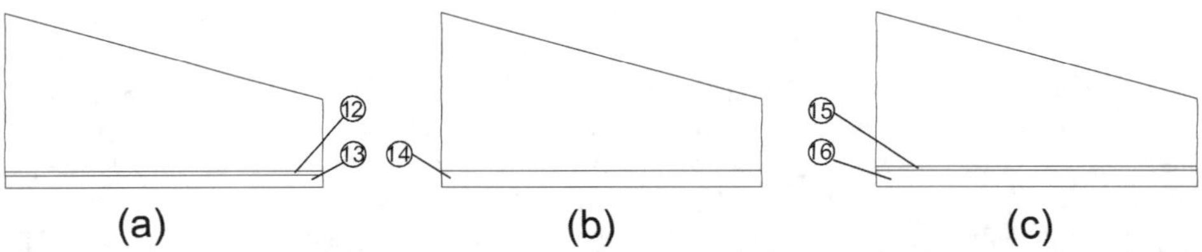

Figure 1. Single slope solar still *contains ;* (a) Iron sand 20mm height and basin water 15mm height namely V1, (b) Iron sand and basin water 20mm height namely V2, (c) Iron sand 20mm height and basin water 25mm height namely V3. 1 = Data Collector, 2 = Solar radiation sensor, 3 = Tgo Sensor, 4 = Tgi sensor, 5 = Tv sensor, 6 = Tw sensor, 7 = Ts sensor, 8 = Gutter, 9 = Water basin buffer, 10 = Fresh water collector, 11 = Weighing, 12 = Iron sand in V1, 13 = Mixed of water and iron sand in V1, 14 = Mixed of water and iron sand in V2, 15 = Basin water in V3, 16 = Mixed of water and iron sand in V3.

3. Result and discussion

Solar radiation and wind speed affect temperature and productivity of solar still [13]. As shown in figure 2, solar radiation and wind speed fluctuate all the time and it affect temperature in solar still. As can be seen in figure 3 to figure 4, temperature inside the solar still is fluctuation and respond to solar radiation, eventhough the response is different for each type of solar still. Temperature profiles are shown on solar still V1, V2 and V3 has a profile like a bell shape even though solar radiation fluctuates this because the heat of solar radiation is absorbed by sensible heat material material and released when solar radiation drop [2][14].

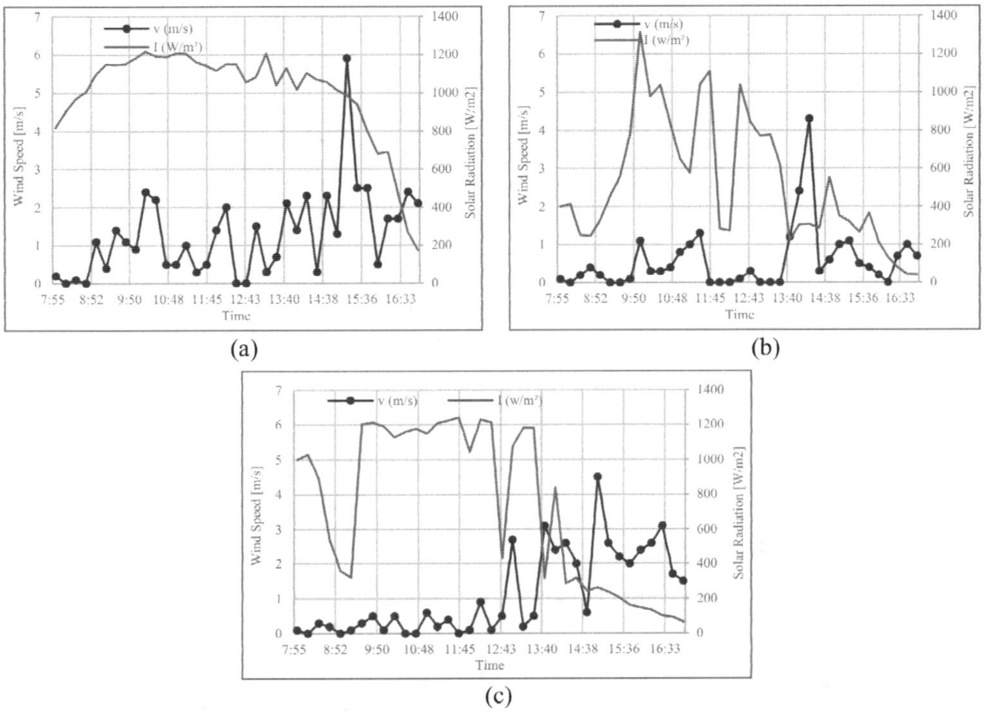

Figure 2. Solar radiation and wind speed at August 19, 2018 (a), August 29, 2018 (b) and September 1, 2018 (c).

In this study, water and sand were mixed. Thus, water temperature (Ta) and sand temperature (Ts) have a very similar temperature profile. At V1, the water surface is below the sand surface and the water temperature is lower than the sand temperature. In V3, the water surface is 5 mm above the sand surface and the water temperature is higher than the sand temperature. These results indicate that there is a gradient temperature due to the propagation of heat in the sand takes some time.

Daily accumulation yield can be seen at Fig 6. Although daily accumulation yield obtained is in accordance with solar radiation, the results follow a certain sequence, from the smallest V1, then V2 and the largest V3.

The level of water in V1 is lower than surface level of sensible heat material. The porosity of sensible heat material is 43.7% and the water rises to the surface with capillarity through the porosity of the material at the speed 0.34 mm/s. Since the surface area of the sensible heat material 1.57 times bigger than the area of solar still and the flow rate 0.0025 ml/s, the heat that accumulated on the surface of sensible heat material is not use to heat and evaporate the water but released. For comparison, the average evaporation rate of V3 is 0.0096 ml/s. That is the reason why the daily yield of V1 is lower than two others. Furthermore, since surface level of water as high as iron sand, part of iron sand particle is higher than water, so that heat accumulated on iron surface transfer to the basin water and air above it. As a result, daily yield of V2 is lower than V3, as shown in **Table 1**.

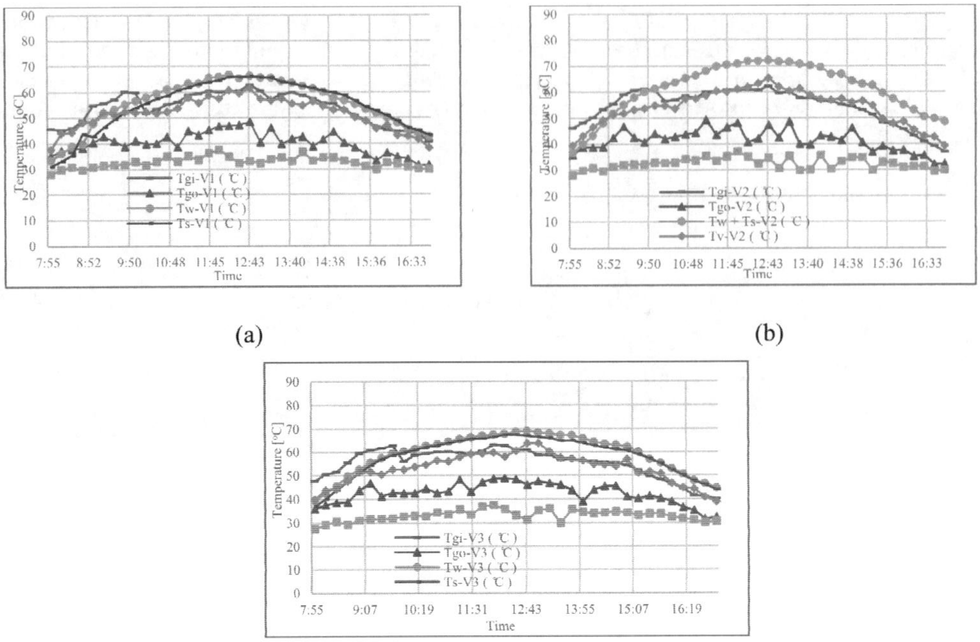

(a) (b)

(c)

Figure 3. Experiment at August 19, 2018. Tgi=inside glass temperature, Tw=water temperature, Ta=ambient temperature, Tgo=outside glass temperature, Tv=vapour temperature, I=solar radiation

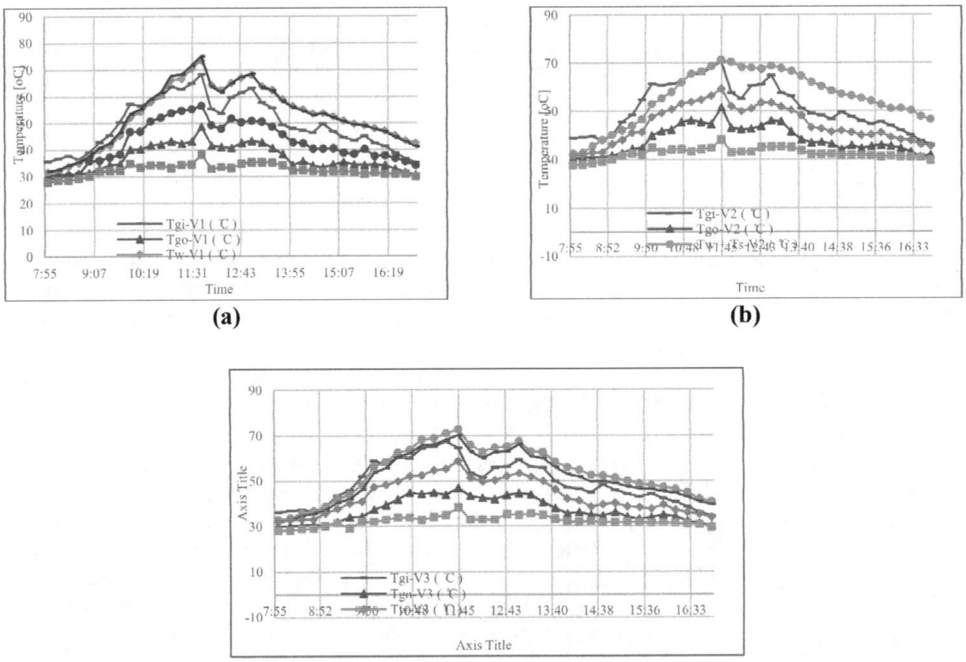

(a) (b)

(c)

Figure 4. Experiment at August 29, 2018

Journal of Physics: Conference Series **1373** (2019) 012014 doi:10.1088/1742-6596/1373/1/012014

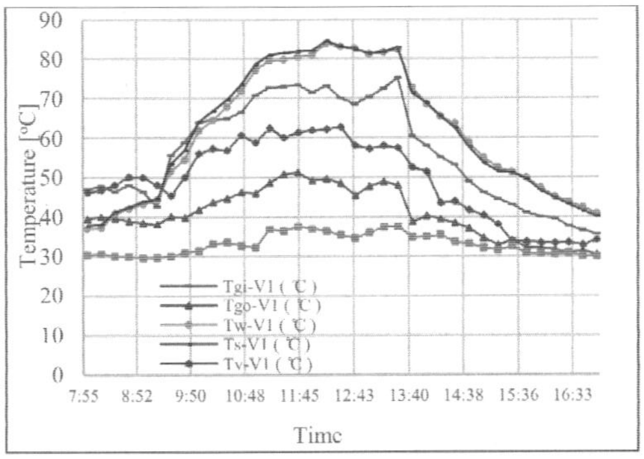

(a)

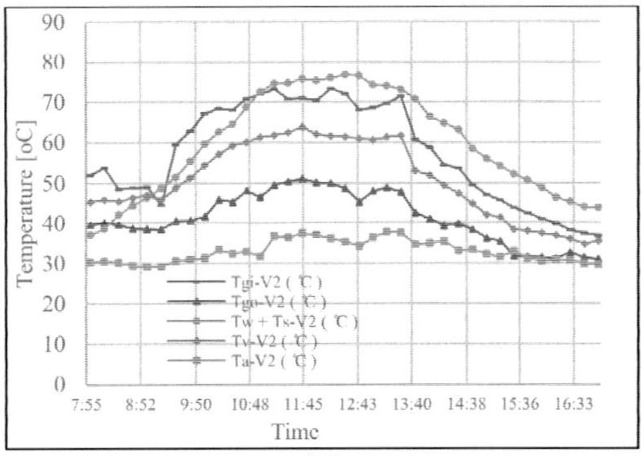

(b)

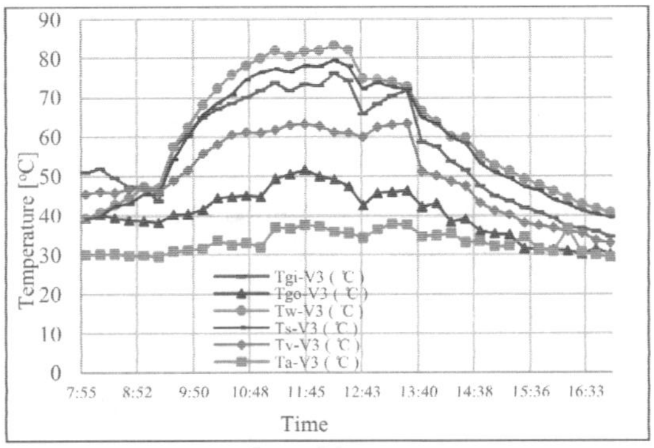

(c)

Figure 5. Experiment at September 1, 2018

TABLE 1. Daily yield of V1, V2 and V3

Date	V1 [gr/m²]	V2 [gr/m²]	V3 [gr/m²]	ΣI [W/m²]
19-Aug-18	925	2,175	2,450	36.933
29-Aug-18	392	667	708	19.056
1-Sep-18	500	958	1,125	26.811

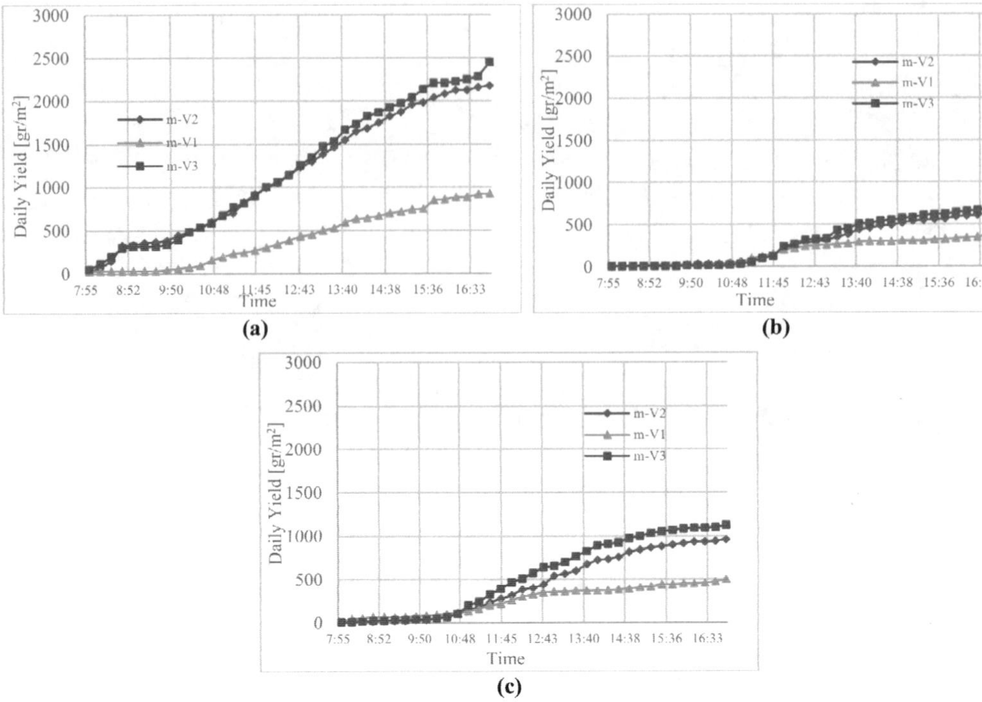

FIGURE 6. Daily yield of P, V1, V2 and V3 at August 19 (a), August 29 (b) and September 1, 2018 (c)

4. Conclusion

The experiment using iron sand as a sensible heat material material and functioned as an absorber was carried out using a single slope solar still. Experiment result shows that daily yield of solar still that using iron sand as absorber affected by basin water surface level. Accumulation of daily yield V2 is on average 49% higher than V1 while V3 is 57% higher above V1 therefore it is recommended to set the water level above the surface of iron sand.

5. References

[1] Naim M M and Abd El Kawi M A 2003 *Desalination* **153** 55-64.
[2] El-Sebaii A A, Yaghmour S J, Al-Hazmi F S, Faidah A S, Al-Marzouki F M, and Al-Ghamdi A A 2009 *Desalination* **249** 699-706.
[3] Deshmukh H S and Thombre S B 2017 *Desalination* **410** 91-98.
[4] Madani A A and Zaki G M 1995 *Appl. Energy* **52** 273-281.
[5] Murugavel K K, Sivakumar S, Ahamed J R, Chockalingam K K S K, and Srithar K 2010 *Appl. Energy* **87** 514-523.
[6] Srivastava P K and Agrawal S K 2013 *Desalination* **311** 198–205.
[7] Srivastava P K and Agrawal S K 2013 *Desalination* **319** 73–78.
[8] Sellami M H, Guemari S, Touahir R, and Loudiyi K 2016 *Desalination* **394** 155–161.
[9] Dumka P and Mishra D R 2018 *Desalination* **437** 15–25.
[10] Elango T and Murugavel K K 2015 *Desalination* **359** 82–91.
[11] Liu K, Li Y, Wang F, Ren J, and Xie H 2018 *Constr. Build. Mater.* **178** 144–152.
[12] Suroso I 2017 *Teknika. STTKD: Jurnal Teknik, Elektronik, Engine* **4** 26–38.
[13] Zurigat Y H and Abu-Arabi M K 2004 *Appl. Therm. Eng.* **24** 1061–1072.
[14] Sellami M H, Touahir R, Guemari S, and Loudiyi K 2016 *Desalination* **398** 180–188.

Journal of Physics: Conference Series **1373** (2019) 012015 doi:10.1088/1742-6596/1373/1/012015

Harmonic Analysis of Ionospheric Total Electron Content (TEC) Using Kalman Filter

Buldan Muslim[1, *], Charisma Juni K[2], Erlansyah[1]

[1]National Institute of Aeronautics and Space, Jl. Dr. Junjunan 133 Bandung, Indonesia 40173
[2]Sepuluh Nopember Institute of Technology,Jl. Raya ITS, Keputih, Kec. Sukolilo, Kota SBY, Jawa Timur, Indonesia 60111

[*]Email: mbuldanr@gmail.com

Abstract. The ionospheric diurnal variation from TEC Global Ionospheric Maps (GIM) data at the epicenter coordinates of the Lombok earthquake on July 29 2018 (-8.4°, 116.5°) has been analyzed using harmonic function which is the sum of the sine and cosine functions with estimated harmonic coefficients obtained using Kalman Filter (KF) method. The initial value of the estimated KF method was obtained from Least Square Estimation (LSE) on the first day data. The estimated TEC value by the KF method shows that there are seasonal variations of the peak values of diurnal variations which is a result of variation in the zenith angle of the sun that change over a year with peaks in March and November in the equator. The results of the first-order harmonic amplitude analysis, namely the diurnal TEC variation (A24) which is the root value of the sum of the squared values of the A1 and B1 coefficients were further analyzed by estimating polynomials with order 17 to eliminate the seasonal effect on A24. The difference in value of A24 from its estimation with order 17 polynomial (delta A24) shows the influence of geomagnetic and seismic activity. A24 analysis based on solar radiation flux at wavelength 10.7 cm (F10.7) does not indicate a strong influence or relationship> Equatorial geomagnetic disturbance index (etc.) shows the influence of geomagnetic storms on diurnal amplitude and earthquake data around Lombok indicating symptoms of pre-earthquake that affects the ionospheric TEC through a decrease in diurnal amplitude a few days before the earthquake.

1. Introduction

Ionosphere is the region of the Earth's upper atmosphere extending from altitudes of approximately 60 km from the surface of Earth to the altitudes beyond 1000 km, where the sufficient amount of charged particles exists. In other words, this is the region of a shell of electrons and electrically charged particles (atoms and molecules), also known as plasma that surrounds the Earth. In plasma, the electrostatic force attracts the negative free electrons and the positive ions to each other, but they are too energetic to stay fixed together in an electrically neutral molecule. The plasma concentration may amount to only about 1% of the neutral concentration and the total ionosphere represents only less than 0.1% of the total mass of the Earth's atmosphere [1].

A theory of the ionization of the upper atmosphere of the earth by the ultra-violet light of the sun is developed based on known laws of pressures and constitution of the high atmosphere, ionic recombination, attachment of free electrons to neutral molecules, and diffusion of ions. It is concluded that the solar ultra-violet light is a necessary and sufficient cause of the Kennelly-Heaviside layer, and that hypotheses of other agencies of ionization, such as charged particles from the sun, penetrating radiation, etc., are uncalled for except perhaps in unusual cases [2].

Because the Sun shines over the equator, the ionospheric density was expected to vary from a maximum at the equator to a minimum at high latitudes. But when the density variation was measured [3]. In [4] it was found to exhibit an unexpected large structure with a trough around the equator, crests near ±15° magnetic latitudes and crest-to-trough ratio about 1.6 in daytime peak electron density (Nmax). This large structure known as the equatorial ionization anomaly (EIA) develops in the morning at around 10 LT, continues to exist well beyond sunset and covers about half the global area in 24 hours; and the position of the crests and crest-to-trough ratio vary with various geophysical conditions.

Solar radiation absorbed by neutral molecules and atomic in the upper atmosphere depends on the angle of arrival of solar radiation as measured by the sun's zenith angle. The greater the zenith angle of the sun, the greater the energy of the absorbed solar radiation. Apart from the intensity of solar radiation, the ionosphere is also affected by the transport process because as charged particles that are in the earth's magnetic field also move as a combined effect between the magnetic field and the electric field in ionospheric charged particles. The ionospheric electric field arises as a result of differences in the warming of the upper atmosphere by solar radiation which gives rise to neutral winds in the upper atmosphere that are westward towards the side of the day and eastward at night. It happens because the response of particles with different electrons and ions to this neutral wind arises the variation of charged particles so that an electric field appears east in the afternoon and westward at night. Therefore, during the day the ionosphere moves upwards and at night the ionosphere moves down.

Variations of the ionosphere can be classified as regular and irregular. Regular variations occur more or less in cycles and therefore can be modeled in advance with reasonable accuracy. Irregular variations, which are mainly due to abnormal solar behavior such as sudden changes in solar radiation, cannot be predicted in advance. Examples of regular variations are diurnal, annual, and solar cycle variations [5]. Possible modulations of the diurnal signal with other frequencies are also identified as regular variations. In this case, the amplitude of the daily signal changes from one season to another. A mathematical foundation is developed to detect and model such occurrences [6].

Since the existence of the ionosphere is due to solar radiation, the relative location of the earth with respect to the sun is responsible for a large part of the ionospheric variations. The daily earth rotation changes the solar zenith angle, which describes a large part of the daily ionospheric variations. The principal diurnal (24 h) component of the daily signal and its higher harmonics (24/n hours, $n > 1$) are due to the day–night variation of the ionosphere and hence of the total electron contents (TEC) values.

Daily variations of harmonic amplitude can be associated with solar activity, geomagnetic activity and seismic activity. Seismic activity can influence the diurnal amplitude variation with the assumption that before an earthquake, anomaly can occur in the electric field above the epicenter of an earthquake that affects the electric field in the upper atmosphere.

The very early versions of seismo-ionospheric coupling models were based on direct calculations of seismogenic electric field effect from the ground surface to the ionosphere [7]; [8]. All of them are based on the methodology of electric field penetration into the ionosphere from thundercloud [9]. This paper describes the results of harmonic analysis of ionospheric TEC above the epicenter of the Lombok earthquake in 2018 to investigate the possible ionospheric precuror before Lombok earthquake.

2. Data and Methodology

The ionospheric TEC data above the Lombok earthquake epicenter was taken from TEC GIM data that can be downloaded from ftp://aiub.unibe.ch in the CODE folder and in 2018 with the CODGDOY0.18I file name, with DOY showing the day to year.

The TEC data above the epicenter of the Lombok earthquake from GIM can be extracted or selected for the TEC data closest to the epicenter of the Lombok earthquake on July 28 2018, namely at latitude 8.4° south latitude and longitude 116.5°. After extracting, the TEC data above the epicenter is arranged in a matrix with the number of rows equal to the number of days from doy 1 to 339, and column 1 to 24 contain the value of TEC from 0 to 23. After that, a TEC harmonic analysis was carried out through the KF method with the initial value of the LSE results of the first day of the TEC in 2018. The sine and cosine harmonic coefficients of the equation (1) the results of calculation using the KF method are filtered from the effect of variations in the sun's zenite angle (seasonal variation) with 17th order polynomial. The filtering results of A24 are correlated with F10.7 which is also filtered with order polynomial 27.

Analysis of the effect of earthquake precursors on the A24 diurnal amplitude can be done by making a graph between delta A24 versus DOY and marking the time of an earthquake with a vertical line on the date of the earthquake. Likewise, the effect of geomagnetic storms is seen qualitatively through delta A24 graph.

2.1 Harmonic Analysis

Twelve years of harmonic TEC analysis used modest harmonic analysis and the most dominant diurnal variation is the main diurnal variation (24-hour period) and half-year period (6 months), in addition to 27 days due to solar rotation and annual variation due to the solar cycle [2]. Based on these results, in this paper the ionospheric TEC diurnal variations were only analyzed by two harmonic components, namely periods 24 and 12. Referring to the formulation of harmonic analysis [10] use different notations, the formulation of harmonic analysis for the two main components of diurnal variation can be given as

$$TEC(t) = A_o + A_1 Sin\left(\frac{2\pi}{24}t\right) + B_1 Cos\left(\frac{2\pi}{24}t\right) + A_2 Sin\left(\frac{2\pi}{12}t\right) + B_2 Cos\left(\frac{2\pi}{12}t\right) \quad (1)$$

with the explanation of the harmonic coefficient as follows:

Ao : TEC average in one day

A1 : Amplitude of the sine component of period 24

B1 : Amplitude of the cos component of period 24

A2 : Amplitude of the sine component of period 12

B2 : Amplitude of the cosine component of period 1

If the harmonic coefficients were obtained, the amplitude of the wave / diurnal variation is

$$A_{24} = \sqrt{A_1^2 + B_2^2} \quad (2)$$

and the wave phase is

$$\Phi_{24} = a\tan\left(\frac{A_{24}}{B_{24}}\right) \tag{3}$$

in the form of a matrix, Equation (1) can be written as follows:

$$\begin{bmatrix} TEC(0) \\ TEC(1) \\ TEC(2) \\ \vdots \\ TEC(23) \end{bmatrix} = \begin{bmatrix} 1 & Sin\left(\frac{2\pi}{24}0\right) & Cos\left(\frac{2\pi}{24}0\right) & Sin\left(\frac{2\pi}{12}0\right) & Cos\left(\frac{2\pi}{12}0\right) \\ 1 & Sin\left(\frac{2\pi}{24}1\right) & Cos\left(\frac{2\pi}{24}1\right) & Sin\left(\frac{2\pi}{12}1\right) & Cos\left(\frac{2\pi}{12}1\right) \\ 1 & Sin\left(\frac{2\pi}{24}2\right) & Cos\left(\frac{2\pi}{24}2\right) & Sin\left(\frac{2\pi}{12}2\right) & Cos\left(\frac{2\pi}{12}2\right) \\ \vdots & \vdots & \vdots & \vdots & \\ 1 & Sin\left(\frac{2\pi}{24}23\right) & Cos\left(\frac{2\pi}{24}23\right) & Sin\left(\frac{2\pi}{12}23\right) & Cos\left(\frac{2\pi}{12}23\right) \end{bmatrix} \begin{bmatrix} A_0 \\ A_1 \\ B_1 \\ A_2 \\ B_2 \end{bmatrix} \tag{4}$$

Furthermore, it can be written in the form of equation:

$$Y = CX \tag{5}$$

with Y containing TEC bonding data from 0 to 23, C is a matrix containing value 1 (column 1), sine and cosine values at 0 to 23 for 24 periods (columns 2 and 3) and 12 periods (columns 4 and 5) X contains unknown harmonic coefficients A0, A1, B1, A2, B2 as explained in equation (1).

The harmonic coefficients in C matrix can be estimated by the LSE method for the first day (January 1, 2018). After that, the harmonic coefficient values become the initial value for the equation in the KF method. Based on the mathematical model in equation (1), a coefficient of harmonic solution can be obtained with the KF method. Furthermore, the application of the KF method to the mathematical model in equation (1) follows.

2.2 Kalman Filter Method

The Kalman Filter (KF) method is a state variable estimation method of a discrete linear stochastic dynamic system that minimizes the estimation error covariance. KF is a technical approach for estimating parameter functions in time series forecasting. The advantage of the KF method is its ability to estimates a situation based on minimal data. The minimum data in question is measurement data (measuring instruments) because KF is a method that combines models and measurements [11].

The KF algorithm consists of four parts, including the first part defining the system model and the measurement model, the second part is the initial value (initialization), then the third and fourth parts of each prediction and correction stage.

The system model used by the KF method can be written as follows [4,11]:

$$x_{k+1} = Ax_k + Bu_k + \eta_k \tag{6}$$

and the measurement model

$$z_k = Hx_k + \varepsilon_k \tag{7}$$

The matrix A in the difference equation (6) relates the state at the previous time step (k) to the state at the current step (k+1), in the absence of either a driving function or process noise. Note that in practice A might change with each time step, but here we assume it is constant. The matrix B relates the optional control input to the state x. The matrix H in the measurement equation (7) relates the state x_k to the measurement z_k. In practice H might change with each time step or measurement, but here we assume it is constant.

Prediction is to predict the state and covariance process at time k + 1 depending on information at time k. Prediction stage:

Forecast real data : $\bar{x}_{k+1} = Ax_k$ (8)

Error covariance forecast : $\overline{P}_{k+1} = AP_kA^t + Q$ (9)

The next phase of correction as follows:

Kalman Gain : $K_{k+1} = \overline{P}_{k+1}H_{k+1}{}^t(H_{k+1}\overline{P}_{k+1}H_{k+1}{}^t + R)^{-1}$ (10)

Error covvariance estimation : $P_{k+1} = (I - KH)\,\overline{P}_{k+1}$ (11)

Real data estimation : $\hat{x}_{k+1} = \bar{x}_{k+1} + K_{k+1}(z_{k+1} - H\bar{x}_{k+1})$ (12)

2.3 Establishment of a System Model

Based on the TEC equation a system model can be formed as follows:

$$x_{k+1} = Ax_k \tag{13}$$

with $x = [A_0\ A_1\ B_1\ A_2\ B_2\ TEC(1)\ \ldots\ TEC(24)]^T$, index 1,2,3…24 represents time (hours) and matrix A is

$$A = \begin{pmatrix} 1 & 0 & 0 & 0 & 0 & 0 & \cdots & 0 \\ 0 & 1 & 0 & 0 & 0 & 0 & \cdots & 0 \\ 0 & 0 & 1 & 0 & 0 & 0 & \cdots & 0 \\ 0 & 0 & 0 & 1 & 0 & 0 & \cdots & 0 \\ 0 & 0 & 0 & 0 & 1 & 0 & \cdots & 0 \\ 1 & a1(0) & a2(0) & a3(0) & a4(0) & 0 & \cdots & 0 \\ \vdots & \vdots & \vdots & \vdots & \vdots & \vdots & \ddots & \vdots \\ 1 & a1(23) & a2(23) & a3(23) & a4(23) & 0 & \cdots & 0 \end{pmatrix}_{29X29}$$

Where:

$a1$(t) is $\sin((2\pi/24).t)$, t= 0, 1, 2, ..., 23)

$a2$(t) is $\cos((2\pi/24).t)$, t= 0, 1, 2, ..., 23)

$a3$(t) is $\sin((2\pi/12).t)$, t= 0, 1, 2, ..., 23)

$a4$(t) is $\cos((2\pi/12).t)$, t= 0, 1, 2, ..., 23) .

The measurement model is

$$z_k = Hx_k \tag{14}$$

Journal of Physics: Conference Series **1373** (2019) 012015 doi:10.1088/1742-6596/1373/1/012015

Where

$$H = \left[0_{24x5}\ \ I_{24x24}\right]$$

The system model and measurement of the KF algorithm were obtained from equations (13) and (14). With the initial conditions A_0, A_1, B_1, A_2, B_2 obtained from the LS method while the initial TEC (1), TEC (2), ..., TEC (24) were obtained from the TEC of first day or the inital TEC. The harmonic coefficients of TEC model is estimated using Kalman filter algorithm as seen in equation (8) to (12).

3. Results and Discussion

By using MATLAB software, software estimation of second order harmonic coefficients has been made as in equation (1) with the KF algorithm as in equation (8) - (12) can be obtained estimation of harmonic coefficients as shown in Figure 1. In Figure 1 can be seen that the coefficient harmonic order 0 (A0) and A1 (sine coefficient period 24) is the most dominant with a value of more than 5 TECU throughout 2018 (blue and red). The peak values of A0 and A1 during 2018 occur at two times, namely at doy between 61 to 91 (March) and between 301 and 331 (November). This shows the peak asymmetry of TEC values in 2018, namely in March and November and not in March and September which is the time at which the sun's zenith is highest. This indicates a delay in the time the TEC reaches its peak in September to November (delayed by 2 months). The delay in the peak value of the TEC in September can be caused by several reasons. First, the epicenter position of the Lombok earthquake in southern latitude could cause changes in the peak value of the TEC due to the presence of meridional neutral winds in the direction of the south which caused an accumulation of ionospheric electron density to peak in November. The second can be caused by the high value of solar radiation in November 2018, but the alleged cause of this asymmetry needs to be supported by the results of previous studies regarding both the occurrence and the explanation of the mechanism.

Use equation (2) and the phase according to equation (3) which is calculated for all days from doy 1 to 339 as shown in Figure 2.

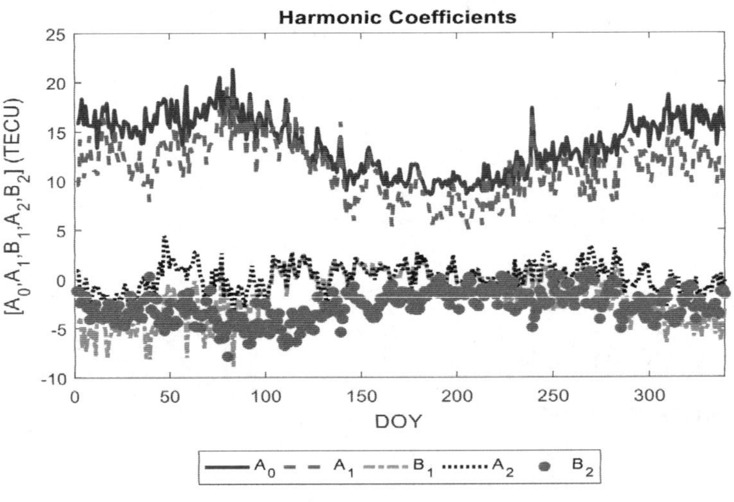

Figure 1. Estimated value of 2nd order TEC harmonic coefficients (A0, A1, B1, A2, B2) with Kalman filters

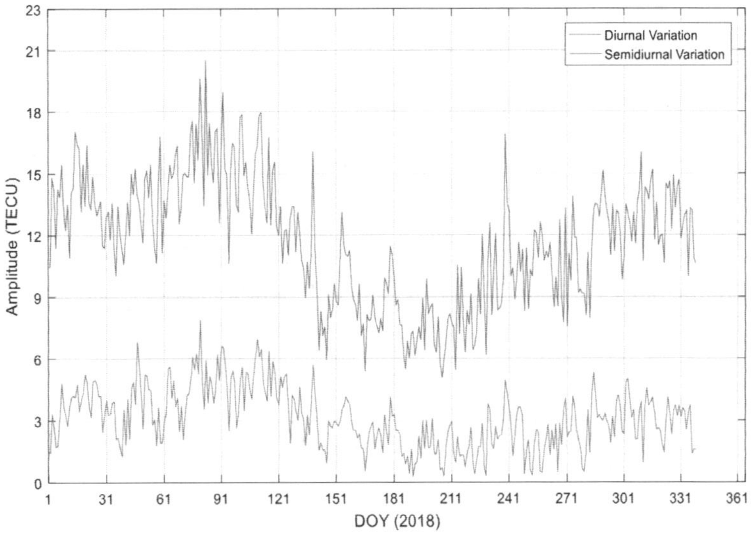

Figure 2. Diurnal (blue) and semidiurnal (brown) amplitude

It can be seen from Figure 2 that the peak values of the second monthly variation in a year (the first in March) of diurnal and semi-diurnal amplitude were slightly different. In October, semi-diurnal amplitude has reached the highest while the amplitude of the diurnal variation has only reached a maximum in November. In addition, in August (doy 240) there was a sharp increase in diurnal and semidiurnal amplitude. To see the diurnal amplitude amplitude daily filtering or separation of diurnal amplitude from the effect of long-term variation with 17th order polynomial as shown in Figure 3. There are semi-annual and 3-month long-term variations in the diurnal amplitude. After reducing the diurnal amplitude value with the estimated value of 17th order polynomial, the diurnal amplitude amplitude is obtained as shown in Figure 4.

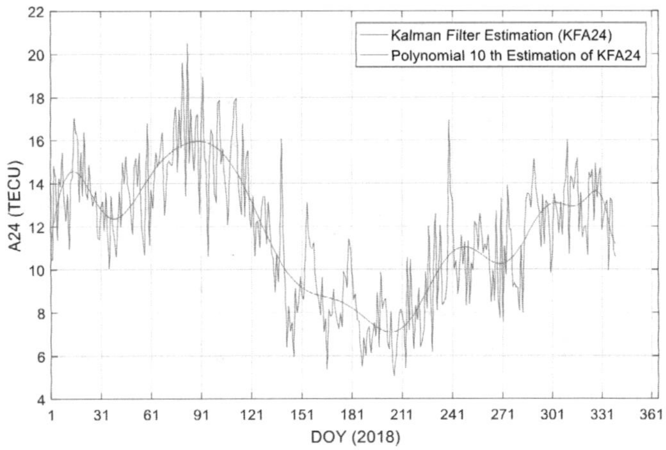

Figure 3. Seasonal variation in amplitude of ionospheric TEC diurnal variation (blue line) and its estimation with 17th order polynomial function (brown line)

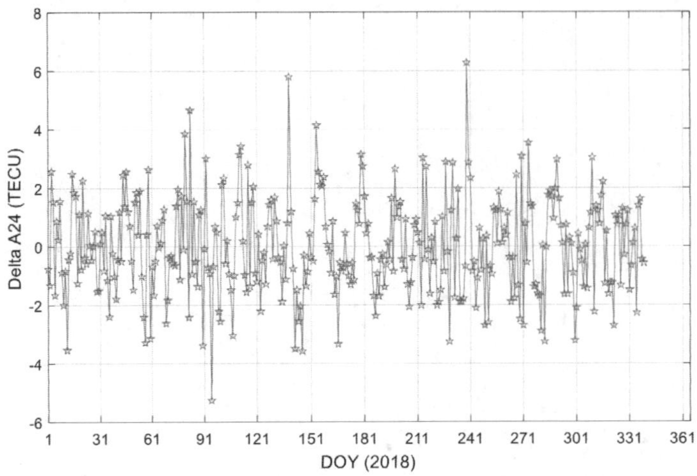

Figure 4. Daily variation of delta A24 ionospheric TEC diurnal variation in 2018 above the earthquake epicenter of Lombok (-8.4°, 116.5°)

Daily diurnal amplitude variations can be caused by daily variations in solar radiation that can be measured by F10.7. Then the daily variation of F10.7 after filtering with long-term seasonal variations can be correlated with delta A24 as shown in Figure 5 which shows very little effect of daily variation F10.7 on diurnal amplitude amplitude daily with a very small correlation indication (0.02) and unpolished relationship between these two parameters.

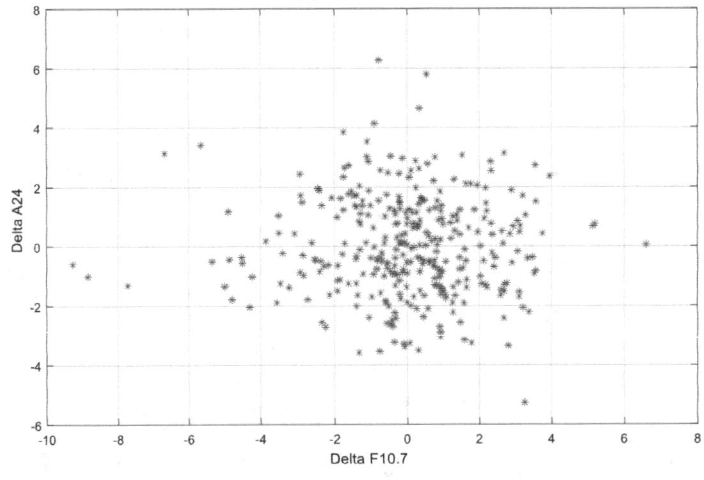

Figure 5. The diurnal delta amplitude relationship with F10.7

Daily diurnal amplitude variations can also be affected by electric fields that appear before earthquakes and geomagnetic disturbances. Then the next analysis is the relationship between prominent diurnal variations in diurnal amplitude and earthquake precursors and geomagnetic storms as shown in Figure 6.

In Figure 6 an earthquake is indicated by a vertical straight line. The first vertical line shows the time of the earthquake on July 29, 2019 to be exact at 6:47 a.m., with a strength of 6.4 M or on July 28 at 22.47 UT according to global time or DOY = 209. In the case of this earthquake, ionospheric TEC anomaly or earthquake precursor in the ionosphere was detected 3 days earlier, namely on July 25 (doy 206) in the form of a decrease in the A24 amplitude until it reaches a minimum with a value of - 2 TECU. In the Lombok earthquake on August 5 (doy 217) the ionospheric earthquake precursor was detected 4 days before that on August 1 (doy 213) in the form of a decrease in amplitude reaching - 2 TECU and in the case of the August 19 earthquake (doy 231) a two-day TEC anomaly was detected previously, on August 17 (doy 229) in the form of a decrease in amplitude A24 to -3.3 TECU. The greatest anomaly caused by the geomagnetic storm August 26 2018 (doy 238) in the form of an increase in diurnal amplitude reached 6 TECU on August 27 (doy 239) one day after the geomagnetic storm.

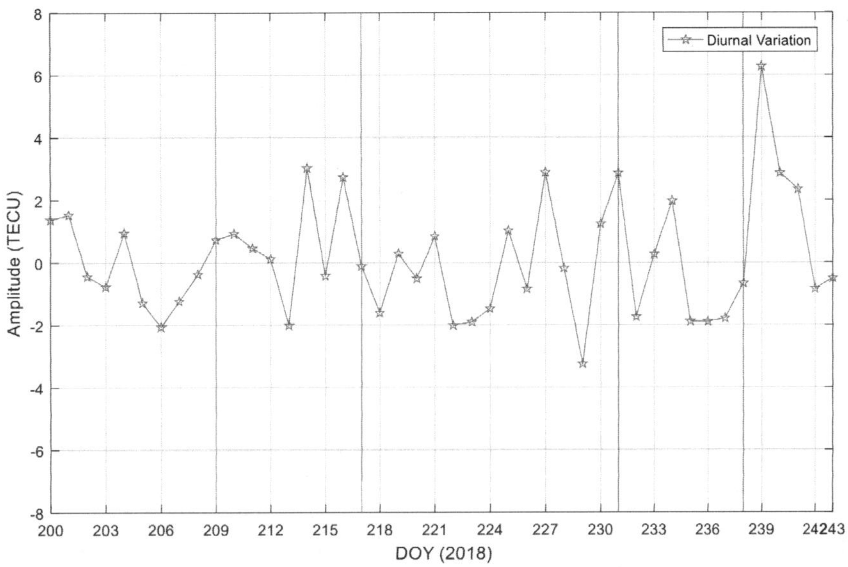

Figure 6. Indication of the precursors of a major earthquake in Lombok in 2018 on July 28 (doy 209) detected on July 25 (doy 206) in the form of a decrease in amplitude A24 to reach a minimum with a value of - 2 TECU, Lombok 5 August earthquake (doy 217) detected on August 1 (doy 213) in the form of a decrease in amplitude reaching - 2 TECU and August 19 (doy 231) detected on August 17 (doy 229) in the form of decreasing amplitude A24 to - 3.3 TECU and the effect of geomagnetic storm August 26 2018 (doy 238) August 27 (doy 239) in the form of an increase in amplitude A24 to more than 6 TECU.

4. Conclusion

The geomagnetic storm affects the amplitude of the diurnal variation, one day after geomagnetism. It can be seen that the amplitude of the variation in the ionospheric TEC is influenced by the strong zenith angle of the sun that changes in a year where in March and October it reaches the highest zenith in Indonesia. The daily variation of solar radiation flux at 10.7 cm does not affect the daily diurnal amplitude variation. Daily variations in diurnal amplitude are more influenced by the portion of transport that varies on a daily basis as indicated by changes in geomagnetic disturbances during geomagnetic storms and electric field anomalies that are suspected as precursors of earthquakes a few days before the earthquake occurred. In this case, to further explore the effects of transport on diurnal variations of TEC, it is necessary to analyze the diurnal TEC variation extracted from variations in the sun's zenith angle by harmonic analysis on a fixed phase ie at 12 LT as the peak of solar radiation.

5. References

[1]. Kelly M C 2009 *The Earth's Ionosphere, Plasma Physics and Electrodynamics* (Boston: Academic Press).
[2]. Hulburt E O 1928 *Phys. Rev.* **31** 1018.
[3]. Namba S, and Maeda K I 1939 *Radio Wave Propagation (*Tokyo: Corona)
[4]. Appleton E V 1946 *Nature* **157** 691.
[5]. Jain S, Vijay S K, Gwal A K 1996 *Adv. Space Res.* **18** 263–266.
[6]. Amiri-Simkooei A R and Asgari J 2012 *GPS Solutions* **16** 77-88.
[7]. Kim V P, Hegai V V, Illich-Svitych P V 1994 *Physics of the Solid Earth* **30** 223-226.
[8]. Pulinets S A, Boyarchuk K A, Hegai V V, Kim V P and Lomonosov A M 2000 *Adv. Space Res.* **26** 1209-1218.
[9]. Park C G, Dejnakarintra M 1973 *J. Geophys. Res.* **84** 960-964.
[10]. Liu J, Chen R, Wang Z, An J and Hyyppa J 2014 *PLOS One* **9** e111497.
[11]. Bishop G 2006 An Introduction to the Kalman Filter (Department of Computer Science, University of North Carolina at Chapel Hill).

Comparative Study of p-type $CuBi_2O_4$ Films and $CuBi_2O_4$ Nanopillars Photocathode for High Performance Photoelectrochemical Water Splitting

Nasori Nasori[1,2*], Agus Rubiyanto[1], and Endarko Endarko[1]

[1]Medical Physics and Biophysics Research Group, Department of Physics, Institut Teknologi Sepuluh Nopember, Surabaya, Indonesia

[2]Occupational and Safety Department UNUSA, Surabaya, Indonesia

*Email: nat.nasori@physics.its.ac.id

Abstract. Traditional thin films and nanostructure are the most reasonable candidates to build the next generation of photoelectrochemical water splitting system with outstanding optical and electrical properties. Especially the use of nanostructure arrays as photoelectrodes might complement the traditional semiconductor photoelectrodes in providing close transfer distance of photoinduced carriers and the increase in the surface reaction sites than thin films. Both of the reasons reduce the probability of carriers recombination and thus enhancing the photoelectrochemical performances. In this work, we demonstrated highly efficient water splitting performance of $CuBi_2O_4$ nanopillars compared to thin film $CuBi_2O_4$ photocathode. The CuBi2O4 nanopillars were fabricated by electrodeposition on anodized aluminum oxide (AAO) template. The $CuBi_2O_4$ nanopillars photocathode gave a notable improvement in photocurrent, from –0.50 to –1.50 mA cm^{-2} at –0.45 V vs. Ag/AgCl by the external quantum yield more than 3 times at wavelength 420 nm. Finally, the result of the study appealed that the photoelectrode based on $CuBi_2O_4$ nanostructure arrays is an encouraging system for showing efficient water splitting system under visible light.

1. Introduction

The conversion of solar energy to fuels cell through water splitting into hydrogen and oxygen evolution in photoelectrochemical (PEC) cell was initiated in 1972 by Fujishima and Honda using TiO_2 electrode [1]. Since then, the technique becoming popular due to its abundance and environmental-friendly nature of hydrogen fuel toward the worldwide sustainable energy questions. Most research on hydrogen energy production has been performed repeatedly up to now, however, the reported efficiencies are still too low to be practical [2-6]. An alternative method to solve this challenge is encouraging novel morphologies of the photoelectrode. This approach has been shown to enormously enhancing the performance of solar water splitting, likely due to their unique size-geometry and shape-dependent features. Nanopillar as a representative nano-structure has been successfully fabricated in various semiconductors for enhancing the PEC performances, such as TiO_2, Cu_2O [7].

Through a direct band gap around 1.6 up to 2.1 eV, $CuBi_2O_4$ (CBO) has long been considered as an assurance nominee for solar energy conversion. Theoretically, the maximum photocurrent of 15 mA cm^{-2} and 18% solar-to-hydrogen conversion efficiency under AM 1.5 illumination is expected to be achieved.[6-10] Furthermore, the p-type materials such as copper are abundant and can be produced at an industrial scale in a cost-effective method [10]. In our previous work, in which the highly active photocathode of CBO thin films was proposed, the photocurrent can reach as high as 1 mA/cm^2 and maintained active for 1h of testing [10]. There have we reported successful treatment for CBO photocathodes using very thin Au and Pt to realize an impressive PEC performance [10]. These remarkable performances determine the unique potential for CBO in PEC water splitting. Furthermore, in this purpose, we synthesize CBO nanopillars arrays efficaciously by the convenient and cost-effective AAO template method. From comparison by the photocathode based on CBO films, the CBO nanopillars photocathode presents a remarkable improvement in photocurrent, from − 0.50 to −1.5 mA cm^{-2} at −0.45 V vs. Ag/AgCl.

2. Experimental section

2.1 Materials and Chemicals

$Bi(NO_3)_3 \cdot 5H_2O$ (≥98.00%), $Cu(NO_3)_2 \cdot 3H_2O$ (≥99.1%), sodium sulfate (≥99.0%), latic acid (≥70%), acetone (≥99.8%), ethanol (≥95%), FTO-coated glass (Product Number: 735213; L×W×D: 100mm×100mm×3mm; Surface resistivity:~10 Ω/sq; Transmittance: ≥83.0% (visible)) were used as received from Sigma-Aldrich. Insulating epoxy (EA 3450) was purchased from Loctite. Copper tape (AT528) was purchased from Advance Tapes.

2.2 Fabrication of Ni nanopore arrays

The production process of pores from Al nickel pore arrays has the same steps as has been written before [12-16]. After the imprinting phase, on a sheet of alumunium by using oil pressure system around 10 kP for 3 minutes, the process of anodization is conducted at 4 ^{O}C temperature for one hour with applied voltage 160 V which later be called as AAO template. To open and broaden AAO pores after the anodized, AAO template is soaked in a 5 wt% H_3PO_4 solution. The duration of the soaking process will control the width of the pore we use. In this work, we made same the condition for deposition time and solvent. After the AAO template was cleansed and dried, the deposition process of Au thin film 25 nm was conducted by using electron beam deposition (Kurt J. Lasker PVD225) procedure with the rate of 0.2 Å s^{-1}.

2.3 Preparation of the CBO films

Firstly, FTO glass was cleaned with acetone, ethanol and DI water, respectively. Secondly, a thin layer of Au was deposited on the cleaned and dried FTO glass as the substrate via a physical vapor deposition (Kurt J. Lasker PVD225) procedure with the rate of 0.2 Å s^{-1}. Then, CBO thin films were cathodically electrodeposited from a bath solution of 8 mM $Bi(NO_3)_3 \cdot 5H_2O$ and 4 mM $Cu(NO_3)_2 \cdot 3H_2O$ in 10% nitric acid. Nitric acid was purposely used to dissolve bismuth nitrate precursor. The pH value of the bath was carefully adjusted to 12.0 by the addition of 3 M sodium hydroxide. The films were grown at a constant potential of −0.40 V vs. Ag/AgCl in a normal three-electrode configuration for a nominal duration of 30 min. The bath temperature was controlled at 45 ° C using a heating plate with an in situ temperature probe. Following deposition, the samples were rinsed with DI water and dried in an air stream. A final annealing procedure at 550 °C for 120 min in the air was performed to achieve better crystallinity. For comparison, we also prepared CBO nanopillars directly using the same procedure.

2.4 Fabrication of CBO photocathodes

A strip of conductive copper tape was stuck on the exposed FTO part of the FTO/Au/CBO to extend the conductor circuit and threaded through a glass tube and then sealed with an insulating epoxy. Electrode areas were optically measured at 0.25 cm^2.

2.5 Photoelectrochemical measurements

External quantum yield was measured with an Oriel 150 W Xe arc lamp (Newport) and a quarter-turn single-grating monochromator (Newport). Sample measurements were recorded with chopped illumination. The output current signal was connected to a Merlin digital lock-in radiometry system, and the output signal from the lock-in amplifier was fed into a computer controlled by TRACQ BASIC software. Current–potential plots and impedance characterization were measured by using the digital BioLogic potentiostat (SP-200) and 0.1 M sodium sulfate aqueous solution served as the electrolyte. A Pt counter electrode and Ag/AgCl reference electrode were used during the measurements. A standard 300 W Xe lamp (Newport) served as the light source, and the light intensity was characterized to 100 mW cm^{-2} by a Si photodiode (Newport). The evolution of hydrogen gas during the water-splitting reaction was measured by the gas chromatograph-mass spectrometer (GC- 2010 plus).

2.6 Characterizations
The powder X-ray diffraction patterns (XRD) were recorded on a Bruker D8 Advance equipped with graphite monochromatized high-intensity Cu Kα radiation (λ = 1.54178 Å). The energy-dispersive X-ray mapping (EDX) mapping was obtained on S4800 HITACHI (Japan). Room–temperature UV-Vis absorption spectroscopy was measured by Varian Cary 5000 UV-Vis-NIR spectrophotometer.

3. Results and discussion

The approach aimed at the outgrowth of the CBO nanopillars was separated into five exhaustive processes that are showed schematically in Figure 1, which here the color of the Al, AAO, Au, Ni, and CBO are described as blue, gray, yellow, orange and black, respectively. The procedure contains anodization (I), gold layer deposition (II), Ni electrodeposition (III), aluminum and barrier layer removal (IV), CBO nanopillars fabrication (V), and template removal (VI). The AAO template was implemented as the dominant templates in our preparation course, the cause of its gorgeous benefits, containing uniform and homogenous oriented nanoporous structures, tunable structural parameters, large area, chive cost, and outstanding thermal and mechanical stability. [11-13]

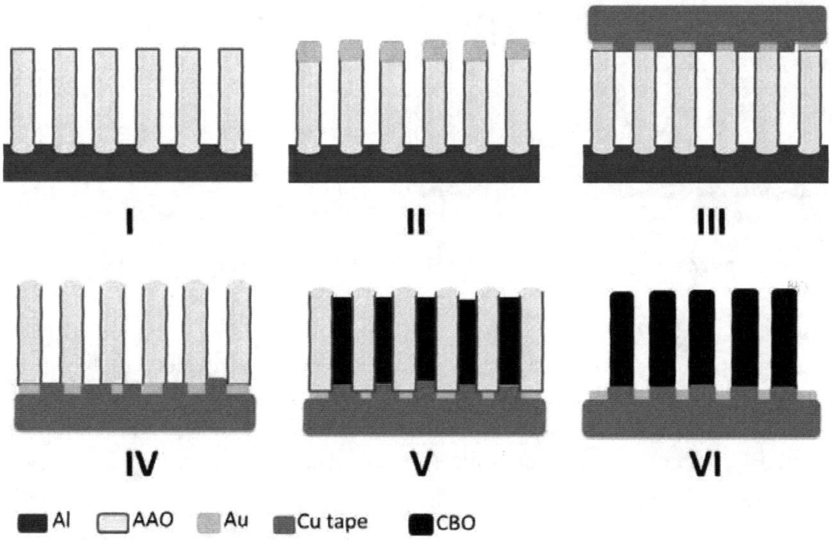

Figure 1. Illustration of the fabrication process of CBO with AAO template: anodization (I), gold layer deposition (II), Ni electrodeposition (III), aluminum and barrier layer removal (IV), CBO nanopillars fabrication (V), and template removal (VI).

Figure 2a shows that the bare AAO template, which was fabricated with one-time oxidization after applied Ni stamp under pressure 10 Pa until 3 minutes, [11,12] gives pretty well-distributed nanopores. It had a diameter of around 150 nm and a length of around 1.5 μm. By succeeding the previously report procedures, the electrodeposition for CBO nanopillars and films were both achieved in a copper sulfate bath holding 3 M lactic acid [14]. Figure 2b demonstrations the representative scanning electron microscopic image of the completed nanopillars on Cu bulk substrates. The nanopillars are vertically joined on the substrate. The size of these nanopillars is assessed as ~ 85 nm in diameter and ~ 1.2 μm in length, respectively. As additional, the CBO films grown on Au/FTO glass are fabricated by unceasingly dispersed polyhedral particles, with the shape chances out to be cuboid, as shown in Figure 2c [14]. It is still in agreement with the reported values,[14,15,16] the band gap of the grown CBO nanopillars and CBO films on Au/FTO glass is characterized as 1.85 and 1.79 eV by the absorption analyses, respectively.

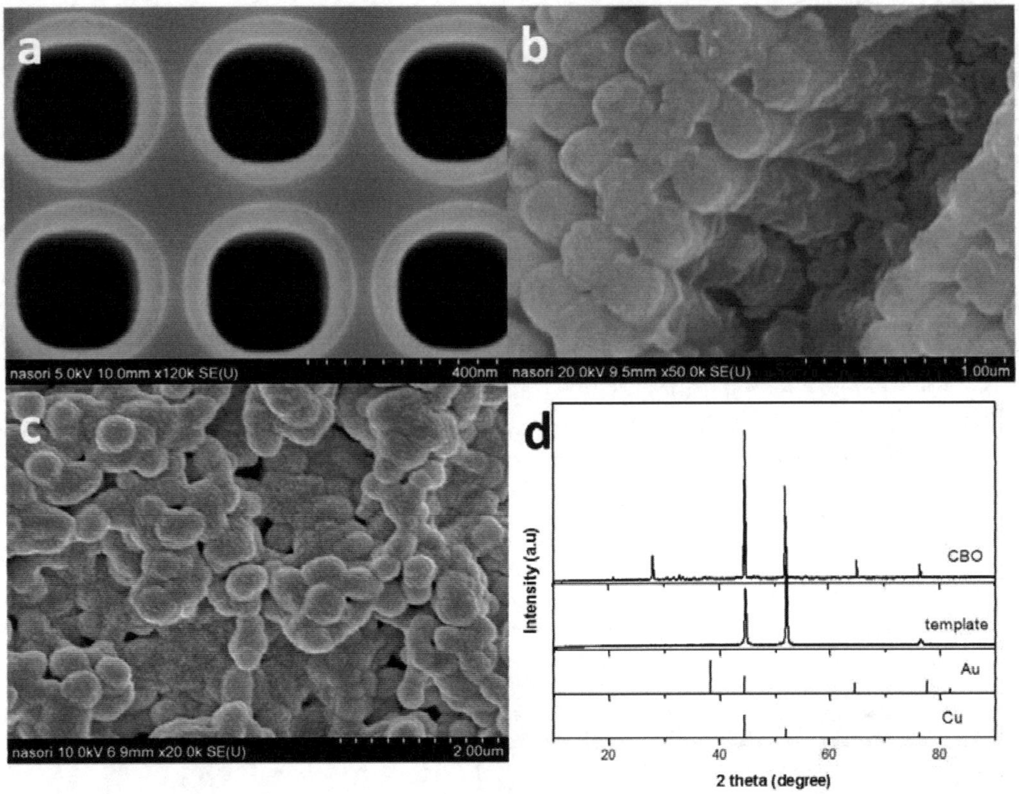

Figure 2. Top view SEM images of the prepared AAO template (a), CBO nanopillars (b) CBO films (c). (d) XRD patterns of CBO nanopillars and substrates.

The XRD patterns of the CBO nanopillars and films are given in Figure 2d, presents the XRD patterns of these two films, where a series of diffraction peaks attributed to Kusachiite CBO phase could be observed (PDF#48-1886), though the peaks of Au and FTO are unavoidable. The highest diffraction peak belongs to the (211) facet [17]. The phases of CBO and Cu are not discernible, indicating a pure CBO structure in the nanopillars and films.

Figure 3a demonstrates the representative steady-state EQY spectra of PEC electrodes of CBO nanopillars and films. The nanopillars photoelectrode on the bulk Cu shows an enhanced EQY below the absorption edge of CBO nanopillars (580 nm) than to CBO thin film photoelectrode. An enhancement of 2-fold is obtained; demonstrating that CBO nanopillars based electrode have a higher photo-to-current efficiency than that with CBO thin films. This result still reliable with those reported in current studies that have focused on the excellent optical and electrical properties of nanopillars and advance in the photocatalytic performance of semiconductor [18,19-25].

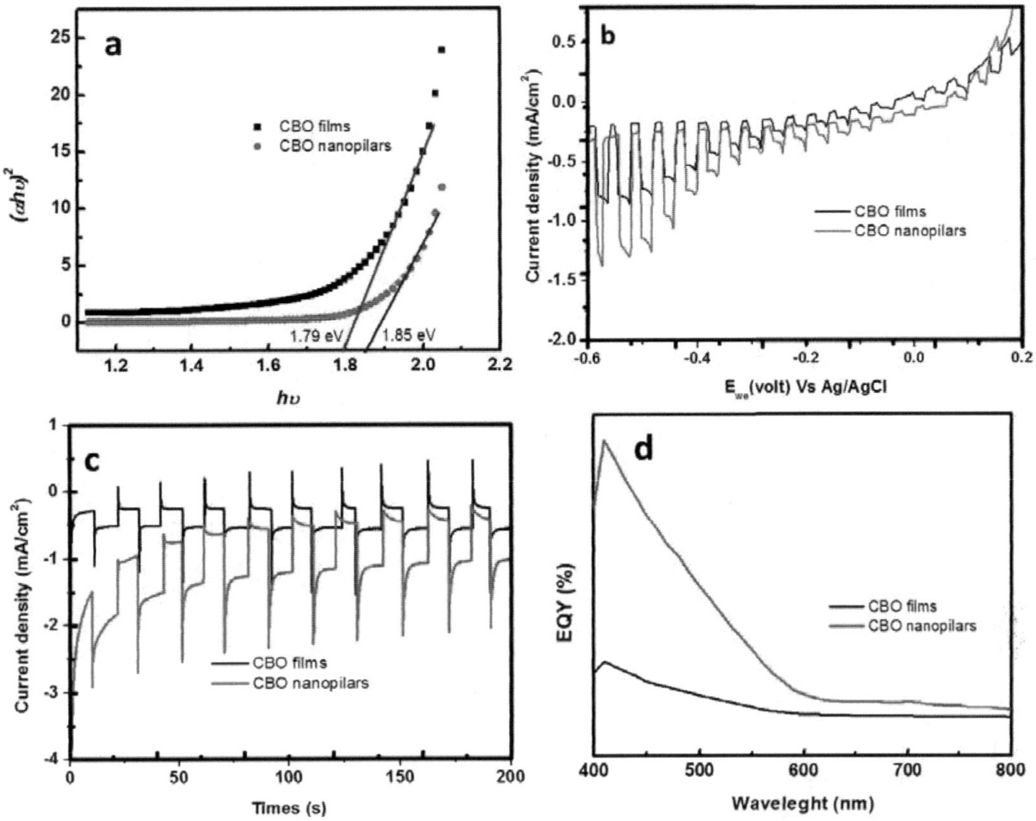

Figure 3. (a) plot of $(\alpha hv)^{1/2}$ vs hv for estimating the band gap value of the film, (b) potential photocurrent profiles, (c) time-dependent photocurrent density spectra and (d) EQY spectra og CBO nanopillars and films photoelectrode

The tested Photocurrent–potential shapes by immersing the photoelectrodes into 0.1 M Na_2SO_4 aqueous solutions, with the Ag/AgCl as the reference electrode and a Pt wire as the counter electrode. Each plot represents a typical photo-response obtained under illumination (AM1.5G, 100 mW cm^{-2}), as shown in Figure 3b. Were J–V curves exhibit a cathodic photocurrent and show a separate p-type feature of two booths CBO nanopillars and films, consistent with our previous report [10]. In work, photo-induced electrons transfer from films and nanopillars to the electrolyte to drive PEC reactions, and in parallel time consistent holes carriage from films and CBO nanopillars to FTO electrode throughout the PEC measurement. To be exciting, likened to the CBO films electrode, the photocathode of CBO nanopillars has a notable improvement in PEC performance, in good bonds with EQY measurement.

As showed in Figure 3c, to examine the photocathode PEC performance and stability under sunlight, the time-dependent photocurrent density was approved out on the CBO nanopillars and films through repetitive chopped on–off sunlight illumination cycles at −0.4 V vs. Ag/AgCl. The result indicates that both samples demonstration fast and reproducible photocurrent reactions upon each illumination. The plateau photocurrent density of the CBO films electrode was 0.45 mA cm^{-2}, and that of the CBO nanopillars was 1.2 mA cm^{-2}, which denotes a more than 2 times increase from the CBO films. Again, the result demonstrates that nanostructure of materials its attain the enhancement of the photoelectric conversion, due to the capability of contribution large surface area and more reaction sites, disassociate light absorption and charge carrier collection, easy carrier diffusion distance compared to the bulk structures [20,21-25].

4. Conclusion

Finally, this study have presented that a novel photocathode based on CBO nanopillars was successfully fabricated via the AAO template method, in comparison with the photocathode of CBO films. The CBO nanopillars photocathode offers notable improvement in photocurrent, from –0.5 mA cm^{-2} to –1.5 mA cm^{-2}, assigning to advantages of the nanostructure with high specific surface area, light-trapping, and small carriers transfer. The results from this study provide an effective-cost, naturally abundant nanopillars material for use in photoelectrochemical cells.

5. Acknowledgments

This work was financially supported by BU-BPKLN Ministry of Education and Culture of the Republic of Indonesia and Universitas Nahdlatul Ulama Surabaya financial support.

6. References

[1] Fujishima A. and Honda K. 1972 *Nature* **238** 37–38.
[2] Bard A. J. and Fox M. A. 1995 *Acc. Chem. Res.* **28** 141–145.
[3] Walter M. G., Warren E. L., McKone J. R., Boettcher S. W., Mi Q., Santori E. a and Lewis N. S. 2010 *Chem. Rev.* **110** 6446–6473.
[4] Chen X., Shen S., Guo L. and Mao S. S. 2010 *Chem. Rev.* 2010 **110** 6503–6570.
[5] Liu R., Zheng Z., Spurgeon J. and Yang X. 2014 *Energy Environ. Sci.* **7** 2504–2517.
[6] Moniz S. J. A., Shevlin S. A., Martin D. J., Guo Z.-X. and Tang J. 2015 *Energy Environ. Sci.* **8** 731–759.
[7] Kudo A. and Miseki Y. 2009 *Chem. Soc. Rev.* **38** 253–278.
[8] Lei Y., W. Cai and G. Wilde, Prog. Mater. Sci, 52, 465–539, 2007.
[9] Paracchino A., Laporte V., Sivula K., Grätzel M. and Thimsen E. 2011 *Nat. Mater.* **10** 456–461
[10] Cao D., Nasori N., Wang Z., Mi Y., Wen L., Yang Y., Qu S., Wang Z. and Lei Y., *Journal of Materials* 440, 2006, 295.
[11] Paracchino A., Laporte V., Sivula K., Grätzel M. and Thimsen E. 2011 *Nat. Mater.* **10** 456–461.
[12] Cao D., Wang Z., Nasori, Wen L., Mi Y. and Lei Y. 2014 *Angew. Chemie Int. Ed.* **53** 11027–11031.
[13] He J., Lindström H., Hagfeldt A. and Lindquist S.-E. 1999 *J. Chem. Phys.* **103** 8940–8943.
[14] Zou Z., Ye J., Sayama K., Arakawa H. 2001 *Nature* 625–627.
[15] Gibson E. A., Smeigh A. L., Pleux L. Le, Fortage J., Boschloo G., Blart E., Pellegrin Y., Odobel F., Hagfeldt A. and Hammarström L. 2009 *Angew Chemie Int. Ed.* **48** 4402–4405.
[16] Siripala W., Ivanovskaya A., Jaramillo T. F., Baeck S.-H. and McFarland E. W. 2013 *Sol. Energy Mater. & Sol. Cells* **77** 229–237.
[17] Zhang Z. and Wang P. 2012 *J. Mater. Chem.* **22** 2456–2464.
[18] Halmann M. 1978 *Nature* **275** 115–116.
[19] Hahn N. T., Holmberg V. C., Korgel B. A. and Mullins C. B. 2012 *J. Chem. Phys.* **116** 6459–6466.
[20] Patil R., Kelkar S., Naphade R. and Ogale S. 2014 *J. Mater. Chem.* **2** 3661–3668.
[21] Arai T., Yanagida M., Konishi Y., Iwasaki Y., Sugihara H. and Sayama K. 2007 *J. Chem. Phys.* **111** 7574–7577.
[22] Da Oh W., Lua S.-K., Dong Z. and Lim T. T. 2016 *Nanoscale* **8** 2046–2054.
[23] Berglund S. P., Lee H. C., Núñez P. D., Bard A. J. and Mullins C. B. 2013 *J.Chem. Phys.* **15** 4554–4565.
[24] Nakabayashi Y., Nishikawa M. and Nosaka Y. 2014 *Electrochim. Acta* **125** 191–198.
[25] Bai S., Ge J., Wang L., Gong M., Deng M., Kong Q., Song L., Jiang J., Zhang Q., Luo Y., Xie Y. and Xiong Y. 2014 *Adv. Mater* **26** 1–7.

Journal of Physics: Conference Series **1373** (2019) 012017 doi:10.1088/1742-6596/1373/1/012017

A Passive Cooling System for Increasing Efficiency of Solar Panel Output

Sunarno A. Rakin[1, 2], S. Suherman[1, *], S. Hasan[1], A. H. Rambe[1], Gunawan[3]

[1]Electrical Engineering Department, Universitas Sumatera Utara, Medan-Indonesia
[2]PT. Inalum, Sumatera Utara-Indonesia
[3]Politeknik Negeri Media Kreatif-Indonesia

*Email: suherman@usu.ac.id

Abstract. Solar cell converts solar energy to electrical energy. As its surface is continuously exposed to sun, solar cell efficiency decreases about 0.5% for 1°C surface temperature increment. Passive active cooling systems have been proposed by researchers. Since active cooling system absorbs the generated current, this paper considers passive system as the cheapest one. The combination of water and heat-sink cooling system is proposed. The experimental implementation shows that the proposed system is able to reduce solar panel surface temperature better than the only water or the only heat-sink cooling system. The output voltage increases in average 21.49 % for average surface temperature decrement of 12.66% compared to basic solar panel. This results significant output power improvement, by more than 40%.

1. Introduction

Alternative energy source is a global issue to anticipate the limited conventional fossil sources. Solar cell is the easiest source to harvest as sun light available whole years. The photon energy comes into two forms: heat and light. Both forms are convertible for electrical source but the latter is the object for solar cell [1]. When converting light into electrical energy, solar cell performance decreases by the appearance of heat form that injected temperature increment on solar cell surface. Solar cell efficiency degrades 0.5% for each 1°C-surface temperature increment as results of heat energy [2]. It is shown in Figure 1 [2].

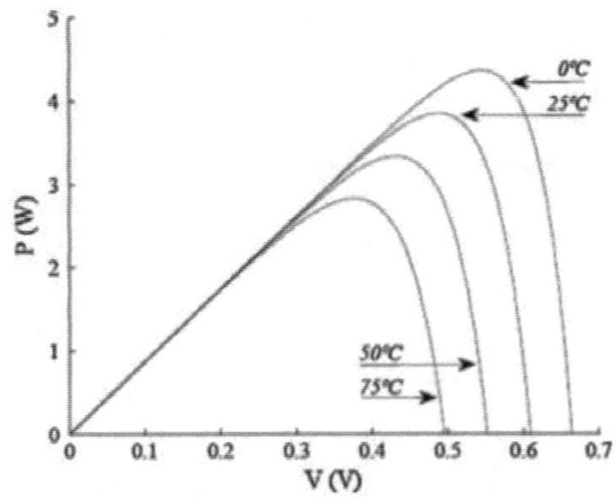

Figure 1. The power and voltage degradation due to surface temperature increment

In order to compensate the efficiency decrement caused by surface temperature, cooling system is utilized. The system can be applied in front and/or at the back of the solar panel. Active cooling system is achieved by flowing a small amount of air or water on solar cell surface (Figure 2 [3]). This is performed by using an electric pump and an electronic controller. Other way is by flowing air at the back of solar panel. Passive cooling system is applied by using heat-sink or a bottle of water that is absorbed by using cottons, other method is by keeping bottom part of solar panel in the water.

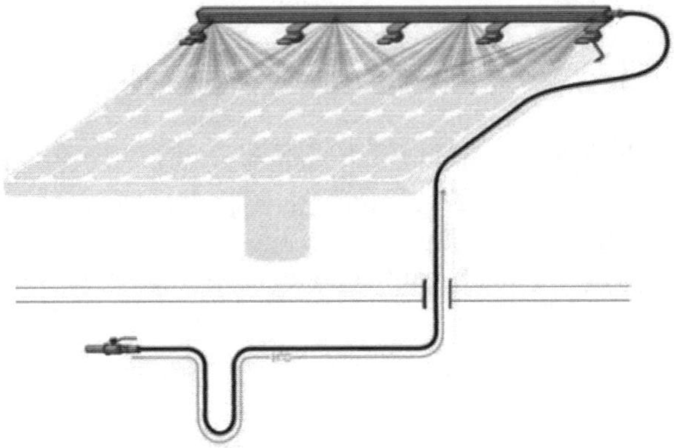

Figure 2. Passive cooling system using water sprinkler

Some researchers proposed active solar panel cooling system such as Haitam et al [4], who used an active cooling system with water and air as the cooler material. The cooling system was attached at the back of solar panel. The proposed system is able to increase solar panel efficiency by about 9% as the surface temperature decreases 20%. Meanwhile, active cooling system with exhaust fan was proposed by Hrica [5], where the surface temperature of twelve solar panels were cooled uniformly. However, the average efficiency was not changed as exhaust fan absorbed energy generated by the solar panels.

The passive cooling system do not consume electrical energy. Arjyadhara et al [2] employed a water-based cooling system to decrease surface temperature of two solar modules of 20 watts 12 volts. Module efficiency increases about 2%. Chen et al [6] used a natural ventilation and heat-sink based passive cooling. The proposed heat-sink has shapes of U and L. The 100 WP solar panel with normal efficiency of 12.9% experienced efficiency increment of about 0.3% and 1.8%. For the higher wind speed, the efficiency improvement achieved 11.8% with output power increment up to 2.4%. Figure 3 shows samples of the effectiveness of passive cooling system of two different heat-sink types, namely Aluminum Round Pins Heat Sink (RPHS) and Aluminum Straight Fins Heat Sink (SFHS) to surface temperature decrement and efficiency increment [7]. SFHS performs better than RPHS. The main problem is that the proposed construction has large solar cell to heat-sink volume ratio that is space inefficient for large solar panel construction (Figure 3 [7]).

Among the passive cooling system, the studies have not found the cooling technique that uses a fixed volume of water placed in an aluminum tube, arranged in array and combined with heat-sink modules. This technique is proposed in this paper. The benefits of the proposed cooling system are that once it is installed, no additional water or energy required, it also saves energy and spaces.

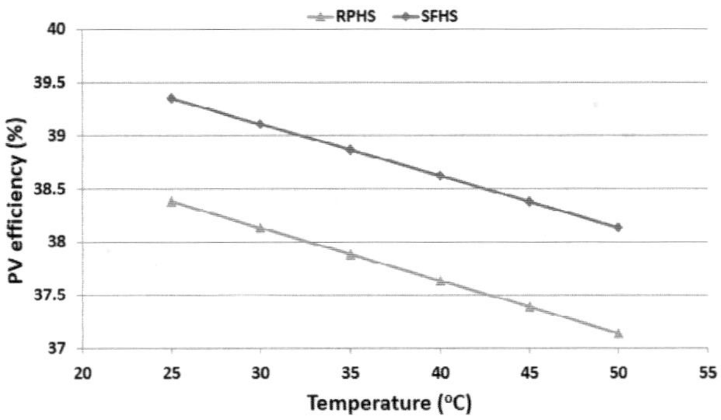

(a) Photovoltaic efficiency

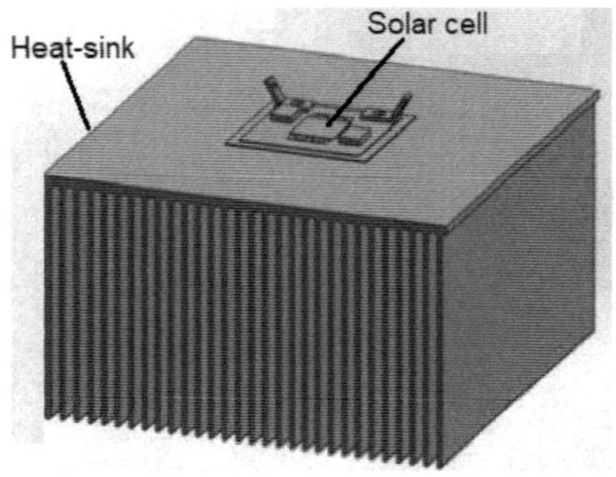

(b) Cooling construction

Figure 3. The round and straight pins heat-sinks performance comparison in cooling solar cell

2. Research Method

The proposed solar panel cooling system is to use both water and Straight Fins Heat Sink (SFHS) in the back side of solar panel. Figure 4 shows the sketch of the proposed system. Water is filled into an array of aluminum beam/cuboids. The cuboids are then attached at the back of solar panel. The SFHS are then attached to cuboids to release heat to the air.

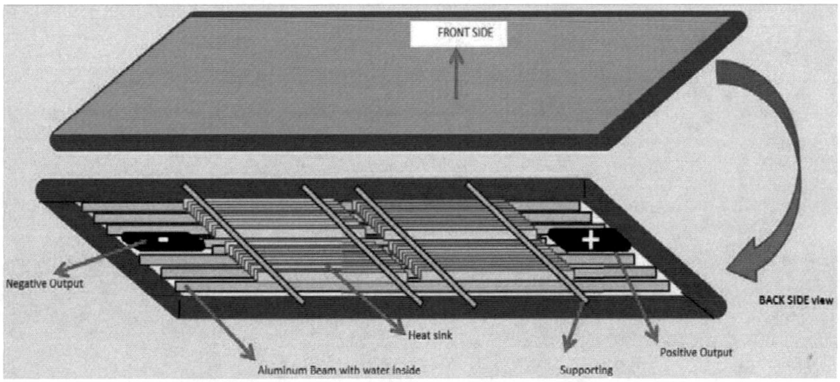

Figure 4. The designed passive cooling system

A solar panel of 50 WP 12 Volts is used as the cooling object (Figure 5a). The designed system will be compared to the baseline un-cooled solar panel, solar panel with water cooling system (Figure 5b), and solar panel with heat-sink (Figure 5c). The measurement is performed by using some sensors, controlled by an Atmega328 microcontroller as shown in Figure 6, a modified version of previous author research [8]. Measurement was done in PT. Inalum, Sigura-gura, Sumatera Utara, Indonesia, located in coordinate of 2.5235980 South and 99.2682700 East.

Journal of Physics: Conference Series **1373** (2019) 012017 doi:10.1088/1742-6596/1373/1/012017

(a) Baseline photovoltage panel

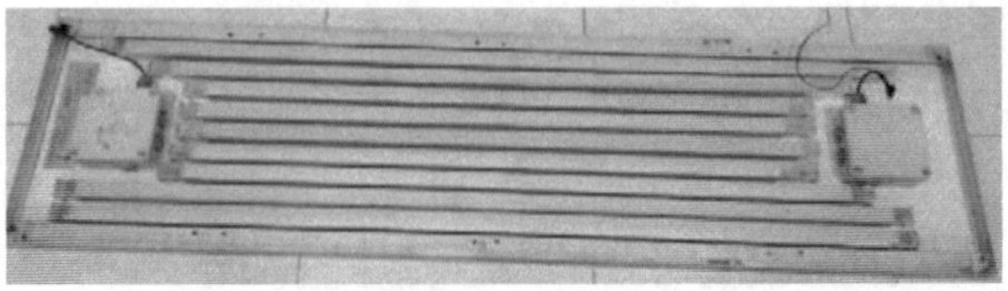

(b)Water cooling system

(a) Heat-sink cooling system

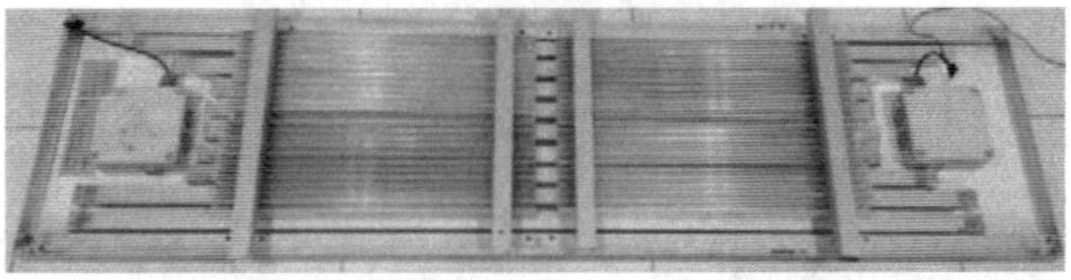

(b) The proposed cooling system

Figure 5. The passive cooling system construction

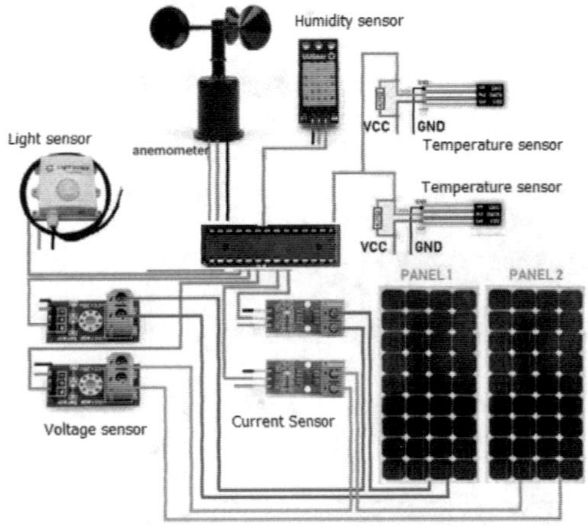

(a) Measurement device

(b) Measurement process

Figure 6. The measurement of the cooling systems performances

3. Results and Discussion

Initial average measurement results are shown in Table 1. The measured values are surface temperatures, output voltages, load currents and output powers of a single solar panel for different cooling systems. The device/sensors measurement results are calibrated using separated multi-tester and thermometer. The passive cooling systems were able to decrease surface temperatures. As results, all the output parameters of cooling systems are better than basic solar panel output without cooling system. This shows that the cooling system is important for solar panel to operate, moreover when solar panel operates in harsh environment.

Parameters	Basic Solar Panel	Heat-sink	Water	Proposed Method
Average Surface Temperature (°C)	53.33	51.83	51.17	46.58
Average Output Voltage (Volts)	9.83	11.42	11.03	11.95
Average Load Current (Ampere)	1.91	2.24	2.16	2.32
Average Power (Watts)	18.82	25.68	23.91	27.80

The measured temperatures, voltages, currents and powers of the proposed cooling system are much better than the compared system. It is able to perform best among the measured panel. In term of average surface temperature, the proposed cooling system is 12.66%, 10.13% and 8.96% subsequently lower than basic solar panel, solar panel with heat-sink and solar panel with water. The percentage is taken from ((53.33°C-46.58°C)/53.33°C)x100%. Consequently, it produces 21.49%, 4.66%, 8.34% higher output voltage. Likewise, the load current increases about 21.27%, 3.73%, 7.50% subsequently than the compared panel. Finally, the output power of the panel with the proposed cooling system achieves 47.71% higher than the plain solar panel. It shows how achievement cooling system can do. The performance comparisons are plotted in Figure 7.

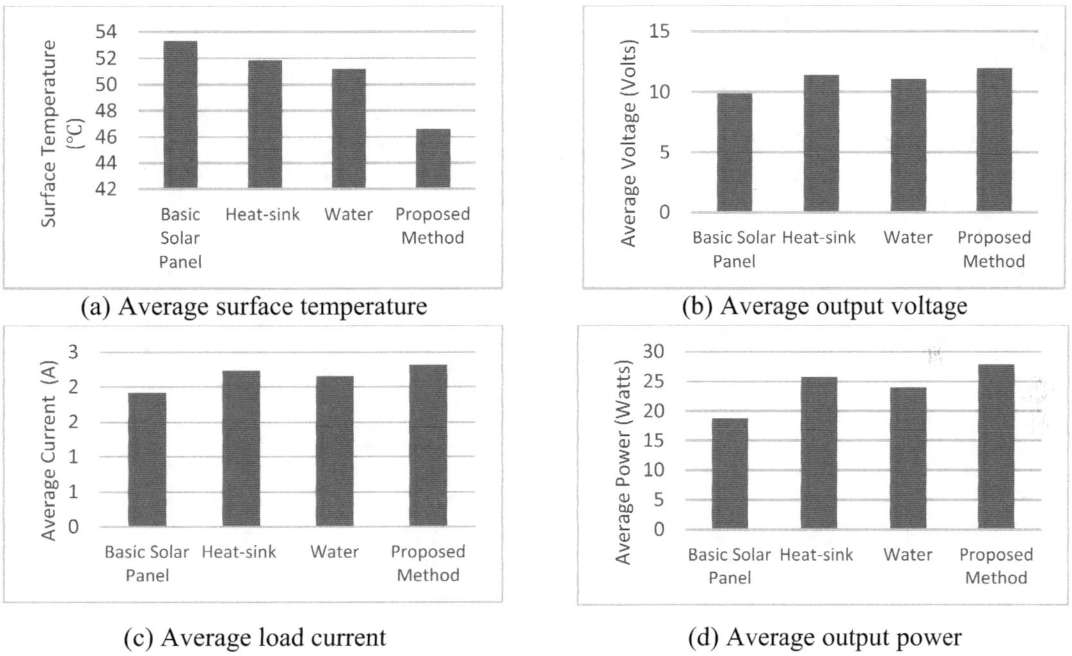

Figure 7. Performance comparisons

4. Conclusions

It is clearly showed that many researches have been carried out to improve solar panel performance though cooling system. Passive cooling system decrease surface temperature slightly; however, it improves the efficiency. This paper reports the performance improvements of solar panel output by proposing the combination of heat-sink and water-cooling system. The experiment shows that the proposed system performs best among the compared system. It reduces the average surface temperature about 12.66%, 10.13% and 8.96% subsequently lower than basic solar panel, solar panel with heat-sink and solar panel with water. As results, the output voltage of the proposed method is 21.49%, 4.66%, 8.34% higher than others. The output power of the proposed cooled panel exerts 47.71% higher power than the basic solar panel.

Acknowledgement
This research has been funded by the TALENTA applied research schema of Universitas Sumatera Utara, 2019.

References

[1] Thedeby A, “Heating and cooling with solar, powered peltier elements”, ISRN LUTMDN/TMHP-14/5308-SE, ISSN 0282-1990, Lund University, 2014.
[2] Pradhan A, Ali S M, Parashar S K S, Paikray P, “Water cooling method to improve efficiency of photovoltaic module”, IEEE 2016 International Conference on Signal Processing, Communication, Power and Embedded System (SCOPES), 2016, IEEE pp, 2017.
[3] Bendra O,“Optimizing solar cells efficiency by cooling techniques,” Capstone report, Al Akhwayn Univ., 2017.
[4] Haitham B M, Rehman S, Gandhidasan P, and Tanweer B. "Experimental evaluation of the performance of a photovoltaic panel with water cooling." *IEEE Photovoltaic Specialists Conference*, 2013 39th, IEEE pp. 2987-2991, 2013.
[5] Hrica J, Chatterjee S, and Samy G, "BAPV array: Thermal modeling and cooling effect of exhaust fan." *Photovoltaic Specialists Conference* (PVSC), 2011 37th IEEE, pp. 003144-003149, 2011.
[6] Chen H, Chen X, Li S, and Ding H 2014 *Int. J. of Smart Grid and Clean Energy*, **3** 374-379.
[7] Aldossary A, Mahmoud S, and Al-Dadah R 2016 *Applied Thermal Engineering* **100** 490-500.
[8] Sani A, Warman E, Depari A P S, Suherman S 2018 *IoP Conference Series: Materials Science and Engineering* 420012051.

Effect of Temperature in Ionic Liquids Pretreatment on Structure of Lignocellulose from Corncob

Umi Rofiqah[1,*], Aditya Kurniawan [1] and Ridwan Widi Nugroho Aji [1]

[1] Chemical Engineering Department, Universitas Islam Indonesia Jalan Kaliurang km 14,5 Yogyakarta 55584, Indonesia

[*]Email: umirofiqah@uii.ac.id

Abstract. Biomass is a potential source of cellulose contained in nature in abundant amounts of either agricultural waste or plant. Corncob is one of biomass derived from waste of agricultural. High content of cellulose from corn cobs have the potential to be processed into glucose which is then fermented to produce bioethanol. Lignin contained in the corn cobs must be removed to produce high glucose levels, through choline acetate as ionic liquid in pretreatment process. Pretreatment process was done by adding choline acetate (ChoAc) as ionic liquid (IL) with a certain ratio between biomass to IL. The pretreatment was carried out at 90°C, 120°C and 150°C of temperature with a reaction time of 20 hours. The highest lignin reduction and index of crystallinity (CrI) is 36.05% and 25.74% respectively, obtained in a pretreatment that uses 0.3 of biomass to IL ratio and 150°C of temperature. This reduction of lignin was aimed to obtain high yield bioethanol in the fermentation process due to high yield of of glucose. And the greater the crystallinity index, the easier the enzymes or microorganisms will degrade glucose into bioethanol.

1. Introduction

Indonesia is one of much country that produce corn as superior commodity [1], it is undeniable that the presence of corn processing waste can also increase. The waste produced includes corn cobs. Corncob is a part of corn fruit that does not contain seeds. Most people only consider corncob as garbage or animal feed that does not have more value. From these problems, the idea arises to utilize corn cobs waste to be processed into bioethanol which is supported by massive cellulose content in the corncob.

In this study, corncob will be made into bioethanol which is a renewable fuel and more eco-friendly than fossil fuel. Corncob is a potential biomass which is one of the agricultural wastes that is rich in cellulose. The components contained in corn cobs are 40-60% cellulose, 20-30% hemicellulose and lignin up to 15-30% [2]. Based on the content, corn cobs have the potential to produce simple sugar compounds that can be further processed to produce alternative fuels in the form of bioethanol. But because the lignin content is not small, treatment is needed in the form of a process to destroy the lignin structure to improve the efficiency of the hydrolysis process which will produce glucose. Glucose is a monosaccharide sugar that can be used as a raw material for the production of fuel, food ingredients, and medicine [3]. Therefore it is necessary to do a delignification process to eliminate the lignin content contained in corn cobs. The aim of this research is to knowing the effect of temperature on lignin degradation by using Ionic Liquid ChoAc. The reason for choosing Choline Acetate as material in this research because choline acetate can be recycled and safe for environment [4].

2. Material and methods

2.1. Material

The material used in this research was 10 grams of corncobs powder with the size 80 mesh and Choline Acetate (ChoAc) used as ionic liquid to decrease the lignin. Corn cobs were obtained from agricultural land in Bantul and Sleman, Yogyakarta. It contain of 37.48 % of hemicellulose, 35.82 % of cellulose, and 15.59 % of lignin.

2.2. Methods

There are two steps in this study were physical and chemical treatment. Corn cob were cuted and grindered until has 80 mesh of size to become powder. Corn cob powder was heated in the oven at 105°C to remove the water content. In chemical treatment, corn cob result from physical treatment was mixed with Choline Acetate (ChoAc) in ratio ionic liquid (IL): corn cob powder 0,3. That sample was added with 500 ml of aquadest then heated in a three-neck rounded glass at 90°C for 20 hours. The sample was filtered using a centrifuge and then the biomass (solids) was washed by entering water so that it becomes suspension. The sample was filtered using a centrifuge again, then ovened the solid until it has a constant weight. Repeat the steps for the heating temperature 120°C and 150°C.

Lignocellulosic component was determined following Chesson method [5] for sample before and after chemical treatment. Sample was analyzed by using X-ray Diffractometry (XRD) to determine the crystallinity of corncob powder. Crystallinity index was defined using Eq. (1) Segal's empirical method [6]:

$$CrI = \frac{(I_{200} - I_{am})}{I_{200}} x100\% \quad (1)$$

Which :
CrI = Crystallinity Index
I_{200} = Crystal intensity at 2θ around 22.5°
I_{am} = Amorphous Intensity at 2θ around 18°

Yield was defined using Eq. (2) :

$$yield\ recovery\ fiber\ (\%) = \frac{m_2}{m_1}\ x\ 100\% \quad (2)$$

Which :
m_1 = mass before ionic liquid treatment
m_2 = mass after ionic liquid treatment

3. Result and discussion

The main component of corn cob powder in this study Table 1 were 37.48% of hemicellulose, 35.82% of cellulose, and 15.59% of lignin. Cellulose content in corn cobs is greater than other agricultural waste sources, rice straw (32.1%) and bagasse (33.4%) [7].

Lignin in corn cob was degraded by using ChoAc ionic liquid at temperature of 90°C, 120°C, and 150°C for 20 hours. ChoAc ionic liquid (IL) was added to biomass in ratio IL: biomass 0.3. From Table 1 and Figure 2 shown that lignin was decreased gradually. The greatest lignin reduction was obtained from treatment at temperature of 150°C that was 36.5%. The greater the temperature used in ionic liquid treatment, the greater of lignin decreasing.

Table 1. Pre-delignified corncob components

Pre-Delignification	
Hemicellulose	37.48 %
Cellulose	35.82 %
Lignin	15.59 %

Table 2. Components of corn cobs in choline acetate pretreatment at various temperature

Ratio IL : Biomass	Temp (°C)	Time (hours)	Composition (%) Hemicellulose	Cellulose	Lignin	Lignin Decreasing (%)
0.3	90	20	36.60	37.00	11.80	24.31
0.3	120	20	35.90	35.70	10.90	30.08
0.3	150	20	34.00	33.60	9.97	36.05

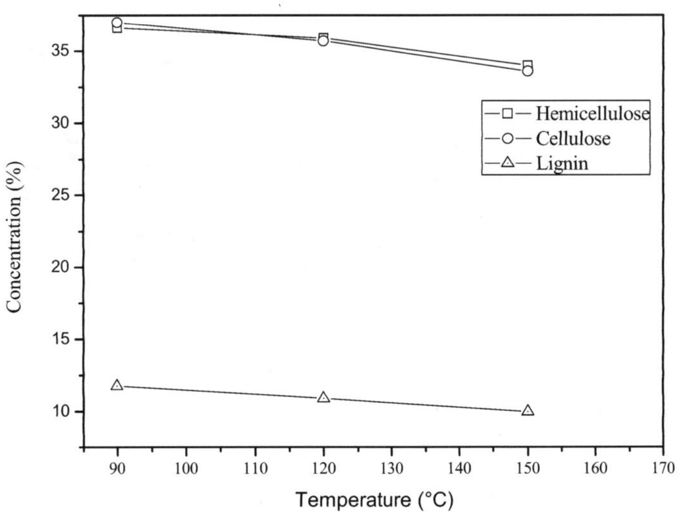

Figure 1. Decreasing Lignoselulosic component levels at various temperature

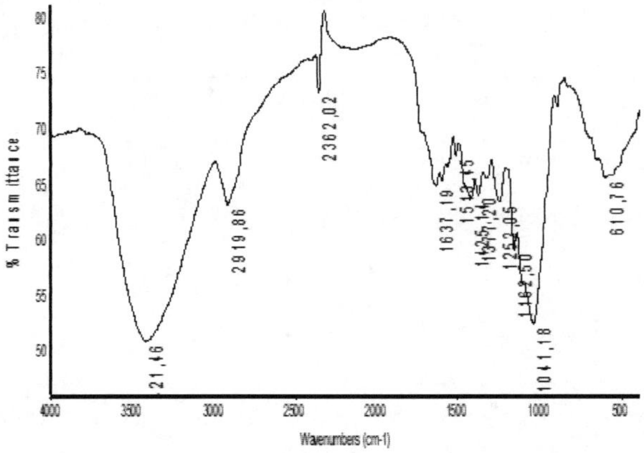

(a)

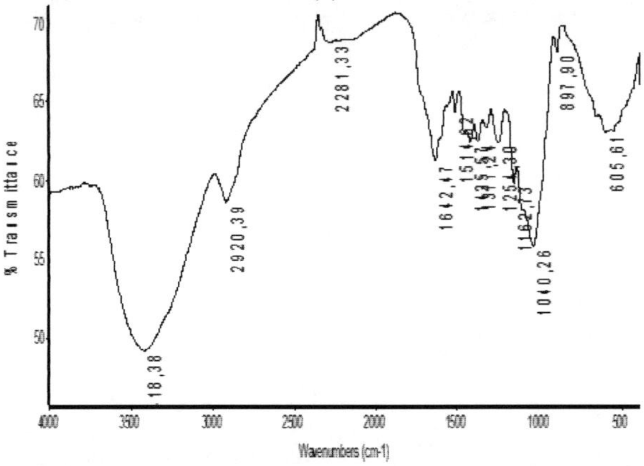

(b)

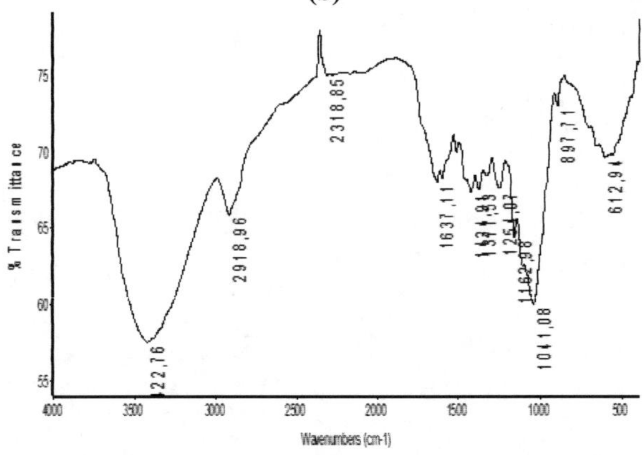

(c)

Figure 2. Graph of FTIR Spectrum at Temperature of (a) 90°C (b) 120°C (c) 150°C

From Figure 2 known that the cellulose and lignin was still in the corncob powder after pretreatment using choline acetate as ionic liquid. Based on Figure 2, cellulose component (peak of OH) exists on wave number 3421.46 at temperature of 90°C, 3418.38 at temperature of 120°C, and 3422.76 at temperature of 150°C as alcohol. And the lignin structure was shown as an aromatic structure [8].

Journal of Physics: Conference Series **1373** (2019) 012018 doi:10.1088/1742-6596/1373/1/012018

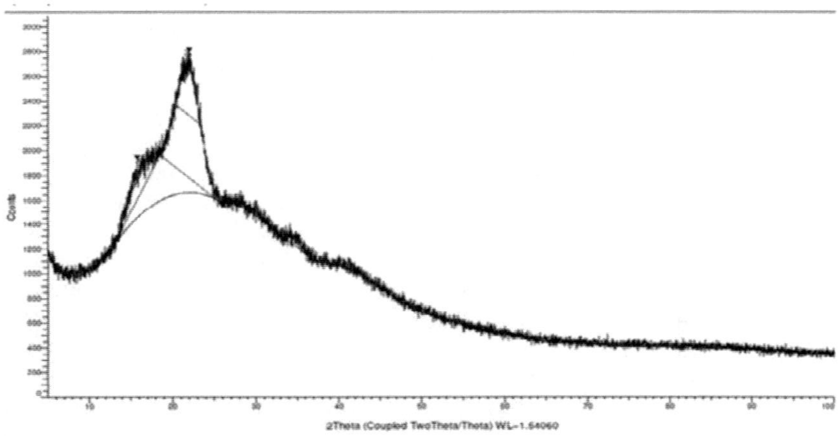

(a)

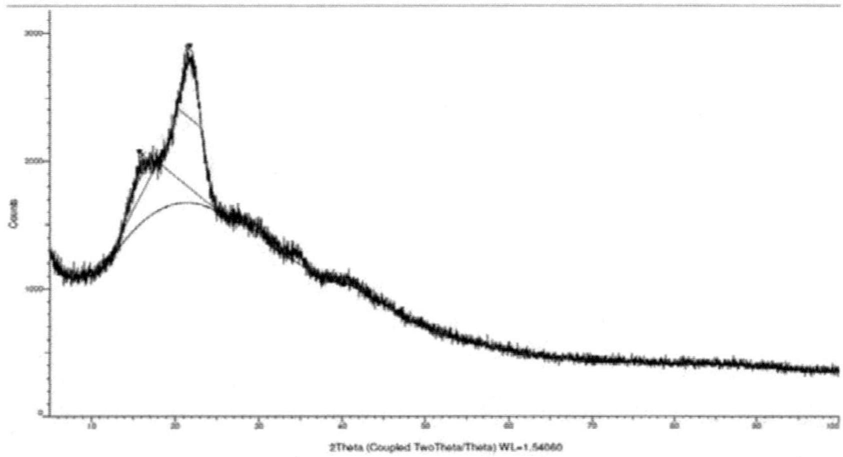

(b)

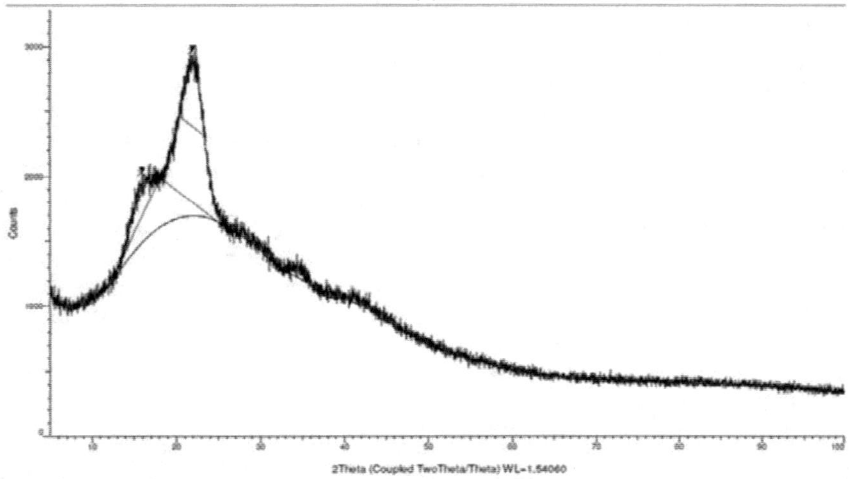

(c)

Figure 3. Graph of XRD at Temperature of (a) 90°C (b) 120°C (c) 150°C

To investigate the cellulose polymorph by using JCPDS ICDD Software 1997 [9].Characteristics of amorphous and crystalline forms in biomass based on 2ɵ values are known that for amorphous read at 2ɵ = 18.7 and crystals at 2ɵ = 22.4. From Figure 3 and Table 3 for XRD analysis at various temperatures kown that there is a change of components (Hemicellulose and Lignin) in form from amorphous to crystalline. In Table 3 it can be seen that there is an increase in the degree of crystallinity of corn cobs as the variable temperature increases. At 90°C the value of Crystallinity Index (CrI) was obtained 21.105%, at 120°C the value of Crystallinity Index (CrI) 23.301% and, at 150°C, the value of Crystallinity Index (CrI) was 25.741%. Data is obtained that there is an upward trend in the Crystallinity Index value explains that the increase in Crystallinity Index is related to the more loss of lignin content from the pretreated sample. Ability of Ionic Liquid Choline Acetate pretreatment can degrade lignin content in corn cobs, thus restructuring amorphous cellose to crystalline cellulose [10]. In the high temperature pretreatment process, the structure which is initially complex and irregular becomes regular, which is indicated by the value of Crystallinity Index. The effect of the structure that changes into crystalline cellulose is that when we process it further the results obtained in the process become more optimal because of its regular structure.

Table 3. XRD Analysis Results for various temperature

Temperature (°C)	Amorf Intensity (2ɵ = 18.7)	Cristal Intensity (2ɵ = 22.4)	CrI (%)
90	2000	2535	21.105
120	1998	2605	23.301
150	2005	2700	25.741

4. Conclusion

The decrease in cellulose content produced from corn cobs in this study was carried out through Ionic Liquid pretreatment. Ionic Liquid pretreatment aims to reduce or eliminate various materials/compounds that can inhibit the hydrolysis rate and increase bioethanol production. The process of destruction of lignin structure is carried out through two stages, namely physical and chemical pretreatment. Physically pretreatment aims to reduce particle size of raw materials and eliminate water content in materials without using chemical compounds. While chemically pretreatment aims to increase the biodegradation of cellulose by removing lignin and/or hemicellulose. In addition to testing delignification samples using the Chesson method, samples were tested using FTIR (Fourier Transform Infrared) tools. Analysis using FTIR is a qualitative functional group analysis of cellulose, hemicellulose and lignin before and after delignification by interpreting absorption peaks from the infrared spectrum. XRD analysis was used to determine the type of cellulose crystals using factometry data, where the data can be obtained to obtain the value of the Crystallinity Index (CrI) of corn cobs in variations of temperature. Pretreatment at temperature of 150°C can reduce lignin in large quantities was 36.05% and can change the structure of lignocellulosic to become crystalline with Crystallinity Index (CrI) of 25.741%. This reduction of lignin was aimed to obtain high yield bioethanol in the fermentation process due to high yield of glucose. And the greater the crystallinity index, the easier the enzymes or microorganisms will degrade glucose into bioethanol.

Acknowledgments

The authors are grateful for the financial support in the project number 009/Dir/DPPM/70/Pen. Pemula/P2/VIII/2018 from the DPPM of Universitas Islam Indonesia as beginner research project. The authors also acknowledge the assistance from the lecturers of Chemical Engineering Universitas Islam Indonesia and all staff of Pharmaceutical Chemistry Laboratory of Universitas Islam Indonesia.

References

[1] McDonnald G and Meylinah S 2018 *Indonesia Grain and Feed Annual Report 2018* GAIN Report Number: ID1808 (USDA Foreign Agricultural Service) pp. 8-13.
[2] Shofiyanto M 2008 Bogor. Fakultas Teknologi Pertanian IPB
[3] Sánchez O J and Cardona C A 2008 *Bioresource Technology* **99** (13), 5270–5295.
[4] Swatloski R P, Spear S K, Rogers J D 2002 *J. Am. Chem. Soc.* **124**, 4974-4975.
[5] Datta R 1981 *Biotechnology and Bioengineering* **23** (9): 2167-2170.
[6] Segal L, Creely J J, Martin A E, and Conrad C M 1959 *Text. Res. J.* **29**(10): 786–794.
[7] Howard R L, Abotsi E J, van Rensburg E L, and Howard S 2003 *J. of Biotech.* **2** 602-619.
[8] Li H Y, Wang C Z, Chen X, Chao X F, Sun S F, and Sun R C 2016 *Bioresource Technol.* **222** 175-181.
[9] Sholikhin A, Hadi Y S, Massijaya M Y, and Nikmatin S 2017 *Makara Journal of Science* **21** 59-68.
[10] Zhou W, Li H, and Wang X 2009 *Bioresource Technology* **100** 5366-5369.

Journal of Physics: Conference Series **1373** (2019) 012019 doi:10.1088/1742-6596/1373/1/012019

Effect of Premium-Butanol Blends on Fuel Consumption and Emissions on Gasoline Engine with Cold EGR System

Firman Lukman Sanjaya [1,*], Syaiful [1] dan Nazarudin Sinaga [1]

[1] Deprtement of Mechanical Engineering, Diponegoro University, Semarang, JL. Prof. Soedarto, Tembalang, Semarang, Indonesia

*Email: sanjaya.firman51@gmail.com

Abstract. Butanol is renewable fuel that can be used as a solution for the increasing fuel consumption and emissions. This is because of the high octane number in butanol increase the effectiveness of the combustion. In addition, high oxygen content in butanol can reduce CO and HC emissions. The purpose of this study is to determine the optimal mixture of premium and butanol on fuel consumption and exhaust emissions on gasoline engines. This study uses a Toyota 7K engine equipped with an Cold EGR system. The percentage of addition butanol in gasoline fuels is 5%, 10%, and 15%. The experiment was carried out at a variation of 2500 rpm to 4000 rpm engine speed. The experimental results show that the addition of butanol in gasoline reduces fuel consumption by 1.9%. The use of the Cold EGR system results in fuel consumption of 0.31 kG.kW/hours in the P85B15 mixture. This value is 16.8% lower than without using the Cold EGR system. Addition of butanol to fuel also increases the EGT. However, the EGT value decreases with the use of the Cold EGR System. The use of a mixture of gasoline and butanol reduced carbon monoxide (CO) and hydrocarbons (HC) by 74.2% and 46.3%. The percentage of CO_2 also showed an increase of 58.6 %. However, the use of the Cold EGR system increases the value of HC and CO. The use of the Cold EGR system also increases the value of CO_2 emissions by 19.04% compared without using the Cold EGR system.

1. Introduction

The growth of motorized vehicles has a serious impact on increasing fuel consumption and exhaust emissions. In addition, poor fuel characteristics such as low heating values lead to increased fuel consumption and exhaust emissions. Therefore, to meet fuel requirements due to an increase in vehicle volume there need to be alternative fuels [1]. Butanol has almost the same characteristics as gasoline. Butanol is usually used as a fuel mixture for gasoline engine vehicles. This is to meet fuel requirements and reduce knocking on the engine so that it makes combustion more efficient. In addition, the relatively high combustion temperature value of butanol helps reduce fuel consumption. Further, higher heat of evaporation in butanol causes better combustion and reduces fuel consumption [2,3]. The high oxygen content in butanol supports the combustion oxidation process. This increases the thermal efficiency of the engine and reduces exhaust emissions [4,5].

Exhaust Gas Recirculation (EGR) is a technology used to reduce NOx and reduce the temperature of the combustion chamber [6,7]. EGR recirculates as exhaust gas has been cooled to the combustion chamber. The EGR brings the unburnt fuel particles back to the cylinder. The exhaust gas contributes to the next combustion cycle so that the engine's performance is better [8]. The EGR system also reduces fuel consumption [9].

2. Methods

In this study, the fuel used is premium and butanol. Premium fuel is obtained from PT. Pertamina Tbk. and is used without further purification. The variation of fuel mixture is premium 100% (P100), premium 95% butanol 5% (P95B5), premium 90% butanol 10% (P90B10) and premium 85% butanol 15% (P85B15). Mixed fuel is prepared before the start of the experiment to ensure that the mixture is homogeneous. A stirrer was used when mixing the fuel.

Table 1. Fuel properties parameters of Premium and Butanol

No	Properties	Premium	Butanol
1	Octane Number	88	98.3
2	Density 15°C (kg/m^3)	744	815
3	Calorific Value MJ/kg	42.7	33.3
4	Water content (%V)	0.003	>5
5	Viscosity (mm^2/s)	0.22	2.63
6	Oxygen content (%)	2.7	21.69

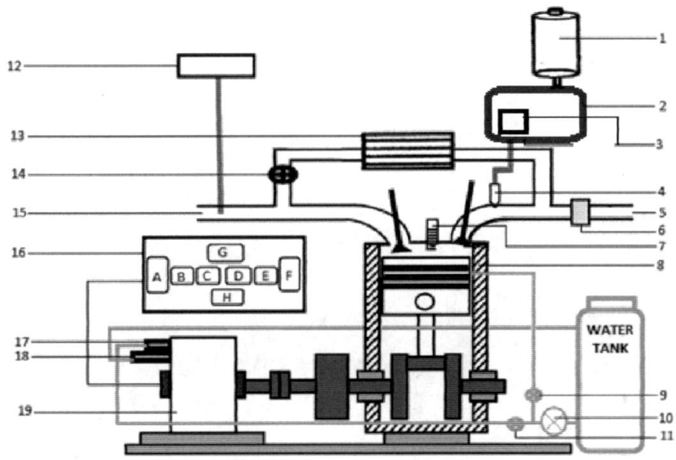

Figure 1. Experimental set-up

The test is carried out with a Toyota 7K gasoline engine. A mixture of premium and butanol, fuel has set the percentage before the experiment begins. Mixer (1) is used to obtain a homogeneous fuel mixture. The location of the mixer is placed higher than the engine so that the fuel mixture flows to the fuel injector (4) based on the principle of gravity and is assisted by a fuel pump (3) installed on the engine. Burette (2) is used to measure fuel consumption, where each volume of fuel is calculated as 90 ml in time. After that, the fuel is sprayed into the combustion chamber using fuel injection.

Experiments were carried out at engine speed of 2500 rpm to 4000 rpm with variations in the percentage of mixed fuels. The dynamometer (19) is applied to measure the brake torque produced by a gasoline engine. Constant engine load is fixed at 25% of maximum engine capability. Engine load is achieved by changing the flow rate of water flowing on the dynamometer. Water flowing from the water tank is pumped by a water pump (10) to the dynamometer. The amount of torque generated on the dynamometer will display a load screen (16).

EGR (13) re-circulates some of the exhaust gas in the intake manifold (5). Type K thermocouples are used to measure temperature. Thermocouples are installed in the exhaust manifold, EGR Inlet, EGR Outlet, intake manifold and engine cylinder. The measurement results are displayed on the thermocouple screen (16). An orifice plate is a tool that can be used to measure EGR air mass flow rates. An orifice plate is applied to the manometer (16) and mounted on the intake manifold and EGR flow. A gas analyzer (12) is used to measure exhaust emissions.

3. Result and Discusion

Figure 2 shown the Specific Fuel Consumption Brake (BSFC) decreases with increasing engine speed (rpm). The highest BSFC value occurs in the P85B15 fuel mixture. Premium produces BSFC 0.291 kG.kW/hour at 4000 rpm. The BSFC value produced by the mixture of P85B15 is 0.296 kG.kW/hour at 4000 rpm. This value increased by 1.9% compared to using premium. Butanol has a low calorific value which decreases the temperature of the combustion chamber. This causes an increase in fuel consumption [3,10,11]. The use of EGR can reduce BSFC. BSFC experienced the largest decrease in the P90B10 mixture of 14.03%. The EGR system reduces fuel consumption and NOx emissions [9].

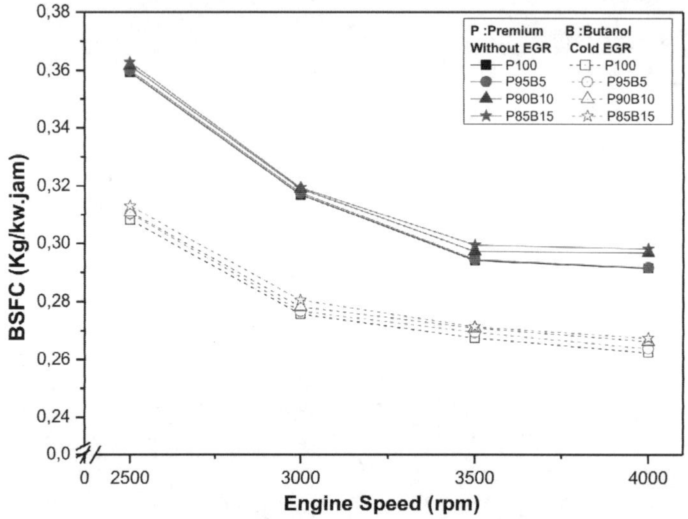

Figure 2. Fuel consumption on gasoline engines uses premium and butanol blend with the Cold EGR system and without the EGR system.

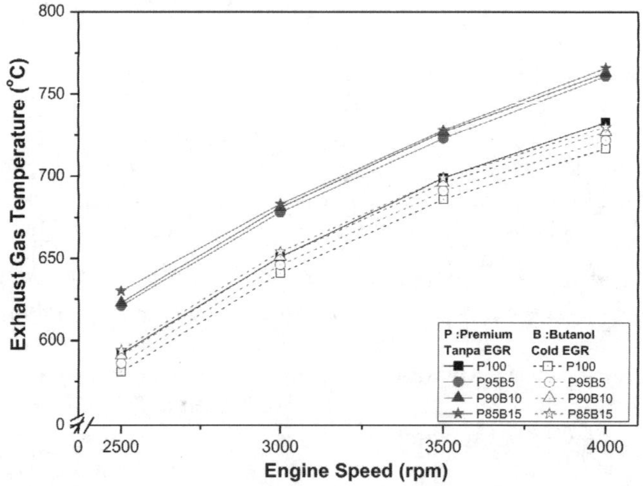

Figure 3. Exhaust Gas Temperature on a gasoline engine uses premium and butanol blend with the Cold EGR system and without the EGR system.

Figure 3 shows the increase in Exhaust Gas Temperature along with the increase in engine speed. The highest EGT occurs in the P85B15 fuel mixture. The premium EGT produced is 592 °C at 2500 rpm. The P85B15 fuel mixture produces EGT of 630 °C at 2500 rpm. EGT has increased by 6.4% compared to premium. Butanol has a higher flame speed than premium. High oxygen content in butanol makes combustion better. this increases the pressure and temperature of the cylinder [12]. Butanol has a high latent heat which increases the compression temperature and exhaust gas temperature [13]. The Cold EGR system decreases the EGT value of 3,2% compared without EGR. The decrease is due to the amount of oxygen circulated into the lower combustion chamber [14].

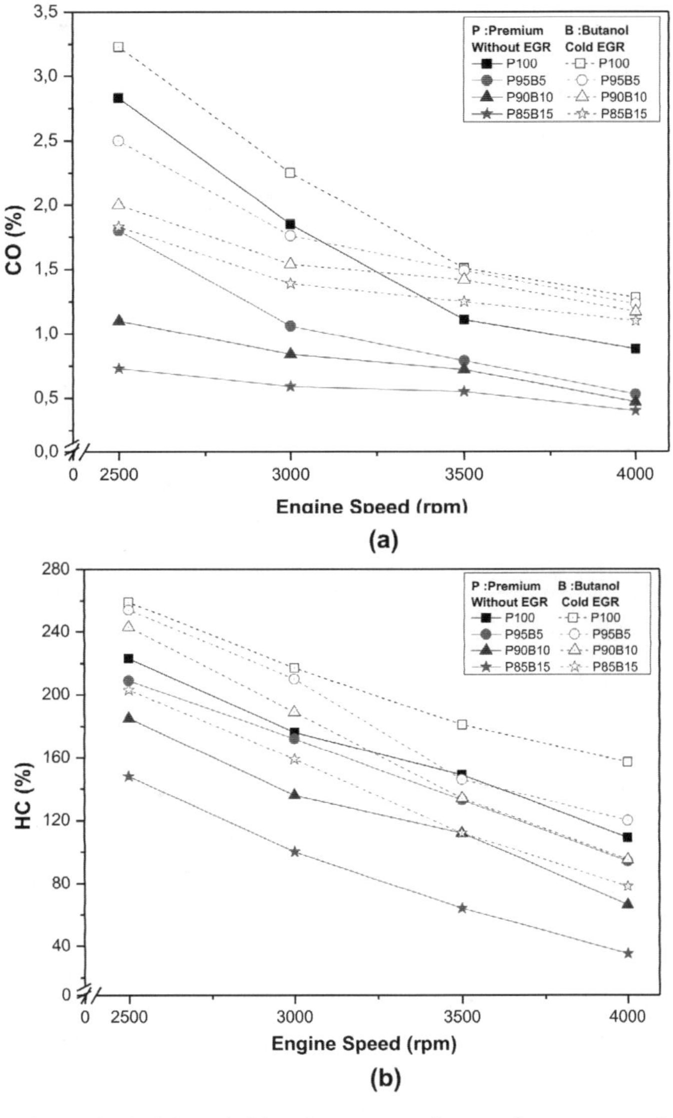

Figure 4. Exhaust emissions of (a) CO and (b) HC on a gasoline engine uses premium and butanol blend with the Cold EGR system and without the EGR system.

Figure 4.(a) show a decrease in CO along with the increase in engine speed. The highest decrease occurred in the P85B15 fuel mixture. The value of CO emissions is 2.83% when the engine uses premium fuel at 2500 rpm. The P85B15 fuel mixture produces CO emissions of 0.73% at 2500 rpm.

This value decreased by 74.2% compared to premium. Butanol has a high oxygen content which causes an increase in leaning effects and more complete combustion. This results in a decrease in CO emissions [15,16]. The use of EGR increases the concentration of CO emissions. Low oxygen concentration produces a rich air-fuel mixture in various locations in the combustion chamber. This heterogeneous mixture does not burn completely and produces higher carbon monoxide [17]. The highest CO emission value occurs in the P90B10 fuel mixture using the Cold EGR system. The value of CO emissions is 0.47% without Cold EGR and 1.47% in the Cold EGR system at 4000 rpm.

Figure 4.(b) also show a decrease in HC along with the increase in engine speed. HC emissions are the highest decreases in the P85B15 mixture. HC emission value uses premium fuel of 149% at a speed of 3500 rpm. The P85B15 fuel mixture has 80% HC emissions at a speed of 3500 rpm. HC emissions decreased by 46.3% compared to premium. The Higher oxygen content of butanol increases combustion efficiency and decreases the concentration of HC emissions. In addition, higher flame speeds in butanol result in more complete combustion and reduced HC emissions [16]. The use of EGR increases the concentration of CO emissions. Low oxygen concentration produces a rich air-fuel mixture in various locations in the combustion chamber. This heterogeneous mixture does not burn completely and produces higher HC [17]. The highest HC emission value occurs in the P85B15 fuel mixture using the Cold EGR system. The value of HC emissions is 47% without Cold EGR and 90% in the Cold EGR system at 4000 rpm.

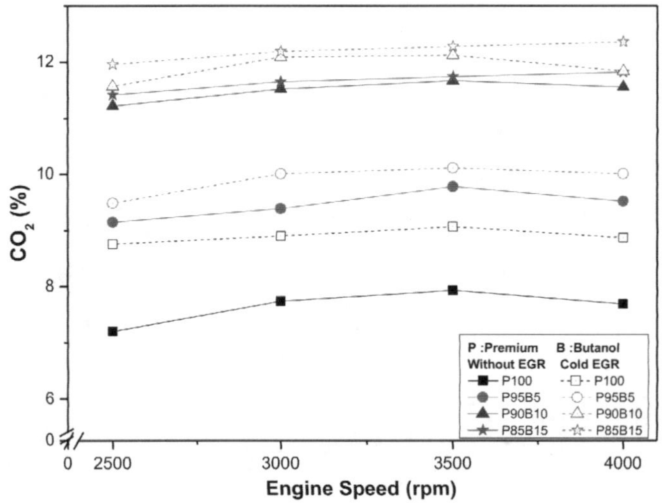

Figure 5. exhaust emissions CO_2 on a gasoline engine uses premium and butanol blend with the Cold EGR system and without the EGR system.

Figure 5 show an increase in CO_2 along with the increase in engine speed. The value of CO_2 emissions experienced the highest increase occurred in the P90B10 fuel mixture. CO_2 emissions of 7.2% when the engine uses premium fuel at 2500 rpm. The P90B10 fuel mixture produces CO_2 emissions of 9.15% at 2500 rpm. The value of CO_2 emissions increased by 58.6% compared to premium fuel. CO_2 emissions increase because oxygen in butanol contributes to increasing the combustion rate. Oxygen in combustion will produce more carbon which can be converted to carbon dioxide. More amount of CO_2 in exhaust emissions shows perfect fuel combustion [12,16]. The addition of the EGR system increases CO_2. The exhaust gas that is recirculated replaces fresh air. Exhaust gases that enter the combustion chamber include CO_2 and H_2O (carbon dioxide and water vapor) produced from the combustion residue [17]. This is what causes an increase of 19.04% CO2 emissions by using the Cold EGR system on the engine.

4. Conclusion

Butanol renewable fuel complementary can be used as a solution to the need for increased fossil fuels and exhaust gas emissions. The use of premium and butanol mixed fuels lowers BSFC but the value is higher than pure premium. The use of the Cold EGR system is proven to reduce BSFC up to 14.04%. EGT has increased with the addition of butanol to fuel. However, the EGT value has decreased when using the Cold EGR system. The high oxygen content in butanol results in reduced CO and HC emissions. The use of the Cold EGR system increases CO and HC emission values. However, CO_2 emissions in the use of premium and butanol mixtures have increased. Increasing CO_2 values shows that combustion is more perfect.

Acknowledgment

The authors acknowledge the assistance from entire Thermofluid Laboratory team, which is one of the laboratories owned by Department Mechanical Engineering, Universitas Diponegoro. This laboratory examines the phenomena that occur in the existing machining process and machine performance testing, especially in the field of Energy Conversion.

Reference

[1] Anderson J E, DiCicco D M, Ginder J M, Kramer U, Leone T G, Raney-Pablo H E, and Wallington T J 2012 *Fuel* **97** 585-594.
[2] Yu X, Guo Z, He L, Dong W, Sun P, Shi W, Du Y, He F 2018 Fuel **229** 1-10.
[3] Feng D, Wei H, Pan M, Zhou L, Hua J 2018 *Energy* **160** 573-581
[4] Liu K, Li Y, Yang J, Deng B, Feng R, Huang Y 2018 *Applied Energy* **212** 13–32.
[5] Chen Z, Zhang Y, Wei X, Zhang Q, Wu Z, Liu J 2017 *Energy Conversion and Management* **152** 57–64.
[6] Edara G, Satyanarayana Murthy Y V V, Srinivas P, Nayar J and Ramesh M 2018 *Case Studies in Thermal Engineering* **12** 188-202.
[7] Wei H, Zhu T, Shu G, Tan L, Wang Y 2012 *Applied Energy* **99** 534–544.
[8] Verma S, Das L M, Kaushik S C, Bhatti S S 2019 *Applied Thermal Engineering* **150** 1090-1103.
[9] Zhang Z, Zhang H, Wang T, Jia M 2014 *Energy* **65** 18-24.
[10] Li Y, Ning Z, Lee C F, Yan J, Lee T H 2018 *Energy* **168** 1157-1167.
[11] Feng H, Zhang H, Wei J, Li B, Wang D 2019 *Fuel* **240** 339–348.
[12] Zaharin M S M, Abdullah N R, Masjuki H N, Ali O M, Najafi G, Yusaf T 2018 *Applied Thermal Engineering* **144** 960–971.
[13] Dhamodaran G, Esakkimuthu G S, Pochareddy Y K, Sivasubramanian H 2017 *Energy* **125** 726-735.
[14] Xie F, Hong W, Su Y, Zhang M, Jiang B 2017 *Energy Conversion and Management* **142** 69–81.
[15] Yusoff M N A M, Zulkifli N W M, Masjuki H H, Harith M H, Syahir A Z, Kalam M A, Mansor M F, Azham A, Khuong L S 2017 *Transportation Research Part D* **57** 23–38.
[16] Sharudin H, Abdullah N R, Najafi G, Mamat R, Masjuki H H 2107 *Applied Thermal Engineering* **114** 593–600
[17] Agarwal D, Singh S K, Agarwal A K 2011 *Applied Energy* **88** 2900–2907.

Experimental and computational study on heat transfer of a 150 kW air-cooled eddy current dynamometer

Nazaruddin Sinaga[1,*] Syaiful[1], B Yunianto[1] and M Rifal[2]

[1]Mechanical Engineering Department, Diponegoro University
Jl. Prof. Soedharto, SH, Tembalang, Semarang 50275, Central Java, Indonesia
[2]Mechanical Engineering Department, Gorontalo University
Jl. Jend. Sudirman No. 247, Limboto, Kota Gorontalo, Gorontalo 96211, Indonesia

Email: *nsinaga.ccfed@yahoo.com

Abstract. The air-cooled eddy current dynamometer is widely used in the measurement of engine performance. However, the load capacity of this device will significantly decrease when operated continuously as a result of temperature rise. To resolve this issue, it needs a high rate air cooling system. Since the heat transfer mechanism is real complicated and has several parameters affecting the cooling rate, it is advantageous to examine the phenomena theoretically, before designing a dynamometer. The primary objective of this study was to validate the numerical simulation modeling against experimental measurement results. The study was conducted with an existing 150 kW eddy current dynamometer that operates at several rotating speeds. Numerical calculation was validated with experimental results by varying turbulence models and dynamometer load. The finite volume method, facilitated by ANSYS FLUENT® software, was used to solve the problem. It has been found that the Spalart-Allmaras model is in good agreement compared to the experimental results. Hence, it is concluded that the computational model might be used as the reference to calculate the effect of dynamometer geometry parameters, and other operational variables, for getting the optimum design of cooling part of the dynamometer.

1. Introduction

To run an energy efficiency program and an environmentally friendly transportation system, a dynamometer is needed to determine the fuel economy, in units of km/liter, and the emissions of a vehicle, in units of gr/km. When measuring these two quantities, the vehicle is running on a chassis dynamometer by following a driving scenario called the driving cycle. To be able to run this driving cycle, the vehicle must be loaded by a dynamometer. Recently, there are almost no workshops or vehicle dealers in Indonesia that use dynamometer for the purposes mentioned above. This is because the price of a dynamometer is relatively expensive, particularly for small-scale workshops, because there has been no dynamometer manufacturer in the country.

The Eddy current dynamometer works according to Faraday's law of electromagnetic induction [1]. According to this law, whenever a conductor cuts magnetic lines of forces, an emf is induced in the conductor, the magnitude of which is proportional to the strength of the magnetic field and the speed of the conductor. If the conductor is a disc, there will be circulated eddy currents in the disc. According to Lenz's law, the direction of the current is in such a way as to oppose the movement of the disc [2].

Essentially the eddy current brake consists of two components, a stationary magnetic field system, and a solid rotating part, which include a metal disc [3, 4] as shown in image 1 and 2. The rotor, generally of mild steel, sometimes referred to as the secondary because the eddy currents are induced. The stator consists of pole core, pole shoe, and winding coil. The coil is wounded at the pole core, where copper or aluminum is used for winding material. The pole core and shoe are made of mild steel and fixed with a bolt and screw. The stator and rotor are separated by a close air gap for torque transmission. During braking, the metal disc is exposed to a magnetic field by the coils and generates eddy currents in the disc. The interaction between the magnetic field and the eddy currents slow down the rotating disc [5]. Consequently, there is no wear as in the friction brake.

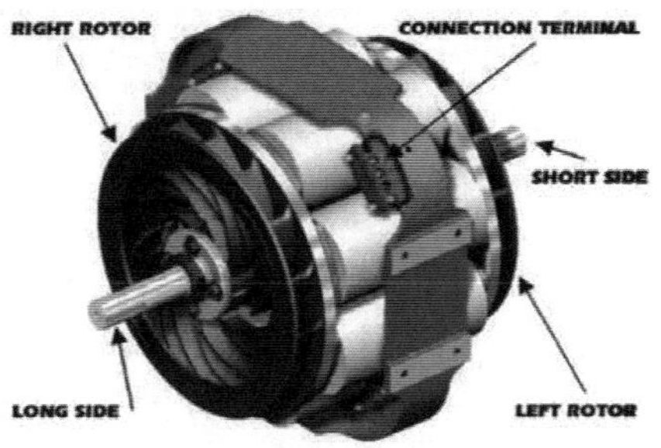

Figure 1. The KLAM CFK-140 Power Absorber Unit produced by Industrial Zelu [3]

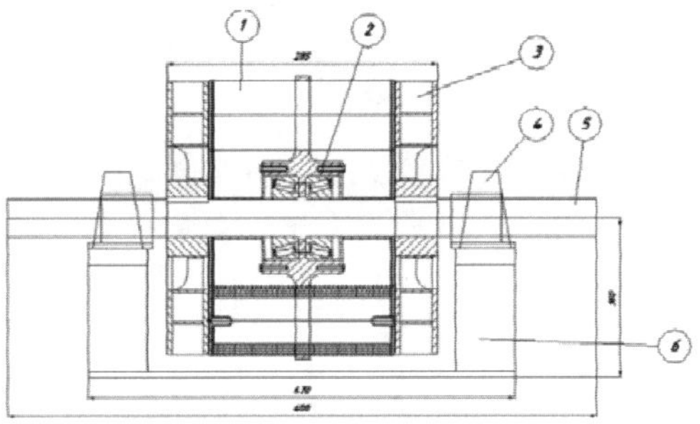

Figure 2. The main parts of dynamometer: (1) solenoid; (2) stator; (3) rotor; (4) pillow block bearing; (5) shaft; (6) frame [4]

Air-cooled eddy current dynamometer is one of the instruments employed to evaluate the power and torque of a rotating engine. Even though there is no surface wear, all of these parts will receive a high amount of heat due to braking. If the cooling system is inadequate, then the stator and rotor surface will experience a high increase in temperature. The temperature rise on both stator and rotor surfaces of the air-cooled dynamometer will drop the dynamometer capacity as low as 70% [6]. To resolve this issue, it needs a system that has a high cooling rate. Therefore, efforts need to be made to increase the cooling rate of the air-cooled eddy current dynamometers. It could be realized by optimizing the dynamometer design using artificial neural networks [7, 8] or other methods. However, the optimization is complicated because it involves many parameters [9, 10, 11].

Since the heat transfer mechanism is very complicated and has several parameters affecting the cooling rate, it is advantageous to study the phenomena theoretically, by using the computational method. With the help of this computational software, calculations can be done quickly. However, the accuracy of the results of these theoretical calculations must first be justified so that the results have low uncertainty. When a precise calculation model has been obtained, the effect of various design parameters from the dynamometer can be searched quickly. This research was conducted in order to obtain a calculation model that can be used to perform optimal design of the air-cooled eddy current dynamometer. The primary objective of this study is to validate the numerical simulation modeling by experimental results. The finite volume method, facilitated by ANSYS FLUENT® software, is used to solve the problem.

2. Methodology

Measurements in the laboratory were carried out by running a dynamometer on a chassis dynamometer at several load values and dynamometer rotational speed. Table 1 shows the main dimensions of the dynamometer being tested, while figure 3 shows the measurement system. Measurements is begun after the dynamometer reached a steady state. At each loading, a rotor surface temperature measurement and flow discharge through the outlet side of the rotor are measured. The next measurement results are compared with the results of numerical calculations. To be able to produce a large enough load, the Toyota Fortuner car has a maximum torque of 400 Nm at 5000 rpm and a maximum power of 125 kW at a speed of 4500 rpm. In figure 2 above shows the construction of a dynamometer in the Efficiency and Energy Conservation Laboratory of Mechanical Engineering Department, Diponegoro University. In this study, eighteen simulations were carried out by varying four types of turbulence models, discretization of first and second order momentum, and three types of discretization of Pressure Velocity Coupling. The turbulence model used is the turbulence model k-ε standard, k-ε RNG, k-ε realizable, and Spalart-Allmaras.

Table 1. Specification of eddy current dynamometer [12]

Parameter	Value
Outer diameter (D_1)	290 mm
Inner diameter (D_2)	140 mm
Inlet blade angle (β_1)	64°
Outlet blade angle (β_2)	26°
Blade number	16
Blade width (b)	50 mm

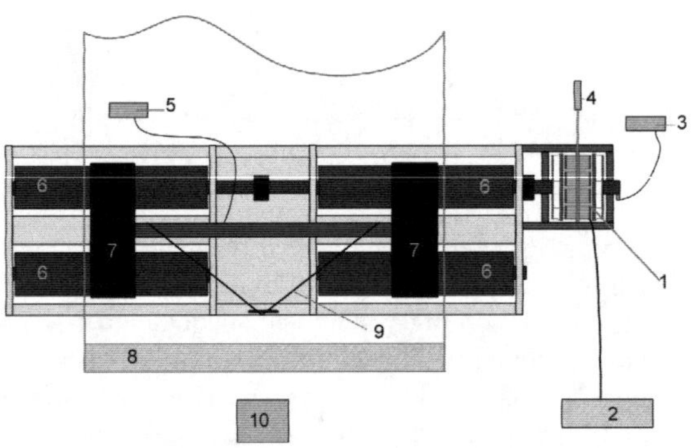

Figure 3. Measurement system: (1) dynamometer; (2) power supply; (3) dyno tachometer; (4) load cell; (5) wheel tachometer; (6) chassis roller; (7) car wheels; (8) car; (9) straps; (10) cooling fan

In this study, the method used is a computational approach, namely the finite volume method. This is a method that is widely used in solving fluid flow and heat transfer problems numerically. In the finite volume method, unknown variables are explained using a specific point on the nodal points. Generally, the basic procedures for solving problems using the finite volume method are as follows: estimating the value of unknown flow variables using a simple function; make a discretization equation by entering the estimation into the general equation of flow, and then doing mathematical manipulation; and finally solve the algebraic equation of the problem at hand [13]. Currently, there is quite several computational fluid dynamic software that can be used to solve flow problems and heat transfer. In this study, ANSYS FLUENT® software is used. One of the advantages of the software is its complete literature and is very widely used in scientific. Besides, this software also has complete features.

3. Turbulence Models

In FLUENT, several turbulence models can be simulated, including the k-ε standard model, k-ε RNG, and k-ε realizable and Spalart-Allmaras. All of these models have a similar shape because they contain transport equations for k (turbulent kinetic energy) and ε (turbulent energy dissipation).

3.1. k-ε Standard

The k-ε standard model is a semi-empirical model compiled based on the transport equation for turbulent kinetic energy and turbulent dissipation energy. In the model k-ε it is assumed that the flow is fully turbulent, and the effect of molecular viscosity is negligible. Therefore this model is only suitable for fully turbulent flow. The turbulent kinetic energy, k and the rate of dissipation ε are obtained from the following equation:

$$\rho\frac{Dk}{Dt}=\frac{\partial}{\partial x_i}\left[\left(\mu+\frac{\mu_t}{\sigma_k}\right)\frac{\partial k}{\partial x_i}\right]+G_k+G_b-\rho\epsilon-Y_M \tag{1}$$

$$\rho\frac{D\epsilon}{Dt}=\frac{\partial}{\partial x_i}\left[\left(\mu+\frac{\mu_t}{\sigma_\epsilon}\right)\frac{\partial \epsilon}{\partial x_i}\right]+C_{1\epsilon}\frac{\epsilon}{k}(G_k+C_{3\epsilon}G_b)-C_{2\epsilon}\rho\frac{\epsilon^2}{k} \tag{2}$$

3.2. k-ε RNG

This turbulence model is derived from the Navier-Stokes equation using a mathematical technique called the renormalization group method. Model k-ε RNG has a shape similar to the k-ε standard model as follows.

$$\rho\frac{Dk}{Dt}=\frac{\partial}{\partial x_i}\left(\alpha_k\mu_{eff}\frac{\partial k}{\partial x_i}\right)+G_k+G_b-\rho\epsilon-Y_M \tag{3}$$

$$\rho\frac{D\epsilon}{Dt}=\frac{\partial}{\partial x_i}\left(\alpha_\epsilon\mu_{eff}\frac{\partial \epsilon}{\partial x_i}\right)+C_{1\epsilon}\frac{\epsilon}{k}(G_k+C_{3\epsilon}G_b)-C_{2\epsilon}\rho\frac{\epsilon^2}{k}-R \tag{4}$$

3.3. k-ε Realizable

The term realizable means that this model meets certain limitations of normal stress, which is consistent with the physical properties of turbulent flow. By combining the Boussinesq relationship and the definition of turbulent viscosity, a normal Reynolds stress can be obtained in an incompressible flow. The transport equation for the k-ε realizable is as follows.

$$\rho\frac{Dk}{Dt}=\frac{\partial}{\partial x_j}\left[\left(\mu+\frac{\mu_t}{\sigma_k}\right)\frac{\partial k}{\partial x_j}\right]+G_k+G_b-\rho\epsilon-Y_M \quad (5)$$

$$\rho\frac{D\epsilon}{Dt}=\frac{\partial}{\partial x_j}\left[\left(\mu+\frac{\mu_t}{\sigma_\epsilon}\right)\frac{\partial \epsilon}{\partial x_j}\right]+\rho C_1 S\epsilon-\rho C_2\frac{\epsilon^2}{k+\sqrt{\nu\epsilon}}+C_{1\epsilon}\frac{\epsilon}{k}C_{3\epsilon}G_b \quad (6)$$

3.4. Spalart-Allmaras

The transported variable in the Spalart-Allmaras model, ν, is identical to the turbulent kinematic viscosity except in the near-wall (viscosity-affected) region. The transport equation for the modified turbulent viscosity ν is

$$\frac{\partial}{\partial t}(\rho\tilde{v})+\frac{\partial}{\partial x_i}(\rho\tilde{v}u_i)=G_v+\frac{1}{\sigma_{\tilde{v}}}\left[\frac{\partial}{\partial x_j}\left\{(\mu+\rho\tilde{v})\frac{\partial\tilde{v}}{\partial x_j}\right\}+C_{b2}\rho\left(\frac{\partial\tilde{v}}{\partial x_j}\right)^2\right]-Y_v+S_{\tilde{v}} \quad (7)$$

where G_v is the production of turbulent viscosity, and Y_v is the destruction of turbulent viscosity that occurs in the near-wall region due to the wall blocking and viscous damping. σ_v and C_{b2} are the constants and ν is the molecular kinematic viscosity. S_v is a user-defined source term. Note that since the turbulence kinetic energy, *k*, is not calculated in the Spalart-Allmaras model, the last term in the equation is ignored when estimating the Reynolds stresses.

4. Simulation Process

4.1. Problem Description

The problem involves modeling the steady-state flow of air through a centrifugal blower running at 1500 rpm. There are 16 blade rows in this blower. To simplify the CFD calculation, the flow is modeled only through double blade rows and uses rotationally periodic boundary conditions on the boundaries between the blade rows. The governing equations are cast in a Single Reference Frame moving at the speed of the rotor. Air is treated as an ideal gas with constant values of specific heat, thermal conductivity, and dynamic viscosity. The schematic of double blade rows and the complete rotor, repeating a single blade row 16 times, are shown in figure 4.

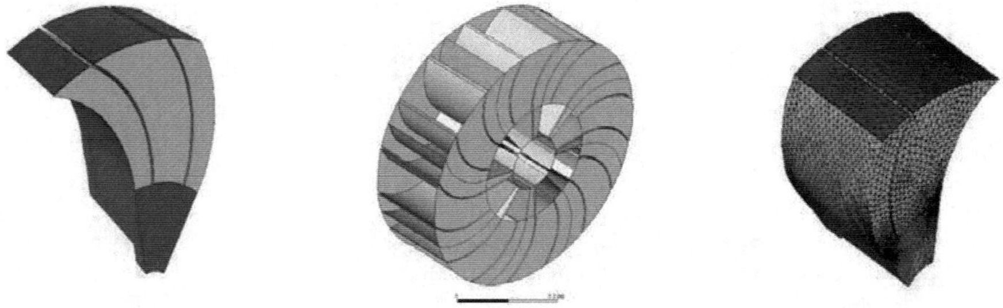

(a) (b) (c)

Figure 4. (a) calculated geometry; (b) full rotor geometry; (c) meshing on the domain geometry

4.2. Domain Geometry

In the present calculation, the rotor plate outer diameter is 290 mm, the inner diameter of the front rotor plate is 140 mm, the number of blades is 16 pieces, and the rotor material is made of St 60 steel. The blade width is 50mm, with the rotor blade entry angle are 64°, while the blade exit angle is 26°.

4.3. Numerical Scheme

There are two numerical schemes provided by FLUENT, namely segregated and coupled solvers. FLUENT solves general integral equations for mass, momentum, energy and other scalar quantities such as turbulence. Both of these numerical schemes use the same discretization process, namely finite volume methods. The only difference lies in the approach used in linearizing and solving discrete equations. The approach taken by coupled methods is to solve continuity, momentum, and energy equations simultaneously.

4.4. Model and Materials

As a first step in the modeling in ANSYS FLUENT, we have to set up the general model as follows: enable the density-based, steady-state solver and retain the default values for the other parameters. The second step is to enable the Spalart-Allmaras turbulence model, and retain the default values for the other parameters. The Spalart-Allmaras turbulence model is an excellent and economical choice for mildly complex boundary-layer flows in turbomachinery. The next step is to set the density of air to ideal-gas. The ideal gas model will automatically enable the solution of the energy equation. For simplicity, we will keep the other fluid properties at their default constant values, though they could be made the function of temperature if desired.

4.5. Boundary Conditions

In figure 5, a cell zone conditions setting is shown in the blade area, which is then given a rotational speed with a specific value in the rotational velocity, for example, 1500 rpm. In this inlet area, the pressure is 1000 Pa gauge, and the initial gauge pressure is 1000 Pa, while the turbulent viscosity ratio is 10. The outlet condition is a pressure outlet of 1000 Pa, and the turbulent viscosity ratio backflow is 10. Figure 5 also shown the boundary conditions at the hub, shaft, shroud-b, and shroud-d areas, where the wall is a moving wall, relative to adjacent cell zone with the rotational axis is X = 1, and rotational speed is zero. For wall boundary conditions of blade1-p, blade1-s, blade2-p, and blade2-s, the settings are the same as before.

4.6. Solver Discretization, Controls, and Monitors

Retain the default solution methods for flow under solve methods, and set the turbulence equation to second order upwind. Activate the pseudo transient method then retain the default solution controls under solving controls. For the residual convergence, set the convergence criterion for continuity to 1e-5. Create three surface monitors as follows: mass flow rate at the inlet, the mass flow rate at the outlet, and mass-averaged total pressure at the outlet.

Table 1. Boundary conditions in inlet and outlet

Name	Boundary Conditions	Label
Inlet-D	Intake Fan	A
Inlet-B	Intake Fan	B
Outlet	Exhaust Fan	C
Hub	wall	D
Shroud-D	wall	E
Shroud-B	wall	F
Blade1-P	wall	H
Blade1-S	wall	I
Blade2-P	wall	J

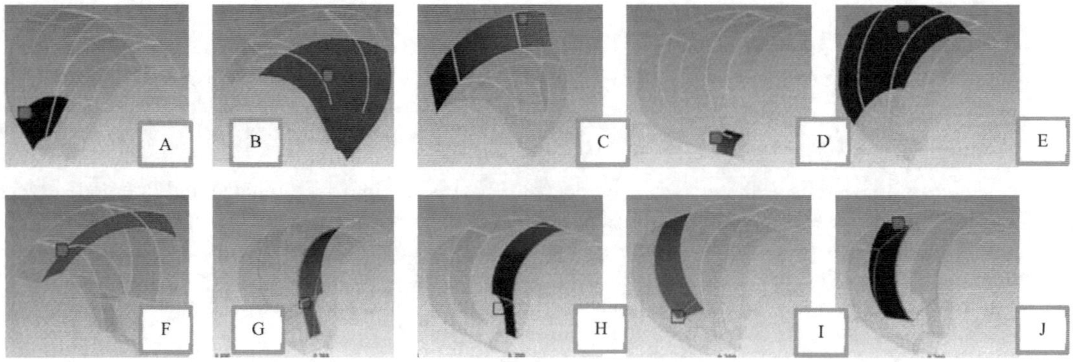

Figure 5. Boundary conditions on inlet-B, D, outlet, and wall

5. Analysis and Discussion

Four turbulence models with standard wall models have been tried in this simulation, using fixed material properties. Initial price initialization is given at the cylinder flow rate. The gathering process is carried out in the following order:

a) The initial grid is simulated until it converges using first-order discretization momentum
b) The results above are refined by using a second-order momentum schematization
c) Smoothing the grid is carried out with adaptive facilities for areas that have a high gradient of velocity, pressure, and kinetic energy. Cells that have Y^+ values smaller than ten are not adapted so that the first cell is not too close to the wall in a standard wall function model.

In this study, 18 simulations were carried out by varying three types of turbulence models, discretization of first and second order momentum, and three types of discretization of PVC (Pressure Velocity Coupling). The purpose of the simulation by varying the three options above is intended to determine the level of convergence of the iteration process and the time required in the calculation. The results of the next calculation are compared with the experimental results. From the comparison of all simulations carried out the results show that the calculation with the Realizable k-ε turbulence model cannot be converged if it uses the second discretization momentum and the PISO. This model can converge with the second order if the SIMPLE method is used. Although the amount and time of the iteration required vary, it turns out that the calculations produced by each simulation process are not much different. The accuracy of the calculation results, according to the properties of this numeric method, will be higher using the second order discretion. When viewed from the number of iterations and the time required, it can be concluded that the use of the PISO method can provide faster calculation results with a smaller number of iterations.

From the comparison of the turbulence model, discretization of momentum, and discretization of PVC that has been done and the description above, in general, it can be said that the three parameters can provide the results of convergent calculations. The effect of discretization of momentum seems not too large, only adds accuracy to the calculation. The influence of turbulence and discretization models of PVC plays a significant role in the correctness of the calculation results. Although not explicitly discussed here, it can be said that the use of the turbulence model that gives the best results is the Spalart-Allmaras model, as seen in figure 6. Discretization of SIMPLE and PISO generally produces almost the same effect even though the number and duration of the iteration process will be affected. Therefore from the results of this study, it is recommended to use the Splart-Allmaras model, which is combined with the PISO discretization and 2nd-order momentum, as a model to study the flow phenomenon in air-cooled dynamometer rotors.

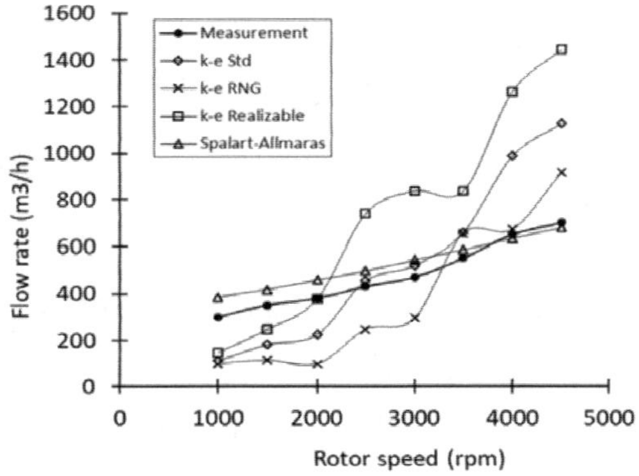

Figure 6. Comparison of experimental and computational results

Figure 7 shows the effect of rotor speed on flow discharge at the front and rear inlet and outlet for optimal blade geometry using the Spalart-Allmaras model. In this case, the blade width is 90 mm, blade inlet angle 73 degrees and the blade outlet angle is 26 degrees. It can be seen that the effect of the rotor speed on the flow rate is linear; that is, the faster the rotor speed, the higher the flow rate. It can be seen that the lowest line gradient is at the flow discharge that enters the rear inlet side. Therefore, it can be concluded that the effect of rotor speed is more effective on the front inlet side. In this study, the maximum flow rate that passes through the outlet side is around 24,200 m3/h at a rotor speed of 3000 rpm. So, to increase the flow rate at both the rear inlet and the outlet side can be done by enlarging the rear inlet rotor diameter.

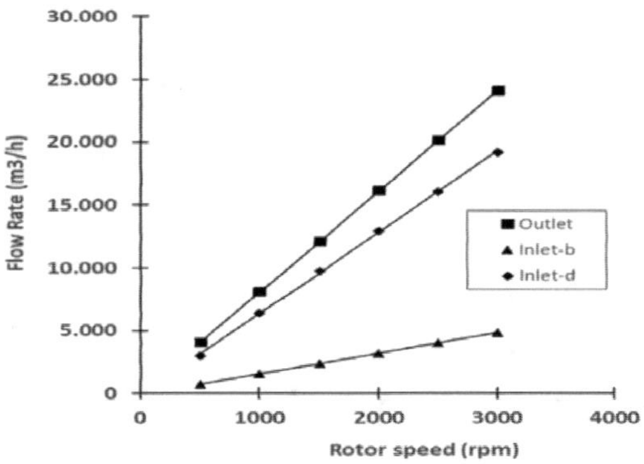

Figure 7. Effect of rotor speed on the air flow rate using the Spalart-Allmaras model

6. Conclusion

In this study, calculations have been carried out using the finite volume method to study the effect of the turbulence models and the discretization scheme on the flow that occurs in the rotor dynamometer. The results of this calculation have been compared with the measurement results. Based on the comparison of volume flow rate from the rotor, the turbulence model that gives the closest results to the measurement results is the Spalart-Allmaras turbulence model. While the best discretization scheme is PISO with second-order momentum. For further model development, research should be conducted to determine the effect of other turbulent models, Reynolds Stress Model, Large Eddy Simulation, and Direct Numerical Simulation.

References

[1] Anwar S 2004 *IEEE Trans. on Control Systems Technology* **12** 422-427.
[2] Habib G 1992 *SAE Technical Paper Series 922450* Telma Retarder Inc., International Truck and Bus Meeting and Exposition (Toledo, Ohio, USA).
[3] KLAM Retarder CFK-140 PAU 2018 *Technical Specification of Power Absorber Unit* (Industrial Zelu, Spain).
[4] Marsono 2013 *Redesign and manufacturing of eddy current dynamometer for testing of a motorized vehicle with a capacity of 150 kW* Thesis (Diponegoro University).
[5] Liu C Y, Jiang K J and Zhang Y 2011 *Int. Journal of Automotive Technology*, **12** 611-616.
[6] Chengye L, Kejun J and Yan Z 2010 *International Conference on Future Information Technology and Management Engineering* pp 551-553.
[7] Syahrullah L I and Sinaga N 2016 *American Journal of Engineering and Applied Science* **9** 222-235.
[8] Haryanto I, Utomo MSKTS, Sinaga N, Rosalia C A and Putra A P 2014 *Applied Mechanics and Materials* **493** 123-128.
[9] Long Z, Li C, Huang B and Wen W 2009 *First International Workshop on Education Technology and Computer Science* pp 521-525.
[10] Ya L, Li D, Ma Y and Jiao B 2011 *IEEE Trans. on Energy Conversion*, **26** 953-958.
[11] Sinaga N and Prasetiyo B 2012 Kaji eksperimental karakteristik sebuah dinamometer sasis arus eddy *Eksergi Journal Teknik Energi* **8** 63-67.
[12] Supriyo and Sinaga N 2011 Perancangan dinamometer arus eddy kapasitas 250 kW *Eksergi Journal Teknik Energi* **7** 46-49.
[13] Ansys Fluent Manuals 2015 *Modeling a Centrifugal Compressor Using the Single Reference Frame Model.*

Numerical simulation of the width and angle of the rotor blade on the air flow rate of a 350 kW air-cooled eddy current dynamometer

Nazaruddin Sinaga[1,*]

[1] Mechanical Engineering Department, Diponegoro University
Jalan Prof. Soedharto, SH, Tembalang, Semarang 50275, Central Java, Indonesia

Email: *nsinaga.ccfed@yahoo.com

Abstract. Design of cooling part of air-cooled eddy current dynamometer plays an important role since the temperature rise will dramatically drop the load capacity. The research aims to investigate the effect of the rotor design on the air flow rate of a 350 kW air-cooled eddy current dynamometer, which has 16 blades and a rotor diameter of 520 mm. The impacts of the rotor blade width and rotor speed, as well as inlet and outlet angle, were analyzed against the air flow rate. The variation of the blade width was 50 - 90 mm, the inlet angle was 55° - 73°, the outlet angle was 17° - 35°, while the rotor speed was 750 to 3000 rpm. The finite volume method was used to solve the problem. It was found that the addition of the blade width would increase the air flow rate so that it could increase the heat transfer rate and the power absorbed capacity. There were an optimum inlet and outlet blade angle, which generated the maximum air flow rate. The optimum inlet and outlet angle was 64° and 26°, respectively. The study concluded that there are optimum parameters which give the best dynamometer performance.

1. Introduction

Eddy currents are electric currents induced in a conductor when exposed to a changing magnetic field, due to the relative motion or variation of the magnetic source and conductor with time. It can cause a circulating flow of electrons, or current, within the body of the conductor. These eddies have inductance and thus induce magnetic fields. Like conventional friction brakes, eddy current brakes are responsible for slowing down an object, such as rotating machinery, a moving train, or even a roller coaster. Another use of this eddy current is to produce lift force on the train (magnetic levitation) or other equipment such as electromagnetic (attractive force) suspension (levitation) and electrodynamic (repulsive) suspension [1]. In the application of electric motors and electric generators, the presence of eddy currents is avoided because it is an irreversible loss, which in turn converts into heat. In this study, eddy currents were made to produce as much as possible to be used as energy absorbers in dynamometers.

The use of a dynamometer to measure the torque and power of the engine has long been known, especially in the automotive machinery industry. Eddy current dynamometer is a type of dynamometer that is widely used in the automotive world, both as a dynamometer and as a braking system (retarder). Some of these advantages are: independent of wheel support, no contact, no wear or tear, no noise or smell, adjustable brake force, high brake forces at high speeds, it uses electromagnetic force and not friction, can be activated at will via an electrical signal, low maintenance, and lightweight. The disadvantages include: braking force diminishes as speed diminishes with no ability to hold the load in position at a standstill, it cannot be used at low-speed vehicles, electromagnetic brakes are used with ordinary mechanical brakes. Nowadays, electromagnetic brakes are used only for safety purpose [2].

Eddy current dynamometer works according to Faraday's law of electromagnetic induction. According to this law, whenever a conductor cuts magnetic lines of forces, an electromotive force, is induced in the conductor, the magnitude of which is proportional to the strength of the magnetic field and the speed of the conductor. If the conductor is a disc, there will be circulatory currents, i.e., eddy currents in the disc. According to Lenz's law, the direction of the current is in such a way as to oppose the cause, i.e., movement of the disc. The eddy current brake consists of two parts, a stationary magnetic field system, and a solid rotating part, which include a metal disc. During braking, the metal disc is exposed to a magnetic field from an electromagnet, generating eddy currents in the disc. The magnetic interaction between the applied field and the eddy currents slow down the rotating disc. Figure1 illustrated an eddy current braking system principle [3].

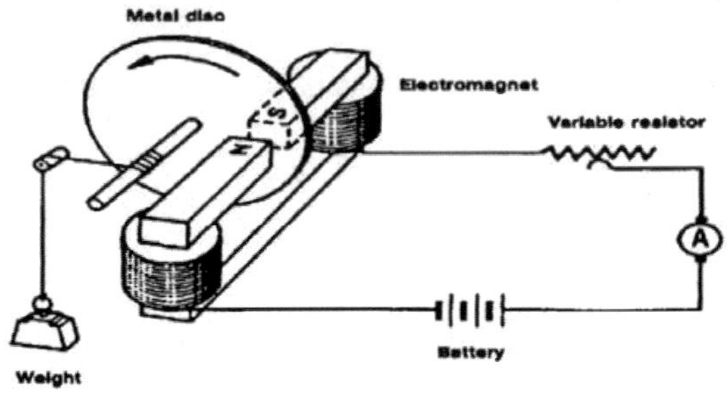

Figure 1. The principle of eddy current braking operation

As explained earlier, besides being a dynamometer, the effect of this eddy current can also be used to braking a rotor, which is called a retarder. Brakes like this are excellent to use as a brake device for a vehicle because these brakes do not experience wear like the brake disk or drum brakes. Even though there is no wear surface, this retarder will receive heat due to braking, so the best cooling system is needed. If the cooling system is inadequate, then the stator and rotor surface will experience a high increase in temperature so that the braking power capacity decreases significantly [4]. The use of dynamometers in laboratories and vehicle workshops is necessary, not only to measure torque and power but also used to measure the brake thermal efficiency (BTE) and the brake specific fuel consumption (BSFC). If this tool is not owned by a workshop, it is not possible to know what the torque capacity and power of the vehicle are being serviced, as well as BSFC and BTE [5]. Dynamometers must also be owned by automobile dealers to know exactly how much torque and power of the vehicle to be sold, or that has undergone repairs or an engine tune-up.

To run an energy efficiency program and an environmentally friendly transportation system, a dynamometer is also needed to determine the fuel economy, in units of km/liter, and the emissions of a vehicle, in units of gr/km. In measuring these two quantities, the vehicle is run on a chassis dynamometer by following a driving scenario called the Driving Cycle. To be able to follow this driving cycle, the vehicle must be loaded by a dynamometer. Until now, there are almost no workshops or vehicle dealers in Indonesia that use dynamometer for the purposes mentioned above. This is because the price of a dynamometer is still relatively high, especially by small-scale workshops, because until now there has been no factory that makes dynamometers in the country.

To be able to make an eddy current dynamometer is indeed not easy, especially by its usage requirements, which are limited by the function of the rotor speed to torque and power, commonly called a dynamometer envelope. Currently, eddy current dynamometers are divided into two types, namely air-cooled and water-cooled. Air-cooled dynamometers have a more straightforward construction than water-cooled ones, so the price is relatively lower compared to water-cooled ones. However, the cooling capacity of the air-cooled eddy current dynamometers is much lower than the water-cooled eddy current dynamometer. This cooling capacity plays a vital role in determining the measuring capacity of the dynamometer after operating in a steady state. Liu et al. [6], and Prasetiyo and Sinaga [7] showed that a decrease in torque measurement capacity could decrease by 70% by using an air-cooled. On the other side, the use of water cooling only reduces the measurement capacity by around 15% [8]. Therefore, efforts need to be made to increase the cooling capacity of the air-cooled eddy current dynamometers, for example by design optimization using artificial neural networks [9, 10]

A water-cooled eddy current dynamometer usually consists of a stator, which produces a magnetic field, and one or two rotors producing eddy currents and at the same time as a cooler, as shown in figure 2. To be able to design a compact dynamometer, the ability to optimize all parameters involved with torque and power capacity. To do this, optimization is complicated because it involves many parameters [11-13]. Optimization involving the process of making a prototype certainly requires quite expensive costs. Therefore, the researchers are currently making efforts using parameter optimization from the simulation results using computational methods. The calculation using computational fluid dynamics software is now very much done because the results are quite close to the measurement results. With the help of this CFD software, calculations can be done quickly. However, the accuracy of the results of these theoretical calculations must first be justified so that the results have low uncertainty. When a precise calculation model has been obtained, the effect of various design parameters from the dynamometer can be searched quickly. This research was conducted in order to obtain a calculation model that can be used to perform the optimal design of the air-cooled eddy current dynamometer.

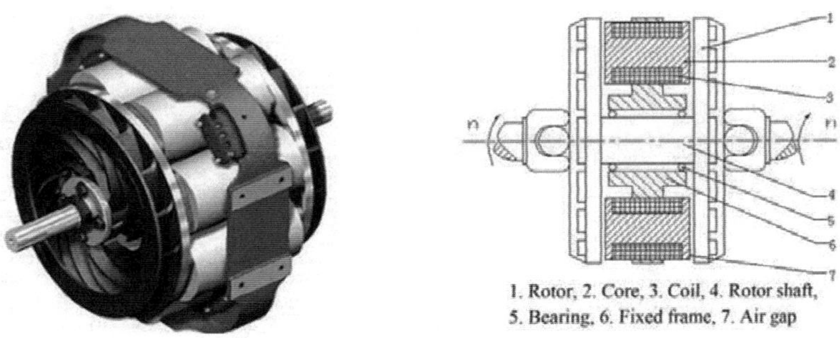

Figure 2. Construction of air-cooled eddy current dynamometer and its components

From the descriptions above, it can be concluded that design of cooling part of air-cooled eddy current dynamometer plays an essential role since the effect of temperature rise will dramatically drop the capacity of power and torque measurement. The air flow rate of the rotor will undoubtedly affect the temperature of the rotor surface and the stator dynamometer. Since the flow and heat transfer mechanism is very complicated and involves many parameters affecting cooling capacity, it is essential to study the phenomena before designing a dynamometer. The study aims to investigate the effect of the rotor blade design to the air flow rate of the rotor of a 350 kW air-cooled eddy current dynamometer, which has the rotor diameter of 520 mm and the number of the blade is 16. In this paper, the impacts of the rotor blade width, as well as inlet and outlet angle, were analyzed against the air flow rate in some speed of the rotor the dynamometer. The finite volume method, facilitated by ANSYS FLUENT software, is used to solve the problem.

2. Methodology

In this study, the method used is a computational approach, namely the finite volume method. This is a method that is widely used in solving fluid flow and heat transfer problems numerically. In the finite volume method, unknown variables are explained using a specific point on the nodal points. Broadly speaking, the necessary steps for solving problems using the finite volume method are as follows: estimating the value of unknown flow variables using a simple function; make a discretization equation by entering the estimation into the general equation of flow, and then doing mathematical manipulation; and finally solve the algebraic equation of the problem at hand [14]. Currently, there is quite a several computational fluid dynamic software that can be used to solve flow problems and heat transfer. In this study, ANSYS FLUENT® software is used. One of the advantages of the software is its complete literature and is very widely used in scientific. Besides, this software also has complete features.

2.1 Geometry and Flow Conditions

In the present calculation, the rotor plate outer diameter is 520 mm, the inner diameter of the front rotor plate is 220 mm, the inner diameter of the rear rotor plate is 460 mm, the number of blades is 16 pieces, and the rotor material is made of St 60 steel. The blade width is 50, 60, 70 and 90 mm, with the rotor blade entry angle are 55°, 58°, 61°, 64°, 67°, 70° and 73°, while the blade exit angle is 17°, 20°, 23°, 26°, 29°, and 32°. The geometry of the dynamometer rotor is shown in figure 3.

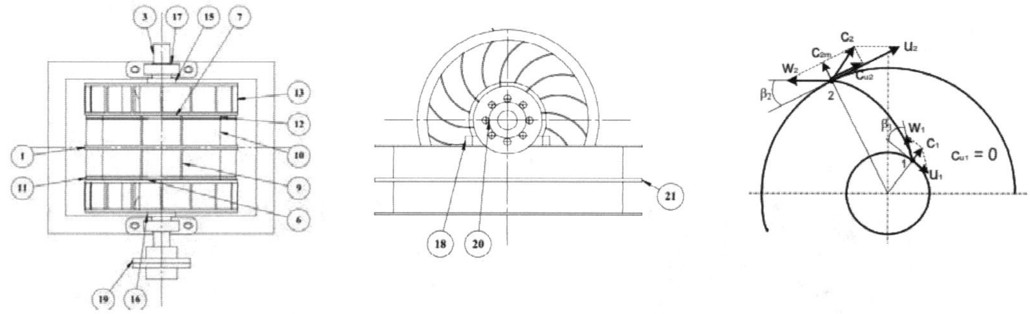

(a) (b) (c)

Figure 3. The geometry of the dynamometer: (a) above view; (b) right view; (c) inlet and outlet angle of the rotor blade

2.2 Problem Description

The problem involves modeling the steady-state flow of air through a rotor running at 1500 rpm. There are 16 blade rows in the rotor. To simplify the CFD calculation, the flow is modeled only through double blade rows and uses rotationally periodic boundary conditions on the boundaries between the blade rows. The governing equations are cast in a Single Reference Frame moving at the speed of the rotor. Air is treated as an ideal gas with constant values of specific heat, thermal conductivity, and dynamic viscosity. The schematic of a single blade row and the complete rotor, repeating a single blade row 16 times, are shown in figure 4.

2.3 Model and Materials

As a first step in the modeling in ANSYS FLUENT, we have to set up the general model as follows: enable the density-based, steady-state solver and retain the default values for the other parameters. The second step is to enable the Spalart-Allmaras turbulence model, and retain the default values for the other parameters. This turbulence model is an excellent and economical choice for mildly complex boundary-layer flows in turbomachinery. The next step is to set the density of air to ideal-gas. The ideal gas model will automatically enable the solution of the energy equation. For simplicity, we will keep the other fluid properties at their default constant values, though they could be made the function of temperature if desired.

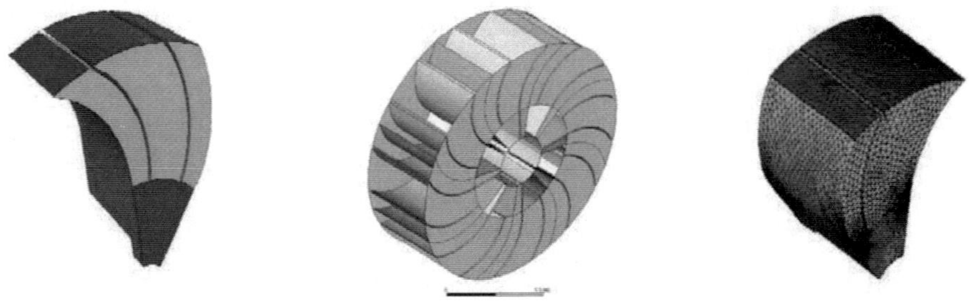

(a) (b) (c)

Figure 4. (a) calculated geometry; (b) full rotor geometry; (c) meshing on the domain geometry

2.4 Boundary Conditions

In figure 5, a cell zone conditions setting is shown in the blade area, which is then given a rotational speed with a specific value in the rotational velocity, for example, 1500 rpm. In this inlet area, the pressure is 1000 Pa gauge, and the initial gauge pressure is 1000 Pa, while the turbulent viscosity ratio is 10. The outlet condition is a pressure outlet of 1000 Pa, and the turbulent viscosity ratio backflow is 10. Figure 5 also shown the boundary conditions at the hub, shaft, shroud-b, and shroud-d areas, where the wall is a moving wall, relative to adjacent cell zone with the rotational axis is X = 1, and rotational speed is zero. For wall boundary conditions of blade1-p, blade1 -s, blade2-p, and blade2-s, the settings are the same as before.

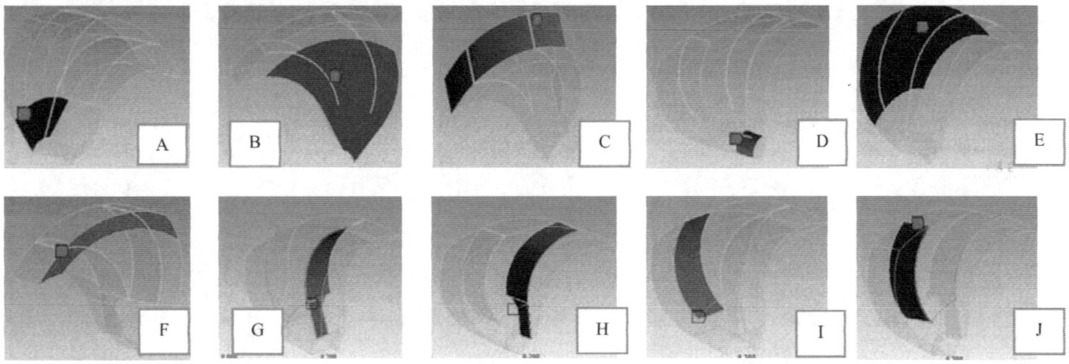

Figure 5. Boundary conditions at inlet-B, D, outlet, and wall

2.5 Solver Discretization, Controls, and Monitors

Retain the default solution methods for flow under solve methods, and set the turbulence equation to second order upwind. Activate the pseudo transient method then retain the default solution controls under solving controls. For the residual convergence, set the convergence criterion for continuity to 1e-4. Create three surface monitors as follows: mass flow rate at the inlet, the mass flow rate at the outlet, and mass-averaged total pressure at the outlet.

Tabel 1. The boundary condition at inlet and outlet

Surface	Boundary Condition	Label
Inlet-D	Intake Fan	A
Inlet-B	Intake Fan	B
Outlet	Exhaust Fan	C
Hub	wall	D
Shroud-D	wall	E
Shroud-B	wall	F
Blade1-P	wall	H
Blade1-S	wall	I
Blade2-P	wall	J

2.6 Solver Discretization, Controls, and Monitors

Retain the default solution methods for flow under solve methods, and set the turbulence equation to second order upwind. Activate the pseudo transient method then retain the default solution controls under solving controls. For the residual convergence, set the convergence criterion for continuity to 1e-4. Create three surface monitors as follows: mass flow rate at the inlet, the mass flow rate at the outlet, and mass-averaged total pressure at the outlet.

3. Results and Discussion

Figure 6 shows the iteration process by observing the residual value and mass flow rate on the outlet side. Figure (a) shows that the iteration process runs quite well towards convergence. In this calculation, the number of cells used is 104,000, which has been tested by the grid-independence test. Although in this study, the residual value is limited to 1e-4, from the residual diagram, it can be seen that up to 3300th the iteration reaches a minimum residual value of 1e-5. This is not very influential on the results of the calculation of mass flow rates, as seen on figure (b). From this value, it can be seen that the calculation has been almost constant since the 1500th iteration. By using a PC with a Pentium-7 processor and 12 GB of memory, the calculation can be completed in 10 minutes.

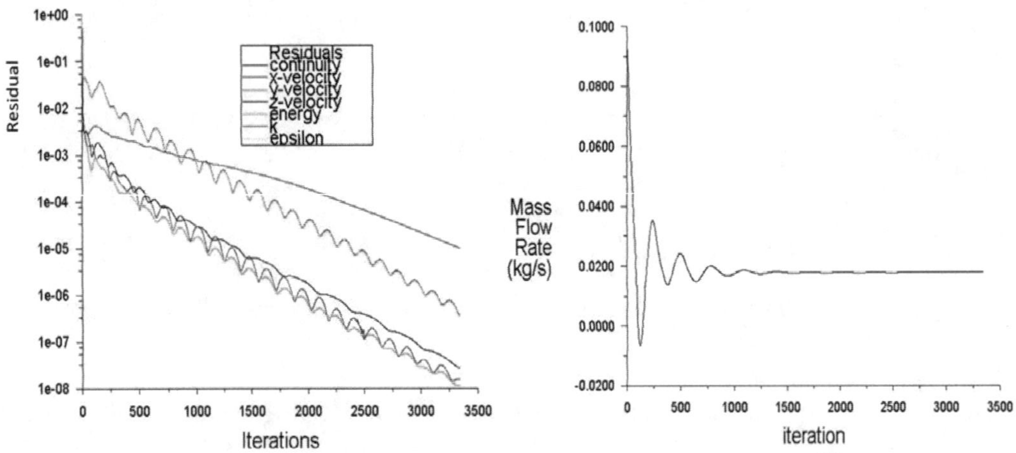

(a) (b)

Figure 6. Residual iteration (a) and iteration of mass flow rate (b) for controlling calculation

3.1 Particle Pathline

To ascertain the direction of flow in the calculation results, figure 7 shows the path lines of the fluid particles in the middle area of the rotor, x = 4.5 mm. From the two pictures, it is clear that the flow is sucked in through the front and rear inlet holes, and is ejected through the outlet side. It is also seen that the movement of fluid flow is quite smooth, and only a few swirls occur. From this pathline, the fluid movements can also be learned from the inlet to the outlet. This observation can give intuition to improve the geometry of the rotor dynamometer so that the resulting geometry is compact but has a more significant flow rate. With an increase in the flow rate, the cooling capacity becomes higher, and ultimately increases the torque capacity and power of the dynamometer.

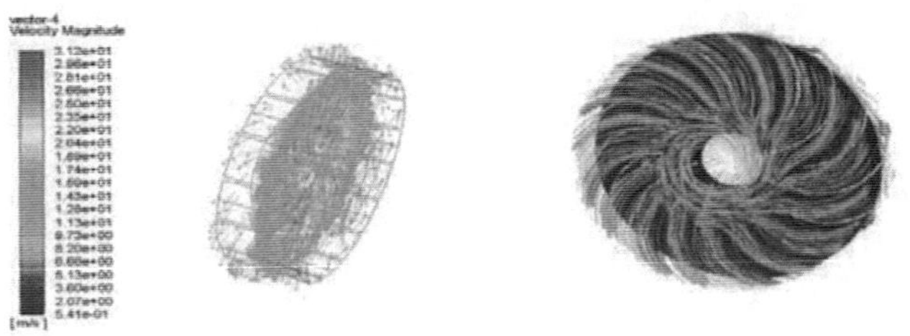

Figure 7. Particle pathline in the middle of the rotor, x = 4.5 mm

3.2 Velocity Contour

Figure 8 shows the tangential, radial, and total velocity contours. It can be seen that the tangential velocity is higher in the area leading to the outlet, while the radial velocity value is relatively more significant in the area of the central part of the rotor. The resultant of these two speeds is the total speed, which is shown in the rightmost image (c). For the rotor cooling process, the dominant speed in conducting convection heat transfer is radial velocity. Based on figure 8 (b), it can be estimated that the area that is effective in heat transfer is the area between the inlet and outlet sides of the rotor. To get a useful heat transfer geometry, it is necessary to find a design that results in a more uniform distribution of radial velocity, starting from the inlet to the outlet side. From figure 7 and 8 it can be said that the flow is reasonable and as expected.

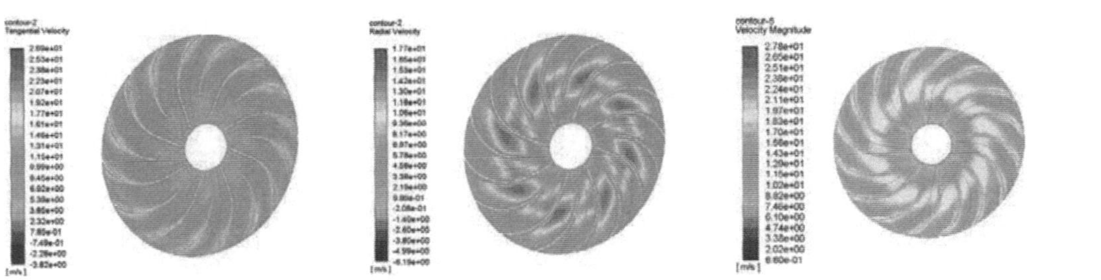

(a) (b) (c)

Figure 8. Tangential, radial and velocity magnitude in the middle of the rotor, x = 4.5 mm

3.3 Pressure Contour

Figure 9 shows the static, dynamic, and total pressure contours. The total pressure is the sum of static and dynamic pressure. The dynamic pressure contour is similar to the speed contour. This is quite clear because dynamic pressure is a pressure that arises due to speed or kinetic energy. The static pressure contour is the opposite of the speed contour. This can be explained using the Bernoulli equation, where the line with a high velocity has low pressure. Therefore the speed and pressure contours generated from the calculation are reasonable and as expected. The value of static pressure on the side of the inlet and outlet is also by the value of the boundary conditions given at the two places, that is 1000 Pa. Observation of the path line, velocity, and pressure contour give the conclusion that the results of this numerical calculation are good enough. The particle pathline, velocity and pressure contour are agree well with the one calculated in Fluent Tutorial [14].

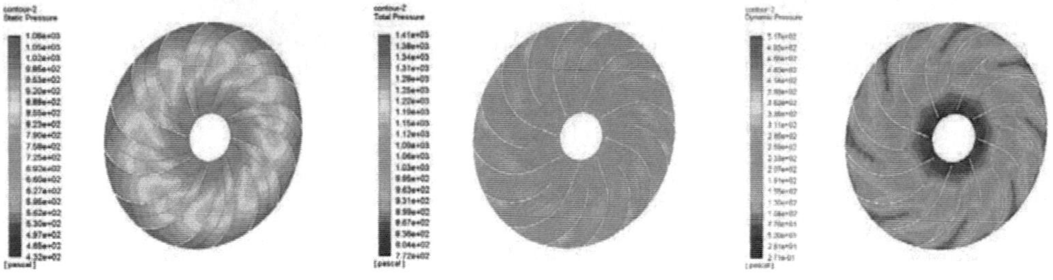

(c) (b) (c)

Figure 9. Static, total and dynamic pressure in the middle of the rotor, x = 4.5 mm

3.4 Effect of the Blade Width

In figure 10 (a), the calculation results of the effect of blade width on the flow rate at the inlet and exit of the dynamometer are shown. It can be seen that the flow rate through the outlet is the same as the sum of the flow rate through the front inlet and the rear inlet. The flow rate through the rear inlet is much smaller than the flow rate through the front inlet. This is because the rear inlet surface area is smaller than the front inlet surface area, while the pressure is almost the same, that is atmospheric pressure. It can also be seen that the blade width is very influential on flow discharge. If a regression equation is made for all three lines, an equation in the form of a parabolic equation is obtained. This indicates that there is an optimum value of the blade width, which gives the highest flow rate. In the simulation carried out in this study, the width of the blade that gives the maximum flow rate is 90 mm.

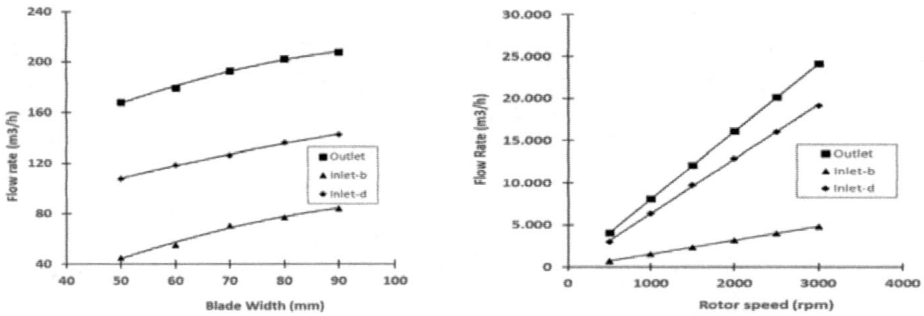

(a) (b)

Figure 10. Effect of blade width at 1500 rpm (a) and rotor speed (b) on the air flow rate

3.5 Effect of the Rotor Speed
Figure 10 (b) shows the effect of rotor speed on flow discharge at the front and rear inlet and outlet for optimal blade geometry. In this case, the blade width is 90 mm, blade inlet angle 73 degrees and the blade outlet angle is 26 degrees. It can be seen that the effect of the rotor speed on the flow rate is linear, that is, the higher the rotor speed, the higher the flow rate. It can be seen that the lowest line gradient is at the flow discharge that enters the rear inlet side. Therefore it can be concluded that the effect of the rotor speed is more significant in the front inlet side. In this study, the maximum flow rate that passes through the outlet side is around 24,200 m3/h at a rotor speed of 3000 rpm. To be able to increase the flow rate at the rear inlet side and the outlet side can be done by enlarging the rear inlet diameter. These results also in accordance with the results of measurements carried out by Fauzi [15].

3.6 Effect of the Blade Inlet Angle
The calculation of the effect of the inlet angle on the flow rate at the front inlet, rear inlet, and outlet side is shown in figure 11 (a). As in the previous explanation, the flow rate at the rear inlet side is smaller than the flow rate through the front inlet side. Unlike the previous pictures, the effect of the inlet angle on the front inlet is different in shape compared to the rear inlet side. It can be seen that with increasing the inlet angle, the flow discharge at the front inlet side will tend to reduce the flow rate. However, the effect of the inlet angle on the flow discharge at the front inlet side is not as strong as the effect on the rear inlet side. Therefore, the flow rate at the outlet side will increase with increasing blade inlet angle. It can also be seen that there is an optimal inlet angle value that gives the highest flow discharge. In this study, the angle is 73 degrees.

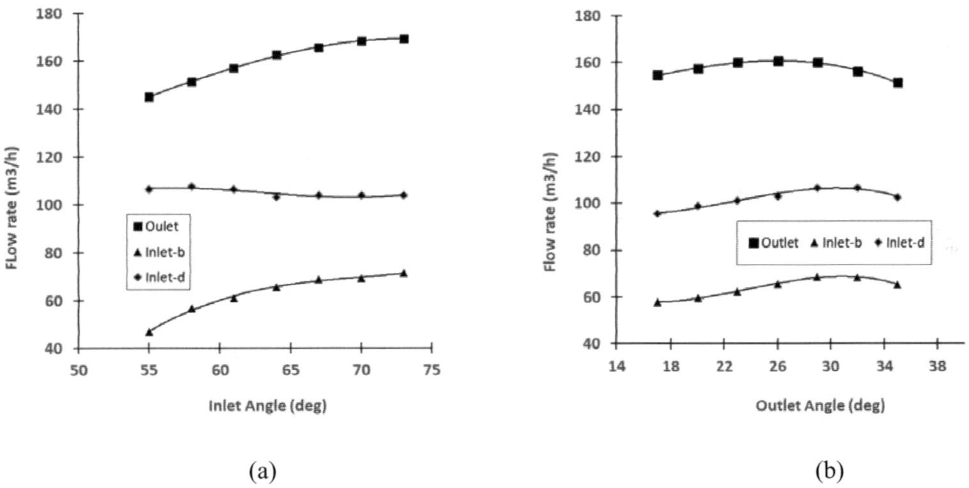

Figure 11. Effect of blade inlet (a) and outlet (b) angle on the air flow rate at 1500 rpm

3.7 Effect of the Blade Outlet Angle
Figure 11 (b) shows the calculation of the effect of the outlet blade angle on the flow rate at the front and rear inlet as well as the outlet. It can be seen that the influence of the blade angle is in the form of the quadratic polynomial equation, wherein the calculated angle range, there is a value that gives maximum discharge. In this study, the blade outlet angle which gives maximum discharge is at 26 degrees. Based on the discussion before, it can be concluded that the blade geometry which provides maximum discharge is at 90 mm blade width, 73° inlet angle, and 26° blade outlet angle.

4. Conclusion

Based on the results of the study, the geometry parameters, namely blade width, blade inlet angle and blade outlet angle have a significant effect on the flow rate produced by the rotor as well as the rotating speed and direction of the blade rotation. Therefore, to get the best design, it is necessary to optimize these parameters. It was found that the higher the speed of the rotor, the higher the air flow rate. It is also found that the addition of the blade width will increase the air flow rate so it could increase the heat transfer rate and the torque and power capacity of the dynamometer. Moreover, there is an optimum value of both the inlet and outlet blade angle which give the maximum amount the air flow rate. The optimum inlet angle is 64°, while the optimum outlet angle is 26° and the blade width is 90 mm. In general, it can be concluded that there are optimum geometric parameters of the dynamometer design that will give the best performance of the dynamometer. This paper provides a useful method for matching and designing the cooling part of the air-cooled eddy current dynamometer.

The speed and direction of the rotor also have a significant effect on the air flow rate generated by the rotor, where the backward type blade produces a higher flow rate than the forward type. The influence of the rotor speed on the flow discharge indicates that with the greater rotational speed, the higher the value of the discharge produced. At 3000 rpm, the forward facing blade generates a discharge of 20,700 m3/h, while at 500 rpm generates a discharge of 1325 m3/h. At 3000 rpm, the backward facing blade generates a discharge of 24,200 m3/h, while at 500 rpm generates a discharge of 1650 m3/h. Therefore, cooling using backward blades would be more effective than the forward blade.

References

[1] Hribar J 2008 *Magnetic Braking* Seminar in University of Ljubljana Faculty of Mathematics and Physics Department of Physics
[2] Karthikeyan B, Goverthana R M, Ganesh K G, Raj M M and Krishnan N 2007 *International Journal of Advanced Research in Management, Architecture, Technology and Engineering*, **3** (9) pp 34-39
[3] Habib G 1992 *SAE Technical Paper Series 922450* International Truck and Bus Meeting and Exposition (Toledo, Ohio, USA)
[4] Ji Y, Wang J, Xu Y and Liu Z 2016 *SAE Technical Paper* 2016-01-0185
[5] King F G, Mitchell W A, Squier W C, Harris D B and Kinsey J S 2002 *Journal of the Air & Waste Management Assoc.* **52** pp 388-395
[6] Liu C Y, Jiang K J and Zhang Y 2011 *International Journal of Automotive Technology* **12** (4) pp 611-616
[7] Prasetiyo B and Sinaga N 2011 *Eksergi Jurnal Teknik Energi* **7** (2), pp 57-62
[8] Ye L, Li, D, Ma Y and Jiao B 2011 *IEEE Transactions on Energy Conversion* **26** (3) pp 953-958
[9] Syahrullah L O I and Sinaga N 2016 *American Journal of Engineering and Applied Science*, **9** (2) pp 222-235
[10] Haryanto I, Utomo M S K T S , Sinaga N, Rosalia C A and Pratama A 2014 *Applied Mechanics and Materials*, **493** (20), pp 123-128
[11] Long Z, Li C, Huang B and Wen W 2009 Simulation research on braking torque of eddy current retarder *First International Workshop on Education Technology and Computer Science* pp 521-525
[12] Anwar S 2004 *IEEE Transactions on Control Systems Technology* **12** (3) pp 422-427
[13] Levin Y, Silveira F L and Rizzato F B 2006 *American J. Physics* **74** (9) pp 815-817
[14] Ansys Fluent Tutorial 2015 *Modeling a Centrifugal Compressor Using the SRF Model*
[15] Fauzi M 2017 *Perancangan, pembuatan dan pengujian rotor dinamometer arus eddy kapasitas 400 kW pada kecepatan putar 3500 rpm* Thesis (Diponegoro University, Semarang) pp 46-48

Enhancements in cure rate index and mechanical properties of silica-filled natural rubber using octadecanol-fatty alcohol

I Surya[1*] and M Ginting[2]
[1]Chemical Engineering Department, Universitas Sumatera Utara, Medan, Indonesia
[2]Chemistry Department, Universitas Sumatera Utara, Medan, Indonesia

Email: *indradanas@yahoo.com

Abstract. The enhancements in cure rate index and mechanical/tensile properties of natural rubber (NR) filled with silica because of the addition of octadecanol (OCD)-fatty alcohol (FA) were investigated. The silica as a reinforcing filler was compounded into the NR at a fixed loading, 30 parts per hundred rubber (phr.). The OCD-FA was derived from the oil of palm kernel and compounded into the NR as an additive agent. The loading variations of OCD-FA were 0 to 4 phr. It was found that the OCD-FA functioned as a curative and plasticizing agent for the silica-filled NR compounds. As a co-curing agent, the OCD-FA enhanced the time of scorches and cure rate index-CRI of the compounds of NR. A higher the OCD-FA loading caused in a higher CRI degree. As a plasticizing agent, the OCD-FA decreased tensile moduli but increased elongation at break. It was also found that the OCD-FA increased both tensile strength and abrasion resistance of the compounds of NR up to a 2 phr. of loading. The addition of OCD-FA with a 2 phr. of was an optimum loading for the compounds of NR.

1. Introduction

Mixing of a type of raw rubber with some rubber additives is defined as rubber compounding operation. Each of raw rubbers and also the rubber additives has a certain function either in the cure, mixing process or rubber products [1]. Raw rubbers will deliver the rubbery behaviour; additives such as accelerator will enhance the cure state; the curing agent will connect chemically the chains of rubber; and plasticizer will reduce the viscosity, enhance deformability and change properties.

In rubber compounding, the additives are defined as curative and non curative. The curative will affect the curing process. The well-known ones of them are zinc oxide, stearic acid, accelerators and sulfur. Non curative such as fillers are utilized in order to enhance mechanical/tensile properties of vulcanized products. Silica is one of the most exploited fillers in the enhancements in mechanical properties of rubber products [2]. Relatively, at higher loading of utilization, the micron-sized particles of the filler that tends to form some bigger aggregates during compounding and causing in a relatively poorer degree of filler dispersion. Besides, the silica filler also has numerous silanol groups on its surface and causes it becomes very polar filler. It can readily interact with zinc oxide during the curing process. The silica-zinc complexes unable to activate the accelerator and as a consequence, the activity of zinc is decreased and hence, the vulcanization or cure process is retarded [2].

As an attempt to solve such problems, a special rubber additive agent which is named as octadecanol (OCT) was added into the compounds of natural rubber (NR) filled with silica. It is fatty alcohol derived from the oil of palm kernel and it was utilized to solve both the cure retardation and filler dispersion problems. The material of OCT-fatty alcohol (FA) is oily with unique polar and non-polar functional groups inside the material; oil can act as a plasticizing agent [3-5] that helps to disperse the filler homogeneously inside the rubber molecules.

This research investigated the effects of OCT-FA on cure rate index and also mechanical/tensile properties of the NR which was filled with silica.

2. Experimental

2.1. Materials of research

A SIR 3L-grade of NR was supplied by the PPTPN VII - Indonesia. IKD, PT locates in Medan of Indonesia prepared IPPD-N-isopropyl-N'-phenyl-p-phenylenediamine, precipitated silica, stearic acid, sulfur, MBTS-mercaptobenzothiazole disulphide, zinc oxide. Ecogreen, PT in Batam, Indonesia supplied OCT-FA. The molecule structure of OCT-FA is $CH_3(CH_2)_{16}CH_2OH$.

2.2. Compounding of NR

A semi EV cure system was applied in making the compounds of NR. Procedures of the making of compounds of NR were conducted according to ASTM D 3184-80. The compounds of NR were made on a laboratory two roll mill. The recipe of compounds of NR is presented in **Table 1**.

Table 1. The formulation of compounds of NR

Chemicals	Content (phr.)
SIR 3L	100
Zinc oxide	5
IPPD	2
Stearic acid	2
MBTS	1.5
Sulfur	1.5
Precipitated silica	30.0
OCT-FA	0; 1; 2; 3 and 4

2.3. Curing of compounds

Cure of compounds of NR was characterised based on an MDR 2000 – Rheometer. The Rheometer was operated to determine the times to scorch and cure according to ISO-3417. The respective compounds of NR were vulcanized at 150 °C.

2.4. The tensile properties

The vulcanized NR was cut from molded sheets based on ISO-37. The tensile properties were tested at a crosshead speed of 500 ($mm.min^{-1}$) and the testing was performed using the Instron 3366-universal tensometer to determine the stresses at 300% and 100% elongations-M300 and M300, tensile strength-TS, elongation at break-EB.

2.5. The abrasion test

The samples of NR compounds were moulded with a KAO – Tech moulding machine with a force 100 kg/cm^2 at 150 °C for optimum cure time plus 5 minutes. The samples were a disc with dimension 12.7 + 0.0 – 0.5 mm in thickness, 63.5 + 0.5 mm diameter with a central hole 12.7 mm diameter. Wallace Abrasion Tester was used to determine abrasion loss of the vulcanized rubber according to BS-903 part A9. The angle between the abrasion wheel and the rubber disc was 15°, the applied weight was 4.5 kg. The abrasion experiment was carried out at room temperature. The test specimen was driven at a constant speed and drives the abrasion wheel which was held against it by a constant force. The test specimen was weighed and mounted on the spindle of the machine by tightening the wing nut finger tight. After 500 revolutions of the trial run, the test specimen was weighted again. The weight of the

rubber loss was converted to volume loss by using the density of the vulcanized rubber. Based on this volume loss, suitable running in period and test run for the sample was determined.
After the running in period, five test runs were carried out and the average was in term of volume loss (cm^3) per-1000 revolutions.

3. Results and discussion

3.1. The cure rate index of compounds of NR
The effects of OCT-FA on scorch time, optimum cure time and cure rate index (CRI) of the NR compounds are shown in **Figs. 1-2**. As shown in **Fig. 1**, the scorch times of silica-filled NR compounds with OCT-FA were higher compared to the compounds without OCT-FA. A higher OCT-FA loading caused a higher in scorch time. In this investigation, the OCT-FA acted as a scorch delay agent. Any material that enhances the scorch time can be defined as a scorch delay material [6-8].

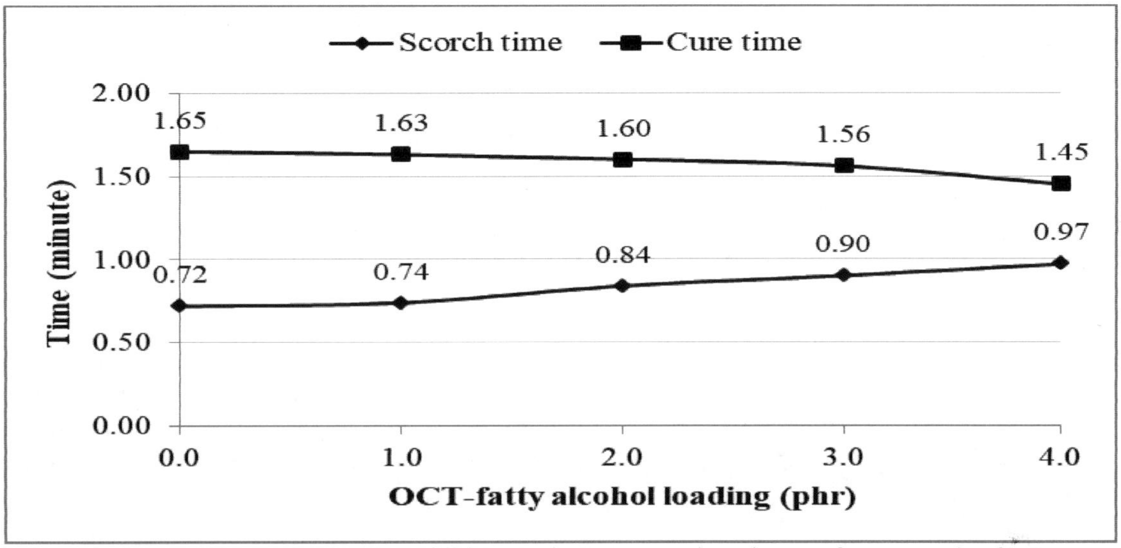

Figure 1. Effects of OCT-FA addition on times to scorch and cure of compounds of NR

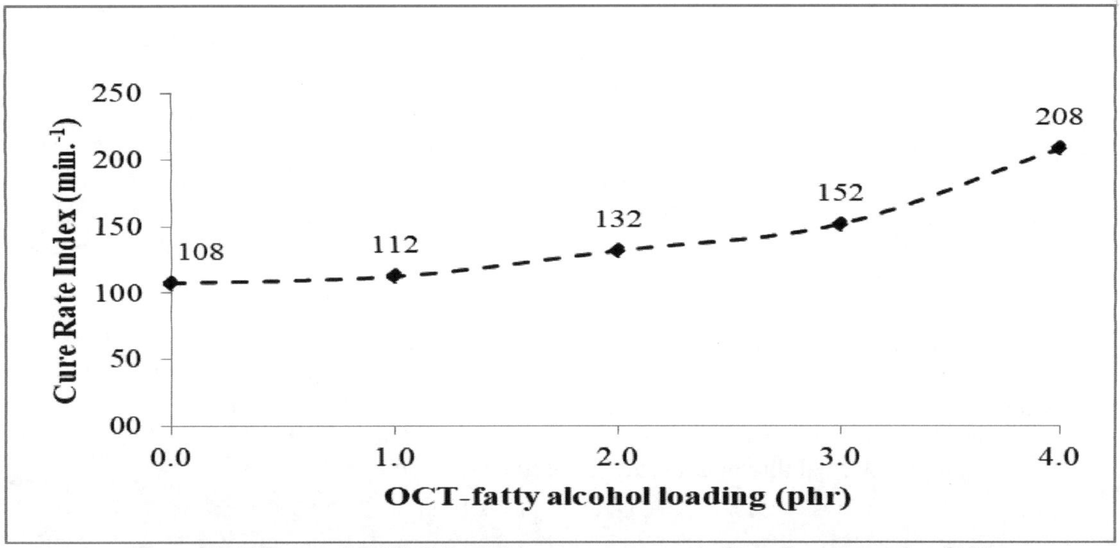

Figure 2. Effects of OCT-FA addition on cure rate index of compounds of NR

From **Fig. 1**, the addition of one phr. of OCT-FA decreased the time of optimum cure. Shorter cure time is defined as a faster cure rate. Therefore, OCT-FA caused in cure improvement. A higher OCT-FA caused in a more pronounced in cure improvement. In this study, the OCT-FA functioned as a cocuring agent for the compounds of NR. Any material that affects the curing properties of a rubber compound is defined as a cocuring or curative agent [9-11].

As shown in **Fig. 2**, the CRI of compounds of NR with OCT-FA were higher compared to the filled NR compounds without OCT-FA. A higher OCT-FA loading caused in a higher in CRI. Again, it was due to the function of OCT-FA as a cocuring agent.

3.2. Tensile properties of the compounds of NR

Effects of OCT-FA on tensile properties of the compounds of NR are visualized in **Figs. 3-5**. As visualized in **Fig. 3**, the OCT-FA decreased the tensile moduli. The higher the OCT-FA, the lower the stresses at 300% and 100% elongations was. The OCT-FA caused in the tensile moduli of compounds of NR with lower values. It was due to the function of OCT-FA as a plasticizing agent which softened filled NR vulcanizates.

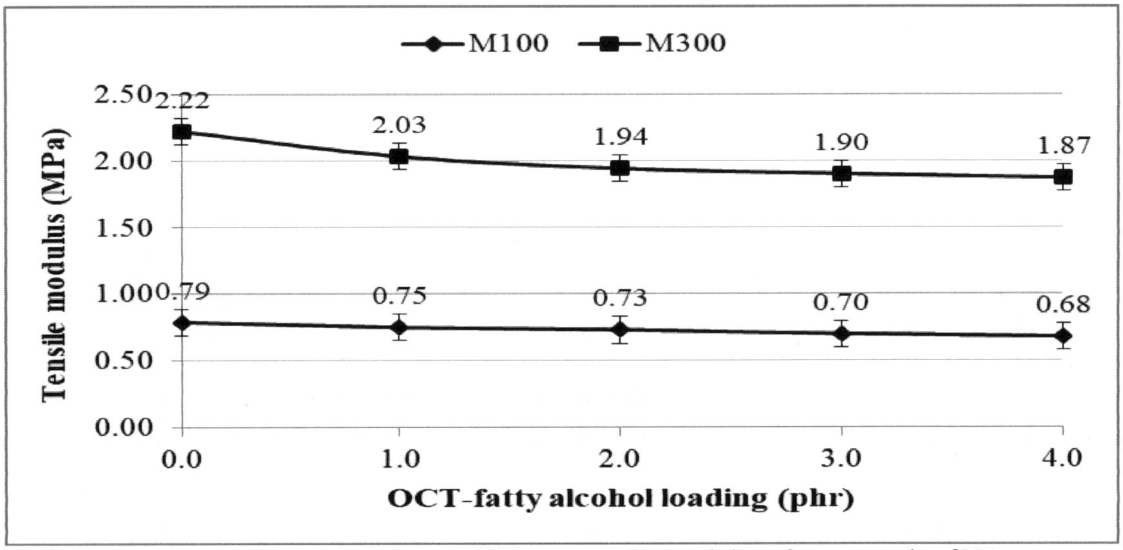

Figure 3. Effects of OCT-FA addition on tensile modulus of compounds of NR

From **Fig. 4**, the TS was increased slightly up to a two phr. of OCT-FA loading and then, was decreased with further increases the loading of OCT-FA. TS enhancement up to a two phr. of OCT-FA might be due to the effect of plasticizing of OCT-FA which increased the degree of silica dispersion and rubber to filler interaction, respectively. Consequently, the crosslink density of the filled NR vulcanizates was enhanced. The deterioration in TS after a two phr. of OCT-FA loading was due to active ingredients absorbing the effect of the excessive OCT-FA that decreased the crosslink density.

As shown in **Fig. 5**, the addition of a one phr. of OCT-FA into NR control compound increased elongation at break (EB). The addition with a more amount of OCT-FA caused further increases in the EB. Again, it was due to the plasticizing effect of OCT-FA to the compounds of NR [12-15]. The additive enhanced the deformability of a polymeric composite. Increasing the OCT-FA loading caused further increases the deformability or extensibility of the compounds of NR.

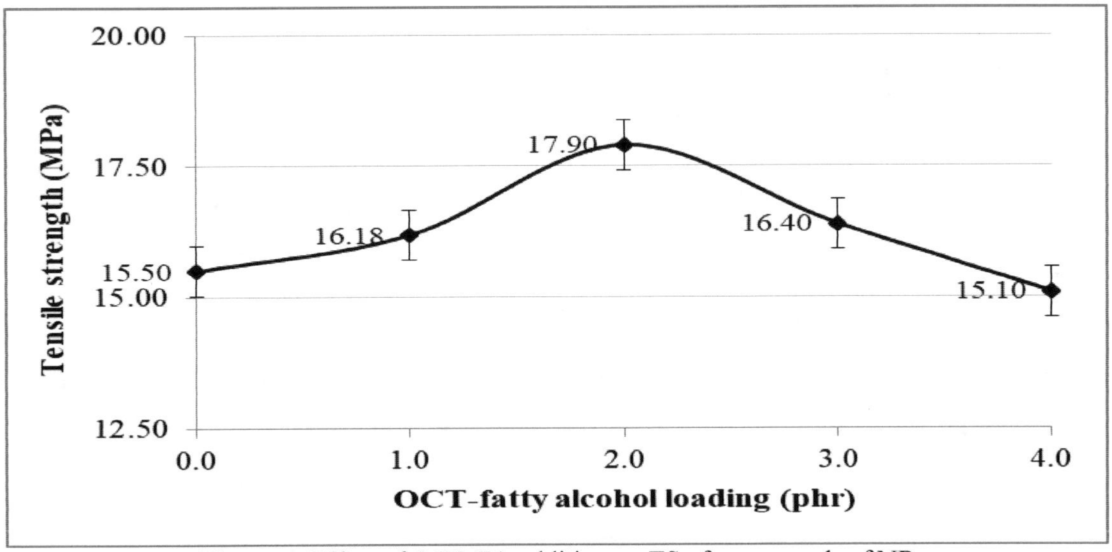

Figure 4. Effect of OCT-FA addition on TS of compounds of NR

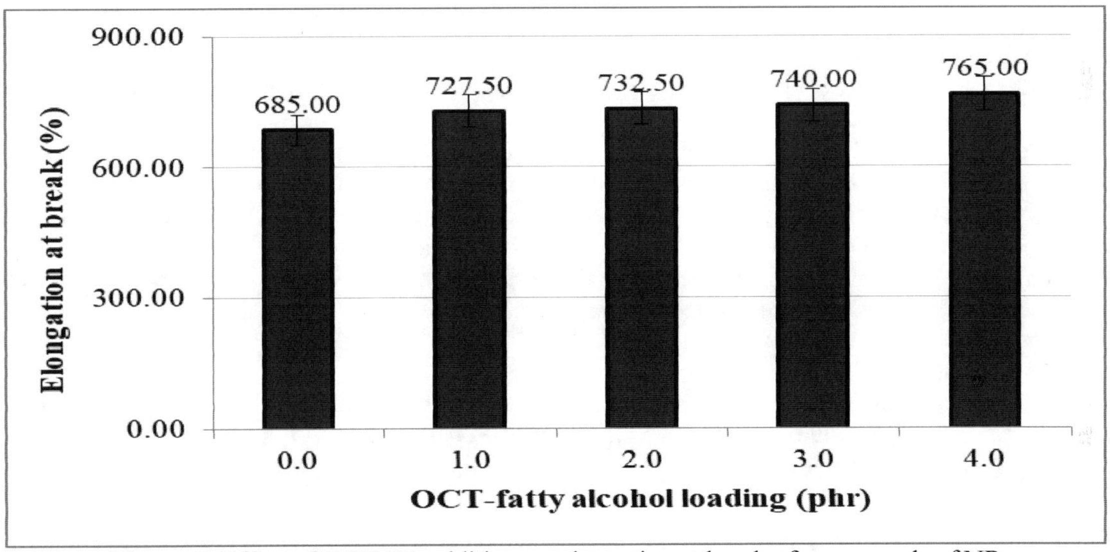

Figure 5. Effect of OCT-FA addition on elongation at break of compounds of NR

3.3. Abrasion loss

The effect of OCT-FA on abrasion loss of the compounds of SIR 3L-NR is visualized in **Fig. 6**. As shown, the abrasion loss of the SIR 3L-NR samples (in grams) was decreased up to an optimum OCT-FA loading (at a two phr.) and then it was increased with further increasing the OCT-FA loading. The abrasion resistance of the SIR 3L-NR was relatively poorer; the OCT-FA improved the resistance. It was due to a higher degree of wetting and silica dispersion inside the OCT-FA plasticized SIR 3L-NR molecules. A higher degree of silica dispersion caused in both a greater degrees of rubber to filler interaction and abrasion resistance [16]. The rubber to filler interaction is a type of physical crosslink which affects the enhancement in overall crosslink density of vulcanized rubber, significantly [17-19]. The enhancement in abrasion resistance was due to the enhancement in crosslink density of the silica-filled with OCT-FA. The deterioration in abrasion resistance after the two phr. of OCT-FA was due to the excessive effect of the additive which decreased the crosslink density of the NR filled vulcanizates.

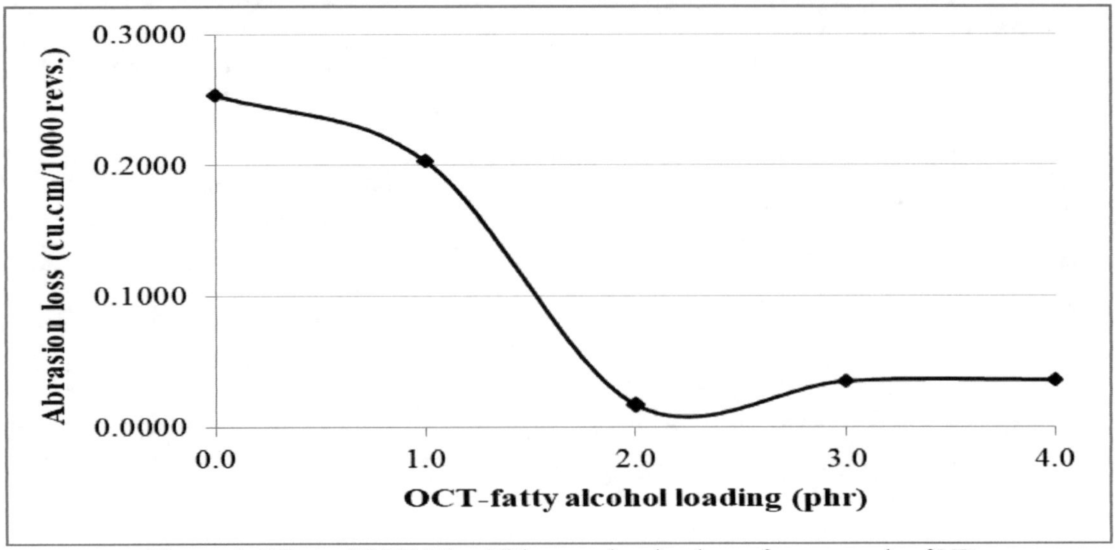

Figure 6. Effect of OCT-FA addition on abrasion loss of compounds of NR

4. Conclusions

The octadecanol-fatty alcohol acted as a co-curing agent for SIR 3L-natural rubber compounds that were filled with silica-filler. It enhanced both times of scorch and cure rate index of the silica-filled SIR 3 L-natural rubber.

The octadecanol-fatty alcohol was also acted as a plasticizing agent. It softened the compounds of SIR 3L-natural rubber and hence, enhanced both the rubber to filler interaction and silica dispersion degree. The octadecanol-fatty alcohol lowered the tensile moduli but increased tensile strength, abrasion resistance and extensibility of the vulcanized SIR 3L-natural rubber.

The enhancements in mechanical properties were due to the crosslink density enhancement as well as cure state enhancement of the compounds of SIR 3 L-natural rubber.

Acknowledgement

Authors appreciate IKD, PT – Medan, Indonesia due to serving the research facilities for conducting the activities of study. Authors thank research grant of PDUPT-2018 with a contract number of No. 062-UN5.2.3.1-PPM-KP-DRPM-2018.

References

[1] Dick J S 2001 *Rubber technology: compounding and testing for performance* Hamser Munich
[2] Fetterman M 1986 *Rubber World* **194 (1)** 38-40
[3] Surya I, Siregar S F and Ismail H 2018 *E3S Web Conf* **34** 01030
[4] Long H 1985 *Basic Compounding and Processing of Rubber* Rubber Division. American Chemical Society Inc USA
[5] Rodgers B 2004 *Rubber Compounding: Chemistry and Applications* CRC
[6] Surya I, Ginting M and Anto J 2018 *AIP Conf Proc* **2024 (1)** 020061
[7] Sianturi R W and Surya I 2018 *J Phys Conf Ser* **1116 (4)** 042033
[8] Andriani F and Surya I 2018 *J Phys Conf Ser* **1116 (4)** 042005
[9] Surya I, Siregar S F and Ismail H 2017 *IOP Conf Ser Mater Sci Eng* **223** 012012
[10] Surya I, Ginting M and Ismail H 2017 *IOP Conf Ser Mater Sci Eng* **223** 012006
[11] Surya I, Ismail H and Azura A 2013*Polym Test* **32 (8)** 1313 - 21
[12] Surya I, Maulina S and Ismail H 2017 *IOP Conf Ser Mater Sci Eng* **299** 012061
[13] Surya I, Sukeksi L and Hayeemasae N 2017 *IOP Conf Ser Mater Sci Eng* **309** 012060
[14] Surya I and Ismail H 2018 *IOP Conf Ser Mater Sci Eng* **309** 012104

[15] Surya I, Ginting M and Ismail H 2018 *MATEC Web Conf* **197** 12005
[16] Byers J T Rubber World 1998 **218 (6)** 38
[17] Hayeemasae N, Surya I and Ismail H 2016 *Int J Polym Anal Ch* **21 (5)** 396 – 407
[18] Surya I and Hayeemasae N 2018 *IOP Conf Ser Mater Sci Eng* **343** 012028
[19] Surya I, Hayeemasae N and Ginting M 2018 *IOP Conf Ser Mater Sci Eng* **343** 012009

The cure and tensile properties of montmorillonite-natural rubber composites in the incorporation of alkanolamide

I Surya[1*] and M Ginting[2]
[1]Chemical Engineering Department, Universitas Sumatera Utara, Medan, Indonesia
[2]Chemistry Department, Universitas Sumatera Utara, Medan, Indonesia

Email: *indradanas@yahoo.com

Abstract. Effects of alkanolamide incorporation on cure and tensile properties of montmorillonite (MMT)-natural rubber (NR) composites were studied. The material of alkanolamide was synthesized from the waste of cooking oil production which was used as a rubber additive and incorporated into the NR compounds at 0, 1, 3, 5, 7 and 9 parts per hundred rubber (phr.). From the results, the alkanolamide decreased the times to scorch and cure but increased elongation at break. It was also found that the alkanolamide improved difference in torque, tensile strength and tensile moduli up to a five phr. of alkanolamide and decreased those properties with further increasing the alkanolamide loading. From the swelling test; the five phr. of alkanolamide loading was the optimum loading which causing the highest density of crosslink of the MMT composite of NR.

1. Introduction

The black and non black fillers are very common rubber additives which are incorporated into rubbers compounds in order to increase the mechanical or tensile properties of the final rubbers goods. The black fillers are carbon blacks (CBs) and non black ones such as $CaCO_3$, montmorillonite (MMT) and precipitated silica. The CB is used to produce some rubbers goods with black colored and non black can produce rubbers products with unwanted black colored. Generally, the mechanical properties of rubbers goods filled with non black are always inferior compared to those of filled with CB. It is due to the surface chemistry of non black fillers which is a more hydrated and polar causing the fillers to be not easy to disperse and react with natural rubber (NR) as a non polar rubber [1].

The MMT, one type of polar fillers, is a nano-sized non black filler. It is very similar to others fillers with micron-sized particles, at relatively higher loading of using, the nano-sized MMT tends to make some bigger aggregates during mixing and causing in a relatively poorer degree of filler dispersion.

As an effort to solve such problems, a special rubber additive which is named as alkanolamide was added into the MMT-filled NR compounds. It is a chemical based on palm oil [2-6] and it was used in order to solve both the low level of mechanical properties enhancement and filler dispersion problems. The alkanolamide is an oily material [7] and oil can be used as a plasticizing agent [8] that improves the degree of filler dispersion.

This research-study investigated effects of the alkanolamide incorporation on the cure and also tensile properties of the composites of MMT-NR.

2. Methodology

2.1. Chemicals
The SMR L grade of NR, MBTS (N-cyclohexyl-benzothiazylsulfenamide), sulphur, IPPD (N-isopropyl-N'-phenyl-p-phenylenediamine), stearic acid and ZnO were supplied by the Rubber Lab of Universiti Sains Malaysia (USM), Malaysia. The alkanolamide, $CH_3(CH_2)_{14}CON(CH_2CH_2OH)_2$ was prepared in the laboratory [3-4].

2.2. Compounding
The semi-EV formulation was applied for the preparation of MMT-NR-alkanolamide composites. The recipe for the making of composites is shown in **Table 1**. The procedures of the making of composites were conducted based on ASTM D-3184-80 using a laboratory two-roll mill.

Table 1. Recipe of the MMT-NR composites

Chemicals	Composition, (phr)
SMR L	100
ZnO	5
MMT	8
Stearic acid	2
Sulphur	1.5
TMTD	1
MBTS	1.5
IPPD	2
Alkanolamide	0; 1; 3; 5; 7; 9

2.3. The cure
The cure data of MMT-filled composites of NR were delivered by operating the MDR 2000 - Rheometer. It was operated for the data of times to cure and scorch and also the differences in torques based on ISO-3417. The composites of NR were vulcanized at a hundred and fifty degrees of Celsius. The composites were compression molded utilizing a stainless steel mold at fifty degree of Celsius, with a 10 MPa of the pressure of a hot press based on their optimum times to cure.

2.4. The swelling percentage
The swelling percentage test was performed in toluene based on ISO-1817. The (30 mm x 5 mm x 2 mm) sized of vulcanized NR was weighed using a balance and was then swollen in toluene for almost 72 hours at room temperature. The samples of composites were cleaned from the residue of toluene and the weight was determined. The change in mass is as follows:

$$\text{Swelling percentage (\%)} = 100\% \, (W_2 - W_1) / W_1 \quad (1)$$

In which W_1 is the beginning mass (gr.) and W_2 is the mass (gr.) after immersion in toluene.

2.5. The tensile properties
The dumbbell-shaped of composites samples were cut from the molded sheets based on ISO-37. Tensile tests were conducted at a crosshead speed of 500 mm/min. Tensile tests were performed on an Instron 3366 tensometer to determine tensile properties as TS (tensile strength), EB (elongation at break) and tensile moduli (M300, M100).

3. Results and discussion

3.1. The cure

The effect of alkanolamide on cure of the MMT-NR composites is shown in **Table 2**. As observed, both of the times to scorch and cure of MMT-NR composite with one phr. of alkanolamide were lower than those of the control composite. The alkanolamide caused improvement in the cure. The higher loading of alkanolamide caused in lowering the scorch and cure times and hence, the greater the affection of cure improvement. It was because of the action of alkanolamide as a cocuring ingredient. In this case, the alkanolamide was an internal accelerator; presumably its amine part increased the rate of cure because amine is one of the accelerator ingredients for rubber compounds. [8-11].

Table 2. Curing data of MMT-filled composites of NR

NR composites	Curing characteristics		
	Scorch, min.	Cure, min.	MH-ML (dN.m)
NR/ Alkanolamide -0	2.79	4.69	5.89
NR/ Alkanolamide -1	2.91	4.98	6.02
NR/ Alkanolamide -3	2.60	4.53	6.69
NR/ Alkanolamide -5	2.16	3.92	6.92
NR/ Alkanolamide -7	1.75	3.20	6.75
NR/ Alkanolamide -9	1.37	2.67	6.70

MH-ML = Max torque minus Min torque

The torque difference or (MH-ML) value of MMT-filled NR composite with one phr. of alkanolamide was higher than that of MMT-filled NR composite with no alkanolamide. The alkanolamide increased the (MH-ML) value. Increasing the alkanolamide loading further increased the (MH-ML) value up to a five phr. of alkanolamide. After the five phr. of alkanolamide, the (MH-ML) started to decrease. The value of torque difference indicates modulus of shear dynamic and it corresponds to composite crosslink density [12-14]. The alkanolamide affected the MMT-NR composites crosslink density. Positively, it was because of the function of alkanolamide as a cocuring agent that improved the cure state.

Decreases in crosslink density beyond the five phr. of alkanolamide loading were because of the excessive effect of alkanolamide which attracted parts of the curatives and as a consequence, a decrease of crosslink density was observed.

3.2. The swelling percentage

Result of swelling percentage test of MMT-NR composites is visualized in **Fig. 1**. Swelling percentage corresponds to the composite crosslink density [15-17]. The toluene with a less penetration into composite means a higher crosslink density. The incorporations up to five phr. of alkanolamide decreased swelling percentage/crosslink density of MMT-NR composites and after the five phr. of alkanolamide loading causing increases swelling percentage/crosslink density. Increases in percentage of swelling were because of excessive amount of alkanolamide that decreased crosslink density [15-17].

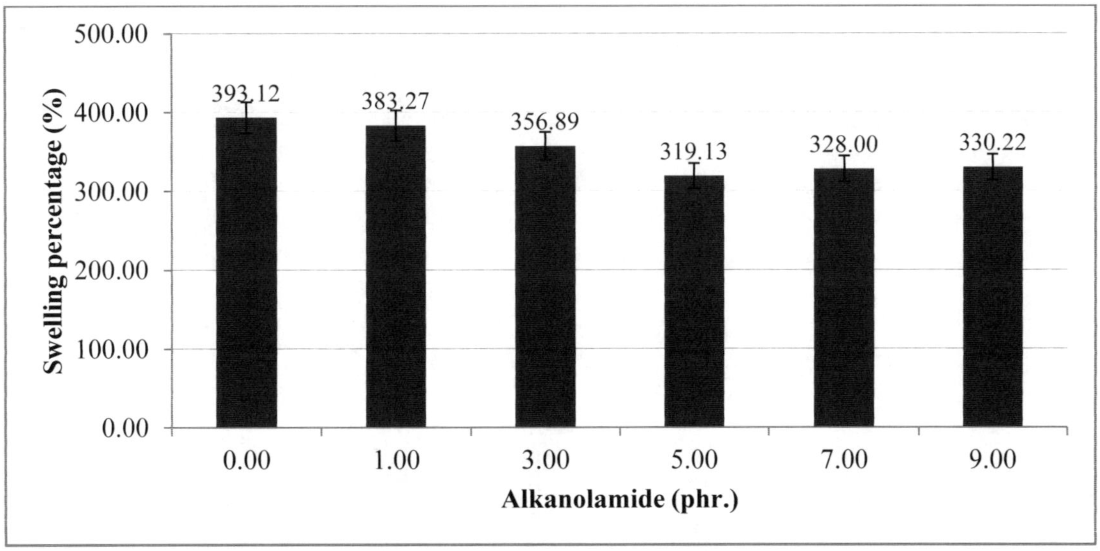

Figure 1. The effect of alkanolamide on swelling percentage of MMT-filled NR composites

3.3. The tensile properties

The TS, tensile moduli and EB of the MMT-NR composites without and with alkanolamide are visualized in **Figs. 2-4**. As observed in **Fig. 2**, alkanolamide affected the EB of the MMT-NR composites. The one phr. of alkanolamide increased EB of the control composite (NR composite with no alkanolamide). The increases the alkanolamide loading caused further increases in the EB of NR composites. It was because of to the action of alkanolamide as a plasticizing material which modified EB of NR composites [18].

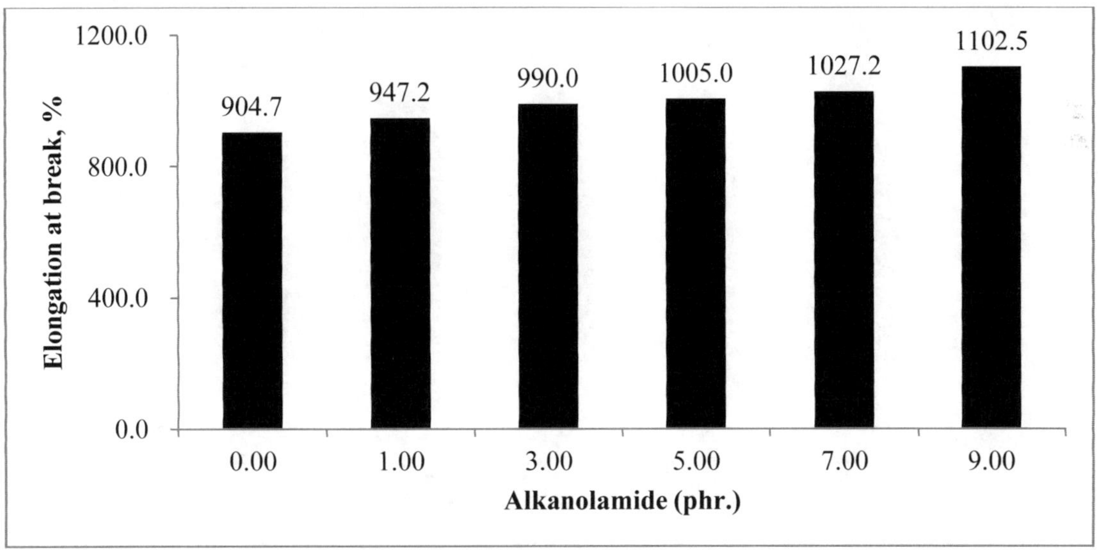

Figure 2. Effect of alkanolamide on elongation at break of MMT-filled NR composites

Incorporation of alkanolamide affected the tensile moduli (M300 and M100) and TS of the MMT-filled NR composites. **From Fig. 3**, the alkanolamide incorporations up to a five phr. improved the

M300 and M100 and further increasing the alkanolamide loading started to decrease those properties. The results of TS showed a similar trend as can be seen in **Fig. 4**. The tensile moduli and TS relate to composite crosslink density [2-4]. The increases in those properties were because of a higher of crosslink density degree (higher torque difference indication in **Table 2**). The decreases in those properties were because of excessive amount of alkanolamide that decreased the composite crosslink density (lower torque difference indication in **Table 2**). See the previous discussion, the crosslink density degree of a composite can be indicated by its torque difference value. A higher in torque difference means a higher is crosslink density degree.

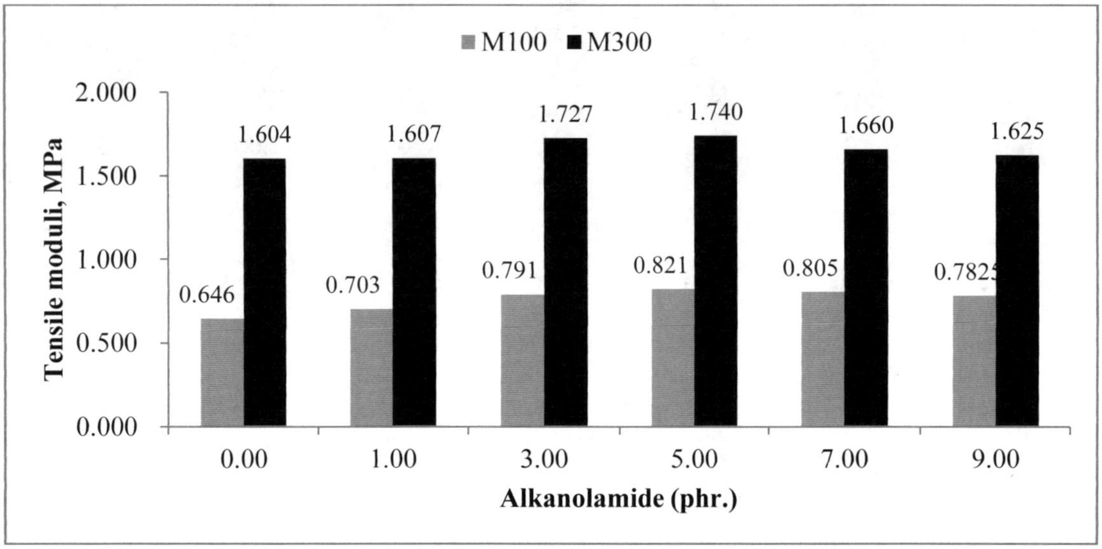

Figure 3. Effects of alkanolamide on tensile moduli of MMT-filled NR composites

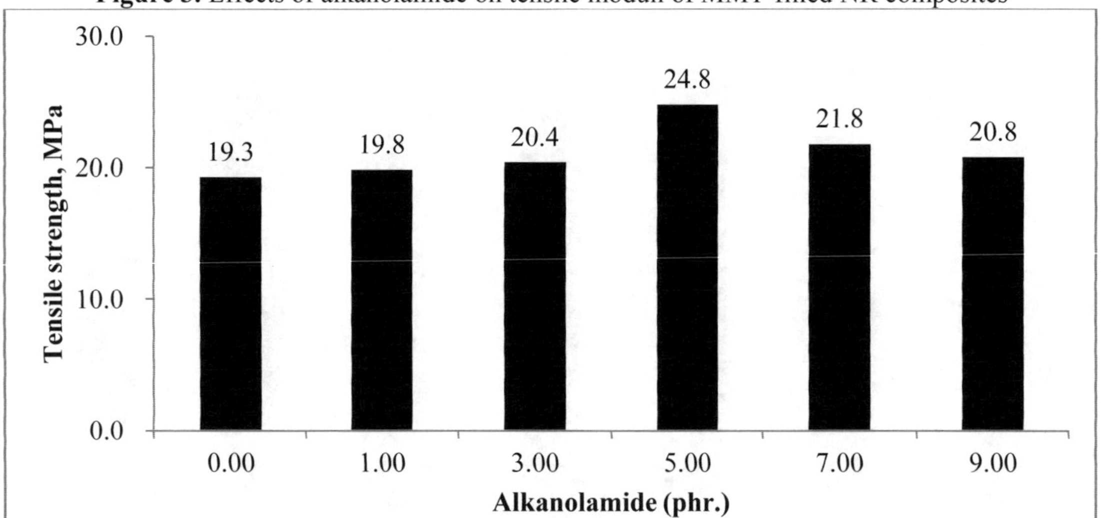

Figure 4. Effects of alkanolamide on tensile strength of MMT-filled NR composites

4. Conclusions

Alkanolamide was a cocurative and internal plasticizing agent in montmorillonite-filled composites of natural rubber. Alkanolamide caused in the decreasing the times to scorch and cure but increasing the elongation at break or extensibility. Alkanolamide improved crosslink density degree and tensile properties of montmorillonite-filled composites of natural rubber up to five parts per hundred rubber of loading.

Acknowledgement
The authors thank the Rubber Lab of Universiti Sains Malaysia-USM, Malaysia for serving research facilities for performing experimental activities.

References

[1] Surya I, Ginting M and Ismail H 2017 *IOP Conf Ser Mater Sci Eng* **223** 012006
[2] I. Surya and H. Ismail: Plast Rubber and Compos Vol. 45 (2016), p. 287
[3] I. Surya, H. Ismail and A. Azura: Polym Test Vol. 32 (2013), p. 1313
[4] I. Surya, H. Ismail and A. Azura: Polym Test Vol. 40 (2014), p. 24
[5] I. Surya, H. Ismail and A. Azura: Polym Test Vol. 42 (2015), p. 208
[6] I. Surya and H. Ismail: Polym Test Vol. 50 (2016), p. 276
[7] I. Surya, M. Ginting and H. Ismail: IOP Conf. Ser., Mater. Sci. Eng Vol. 223 (2017), 012012
[8] B. Rodgers: *Chemistry and Applications* (CRC, 2004).
[9] Surya I, Ginting M and Anto J 2018 *AIP Conf Proc* **2024 (1)** 020061
[10] Sianturi R W and Surya I 2018 *J Phys Conf Ser* **1116 (4)** 042033
[11] Andriani F and Surya I 2018 *J Phys Conf Ser* **1116 (4)** 042005
[12] Hayeemasae N, Surya I and Ismail H 2016 *Int J Polym Anal Ch* **21 (5)** 396 – 407
[13] Surya I and Hayeemasae N 2018 *IOP Conf Ser Mater Sci Eng* **343** 012028
[14] Surya I, Hayeemasae N and Ginting M 2018 *IOP Conf Ser Mater Sci Eng* **343** 012009
[15] Surya I, Sukeksi L and Hayeemasae N 2017 *IOP Conf Ser Mater Sci Eng* **309** 012060
[16] Surya I and Ismail H 2018 *IOP Conf Ser Mater Sci Eng* **309** 012104
[17] Surya I, Ginting M and Ismail H 2018 *MATEC Web Conf* **197** 12005
[18] Surya I, Maulina S and Ismail H 2017 *IOP Conf Ser Mater Sci Eng* **299** 012061

Study of Desiccant Rice Husk Ash for Air Conditioning System

Syahbardia[1*], Berkah Fajar TK.[2], SH. Winoto[3], MSK.Tony Suryo[4], and L.Abdurrachiim[5]

[1,5]Mechanical Engineering Department, Universitas Pasundan, Bandung, Indonesia
[1,2,3,4]Mechanical Engineering Department, Universitas Diponegoro, Semarang, Indonesia

*Email: syahbardia@student.undip.ac.id

Abstract. Indonesia is one of the producers and consumers of paddy or rice due to more than 95% of the Indonesian population consume rice as staple food. The by-product of paddy processing as to rice is also hush or rice hush. Up until now, in Indonesia, it is widely used as animal feed, fertilizer, mushroom planting media and fuel for making bricks, tiles and others. To increase the benefits of rice husk, the study of the usage of rice husk as an active ingredient in the SDC cooling system was conducted. The research method begins with the making of rice husk test samples and silica gel as a comparison. Rice husk was burned and heated using an oven until all carbons became ash. Then, the rice husk ash was blended and filtered, these applies to silica gel as well. Then these two samples are tested for characteristics with GSA and XRF instruments. From this test it can be concluded that the adsorption / desorption ability of ash is 1/6 time silica gel. If rice husk ash is used as an alternative for desiccant materials, it is necessary to modify the material.

1. Introduction

Indonesia is one of the producers and consumers of paddy or rice due to more than 95% of the Indonesian population uses rice as staple food. The by-product of paddy to rice, is hush or rice hush. Indonesia's dry paddy production in 2018 was estimated around 82 million tons. Rice mills produce 72% of paddy, 5-8% of bran, and 20-22% of rice husk. If the content of Hush is 20%, then rice production produces around 16.4 million tons per year. So far, rice husk is a by-product of agriculture. If there is no use, it can become soil pollution. Up until now in Indonesia, it is widely used as animal feed, fertilizer, mushroom planting media and fuel for making bricks, tiles and others.

Today there has been a lot of research that has been done about the usage of rice husk as an economical raw material in making silica. Because manufacturing silica from quartz sand, may causes damage to the environment as well as global warming effect. The melting of quartz sand into sodium carbonate requires a furnace temperature of 1300 ◦C [1].

SDC cooling system (Solid Desicant Cooling), is one type of AC machine that uses silica gel as a natural active ingredient in energy and mass conversion in cooling systems. The advantages of *SDC* silica gel compared to silica gel from quartz sand are: (1) Increase the usefulness of rice husk, (2) The process of making silica gel will be more economical, (3) Reducing global warming and soil pollution,

(4) Raw materials are abundant, and (5) Some treatments to rice hush in the process of making silica gel, may improve product quality.

The adsorption / desorption phenomenon was discovered more than two centuries ago by C. W. Scheele in 1773 and by F. Fontana in 1777. In 1785, they discovered that when they heat the charcoal contained in the test tube the gas is decomposed. The gas is then adsorbed again when the charcoal is cooled [2]. Adsorption is classified into (1) physical absorption, where attraction that occurs between the adsorbate molecule and the porous surface of the material, this is the Vander Waal style. In physical adsorption there is a weak pull between vapor or gas and the surface of the porous material, then the type of adsorption can be easily reversed ie reversible; and chemical absorption, where attraction between steam or gas and the surface of a porous material has the same strength as a chemical bond, this type of absorption is called chemical adsorption. In chemisorption, the attraction is so strong that chemisorption adsorption cannot be easily irreversible [3,4].

Adsorption isotherms measured on various gas-solid systems have various forms, and can be grouped into one of six types according to the International Union of Pure Applied Chemistry (IUPAC) Classification 1994 [5] with the majority produced from physisorption, Fig. 1 showing six types of isotherms, which only apply to the adsorption of a single component gas in a condensable range, and are very useful for studying porous materials.

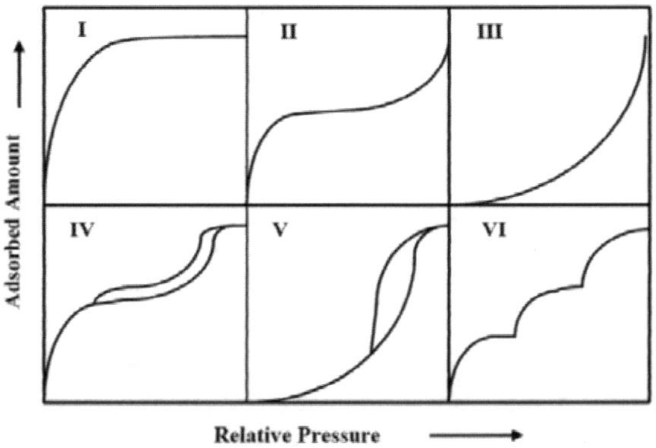

Figure 1. Six types of isotherms, based on the IUPAC classification standard [6]

Type I represents a highly hydrophilic material with a steep increase in gas absorption at very low relative pressure (P / Po). Substances with type II or IV isothermic forms are also classified as hydrophilic materials. For these materials, the water adsorption capacity is very high measured at low and medium P / Po. In addition, hydrophilic materials include those which exhibit unusual type-like isothermic types. In contrast, materials with type III isotherms are recognized as hydrophobic or low hydrophilic, indicating low water adsorption at low P / Po and a sudden increase in water adsorption at high P / Po. Similar to type III isotherm, type V isotherm also shows low water adsorption at low P / Po. However, the sigmoidal or S-shape curve on medium or high P / Po is studied further, this shows an increase in sudden water absorption between narrow P / Po. The optimum desiccant for the SDC system must be like type V isotherm. This is because materials that have type I, II or IV isotherms are more difficult to regenerate because they are more hydrophilic and have a higher amount of water adsorption at low P / Po. In addition, water adsorption occurs in P / Po which is too high for material that has type III isotherm [7]. Within the SDC system, the active material desiccant is a material that has properties, capable of absorbing as much water vapor as possible by adsorption, if it is in the air flow containing water vapor and is able to release absorbed water vapor (desorption) if it is in the womb low water vapor. Thus desiccant has properties as a material capable of alternating adsorption and desorption (regeneration) of steam, which is ideally reversible. The characteristic parameters of a solid desiccant material can be concluded are (1) Chart of adsorption / desorption, (2) The main dimensions of the desiccant pore, and (3) Composition of desiccant compounds. This diagram is also known as the isotherm diagram. From sub-chapter of types of adsorption isotherm, the types have been explained. For solid desiccants that have good adsorption / desorption characteristics are types II, III and IV. For type I the adsorption power is very good but the desorption is very low.

The adsorption / desorption characteristics can be known easily using the Vapor Sorption Analyzer (VSA) or Gas Sorption Analyser (GSA). VSA is very suitable to be used for the analysis of moisture adsorption / desorption properties in solid materials. While GSA generally applies to know the characteristics of adsorption / desorption of various kinds of gases on various adsorbent media. The main difference between the two types lies in the adsorbent media. DVS uses steam + nitrogen gas as an adorbent medium. GSA instruments use liquid nitrogen and gas media as adsorbent media. Based on the specifications of the Silica Gel Standard JISS-0701, the characteristic parameters are: Density, Surface Area, Pore Volume and pore diameter. This parameter greatly influences the desiccant's ability in gas adsorption and desorption characteristics. Pore diameters based on the International Union of Pure Applied Chemistry (IUPAC) Classification 1994 [627] are divided into three groups, namely (1) Micropore material: 0.2-2 nm, (2) Mesopore material: 2-50 nm, and (3) Macropore material: 50-1000 nm. Desiccants that have the size of most micropores, generally have characteristic adsorption isotherms curves such as type I. Examples are some types of activated carbon material included in this type. The advantages of this type of gas adsorption power and solid particles are very high and the adsorption capacity is large because of the large porous volume. But the ability of desorpsis very low. This type is suitable for use in media to reduce the levels of toxic gases and particles in air and water. This media is used once, it cannot be regenerated.

Mesopore and pore macro materials have the properties of type IV and V isotherm curves, which have a high adsorption and desorption power. This can be seen also from the curve there is a hysteresis area. But if the amount of adsorption power and desorption is equal, hysteresis on the isotherm curve is not visible, even though the desiccant material has desorption properties. The hysteresis area that is too wide is also not well used as an active medium in SDC, because it shows that the desiccant has low regeneration properties. The vapor / gas absorbed by the desiccant remains in the pore of the media, even though it has passed through the regeneration phase. Some types of desiccants also have an enlarged hysteresis area over time. Indeed, the desiccant material has a useful life. One indication of the ability of adsorption / desorpsion is that the hysteresis area is enlarged. The following table 1 is the main dimension of pore desiccant for standard silica gel and silica gel in many markets.

Table 1. The main dimensions of the desiccant pore standard JIS and Non-Indicating Silica Gel

Adsorbent	Surface Area (m^2/g)	Pore Volume (cm^3/g)	Average pore diameter (nm)	Average pore width (4V/A)	Spesific weigh (g/cm^3)
Silica gel standard JISS-0701	650	0.36	2.2	2.21	0.7
Non-Indicating Silica Gel	600 - 800	0.35 – 0.45	2-3	-	0.7-0.77

In general, solid desiccant materials based on their sources can be divided into two, namely natural and synthetic types. The first type which is already available in nature can be directly utilized, while the second type needs to be synthesized with raw materials from nature. Adsorptivity / desortivity of a solid material is determined by its surface structure. The surface of a solid desiccant containing pores is one of the main factors that determines adsorptivity / desortivity. The more pores formed, the higher the adsorptivity / desortivity. Rice husk on the surface has pores. The amount of pore can be increased, one of which is by combustion into rice hush ash. This combustion not only changes the surface structure but also the composition of compounds formed in solid desiccants. One parameter that affects adsorptivity / desortivity is the composition of the compound. The composition of the rice husk ash compound is also determined from the source or location of the rice field. To determine the composition of desiccant compounds XRF was tested.

The purpose of this research is therefore to explore the feasibility of rice husk that is used as an active ingredient of solid wheel desiccant on cooling machines of Solid Desiccant Cooling type. The desiccant performance of rice husk is compared to desiccant silica gel that is produced from quartz sand. This study uses raw materials for making silica gel samples from rice husk obtained from the rice fields in Sumedang district, west java, Indonesia. Characteristics testing uses laboratory facilities at: ITB, Unpad and UNY. The production of silica gel from rice husk is done simply at the Chemical Laboratory of Unpas Food Engineering Department. Pure silica gel that is used for comparable material, comes from many that are sold freely.

2. Experimental Method

Two different samples, namely the RRC silica gel which was widely marketed, but the specifications were not known. This sample is used as a comparison for the second sample from the burnt straw. This rice husk from the rice fields of Sumedang district (West Java). Stages of making rice husk ash samples include (1) Rice Husk gradation (sieving, filtering) by using 20-200 mesh sieves. The main part, which usually consists of a size of 20 mesh, is used for experiments, (2) Washing and drying with distilled water for the purpose of cleaning and drying the sun for 48 hours.

The characterization of desiccant materials uses two instruments, namely: Gas Sorption Analyzer (GSA) and X-Ray Fluorescence (XRF). GSA is used to determine the adsorption / desorption properties and the main dimensions of the desiccant pore sample material (pore surface area, pore diameter and pore volume). While XRF is used to determine the composition of desiccant compounds.

3. Results and Discussions

3.1. Adsorption / Desorption

Figure 5 illustrates the isothermic adsorption / desorption diagram N_2 at -196oC on SG (silica gel) and RHA (rice hull ash). The adsorption / desorption power of silica gel is much higher than that of rice husk ash. The hysteresis curve shows that more N_2 ash is left in the media pore compared to silica gel during the desorption process. Silica gel is more hydrophilic than ash. The silica gel isotherm curve seems to be included in the mixed type in the IUPAC classification; type II and type IV, while the isotherm curve of rice husk ash is a mixed type III and type IV. Thus, from the adsorption / desorption diagram, silica gel is better than rice husk ash as an SDC cooling system.

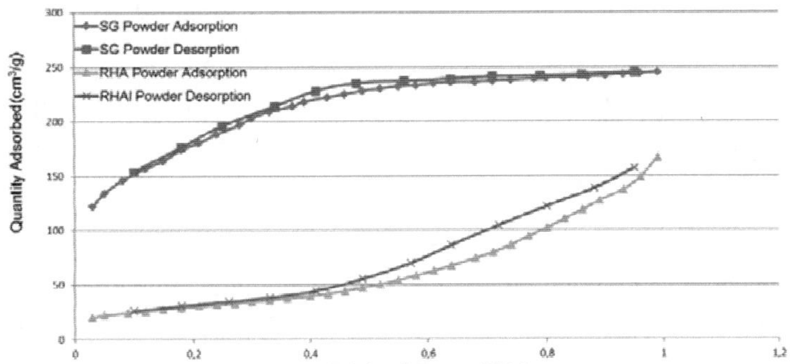

Figure 2. Graph of N_2 adsorption-desorption isotherm from silica gel and rice husk ash

3.2. Analysis of Pore Structure

To compare the texture properties of the adsorbent, the parameters obtained from the BET surface area, pore volume, pore diameter and micropropicity are tabulated in Table 2. This table shows that the porous surface area of silica gel powder is 6x the surface area of pore rice husk ash, powder

Table 2. Characteristics of the pore texture of the adsorbent

Adsorbent	Surface Area (m^2/g)	Pore Volume (cm^3/g)	Average pore diameter (nm)	Average pore width (4V/A)	Spesific Weigh (g/cm^3)
Silica Gel RRC	632	0.38	2.4	2.41	0.7
Rice husk ash	106	0.26	9.7	9.81	0.7

Rice husk ash pore volume is 2/3 parts volume of silica gel pore. The average diameter of the silica gel pore is 3x the average diameter of the porous ash. But based on pore size, silica gel and rice husk ash are included in the mesopore group

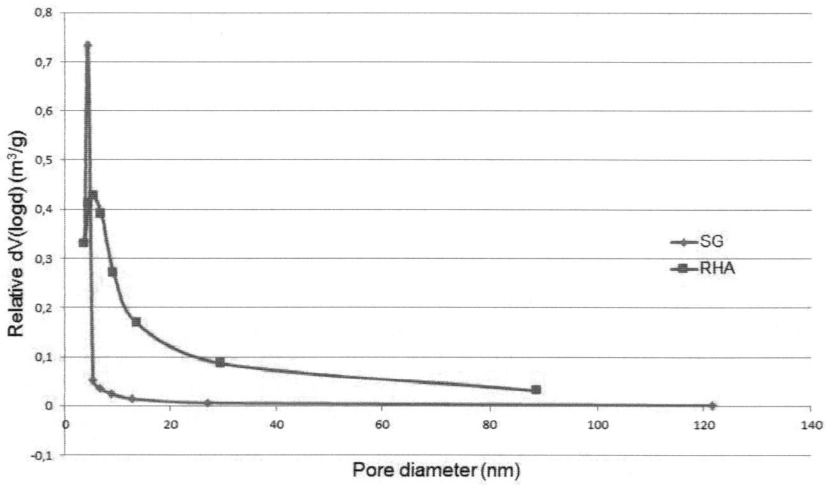

Figure 3. Pore size distribution desorption

The pore size distribution of silica gel and rice husk ash samples can be seen in Fig. 6. calculated from the N_2 adsorption isotherm following the method Barrett-Joyner-Halenda (BJH) with calibration. This is clearly shown in Fig. 6, Silica gel reveals a relatively narrow pore distribution centered around 2.4 nm, while the ash distribution is relatively wide at around 9.7 nm.

3.3. Desiccant Compound Composition Analysis

X-ray fluorescence (XRF) is an analytical technique that can analyze the compounds that make up a solid material. Nowadays with the development of XRF instrument technology, there is already a portable type, and its use is easier.

We are testing the composition of silica gel and ash compounds using ITB Mining Engineering laboratory facilities.. Composition of adsorbent compounds can be seen in table 3. From the table it can be seen that the purity of silicon oxide in rice husk ash is lower than silica gel. Rice husk ash content of silicon oxide is 94.95% while silica gel is 99.83%.

Table 3. Composition of the adsorbent compounds Silica gel and rice husk ash

Components expressed as oxides	SiO_2	Al_2O_3	Fe_2O_3	CaO	Na_2O	K_2O	MnO	TiO_2	MgO	P_2O_5
Silica Gel RRC	99.83	0.61	0.02	0.54	0.25	0.25	0.94	0.16	0.04	0.13
Rice husk ash	94.95	0.39	0.26	0.54	NA	0.02	0.94	0.16	0.02	0.9

The shape of the surface texture of the adsorbent material is very determined from the purity of silicon oxide and the temperature of treatment [8]. Surface texture parameters: pore surface width, pore diameter, pore distribution and pore volume. High purity adsorbents, wider pore surface and, average pore size including mesoporous groups. Temperature treatment process too high, above 1000 °C the phase structure formed is crystal. Crystal structure is not desirable, because the surface textures have no pores.

The results of the analysis showed that the rice husk ash or RHA in this study showed that the properties of the adsorbent were lower compared to silica gel. If it is used as a desiccant media on SDC its performance is 1/6 (if seen from the pore surface area) the performance of silica gel. Thus, technically, rice husk ash without treatment is not suitable to be used as SDC desiccant media. Some research results in the form of synthesis or modification of RHS [8-10] can improve the performance of the adsorbent. Thus the RHS is potential as a substitute for silica gel for desiccant media in SDC cooling systems.

4. Conclusion

Rice husk ash which is processed simply without a specific treatment, it is concluded that it has not yet fulfilled the technical requirements as SDC desiccant media. The absorptivity performance of rice husk ash can be increased in several ways, namely: treatment with the aim of increasing the purity of silicon oxide and making composite desiccant with basic ingredients of rice husk ash.

References

[1] Priombodo A 2008 *The Usage of Rice Husk Waste in Silica Gel Production* Bechelor Thesis FTUI Depok Indonesia.
[2] Aranovich G, and Donohue M 1998 *Journal of Colloid and Interface Science* **200** 273-290.
[3] Meunier F 2002 *Proceedings of the International Sorption Heat Pump Conference (Shanghai)* (Shanghai: Science Press).
[4] Kluppel RP and Gurgel JMAM 19 *Sunworld* **12** 113-4.
[5] Rouquerol J, Avnir D, Fairbridge CW, Everett DH, Haynes JM, Pernicone N, Ramsay JDF, Sing KSW and Unger KK 1994 *The Scientific Journal of IUPAC* **66** 1739-1758.
[6] Tanaka S, Hasegawa T and Waki T 1983 *I.S.E.S. Solar World Congress (Perth)*.
[7] Bakar RA, Yahya R and Gan SN 2016 Procedia Chemistry **19** 19-195.
[8] Nawaz K, Schmidt SJ and Jacobi AM 2016 *International Refrigeration and Air Conditioning Conference* (Purdue University-Purdue e-Pubs).
[9] Fernandes IJ, Calheiro D, Sanchez FAL, Camacho ALD, Rocha TLAC, Moraes CAM, Sousa VC 2017 *Materials Research* **20** 512-518.
[10] Zheng X, Ge TS and Wang RZ 2014 *Energy* **74** 20-294.

Journal of Physics: Conference Series **1373** (2019) 012025 doi:10.1088/1742-6596/1373/1/012025

The Effect of Dust on Solar PV System Energy Output Under Urban Climate of Surabaya, Indonesia

Elieser Tarigan[1,*]

[1] Department of Electrical Engineering and Center for Renewable Energy Studies, PSET, University of Surabaya, Jl. Raya Kalirungkut, Surabaya 60292, Indonesia

*Email: elieser@staff.ubaya.ac.id

Abstract. Solar Photovoltaic (PV) electricity is one of the most effective, eco-friendly, and sustainable source of energy. The amount of electricity produced by a PV system is directly affected by the level of solar irradiation incident to the PV panels. In the real operation, the dust from the surrounding environment falls and accumulates on PV panels. The accumulation of dust causes a decrease in the energy output of the PV system. This study was aimed to investigate the impact of dust on PV operation to its energy output under the urban climate of Surabaya, Indonesia. The study was conducted by comparing two identical PV panel systems, each of 300 Wp. For one system, the PV panels were manually cleaned regularly for the dust, while for another system is remained without any cleaning. Both systems were measured for their energy output, and the results are compared. The results showed that, during the dry season in Surabaya, the energy output from the cleaned PV panels system is higher by 20% in comparison with the uncleaned PV panels system. While, during the rainy season, both systems produce the same amount of energy. In total throughout the year, cleaning of PV panels system would increase the energy output of about 8%.

1. Introduction

The use of solar energy can reduce the need for fossil energy and in turn reduce the number of greenhouse gasses being released into the atmosphere. Solar energy which is available and renewable in the form of the photon can be converted into electricity form directly using devices named photovoltaic (PV) system. The efficiency of energy conversion is affected by many factors such as temperature, radiation, and dust. The presence of dust on the surface of PV modules significantly affects the performance of a PV system[1,2].

A quite number of studies on the subject of dust impact of PV system had been reported from the different site of the world with different climates [2-6]. Study of dust effect on photovoltaic module for photovoltaic for various type of PV modules in Thailand was reported by Ketjoy and Konyu [1]. The results were used to derive a mathematical model for the local site. A study was reported by Mejia, et. al.,[5] from the investigation of one year of power output from a PV site in Santa Clara, CA demonstrated how soiling decreases the efficiency of solar PV plants. Soiling effects were found to strongly depend on the dry exposure time of the panel. Soiling losses have their largest impact during the long dry summers. The losses caused by the accumulation of dust were estimated to be -0.0021 per day in relative solar conversion efficiency.

The effect of the dust depends on the local area where the PV system is mounted, therefore it might difficult to apply a general model or calculation in all cases [3]. From the literature review, it was found that a few studies for this subject under the condition of Indonesia have been reported. Wibowo et., el.[7] reported a study on the cleaning system of PV panel, however, it focused on the device itself instead of energy gain from the cleaning of panels. A similar study reported by Agus et. al., [8] where the cleaning system device was designed to be operating using wifi and SMS.

At the time of conducting studies in this paper, none of research has been found in the literature regarding the dust effect on the output of PV systems in Surabaya. The objective of this study, therefore, is to investigate the effect of dust on the energy output of solar PV system deployed in the specific area of Surabaya, East Java, Indonesia. As mentioned above, the impact of the dust depends on the local area where the PV system is mounted. As situated in around the equator line, i.e., 07° 19' 17.83" South and 112° 46' 3.19" East, Surabaya has only two major seasons around the year, dry and rainy season. The dry season commonly occurs from July to October, while October – July is usually rainy season. The annual temperature varies from 23 – 36°C, and wind speed of between 5 – 19 km/h [9]. The tested PV system in this study was installed on the roof of a house and there was no any other higher objects or buildings surround the system that could shade the panels from the sun. Based on the effect of dust on PV modules every season, the impact of action for cleaning modules is analyzed. The results and information from this study are expected to be useful for PV application development for larger scale in Indonesia.

2. Methods

Two identical PV systems named system 1 and system 2, each 400 Wp, are mounted side by side on the roof of a building. Both systems are identical in term of the system components, size, and configuration. Each system consisted of PV modules, a solar charge controller (CC), battery, and inverter. The energy produced by the PV system is used daily for powering electricity appliances. The system configurations are as shown in Figure 1, while the specification of the components is shown in Table 1. The energy produced by each system is recorded using a power meter with data logger attached/connected right at the PV modules output (before the charge controller).

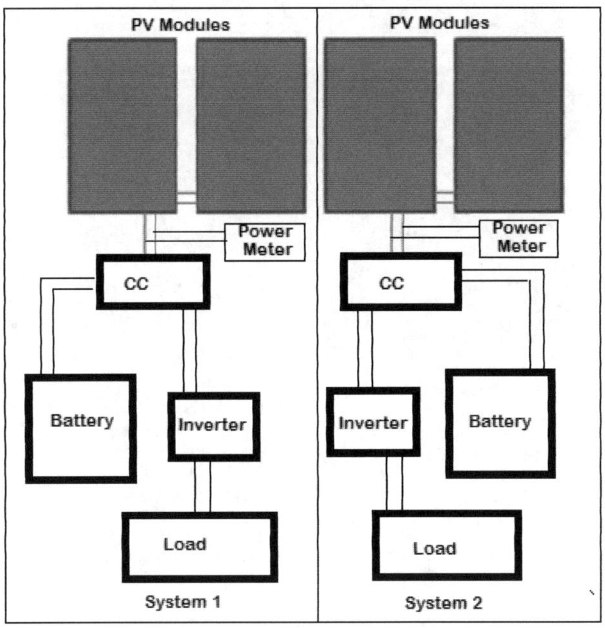

Figure 1. Experimental PV systems configuration

Table 1. System component and specification

Component	No	Brand	Model	Size	Other
PV Modules	2	I- Solar 1	SPU-180M	150 W Max Power	Mono Crystalline Module
Charge controller (CC)	1	S series	SC-20	Load Current 20 A	Self-consumption 6 mA
Battery	2	CIT BAT	CT12 100FR	12 Volt 100 Ah	No Data Sheet Found
Inverter	1	Intelligent	SP-1500	1500 W /Surge 3000 W 24 V DC => 220 V AC	Modified Sine Wave

During the dry season when the intensity of dust from surrounding commonly high, the PV modules for system 1were regularly and manually cleaned weekly using water, while PV modules for system 2 were left without any cleaning treatment (naturally). During rainy season both systems do not need any cleaning treatment as the PV modules were naturally cleaned by rainwater. Data for energy output were collected for both rainy and dry season throughout the year 2018, Monday – Sunday from 06.00 – 18.00. Figure 2 shows the photograph the typical appearance of PV modules for both systems during a dry season.

Figure 2. Photograph of modules system 1 and system 2 during the dry season.

3. Results and Discussion

During the rainy season, the result of measurements for power and energy output from both system 1 and system 2 are similar. The PV modules for both systems were naturally cleaned by rainwater. The surface of the modules is relatively clean, and it can be clearly visible. In this period, the impact of dust on the power output of the PV systems is considered very small. Solar irradiation during the rainy season varies between 2.0 – 4.5 kWh/day.m^2 and the daily energy output ranged from 500 Wh to 1300 Wh per day.

Table 2. Measurement results during the dry season

Day no.	Output Energy (Wh)	
	System 1	System2
1	1,279	1,097
2	1,261	1,083
3	1,285	1,028
4	1,097	799
5	1,353	1,040
6	1,286	997
7	1,240	1,040
8	1,300	1,050
9	1,305	1,005
10	1,286	1,011
11	1,290	1,003
12	1,275	1,010
13	1,099	809
14	1,311	1,000
15	1,120	845
16	1,089	834
17	1,350	1,152
18	1,300	1,092
19	1,288	1,069
20	1,199	973
21	1,245	1,016
22	1,190	889
23	1,245	1,039
24	1,278	981
25	1,300	1,069
26	1,353	1,107
27	1,210	930
28	1,098	792
29	1,088	888
30	1,050	852

Table 3. Measurement results during the rainy season

Day no.	Output Energy (Wh)	
	System 1	System 2
1	915	918
2	834	838
3	756	753
4	1,112	1,111
5	856	853
6	743	745
7	619	622
8	638	640
9	723	728
10	472	475
11	629	636
12	721	724
13	890	894
14	567	562
15	638	636
16	567	564
17	719	714
18	900	893
19	1,035	1,038
20	1,120	1,125
21	589	587
22	893	896
23	900	906
24	500	507
25	673	675
26	598	602
27	890	888
28	783	779
29	590	585
30	800	804

During the dry season, the daily solar irradiation is relatively high and constant in Surabaya, i.e., within 4.2 – 5.8 kWh/day.m^2. With this level of radiation, system 1 (cleaned modules) produced electricity ranged from 1200 – 1400 Wh/day. During 15 – 20 days of the first month of the dry season, the energy output by system 1 (cleaned modules) relatively similar to the system 2 (uncleaned modules). The dust commonly falls onto the modules gradually, and it is accumulated day by day. The measurement indicated that the impact of the dust was significantly affected the energy output after about 25 – 30 days of dust accumulation without any rain. Therefore, during the dry season, whenever possible, a cleaning process for a PV system module in Surabaya is recommended at least once per month to lower energy lost by dust impact. In this study, however, the cleaning process is applied every week (for system 1) to investigate the impact of dust on the energy output from the PV system. Table 2 shows the daily energy output for both systems during 30 consecutive days of measurement results, recorded from 0.6.00 – 18.00 during fully month of the dry season (August 2018), while Table 3 shows the results for 30 consecutive days of the rainy season (February 2018).

The impact of the accumulated dust was significant and relatively stable (the daily energy lost was relatively constant) for system 2 after 30 days of the dry season until the next rainy season arrived. The comparison of output energy by system 1 and system 2 for 6-consecutive days during the dry season in the form of the bar chart is shown in Figure 3. The data was from measurements during the days after 2 months of the dry season. Calculation results showed that during the dry season in Surabaya, the energy output from the cleaned PV panels system is higher about 20% in comparison with the uncleaned PV panels system as can be seen day by day in Figure 3.

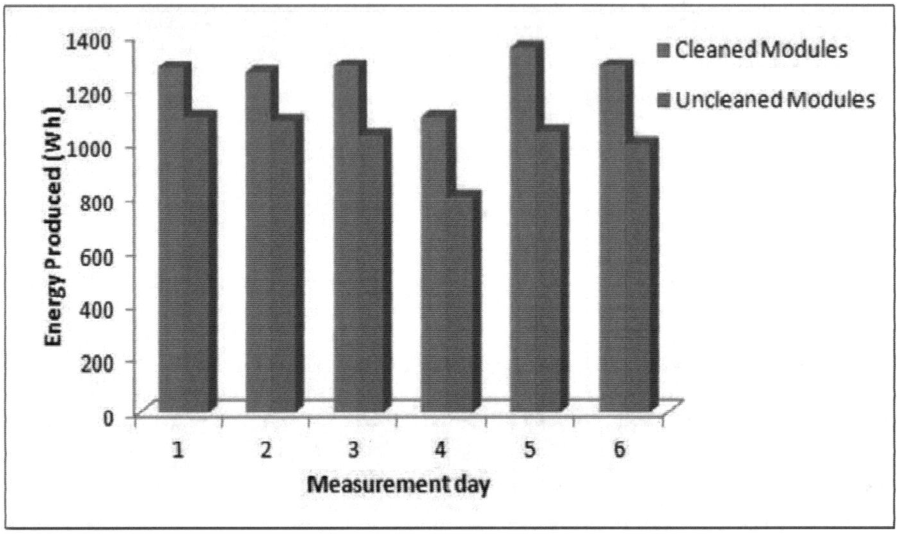

Figure 3. Daily energy output by system 1 (cleaned) and system 2 (uncleaned)

The daily energy output from PV systems is from the accumulation of energy produced since morning to evening, i.e. the accumulation of the power produced by time. The comparison of power output between system 1and system 2 for one day is shown in Figure 4. During higher solar radiation, the power output by system 1 was reached 296 Watt, at the same time system 2, produces power about 170 Watt. However, for lower radiation the different power output by both systems was lower, means that the impact of the dust is significant during high solar radiations.

As previously mentioned, the dry season in Indonesia, including in Surabaya where this study was taken place, commonly occurs from July to October, while October – July is commonly rainy season. However, in the recent past years, the dry or rainy seasons became more unpredictable. A particular study should be attempted to find the reasons for this situation. A rainfall intensity for a typical year (2018) in Surabaya, taken from Meteorology, Climatology, and Geophysical Agency (Indonesian: Badan Meteorologi, Klimatologi, dan Geofisika, abbreviated BMKG) [10] is shown in Figure 5.

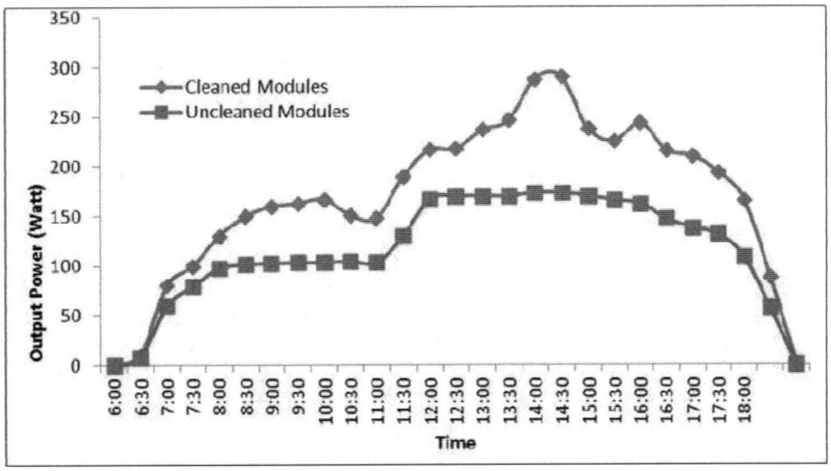

Figure 4. Daily power output by system 1 and system 2

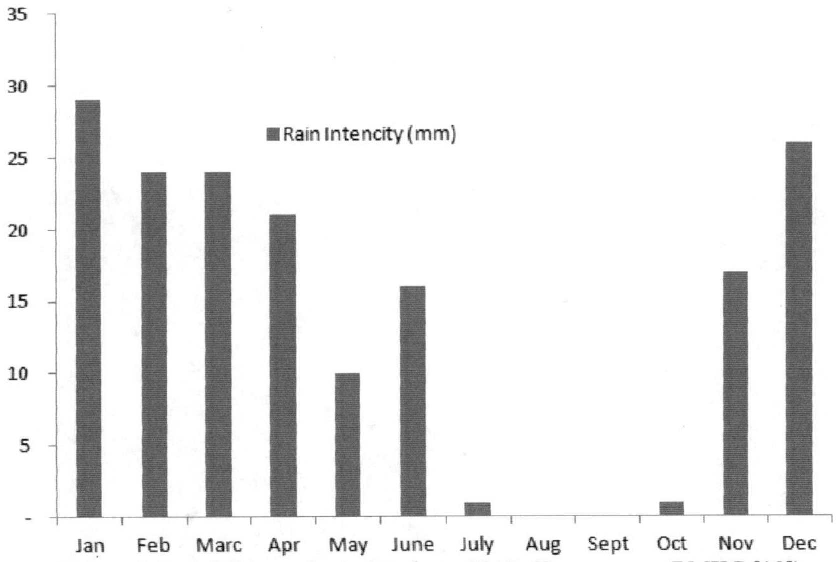

Figure 5. Rainfall intensity in Surabaya 2018. (Data source: BMKG [10]).

With the reference of data in Figure 5, it can be seen that rain occurs from November to June (with different intensity). During this period the dust would not impact the PV system power output. However, a long dry season occurs during June – November, means that the dust will impact on PV system for about 4 months. Using this scenario, cleaning of PV panels system would increase the energy output of about 8% in total for one year.

4. Conclusions

The presence and accumulation of dust on the PV modules significantly affect the performance of a PV system. The dust commonly falls onto the modules gradually, and it is accumulated day by day. It was found that dust was significantly affected the energy output after about 25 – 30 days of dust accumulation without any rain. Therefore, during the dry season, whenever possible a cleaning process for a PV system module in Surabaya is recommended at least once per month to lower energy lost by dust impact. Cleaning of PV panels system would increase the energy output of about 8% in total for one year.

References

[1] Ketjoy N and Konyu M 2014 *Energy Procedia* **52** 431–437.
[2] Rahman M M, Hasanuzzaman M, and Rahim N A 2015 *Energy Convers. Manag.* **103** 348-358.
[3] Kaldellis J K and Kapsali M 2011 *Energy* **36** 5154–5161.
[4] Tanesab T, Parlevliet D, Whale J, and Urmee T 2016 *Energy Procedia* **100** 65–68.
[5] Mejia F, Kleissl J, and Bosch J L 2013 *Energy Procedia* **49** 2370-2376.
[6] Menoufi K, Farghal H F M, Farghali A A, and Khedr M H 2017 *Energy Procedia* **128** 24-31.
[7] Wibowo E P, Notosudjono D, and Fiddiansyah D B 2018 *Jurnal Online Mahasiswa* **1** 1-11.
[8] Agoes S, Mardian D, Latupapua H, and Susila T 2015 *J. ELEKTRO* **8** 109-122.
[9] worldweatheronline.com, "Surabaya Monthly Climate Averages." [Online]. Available: https://www.worldweatheronline.com/surabaya-weather-averages/east-java/id.aspx. [Accessed: 15-May-2019].
[10] BMKG Indonesia, "BMKG | Badan Meteorologi, Klimatologi, dan Geofisika." [Online]. Available: https://www.bmkg.go.id/. [Accessed: 28-Dec-2018].

Sensor Fabrication for Determination of Cadmium Using Nafion/Ionic Liquid/Graphene Composite with Thick Film Technology

Teni Tresnawati[1], Aminuddin Debataraja[2] and Robeth Viktoria Manurung[2]

[1] Electrical Engineering, State Polytechnic of Jakarta, Jalan Prof. Dr. G. A.Siwabessy, Kampus UI, Depok, Indonesia.

[2] Research Center for Electronic & Telecommunication, Indonesian Institute of Sciences, Jl. Sangkuriang - Komplek LIPI Gedung 20, Lt. 4. Bandung, Indonesia.

Email: [1*]teni.tresnawati@gmail.com
[2*] adebataraja@yahoo.com
[3*] robeth_vm@yahoo.com

Abstract. The cadmium (Cd (II)) detection sensor consists of working electrode, counter electrode and reference electrode made with thick film technology on sensitive material-based carbon electrodes (G / N / IL / SPCE) and coated with Bismuth film. Carbon (C) paste, Ag | AgCl is used as working electrode, counter electrode and reference electrode material on alumina substrate by screen printing. The working electrode was doped with Graphene (G), Nafion (N) and Ionic liquid (IL). Furthermore, Bismuth-Film (BiF) is coated with electrodeposition on the surface of the working electrode. Electrode characterization with SEM is conducted to see the morphology of sensitive materials. It showed that the particles were distributed evenly on the surface of the electrode with spherical particles. Analysis using EDS shows the element's atomic composition is C 91.57%, N 3.48%, O 3.23%, Bi 0.01%. The performance of the cadmium sensor was tested with a potentiostat in a standard Cd solution with a concentration of 0.005 mgL^{-1} – 1 mgL^{-1} indicating an average output voltage range of 10.36 volts - 10.37 volts. This can be explained that the sensor can detect Cd (II) content properly according to performance.

1. Introduction

Cadmium (Cd) is one of toxic elements that can be found in the environment. Therefore, detection of cadmium levels in the environment, food and biological samples is of concern to researchers. Methods and analysis in determining the content of Cd continue to be efficiently developed such as Atomic Absorption Spectrophotometry [3] [4] [5], differential pulsed anodic stripping voltammetry (DPAS) [6], square wave anodic stripping voltammetry [7].

Electrochemical methods, especially electro polymerization have been widely recognized as electrode coatings for heavy metal detection sensors because of their low cost, easy operation, excellent stability, high sensitivity, and low detection limits [8] [9] [10] [11]. The use of mercury film-based electrodes (MFE) [12] [13] has been used in electrode coating techniques because of advantages such as surface purity, reproducibility, high sensitivity, high hydrogen potential [14]. However, the disadvantage is that mercury is a toxic material. The alternative developed in the study was using bismuth film. The characteristics of electrode coating are almost the same as coating with mercury electrodes. The formation of bismuth films was carried out by electrodeposition such as at the Glass Carbon Electrode (GCE) [15] [16], Carbon Paste Electrode (CPE) [17], Boron Doped Diamond Electrode (BDDE) [18] and Screen Printed Carbon Electrode (SPCE) [19] [20] [21] In this study the SPCE method based on thick film was chosen in the manufacture of electrodes because the process is easy and fast with low production costs.

Graphene (G) has been used in many studies because of its properties like extremely high thermal conductivity, high mobility of charge carriers, good mechanical strength, large specific surface area and outstanding electrical properties [22]. *Ionic Liquid* (IL) is a high ionic conductivity material, widely used as an electrochemical solvent and electrode modifier for the manufacture of sensors [23] [24]. The use of *Nafion* (N) [25] which has excellent anti-pollution capacity properties, chemical inertia, and high permeability to relevant cations to increase detection sensitivity to heavy metals.

In this study, the SPCE electrode was modified by G, IL and N composite coating on a working electrode by electropolymerization technique. Integration of IL and N composites can cause synergistic effects in electrochemical applications. Furthermore, the addition of Bismuth-film to the working electrode makes the electrode strong and good for the determination of heavy metals. Finally, this simple, inexpensive and sensitive sensor is applied to determine Cd (II) content by using potentiostat analysis and morphological analysis using Scanning Electron Microscope (SEM).

2. Experiment

2.1. Tools and Materials

The printing process of Carbon paste and Ag | AgCl uses the Accu-coat 3230 (Aremco Products Inc.) screen printer machine. After the printing process is dried using a National laboratory oven at 150°C for 10 minutes. Then proceed with the combustion process on a conveyor furnace machine (RTC Radiant Technology Corp.) at a temperature of 800 °C for 30 minutes. Measurement and determination of the sensor response to Cd concentration is carried out using a home made potentiostat (SOLITON EM 10/20), APPA 505 digital multimeter connected to a computer.

The electrode material uses carbon paste (Dupont, BQ242) and Ag AgCl material (Dupont 5874). Graphene was obtained from ACS Material, LLC (Medford, USA). As an anion Amoniumhydrogen phosphate (MERCK) is used. Ionic Liquid uses types 1-butyl-2, 3-dimethylimidazolium tetrafluoroborate and Nafion D-521 5% (Alfa Aesar), N-dimethylformamide (DMF) from Sigma-Aldrich (Buchs, Switzerland). Buffer solution is obtained from Merck (Darmstadt, Germany). Cd standard solutions in 1000 mg L^{-1} concentration were obtained from Certipur (Germany). Aquadesk 17MΩ is purified by Ultrapure Water Purification Systems (USA) Ultrapure Water Purification Systems (USA) water purification systems (Thermo Scientific ™ E-Pure ™).

2.2. Fabrication of Electrodes

Thick film technology screen printing method is used to make cadmium detection sensor electrodes. This technology was first used in the mid 1960s and became popular in 1970. This process was chosen to obtain a microelectronics circuit with the advantage of the size of the sensor can be diminiatur and conductor lines in small sizes (mm). Layer thickness in thick film technology ranges from 100 micrometers. Printing carbon paste material (Dupont, BQ242) is applied to make working electrodes and counters. Whereas silver | Siver chloride (Ag | AgCL) paste is applied to make reference electrodes. The stages of thick film processing include screen film making, printing of paste materials, drying, combustion, assembly, packaging and other additional processes. Figure 1 shows the flow of the electrode fabrication process.

Figure 1. Electrode Fabrication Process

2.3. Preparation of the N/IL/G composite

First, 2.0 mg of graphene was dispersed in 2.0 mL of DMF by stirring for about 2 hours. Next, 0.5% (m:v) of ionic liquid and 0.1% (v:v) of Nafion solution were added to the graphene dispersion and stirred for further 30 minutes.

2.4. Nafion / Ionic Liquid / Graphene Coating

The N/IL/G composite is applied to the surface of the working electrode which has been evenly polymerized and left to dry. Then reference voltage measurement is conducted in the KCL 3M solution to see the voltage stability of the sensor electrode.

2.5. Bismuth-film Electropolymerization

The bimuth film coating on the surface of the working electrode is done by electropolymerization. The electropolymerization solution used consists of 200.0 ng mL^{-1} Bi (III), 0.1 M acetate buffer solution with pH 4.5. The solution is mixed and stirred and at the same time given a voltage of -1.4 V for 2 minutes. The sensitive membrane dopping process is shown in Figure 2. After that amperometrik measurement is conducted on the sensor electrode to evaluate and characterize the sensor according to different Cd standard solutions.

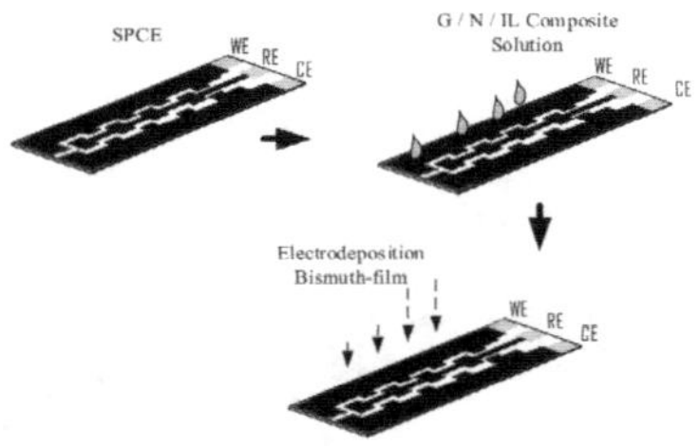

Figure 2. The process of sensitive membranes doppin

3. RESULT AND DISCUSSION

3.1. Electrode Characterization Using SEM

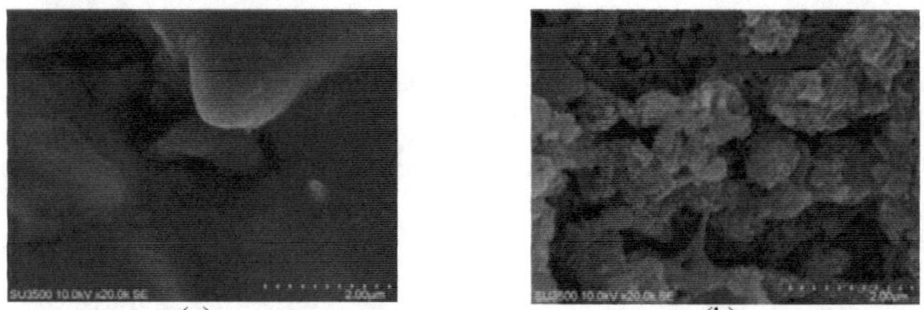

(a) (b)

Figure 3. SEM Photograf (a). Electrode carbon before electrodeposition. (b). Electrode carbon after electrodeposition

Morphological characteristics were tested using Scan electron microscopy (SEM SU3500), 10kV with an enlargement of 20k secondary electron. Tests were carried out on carbon electrode samples (SPCE) before and after conductive material coating. The morphological characteristics of SPCE before coated with conductive material are shown in Figure 3A. The carbon layer dominates on the surface of the SPCE. Whereas in (Figure 3B) is the G / N / IL / Bi / SPCE SEM morphology shows the particles are distributed evenly (uniformly) at all surfaces of the working electrode. So it shows that G / N / IL / Bi / SPCE conductive performance has increased due to carbon and other composite contributions.

Furthermore, testing is done to see the composition of each element in the electrode layer which consists of carbon (C), Nitrogen (N), Oxygen (O) and Bismuth (Bi). Based on the results of testing with Energy Dispersive Spectroscopy (EDS) the atomic composition of elements is C 91.57%, N 3.48%, O 3.23%, Bi 0.01% as in Table 1 according to the material specified. Carbon elements are more dominant than other elements because they use thick films with a thickness of 100 micrometers. The composition of each element shows a good relationship to the specifications of the electrode.

Table 1. Energy dispersive spectroscopy testing (EDS)

Elemen	**Weigth %**	**Atomic %**	**Error %**	**Net Int.**	**K Ratio**	**Z**	**A**	**F**
C	89.16	91.57	2.75	3140.86	0.7873	1.0064	0.8774	1
N	3.95	3.48	31.35	19.16	0.003	0.9792	0.0774	1
O	4.19	3.23	20.4	51.86	0.0044	0.9559	0.1092	1
F	2.65	1.72	15.43	63.65	0.0035	0.886	0.149	1
Bi	0.05	0.01	73.23	14.32	0.0022	0.5294	8.0027	0.9969

3.2. *Electrochemical Characterization of Electrodes*

3.2.1. *Stability of reference electrodes*

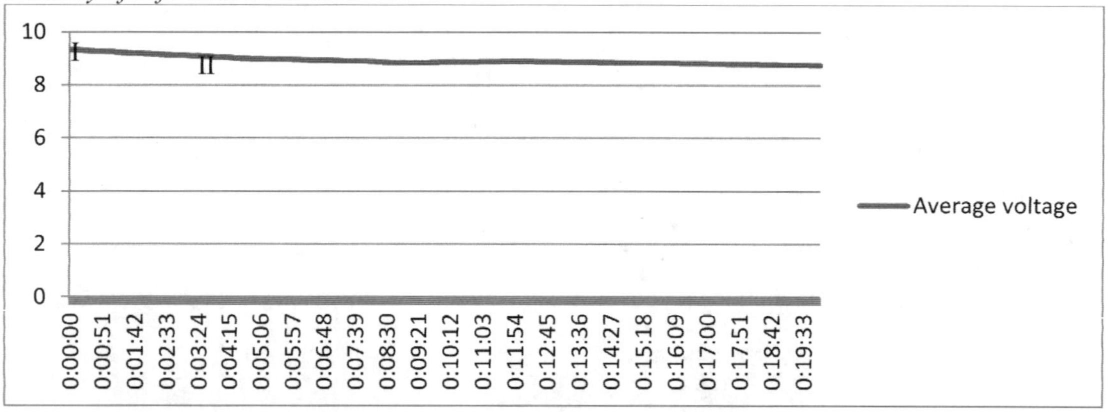

Figure 4. Output response of Ag|AgCl electrode vs Ag|AgCl comercial electrode

Reference electrode Ag | AgCl is used to maintain the voltage stability produced by working electrodes and counter electrodes. The reference voltage test results are shown in Figure 5. Testing is conducted in KCL electrolyte solution with a concentration of 3 M. In quadrant I there is a decrease in voltage affected by the spread of KCL solution which is not evenly distributed on the electrode surface. Whereas in quadrant II it is relatively stable. This result shows that the output voltage is quite good (stable) with a range of 8.5 - 9 mV and can be used as a detection sensor device for Cd (II) content.

3.2.2. • Cadmium (Cd) Electrochemical Detection

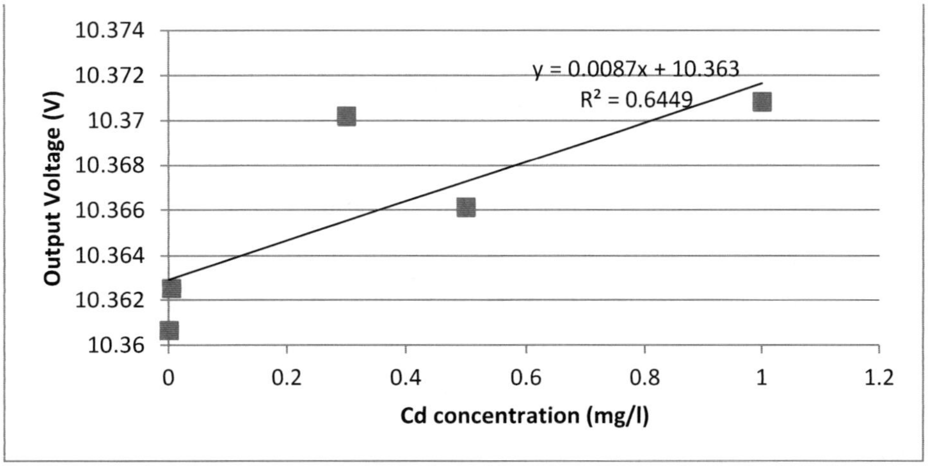

Figure 5. Linearity response of Cd sensor

The performance of the sensor electrode was tested with a potentiostat to control electrodes (WE, RE, and CE) and measure the experiment electroanalytically. The source voltage used was 700mV with 10KΩ resistance. Figure 5 shows the output voltage of the Cd content test using a standard Cd solution. The prepared test concentration consisted of 0.001 mgL^{-1}, 0.005 mgL^{-1}, 0.3 mgL^{-1}, 0.5 mgL^{-1}, and 1 mgL^{-1}. The average output voltage for each sample test solution is 0.001 mgL^{-1} of 10,360 volts, 0.005 mgL^{-1} of 10.362 volts, 0.3 mgL^{-1} of 10.37 volts and 0.5 mgL^{-1} of 10.366 volts, and 1 mgL^{-1} of 10.37 volts . The biggest voltage difference occurs between 0.001 mgL^{-1} test solution and1 mgL^{-1} test solution with a difference of 1.02 volts. When viewed from the sensor liner of 64.5%, the results indicate that the sensor can detect Cd (II) content properly.

4. Conclusion

Based on this research, G / N / IL was successfully deposited on carbon electrodes plus Bismuth films. The morphology of the surface of the working electrode using SEM shows that the particles are distributed evenly (uniformly) on all surfaces. So this shows that G / N / IL / Bi / SPCE conductive performance has increased due to carbon and other composite contributions. The voltage stability of the reference electrode shows good results with a voltage range of 8.5 - 9 mV. The sensitivity of the sensor electrode increases by depositing bismuth film by electrodeposition for 2 minutes. The electrode was applied to galvanostatic potential with the measurement results showing that the electrode had a linear response (R = 64.5%). Measured detection limits are 0.01mgL^{-1} - 1mgL^{-1} concentrations indicating an average output voltage of 10.36 volts - 10.37 volts. This can be explained that the sensor can detect Cd (II) content well according to performance. The electrode has been used as a working electrode for Cadmium sensor.

References

[1] Kwon J. Y., Jang Y. J., Lee Y. J., and Kim K. M., 2005 *J. Am. Chem. Soc* **127** 10107–10111.
[2] Ma R., Van Mol W., and Adams F., 1994 *Anal. Chim.* **285** 33–43.
[3] Tu Y., Ju S., and Wang P., 2016 *Spectrosc. Lett.* **49** 249–256.
[4] Flores É. M. M. *et al.*, 2009 *Int. J. Environ. Anal. Chem.* **89** 129–140.
[5] Silva E. L. and Dos P., Roldan S., 2009 *J. Hazard. Mater.* **161** 142–147.
[6] Li N. B., Duan J. P., and Chen G. N. 2004 *Chinese J. Chem.* **20377007** 553–557.
[7] Anastasiadou Z. D., Sipaki I., Jannakoudakis P. D., and Girousi S. T. 2011 *Anal. Lett.* **44** 37–41.

[8] Guell R., Aragay G., Fontas C., Antico E., and Merkoc A. 2008 *ScienceDirect* **7** 219–224.
[9] Rojas C., Segura R., and Nagles E. 2012 *Talanta* **99** 119–124.
[10] Liu X., Venkatraman K., and Akolkar R. 2018 *J. Electrochem. Soc.* **165** B9–B11.
[11] Dai Y. and Liu C. C. 2017 *Sensors* **17** 950.
[12] Palchetti I., Majid S., Kicela A., Marrazza G., and Mascini M. 2003 *Int. J. Environ. Anal. Chem.* **83** 701–711.
[13] Lara J., Torres J. F., Beltrán O. G., Nagles E., and Hurtado J. 2017 *Int. J. Electrochem. Sci.* **12** 6920–6929.
[14] De Oliveira M. F., Saczk A. A., Okumura L. L., Fernandes A. P., De Moraes M., and Stradiotto N. R. 2004 *Anal. Bioanal. Chem.* **380** 135–140.
[15] Guzsvány V., Papp Z., Zbiljic J., Vajdle O., and Rodic M. 2011 *Molecules* **16** 4451–4466.
[16] Wang Z. M., Guo H. W., Liu E., Yang G. C., and Khun N. W. 2010 *Electroanalysis* **22** 209–215.
[17] Ashrafi A. M. and Vytřas K. 2013 *Int. J. Electrochem. Sci.* **8** 2095–2103.
[18] Toghill K. E., Wildgoose G. G., Moshar A., Mulcahy C., and Compton R. G. 2008 *Electroanalysis* **20** 1731–1737.
[19] Chaiyo S., Mehmeti E., Žagar K., Siangproh W., Chailapakul O., and Kalcher K. 2016 *Anal. Chim. Acta*.
[20] Paukpol A. and Jakmunee J. 2016 *Chiang Mai Univ. J. Nat. Sci.* **15** 77–88.
[21] Lu Z., Zhang J., Dai W., Lin X., and Ye J. 2017 *Microchim. Acta* **184** 4731–4740, 2017.
[22] Guo S.J. and Dong S.J. 2011 *Journal of Materials Chemistry* **21** 18503-18516.
[23] M. Armand, F. Endres, D. R. MacFarlane, H. Ohno, and B. Scrosati, "Ionic-liquid materials for the electrochemical challenges of the future," *Nat. Mater.*, vol. 8, no. 8, pp. 621–629, 2009.
[24] Niu X., Zhao H., and Lan M. 2011 *Electrochim. Acta* **56** 9921–9925.
[25] Segura R. A., Pizarro J. A., Oyarzun M. P., Castillo A. D., Díaz K. J., and Placencio A. B. 2016 *Int. J. Electrochem. Sci.* **11** 1707–1719.

The Crystallization Process of Zr-Based Metallic Glass by the Influence of Heating Rate and Alloy Composition

Triwikantoro[1,*], Y Kriswandono[2], and F Munawaroh[3]

[1] Physics Department, Faculty of Science, Sepuluh Nopember Institute of Technology, Surabaya 60111, Indonesia.

[2] Natural Science Education Study Program, Faculty of Education, University of Trunojoyo Madura, Bangkalan 69162, Indonesia

Email: [1*] triwi@physics.its.ac.id

Abstract. A study of the crystallization process of Zr-based Metallic Glass by the influence of the heating rate and variation of alloy composition was investigated. The alloys used were $Zr_{68}Cu_{14}Ni_{11}Al_7$, $Zr_{64.5}Cu_{17}Ni_{11}Al_{7.5}$, and $Zr_{69.5}Cu_{12}Ni_{11}Al_{7.5}$. Thermal stability was analyzed using DSC (*Differential Scanning Calorimetry*) with a heating rate of 5°C /min and 10°C / min until the temperature reached 500°C. The three samples were heated with temperature variations of 400°C, 425°C, and 450°C for one hour in air. Crystal growth was observed by XRD test. The DSC curve of the three samples shows that the higher the rate of heating, the shift of the crystallization peak temperature shifts to a higher temperature. At higher temperatures, the crystallization process occurs faster than at lower temperatures. The composition of zirconium contributes to the resistance properties of crystallization. The alloy with the highest zirconium composition has a high resistance to crystallization. The crystalline phase formed is ZrO_2 tetragonal in all samples heated at a temperature of 450°C, and there is an intermetallic phase of Zr_2Ni identified on $Zr_{64.5}Cu_{17}Ni_{11}Al_{7.5}$

1. Introduction

These guidelines, Metallic glass alloys are metal alloys that have an amorphous structure, and in recent years, research related to metallic glass still receives significant attention [1], [2], [3], [4], [5]. Its amorphous structure makes metallic glass have excellent properties, including oxidation resistance, corrosion resistance, ductility, toughness [6],[7]. Related to usage, besides having superior mechanical strength, the stability of the amorphous structure is also an important concern.

Meanwhile, the structure will change if the amorphous alloy undergoes heat treatment and is in an environment that does not vacuum. Oxygen has a very significant role and maintains an amorphous structure. The amorphous structure will be damaged and transformed into crystals when the amorphous alloy is heat treated at the crystallization temperature and in a high oxygen-containing environment [8], [9], [10]. Structural resistance and resistance to the environment can be identified through structures formed after the alloy has experienced heating, for a long time and in an environment that does not vacuum. Koster has reported that crystallization and oxidation are two phenomena that compete with each other when amorphous alloys are subjected to heat treatment, held for a certain period and not vacuum conditions [11,12], [13].

In this paper, the effect of heating rate and composition on the amorphous structure of Zr-based metal has been studied. Also, it also examines the effect of sample storage time on structural stability and oxidation resistance, when the alloy is heated around the crystallization temperature in the air for 1 hour.

2. Experimental Methods

Zr-based amorphous alloys with variations in composition derived from arcs melting Zr, Cu, Ni and Al (> 99.9%) made by melting spinning as reported elsewhere in detail [12,13]. Amorphous alloys have been stored in boxes made of cardboard for more than 20 years. Thermal characteristics of the sample were tested by DSC (*Differential Scanning Calorimetry*) Type DSC-60 Schimadzu with the variation of heating rate. Stability of amorphous alloy structure investigated by XRD test. Amorphous alloy transformation kinetics can be seen from the DSC curve. And the phase formed in the amorphous alloy during heating around the crystallization temperature was studied from the XRD results.

3. Result and Discussion

Fig. 1 shows the X-ray diffraction patterns for the initial condition of the amorphous Zr-based alloys before heat treatment. In general, all samples are amorphous, even though there is an unidentified small peak in $2\theta = 26.48^{o}$ for $Zr_{64.5}Cu_{17}Ni_{11}Al_{7.5}$ sample.

The resulted DSC measurement during thermal characterization with different heating rate is shown in fig 2. Each curve in Fig. 2 expresses the heat flow to the samples that the energy released by the sample to transform into an order of atomic arrangement. On the DSC curve, there does not appear to be a noticeable change in the curve of the baseline, as observed (see insert on Fig.2) by Inoue A. et.al [14]. It indicates that the glass transition temperature does not observe. The initial condition of the sample is not completely amorphous (see Fig.1). Sample storage for a long time does not guarantee that it does not react with oxygen, even if only on the surface. In this alloy, Zirconium has the largest composition, which has a relatively high affinity for oxygen, so oxidation is easy. The oxide formed causes unclear glass transition regions. This one of the reasons for the glass transition on the DSC curve is not obvious. The complete data analysis was shown in Tab. 1.

Table 1. DSC data for amorphous Zr-based alloys with the different heating rate

No	Samples	Heating rate (˚C/min.)	Crystallization (ºC)		
			Start	Peak	end
1	$Zr_{68}Cu_{14}Ni_{11}Al_{7}$	5	393.07	419.11	425.74
		10	404.42	427.53	435.04
2	$Zr_{64,5}Cu_{17}Ni_{11}Al_{7,5}$	5	402.24	416.87	422.96
		10	418.03	426.20	431.84
3	$Zr_{69,5}Cu_{12}Ni_{11}Al_{7,5}$	5	422.96	433.60	437.63
		10	436.92	442.26	446.51

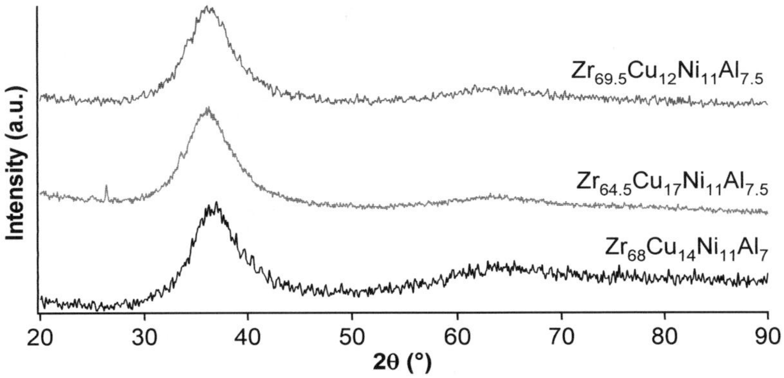

Figure 1. X-ray diffraction patterns of the initial condition of $Zr_{64.5}Cu_{17}Ni_{11}Al_{7.5}$, $Zr_{68}Cu_{14}Ni_{11}Al_{7}$, and $Zr_{69.5}Cu_{12}Ni_{11}Al_{7.5}$

Crystallization temperatures were observed for the heating rate of 5° C/ min at the peak of the DSC curve for $Zr_{64.5}Cu_{17}Ni_{11}Al_{7.5}$, $Zr_{68}Cu_{14}Ni_{11}Al_{7}$, and $Zr_{69.5}Cu_{12}Ni_{11}Al_{7.5}$ of 416, 419 and 433°C, while for the heating rate of 10°/min of 426, 427 and 442°C, respectively. From this data, amorphous Zr-based alloys were then heat treated with the temperature around the crystallization temperature, namely: 400, 425, and 450°C in the air.

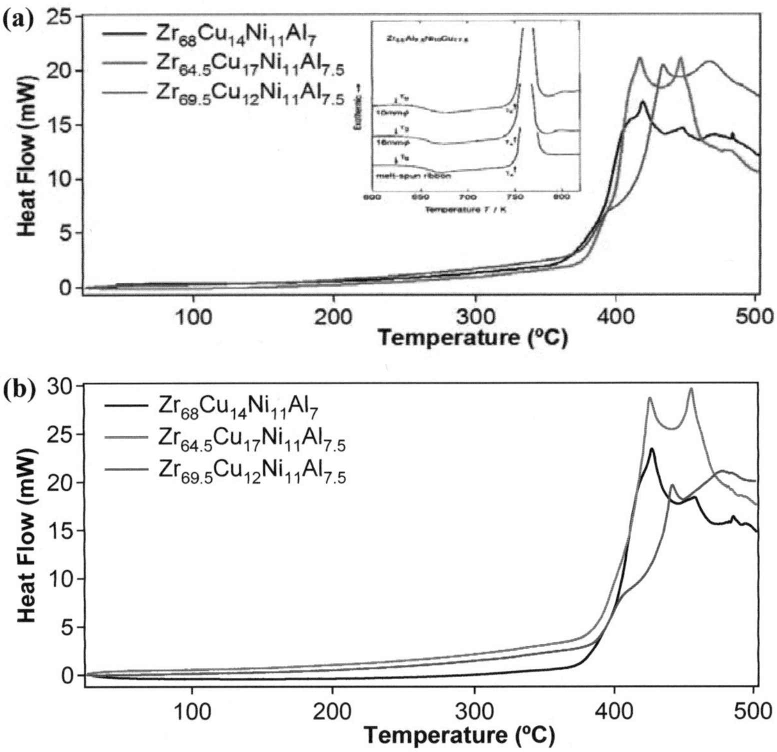

Figure 2. The DSC curves of $Zr_{68}Cu_{14}Ni_{11}Al_{7}$, $Zr_{64,5}Cu_{17}Ni_{11}Al_{7,5}$, and $Zr_{69,5}Cu_{12}Ni_{11}Al_{7,5}$ with different heating rate (a) 5°C/min and (b) 10°C/min.

The DSC curve can also be used to analyze the transformation process. In metallic glass alloys, the kinetics of phase transformation from amorphous to the crystal can be carried out by providing a transformed fraction model. The DSC peak represents the energy released when there is a transformation. Based on the DSC test, data can be described as the transformation kinetic curve model (S curve). The S curve is a plotting of material transformed as a function of time. Fig. 3 shows the transformed fraction curve vs. time for amorphous Zr-based alloys at the different heating rate. From the transformed fraction curve with time shows that the magnitude of the heating rate causes in a curve shift to the right. This is related to the heat transfer or heat flow to the sample. This phenomenon was classified by Wu.

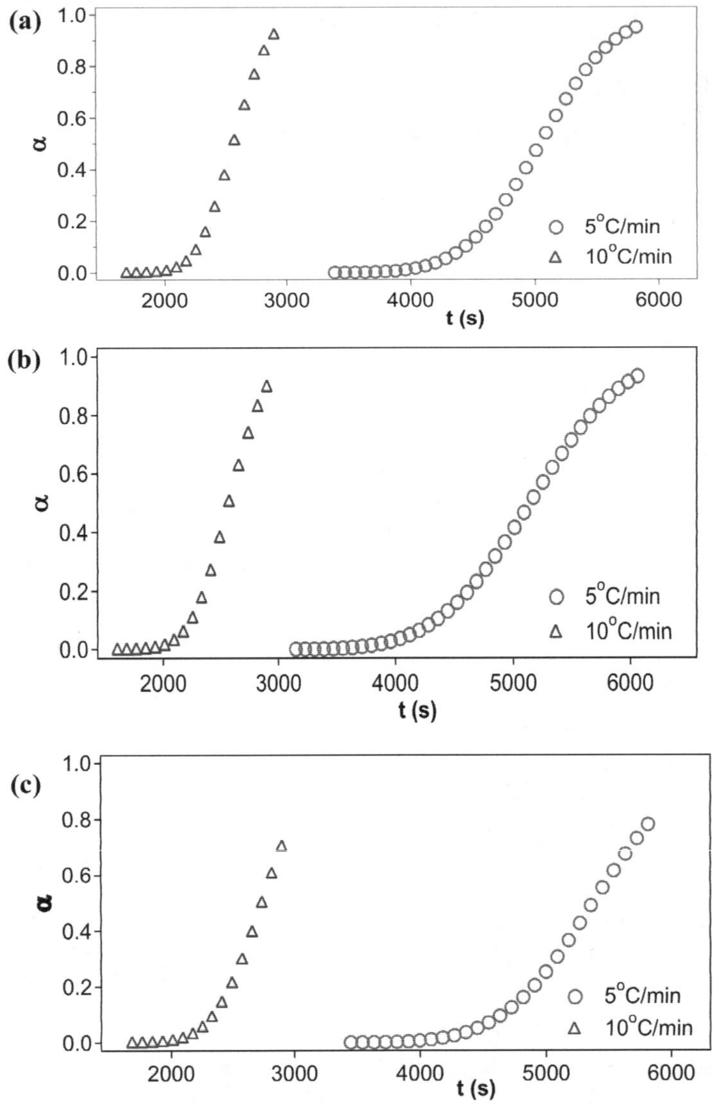

Figure 3. Curve of transformed fraction vs time_for (a) $Zr_{64.5}Cu_{17.5}Ni_{11}Al_{7.5}$, (b) $Zr_{68}Cu_{14}Ni_{11}Al_7$ and (c) $Zr_{69.5}Cu_{12}Ni_{11}Al_{7.5}$

In the other side, it has conducted a study related to the effect of heating rate variations on the shift of the DSC curve. It has been reported that the shape of the DSC curve (S curve) is influenced by the heating rate. The peak of the DSC curve will shift to a higher temperature for a higher heating rate. Besides, the rate of crystallization will be faster if the heating rate is relatively high. At a heating rate of 20 - 30°C/min. It produces a more stable crystallization, compared to a heating rate of 5 - 10°C/min. This informs that the increased heating rate causes the crystallization process to occur earlier and in a short time. The S curve in figure 3 - 5 shows that for the crystallization process, up to 80% of the transformed fraction requires a shorter time for the heating rate of 10°C / min than 5°C / min.

Fig. 4 - 6 shows the pattern of the diffraction pattern of Zr-based amorphous alloys composition varies after being heated in the vicinity of the crystallization temperature of 400, 425, and 450°C. At temperatures of 400°C, all samples were still in an amorphous state, except that $Zr_{64.5}Cu_{17.5}Ni_{11}Al_{7.5}$ began to emerge crystalline oxides identified as t-ZrO_2. In general, after being given heat treatment for 1 hour with varying temperatures, the phase formed is, tetragonal ZrO_2 (t-ZrO_2), while phases other than oxide crystals were only identified on $Zr_{64.5}Cu_{17.5}Ni_{11}Al_{7.5}$. Storage for a long time makes Zr-based metallic glass material, an alloy that has excellent structural stability. There is no phase intermetallic phase like Zr_2Ni when it has been heated for 1 hour.

Meanwhile, it has been reported that the super-Zr-based alloy with the composition $Zr_{69.5}Cu_{12}Ni_{11}Al_{7.5}$ has been crystallized at a temperature of 370°C for 4 hours in the air [5]. The greater the Zr content, the higher the crystallization, which means the stability of the structure is getting better, while at a temperature of 400°C these alloys are still amorphous. $Zr_{69.5}Cu_{12}Ni_{11}Al_{7.5}$ is an alloy that has the best properties for structural stability and oxidation resistance.

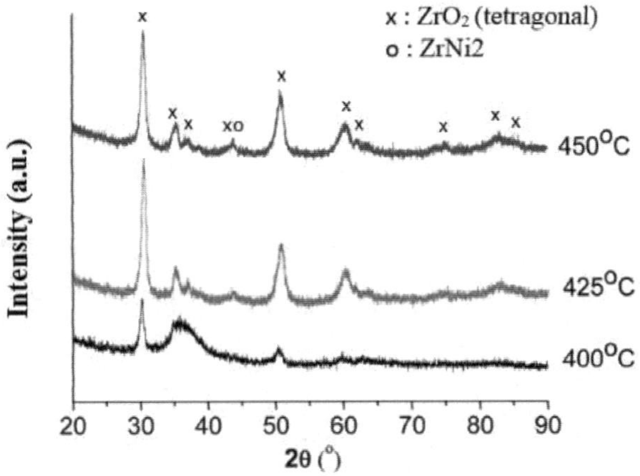

Figure 4. The XRD patterns of $Zr_{68}Cu_{14}Ni_{11}Al_7$ during heating at different temperature for 1 h

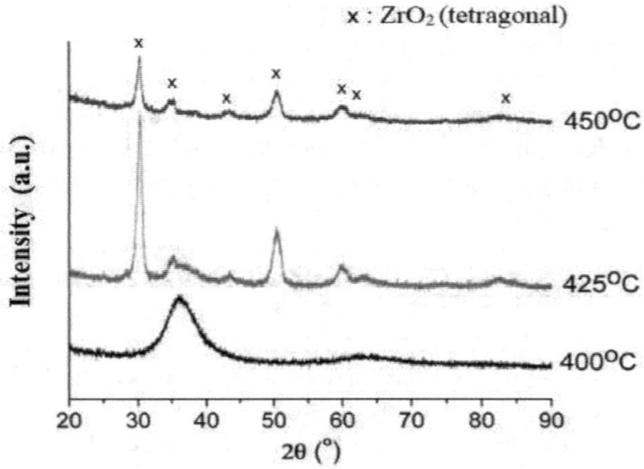

Figure 5. The XRD patterns of $Zr_{64.5}Cu_{17}Ni_{11}Al_{7.5}$ during heating at different temperature for 1 h

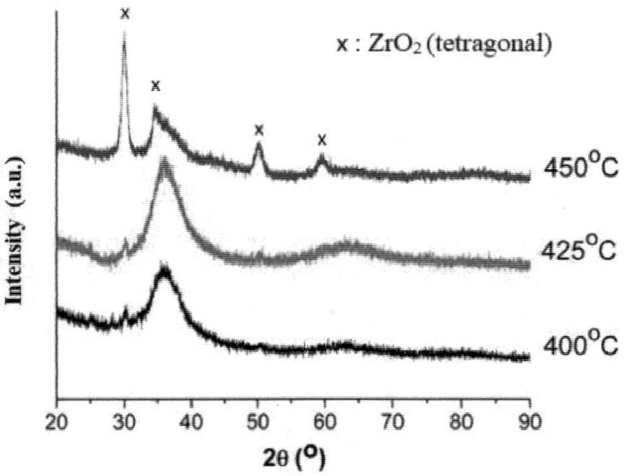

Figure 6. The XRD patterns of $Zr_{69.5}Cu_{12}Ni_{11}Al_{7.5}$ during heating at different temperature for 1 h

4. Conclusion

Storage for a long period of more than 20 years can be identified by the structural stability and oxidation resistance of amorphous Zr-based alloys. The initial oxide layer formed can inhibit its crystallization or can maintain an amorphous structure and oxidation resistance. After heating 400 - 450°C, generally the alloy only undergoes oxidation. The best crystallization and oxidation resistance was found in the $Zr_{69.5}Cu_{12}Ni_{11}Al_{7.5}$ alloy, while the worst oxidation and oxidation resistance was identified on $Zr_{64.5}Cu_{17}Ni_{11}Al_{7.5}$.

Acknowledgments

The author would like to thank the BCI Research Group, Faculty of Chemical Engineering, TU Dortmund, Germany.

Reference

[1] Churyumov A Y, Bazlov A I, Tsarkov A A, Solonin A N and Louzguine-Luzgin D V 2016 *J. Alloys Compd.* **654** 87–94.
[2] Park S H, Lim K R, Na M Y, Kim K C, Kim W T and Kim D H 2015 *Corros. Sci.* **99** 304–12.
[3] Nie X P, Yang X H, Ma Y, Chen L Y, Yeap K B, Zeng K Y, Li D, Pan J S, Wang X D, Cao Q P, Ding S Q and Jiang J Z 2012 *Intermetallics* **22** 84–91.
[4] Singh D, Mandal R K, Tiwari R S and Srivastava O N 2015 *J. Alloys Compd.* **648** 456–62
[5] Zhang M, Yao D, Wang X, and Deng L 2014 *Corros. Sci.* **82** 410–9.
[6] Kang CC and Jordi S 2014 Metallic Glasses *Phys. Metall. Fifth Ed.* **1** 305–85
[7] Chatterjee R 2017 *Adv. Mater. Manuf. Charact.* **7** 24
[8] Zhang Q C, Pang S J, Li Y and Zhang T 2011 *Sci. China Physics, Mech. Astron.* **54** 1608–1611
[9] Munawaroh F and Triwikantoro 2012 *Berkala Fisika Indonesia* **4** 27–32
[10] Triwikantoro and Sukendar S 2007 *Jurnal Fisika Dan Aplikasinya* **3** 5–8
[11] Yamamoto M 1991 Mechanism of Oxygen Absorption by Zirconium *Dissertation*
[12] Triwikantoro, Toma D, Meuris M and Köster U 1999 *J. Non. Cryst. Solids* **250-252** 719-723
[13] Köster U, Zander D, Triwikantoro, Rüdiger A and Jastrow L 2001 *Scr. Mater.* **44** 1649–54
[14] Inoue A, Zhang T, Nishiyama N, Ohba K and Masumoto T 1994 *Mater. Sci. Eng. A* **179** 210–4

The Real-Time Vital Sign Monitor for Heart Rate and SPO_2 Parameter Using Internet of Things Technology

Bedjo Utomo[1,a], Triwiyanto[1], Sari Luthfiyah[1], Urip Mudjiono[2], Torib Hamzah[1]

[1]Department of Electromedical Engineering, POLTEKKES KEMENKES Surabaya

[2]Shipbuilding Institute of Polytechnic Surabaya

[*]Email: bedjoutomo123@gmail.com

Abstract. Recently, the internet of thing (IOT) has been growing rapidly in many fields including in the biomedical vital sign monitoring of a human. Heart rate and SPO2 are vital sign parameters for human life. Those parameters needed to be observed for patients in an emergency situation. In patient care at hospitals, doctors need to monitor the patients' heart rate and SPO2 parameters continuously. Therefore, the objective of this study was to develop the heart rate and SPO2 parameters based on IOT technology. The heart rate and SPO2 parameters were detected using a finger sensor and an oximeter sensor, respectively. The digital data from those sensors, further, was processed using ESP8266 microcontroller. By using a built-in Wi-Fi module in ESP8266, the vital sign was sent to the user wirelessly to handphone or central monitor using Think Speak application, remotely. In this experiment, the sensors were placed on the finger while the patient was in the standing, lying or sitting position. In the same position, the parameters were acquired for ten times. Some analyses were conducted to find the accuracy of the measurement. The results showed that the uncertainty were 0.26 and 0.04 for heart rate and SPO2, respectively. Based on the medical standard calibration (<3%), those parameters are feasible to be used in the medical system measurement. The T-test analysis statistics for one independent sample was performed to obtain a significant difference among the data. The p-values were 0.118 and 0.000 for heart rate and SpO2, respectively. Finally, this study suggested that the IOT system was able to monitor the parameters effectively.

1. Introduction

Exercise is an activity that cannot be separated from health, the more people diligently doing exercise, the healthier their body and the stronger their muscles would be [1]. When doing exercise, the cardiovascular system function will experience an increase and decrease in pulse rate followed by an increase in stroke content, lung capacity and lactic acid accumulation, collateral blood vessels, HDL cholesterol and also experience decrease in Atherosclerosis [2]. In resting state the heart beats 70 times/minute. When doing many movements, the heartbeat can reach 150 times/minute with a pump power of 20-25 liters of blood/minute [3]. Cardiac output is the volume of blood pumped by each ventricle per minute. While the normal speed of the heart rate is the amount of beat every minute, in adults, it is 60-80 per minute. According to Nutrition Guidelines for Sports Achievement, the heart rate can be calculated in a way namely cardiac output = heart rate x stroke volume, for example 70 beats/minute x 70 ml / pulse = 4,900 ml/minute or 5 liters/minute (average volume of 70 ml/pulse) [2].

With the increase in a person's physical activity, the need for blood containing oxygen will be even greater. The heart will fulfill the need for oxygen by increasing the blood flow. This will also be responded by the blood vessels by widening the diameter of the blood vessels so that it will have an impact on the individual's blood pressure. When this process takes place, the condition of the blood will increasingly thicken and inhibit blood circulation in transporting digested food and oxygen throughout the body. The role of body fluids (water) is as balancer or blood thinner so that the blood is normal and can circulate the extracts of food and oxygen throughout the body [2]. To determine the speed of a person's pulse, it can be done by knowing the pulse rate by calculating the sudden changes in pressure. This pressure are propagated as waves on the blood wall while measurements can be made on carotid artery (neck area), located in the neck below the ear lobe, where there is a carotid artery between the trachea and the sternocleidomastoid muscle that is often used for babies in cases of cardiac arrest and for monitoring blood circulation. The frequency of the human heart rate varies depends on many factors that influencing it during normal activities. The radial artery (wrist) is located along the radial bone. It is more easily palpated above the wrist on the side of the thumb relatively easy and is often used routinely. BPM or Heart Rate is a representation of the pulse per unit time of an object. Usually, an adult's heart beats 60 to 100 times per minute in resting conditions. If the heart rate exceeds 100 beats per minute, the person is indicated to have Tachycardia heart abnormalities. On the other hand, if the heart rate is less than 60 beats per minute, the person is indicated to have a Bradycardia heart disorder especially if accompanied by other symptoms such as dizziness, shortness of breath or frequent fainting.

Heart Rate or BPM (Beat per Minute) and Oxygen Monitoring is continuous monitoring of a patient's condition that helps medical staff and paramedics in monitoring the progress of the patient's condition [4]. In general, monitoring of BPM and Oxygen can also be done non-invasively by practitioners and the general public by using finger sensors namely photodiode sensors and infrared clipped to the index finger of the hand. In designing the tool by using a sensor oximeter, oxygen level reading can be done easily, but with advances in technology in the field of informatics, the development of Internet of Things (IoT) – it will be much easier to do[5]. In a previous study, detection of conventional heart rate signals still found many obstacles in analyzing, so that by applying this application it would be easier for everyone to analyze the results of data recording in real time during sports activities. The benefits of doing adequate and regular exercise have been widely informed in various health articles, popular articles, and health journals, exercise can prevent obesity, diabetes mellitus, hyperlipidemia, stroke, and hypertension. In their research in Belgium concluded that aerobic exercise can be applied as management of hypertension not only for prevention. Still, in the same study, it was also mentioned that blood fats can be reduced by doing exercise, especially aerobics. Blood fat is what will cause atherosclerosis later if the level is high.[6]

This study aimed to analyze the effectiveness of heart rate (BPM) and blood oxygen level (SPO2) monitoring based on Internet of Things (IoT). The use of computers in the future can dominate human work and defeat human computing capabilities such as controlling electronic equipment remotely. Internet of things (IoT) is designed to connect astronomical numbers of ordinary and everyday objects to the Internet and consequently expand the Internet into the human physical world. This device is usually limited to resources, which have low computing and storage capabilities and are battery-powered [7]. In the coming years, the IoT is expected to bridge diverse technologies to enable new applications by connecting physical objects together in support of intelligent decision making [8].

2. Materials and Methods

2.1. Model Design

Finger sensor and oximeter sensor send analog data to Node MCU ESP8266 via Arduino [9]. From this Arduino module, the input will be processed into data. Furthermore, NoteMCU ESP8266 from the module receives data from Arduino in the form of Wi-Fi data. [10]. After connected with Wi-Fi, this data will be sent into smartphone media through an application to monitor the results, the data update is carried out every 5 seconds.[11]

2.2. Type of Research

This research is experimental research with the design of one sample one group. The population and samples in this study were students majoring Electromedical Engineering. Each sample underwent ten treatments. The dependent variable in this study is the description of heart rate (BPM) signals and oxygen levels in the blood. The independent variable is aerobic exercise. The materials and equipment used in this study were: a) stopwatch, time measurement tool, b) module for designing heartbeat (BPM) and oxygen levels in the blood (SPO2) devices, c) smartphones as media to monitor the data that were then processed by using Excel and displayed in graphical form per 30 seconds.

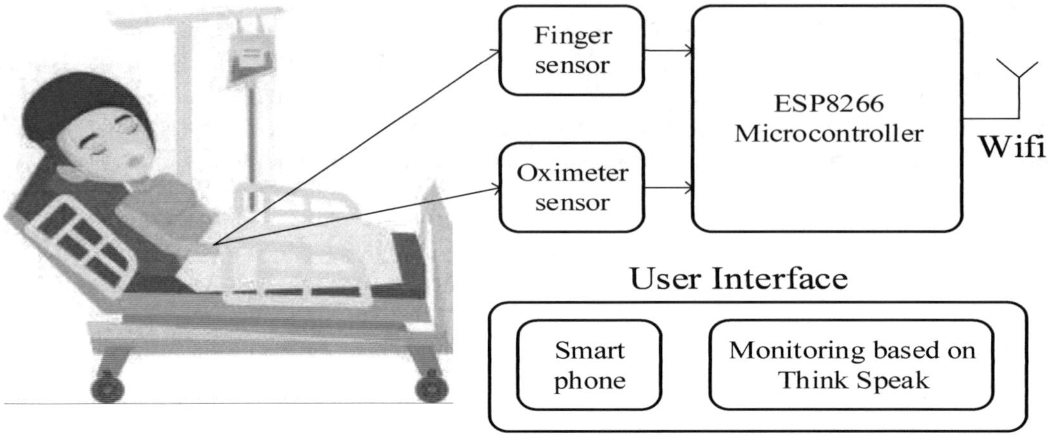

Figure 1. The Experimental Setup on the Development of IOT System for Heart Rate and SPO2 Parameters Monitoring

2.3. Data Acquisition and Processing Techniques

Data retrieval was done through some stages, namely: 1) Retrieval of Data, as follows: a. attaches the index finger to the sensor b. data will be recorded and monitored in the Thing Speak application on Android c. then the data is downloaded on the application on the PC. 2). Data Analysis, the data recorded on the device will be transferred through Wifi media through the IoT web by using the Thing Speak application on the smartphone. From the Thing Speak data web, the data can be downloaded with the excel data extension (.cvs). The results of the data on the web on thing speak application can then be monitored in real time in accordance with the amount of value that is on the sensor readings on the equipment.[8]

3. Result and Discussion

3.1. Statistic Analysis

The statistic test was done to compare the data tools with the standard data. The aim was to determine the level of significance of the data from the design of the tool with the standard value data. The results of the independent T-Test on the value of the heart rate (BPM) of 10 samples obtained a significance level of $\alpha = 0.118$ (> 5%) so that, statistically, there was a difference between the value of the heart rate (BPM) module and the value standard heart rate BPM. According to the calibration standard for differences, values <5% is considered close to the threshold, which is obtained by measuring the heart rate (BPM), average = 64.8, the standard deviation value was 2.84 and the uncertainty value was = 0.28. Furthermore, the measurement of SpO2 level obtained an average of 95, the standard deviation was 0.44 and the uncertainty value = 0.04. The results of the Independent statistical T-test for SpO2 level obtained a significant value of $\alpha = 0.000$ (<5%) so that it can be said that there was no difference between the value data from the Design Module and the standard value data, the result of the data monitoring can be seen in the table below

Table 1. Measurement of Heart Rate (BPM) and SPO_2 using the design

Time of Measurement (minute)	1	2	3	4	5	6	7	8	9	10	Mean	SD
Heart Rate (BPM)	64	64	65	63	66	64	62	67	66	67	64.8	1.6
SPo2 (%)	94	95	95	95	96	95	94	96	95	95	95	0.6

Table 2. Measurement of Heart Rate (BPM) and SPO_2 using a calibrator

Time of Measurement (minute)	1	2	3	4	5	6	7	8	9	10	Mean	SD
Heart Rate (BPM)	67	65	64	66	66	65	65	66	65	65	65.4	0.84
SPo2 (%)	96	97	97	97	96	96	96	95	97	97	96.4	0.69

3.2. Test Equipment

Test equipment was grouped into two types of measurements, namely determining the value of the heart rate (BPM) and oxygen level in the blood taken from the Thing Speak application data. The result obtained can be seen in the graph below [6].

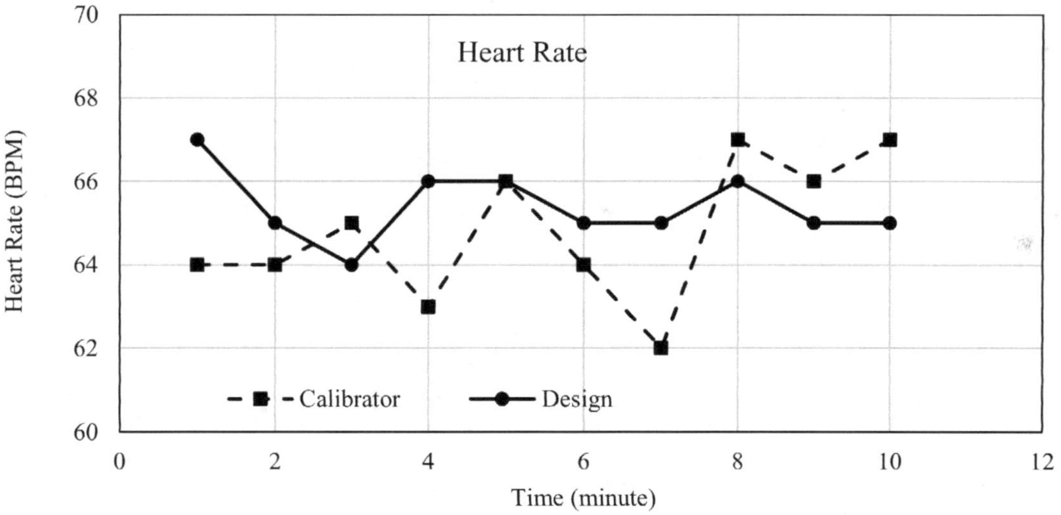

Figure 2. Result of heart rate (BPM) monitoring when subject perform activity

The result of the heart rate (BPM) monitoring of the equipment was 62.5, the standard deviation was 5.7, and the standard error was 1.8. The average result of the standard BPM was 65.4, the standard deviation was 0.84, and the standard error was 0.26. The data collection obtained from measuring the value of oxygen in the blood (SPO2) can be seen in the graph below.

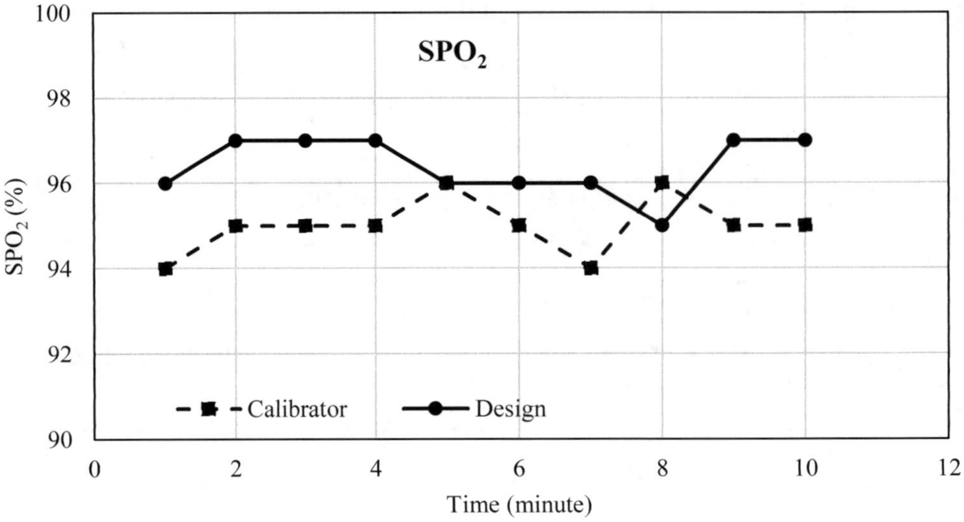

FIGURE 3. Result of Monitoring SPO_2 Level When Doing Activity

The average result of monitoring SPO_2 level on the equipment was 95.1, the standard deviation was 0.56 and the standard error was 0.17. The result of the standard SPO_2 equipment obtained was 96.4, the standard deviation was 0.69, and the standard error was 0.2

4. Conclusion

The results of this study can be summarized as follows: 1) the design of the heart rate (BPM) monitoring and SPO_2 level can be used on subject to monitor the BPM and SPO_2 parameters, 2). the monitoring of those parameters can be done by using the IoT system using ThingSpeak application and it can run normally with a 1-minute period to read on a smartphone. Over all results showed that the design is feasible to be used at home or medical clinics. The researcher suggested and recommended to develop an application on equipment that can be used for multi-modal monitoring.

References

[1] Hargreaves, M. 2008 Physiological limits to exercise performance in the heat (*Journal of Science and Medicine in Sport* vol 11) pp. 66–71.
[2] Kementrian Kesehatan Indonesia 2014 *Pedoman Gizi Olahraga Prestasi* vol 5201590(021) pp 5201590.
[3] Loria1, P.,Ottoboni1,S.,Michelazzi, L., Giuria1, R., 2014 Salivary Cortisol in an Extreme Non-Competitive Sport Exercise: Winter Swimming (Natural Science vol 6) pp 387-398.
[4] Meddeb, M., Dhraief, A., Belghith, A. 2018 A Promising Architecture for the Internet of Things (IoT) (*Int. J. Semant. Web Inf. Syst.* vol. 14)pp 86-112.
[5] Rozik, A. S., Tolba, A., Eldosuky, M., 2016 Design and Implementation of the Sense Egypt Platform for Real-Time Analysis of IoT Data Streams (*Adv. Internet Things* vol. 6) pp 65-91.
[6] Sherwood, L. 2011 *Fisiologi Manusia Dari Sel ke Sistem edisi ke 6* (Jakarta: ECG) pp. 287.
[7] Bormann, A. K. C. , Ersue, M., and Ericsson 2014 Internet Engineering Task Force (IETF) (*Terminology for Constrained-Node Networks*).
[8] Al-Fuqaha, M. A. Ala, Guizani, M., Mohammadi, M., Aledhari, M., 2015 Internet of Things: A Survey on Enabling Technologies, Protocols and Applications (*IEEE Commun. Surv. Tutorials).*
[9] Schwartz,M. 2018 Internet of Things dengan ESP8266 (*Internet of Things dengan ESP8266*) pp 226.
[10] Team, E.I., 2018 ESP8266EX Datasheet.
[11] Fisher, D.K., Fletcher,R.S., Anapalli1, S.S., Pringle III, H.C. 2018 Development of an Open-Source Cloud-Connected Sensor-Monitoring Platform (*Adv. Internet Things* vol 8) pp 1-11.

Journal of Physics: Conference Series **1373** (2019) 012029 doi:10.1088/1742-6596/1373/1/012029

Thickness and Resistivities of Cu/Ni Film Resulted by Electroplating on the Various Electrolyte Temperature

Jumratul Wustha[1,*], Moh. Toifur[2] and Azmi Khusnani[3]

[1,2,3]Dept. of Physics Education, Ahmad Dahlan University, Jl.Pramuka 42 Sidikan, Yogyakarta 55161, Indonesia

Email: [1,*] jumratulwustha423@gmail.com

Abstract. Synthesis of Cu/Ni thin films has been carried out by electroplating method assisted by magnetic fields at variations in the temperature of the solution. The use of magnetic fields in the deposition process is to accelerate mass transport, reduce the reaction effect of hydrogen evolution, improve surface morphology. An increase in the temperature of the solution can accelerate the rate of Ni ions from the anode to the cathode. The electroplating process is carried out at a DC voltage of 1.5 volts, a magnetic field of 200 gauss that is perpendicular to the electric field, within 5 seconds, and the electrode distance is 4 cm. The solution temperature varied from 40°C — 80°C at intervals of 10 °C. Characterization is done by testing the thickness and resistivity of the Ni layer. Thickness test is done with the help of calculations from the results of weighing Ni. The test of sheet resistivity is done by means of a 4 point probe

The results showed that the greater the temperature of the solution the thicker the Ni layer. The thickness of the Ni layer ranges from (0.09 ± 0.001) μm to (0.38 ± 0.006) μm which is the opposite to the resistivity of the chip, namely the higher the temperature of the solution the smaller the sheet resistivity. The size of the sheet resistivity is from (1.31 ± 0.02) Ω/sq. To (1.38 ± 0.06) Ω/sq to identify relevant articles in literature searches, great care should be taken in constructing both

1. Introduction

The progress of research in the field of material physics is in line with the times. Some materials are considered to have advantages such as easy to get, resistant to corrosion, resistant to wear, have strength, hardness, and low cost.

Thin layers are increasingly needed in various fields such as in industry, health, education, security and in other fields [1-3]. The thin layer is a layer of metal whose thickness ranges from the order of angstroms to microns [4,5]. In the medical field to preserve vaccines so that they remain stable and can last for a long time, preservation is carried out at low temperatures, namely cryogenic freezing below -150°C [6]. Similarly, cryonic methods have been developed to preserve organs so that for a long time the tissue is still alive and can be used at any time needed.

At this temperature, objects are preserved on the freezing unit. For that, we need a thermometer that can measure these low temperatures. One material that can be used as a low-temperature sensor is a combination of Cu and Ni so CuNi or Cu/Ni layers are produced

One of the benefits of Ni, when combined with Cu, is to increase the resistivity. Resistivity is one parameter that play an important role in determining sensitivity. The higher the resistivity of the material the more sensitive the material, because of the greater the resistivity range available to hold the temperature. This is because the material has a greater resistivity range to accommodate the temperature changes.

Ni has a higher resistivity (7.3 μΩcm) than Cu (1.7 μΩcm) [7] so that the combination Ni with Cu will increase the resistivity of Cu. In the study of metal, the Cu and Ni are two materials that is not difficult to combine because they are both transition metals that have almost the same radius of the atom, which is 1.25Å for Ni and 1.28Å for Cu so that if Ni is coated with Cu both of them will produce a good microstructure [8].

Cu and Ni did not difficult to be combined because both included transition metals that have almost the same atomic radius, namely 1.25Å for Ni and 1.28Å for Cu. If Ni is coated to Cu, a good microstructure will be produced [8], that is a parameter to produce advantage material properties.

One of the methods to make the thin film is electroplating [9] is the process of metal deposition over other metals using electrolysis [10]. Giving direct current to the solution causes a reduction process in the cathode and oxidation at the anode. In the process of electroplating, the temperature and magnetic field are factors that can affect the quality of the coating. The increase in temperature will cause an increase in the diffusion of ions to the cathode. This can prevent the occurrence of uneven coating which appears as a vacuum of ions in the cathode. The appropriate temperature will tend to improve the quality of the coating.

Furthermore, using a magnetic field in a direction perpendicular to the electric field, namely in the direction of the anode-cathode, will increase the chemical reaction. This is because there is a flow of magneto hydro dynamics (MHD) induced in the solution. MHD is a conductive fluid that flows due to the effect of using an external magnetic field. The Lorentz force is one of the effects of MHD.

In Cu-Ni coating the presence of Lorentz force causes the direction of the Ni ion flow in electrolyte deviates from the original direction. Near the cathode, the presence of Lorentz effect leading infiltration of Ni ions in the oblique direction so that vacancies in substrate Cu which is not visible from the direction of the Cu surface becomes visible when viewed from the oblique direction. Therefore the Ni ions can occupy these vacancies. This makes the layer on the surface becomes more solid and microscopically it will be better than Cu surface at normal conditions. [11, 12].

2. Research Methods

2.1. Material research

The material used in this study consisted of copper plates with a size of 10×1.3 cm^2 of 5 pieces. This plate was used as a substrate and installed as a cathode. Nickel with the same size was used as an anode. Electrolyte solutions were made from a mixture of boric acid (H_3BO_3), nickel sulfate ($NiSO_4$), and nickel chloride ($NiCl_2$). Boric acid was used as a pH controller for electrolyte solutions consisting of nickel sulfate and nickel chloride. The distilled water was used to dilute the solution. The equipment consisted a plating bath which contains 300 mL electrolyte, a TCA-BTA thermocouple was used to control the temperature of the solution, electric stove to heat the electrolyte solution, the DC voltage source that was set at 1.5. volt, a PA214 Ohaus balancing to measure the mass of the plate before and after deposition, ultrasonic cleaner to clean the sample, and Multimeter to display the voltage and current on the four points probe.

The title is set 17 point Times Bold, flush left, unjustified. The first letter of the title should be capitalized with the rest in lower case. It should not be indented. Leave 28 mm of space above the title and 10 mm after the title.

2.2. Method of experiment

In the electroplating process electrolyte solutions were prepared with the composition of H_3BO_3 (30 g), $NiSO_4$ (195 gr), NiCl2 (450 gr) and H_2O (750 ml). The solution is heated with Bunsen burners up to 40 °C, then the equipment is assembled according to Figure 1. The electroplating process is carried out at a voltage of 1.5 volts for 5 seconds with an electrode distance of 4 cm, and a magnetic field of

200 gausses. The electroplating process is repeated by varying the electrolyte temperature of 50 °C, 60 °C, 70 °C, and 80 °C.

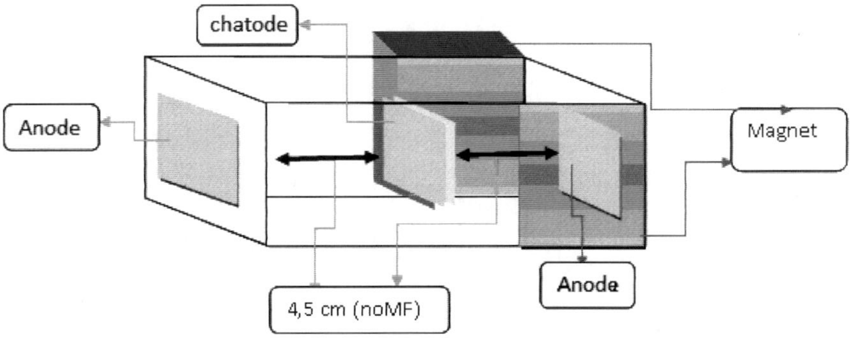

Figure 1. Scheme of equipment in the electroplating process

The mass of the substrate before and after electroplating was weighed to determine the thickness of the Ni layer following the equation:

$$d = \frac{\Delta m}{\rho A} = \frac{m_{\text{Cu/Ni}} - m_{\text{Cu}}}{\rho_{\text{Ni}} A} \tag{1}$$

where d is the thickness of the Ni layer (cm), Δm is mass difference before and after coated (gr), m_{Cu} is mass of substrate Cu, $m_{Cu/Ni}$ is mass of Cu/Ni sample mass, ρ is Ni density (gr/cm3), and area A is Cu/Ni surface (cm^2). The errors of d are obtained from the propagation of errors from $m_{Cu/Ni}$, m_{Cu}, and A so that they become:

$$s_d = \sqrt{\left(\frac{1}{\rho A} s_{m_{\text{Cu/Ni}}}\right)^2 + \left(\frac{1}{\rho A} s_{m_{\text{Ni}}}\right)^2 + \left(\frac{m_{\text{Cu/Ni}} - m_{\text{Cu}}}{\rho A^2} s_A\right)^2} \tag{2}$$

Sheet resistivity was obtained by means of a four-point probes with the equation:

$$R_s = \frac{\pi}{\ln 2}\frac{V}{I} \tag{3}$$

If the measurement is done by variating I and recording the value of V, then Eq. (2) can be written as:

$$V = \frac{\ln 2}{\pi} R_s I \tag{4}$$

If it is assumed that $y = V$ and $x = I$, then by doing a linear regression on the data set (Ii, Vi) by following the equation $y = ax + b$ obtained:

$$a = \frac{\ln 2}{\pi} R_s \tag{5}$$

And then sheet resistivity is given,

$$R_s = \frac{\pi}{\ln 2} a \tag{6}$$

And the error of R_s is

$$s_{R_s} = \frac{\pi}{\ln 2} s_a \tag{7}$$

Where

$$s_a = s_{\hat{V}} \sqrt{\frac{N}{N\Sigma I_i^2 - (\Sigma I_i)^2}} \tag{8}$$

and

$$s_{\hat{V}} = \sqrt{\frac{\Sigma\left(V_i - \hat{V}_i\right)^2}{N-2}} \tag{9}$$

Where N the amount of data [12].

3. Results and discussion

The weighing Cu substrate and Cu/Ni sample are listed in table 1. Then in column 4 the data of weight of Ni was obtained using equation (1) with data ρ_{Ni} =8.91 g/cm^3 and the cross-sectional area of sample A = 7.61 cm^2.

Table 1. Nickel mass and the thickness at various electrolyte temperatures

Electrolyte Temp. (°C)	Mass of Cu (g)	Mass of Cu/Ni (g)	Mass of Ni (g)	Thickness of Ni (μm)
40	4.3531	4.3591	0.0060	0.09 ± 0.43
50	4.1126	4.1292	0.0166	0.24 ± 0.43
60	4.111	4.1323	0.0213	0.31 ± 0.43
70	3.5062	3.5322	0.0260	0.38 ± 0.43
80	4.0827	4.1144	0.0317	0.47 ± 0.43

In Figure 1 a curve of the relationship between the thickness of the Ni layer and the temperature of the solution is displayed.

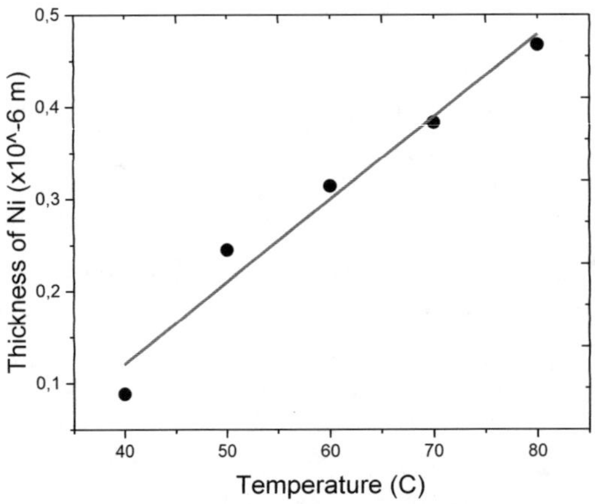

Figure 2. The thickness of the Ni layer at various electrolyte temperature

The thickness and the electrolyte temperature have a linear relationship

$$d = 0.009\ T - 0.243\ (\mu m) \tag{10}$$

with index terminated $R^2 = 0.97$.

Figure 1 shows that the thickness of the Ni layer increases with a higher electrolyte temperature. This is because the temperature has an influence on the rate of reaction. The higher the temperature, the faster the reaction rate. The increase in the reaction rate causes an increase in the kinetic energy of the ion so that the Ni ions move towards the cathode faster. With the high kinetic energy, the layers formed tend to be thicker [13,14]. The magnetic field here has a role to produce a layer with morphology that is formed to become finer.

Furthermore, the data of sheet resistivity of Cu before and after plated with Ni, and the resistivity difference between Cu and Cu/Ni is showed in table 2.

Table 2. The increasing sheet resistivity of Cu after plated with Ni

Electrolyte Temperature (°C)	R_s Cu (x10^{-3} Ω/sq)	Rs Cu/Ni (x10^{-3} Ω/sq)	ΔRs (x10^{-3} Ω/sq)
40	1.21 ± 0.03	1.34 ± 0.02	1.33 ± 0.02
50		1.33 ± 0.03	1.23 ± 0.03
60		1.32 ± 0.03	1.10 ± 0.03
70		1.30 ± 0.02	0.98 ± 0.02
80		1.29 ± 0.04	0.83 ± 0.04

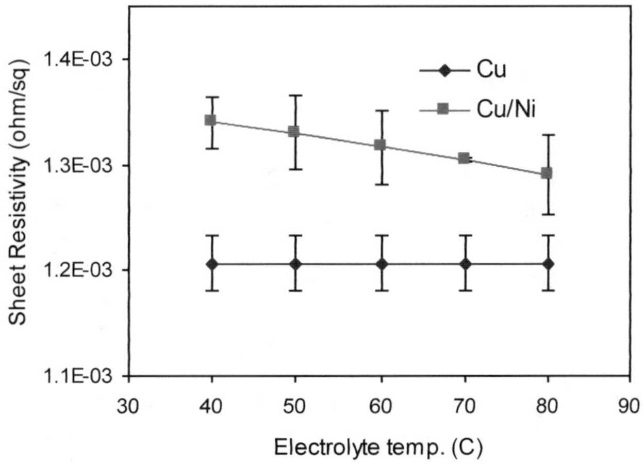

Figure 3. The sheet resistivity of Cu and Cu/Ni

Furthermore, Figure 3 shows the resistivity of Cu/Ni and Cu. The sheet resistivity of Cu/Ni decreases with the higher electrolyte temperature. The decreasing resistivity is caused by the increasing thickness of the Cu/Ni layer. Because nickel is a metal, the thicker the Ni layer that attaches to the Cu substrate, the Ni layer becomes more continuous, so that the sheet resistivity becomes smaller. Here, the electrolyte temperature significantly affects the diffusion of Ni ions that leading to the Cu cathode. If the temperature of the electrolyte solution gets higher, the solution becomes runnier.

From the curve in Fig. 3 it appears that sheet resistivity (R_s) from Cu/Ni has a linear relationship to the temperature of the solution T according to

$$R_s = -1.259 \times 10^{-6}\, T + 0.001\ (\Omega/\text{sq}) \tag{11}$$

where the determination index $R^2 = 0.99$. With this equation, it can be estimated the value of R_s which is related to the temperature of the electrolyte solution T.

4. Conclusion

Based on the results of the study, it was found that the Ni layer on the Cu substrate had succeeded to be synthesized in the variation of the electrolyte solution temperature. In this study, it is also known that the thickness of the Ni layer is proportional to the temperature of the electrolyte solution and inversely proportional to its resistivity.

Acknowledgments

This is the preliminary research from the main research concerning with Cu/Ni temperature sensor. The researcher expresses his deepest gratitude to the Ministry of Research, Technology and Higher Education of the Republic of Indonesia which has provided research funding through the 2019 Higher Education Advanced Applied Research (PTUPT) scheme.

References

[1] Soto VM, Leandro D, Aldaba AL, López JJB, Landazábal JIP, Auguste JL, Jamier R., Roy P and Amo ML 2017 *Sensors* ***17*** 2773.
[2] Yeager CJ and Courts SS 2001 *IEEE Sensors Journal* **1** 352-360.
[3] Ukil A, Braendle A and Krippner P 2012 *IEEE Sensors Journal* **12** 885-892.
[4] Oliveira Jr. ON 1992 *Brazilian Journal of Physics* **22** 60-69.
[5] Abass M, Al-Fawadi E and Al-Samuraee AK 2009 *Materials Science and Technology* 2637-2646.
[6] Cron, R 2014 *Is Cryonics an Ethical Means of Life Extension? Dissertation* (England: University of Exeter).
[7] Boylestad, R.L 2007 *Introductory circuit analysis, Eleventh edition* (Parential Hall Pearson Education International).
[8] Chang, R 2003 *Kimia Dasar Konsep-konsep Inti Edisi Ketiga Jilid 1* (Jakarta: Erlangga-Alih Bahasa Departemen Kimia ITB).
[9] Eckertova, L. 1986. *Physics of Thin Film, 2nd*. (London: Plenum Press).
[10] Indarwati SHN, Gunadi I and Suryono 2017 *Youngster Physcs Journal* **6** 339–347.
[11] Koza JA, Uhlemann M, Gebert A, Schultz L 2008 *Journal of Solid State Electrochemistry* **22** 181-192.
[12] Prioteasa P, Codescu M, Patroi E and Patroi, D 2013 *Optoelectronics And Advanced Materials – Rapid Communications* **7** 90 – 95.
[13] Toifur M 2014 *Prosiding Pertemuan Ilmiah XXVIII HFI Jateng & DIY* (*Indonesia*) vol 26 (Indonesia: HFI Jateng & DIY) p 1-7.
[14] Park JM, Kim DY, Baek JD, Yoon YJ, Su PC, and Lee SH 2018 *Energies* **11** 473.

Potential of Carbon Stocks and Its Economic Values in Tropical Karst Landscape (Case Study in Biduk-Biduk Karst, East Kalimantan, Indonesia)

Danardono[1,3,*], Eko Haryono[2,3] and M. Widyastuti[2,3]

[1] Faculty of Geography, Universitas Muhammadiyah Surakarta, Jl. A.Yani Tromol Pos 1, Surakarta 57162, Indonesia.
[2] Faculty of Geography, Universitas Gadjah Mada, Jl. Kaliurang, Bulaksumur, Yogyakarta 55281, Indonesia.
[3] Karst Research Group, Faculty of Geography, Universitas Gadjah Mada, Jl. Kaliurang, Bulaksumur, Yogyakarta, Indonesia.

[*]Email: danardono@ums.ac.id

Abstract. Karst especially tropical karst landscape has a potential value to store an amount of both organic and inorganic carbon. Tropical karst landscape can store organic carbon derived from forest ecosystem biomass that covers this karst landscape. While inorganic carbon stocks in tropical karst landscape are stored on carbonate rock (CaCO3). This potential can be calculated to identify the economic carbon value to support carbon trade program in East Kalimantan, Indonesia. This research aims to (1) identify carbon stocks value from organic and inorganic carbon in the study area; (2) calculate economic carbon value in the study area. Organic carbon stocks were measured using field measurement to calculate aboveground, belowground, and litter biomass. Calculation of aboveground, belowground, and litter biomass was measured based on the standard method in SNI 7724: 2011. X-Ray Fluorescence (XRF) method was used to identify the CaO concentration in carbonate rock. This CaO concentration was measured to calculate carbon (C) concentration in carbonate rock formation namely inorganic carbon stocks. Economic carbon value was calculated using cost-benefit transfer method. Biduk-Biduk Karst Region has a big value of organic carbon stocks of 0.007 Pg-C and inorganic carbon stocks of 9.02 Pg-C. Therefore, the total carbon stock in Biduk-Biduk Karst Region is 9.037 Pg-C. Based on the total value of carbon stocks, Biduk-Biduk Karst has a total economic value of US$ 45,171,544,296.

1. Introduction

The potential of global carbon storage has become one of the interesting topics in a few decades ago. The impact of global climate change that is increasingly felt in the last few decades makes efforts to reduce factors causing this phenomenon need to be done. One of the biggest factors that occur global climate change is greenhouse gases accumulation in the atmosphere. One of greenhouse gas constituents is carbon dioxide which has a significantly increased. An effort to reduce the amount of carbon in the atmosphere is initiated to inventory the potential of carbon stocks in various ecosystems [1], [2], [3]. The inventory of carbon stocks potential is intended to protect ecosystems that have ability to store carbon. The preservation of these ecosystems can reduce the number of carbon emissions into the atmosphere. Therefore, research to find and calculate the potential of carbon stocks in various ecosystems become one of the interesting topics for climatic researchers.

Until now, research on carbon stocks that have developed in Indonesia is focused in vegetation (green carbon), soil, and ocean (blue carbon) stocks [4], [5], [6]. Even though, there are still a lot of carbon stocks potential that have not been widely studied, such as the carbon stocks potential in rock formations. According to previous research, the potential value of carbon stocks in rock formations is greater than the potential of carbon storage in vegetation and soil. Estimation of carbon storage in rocks is 10.000 Pg-C. This value is just lower than the potential of carbon stocks in the ocean of 38,000 Pg-C [7], [8], [9].

One of the rock formations that have the ability to store the number of carbon is carbonate rocks or limestones [10]. Carbon is stored inside the limestone body which has the chemical formula of CaCO3. Carbon storage in the limestone body is known as inorganic carbon stocks. Carbon in limestone will be stored forever unless there is a limestone extraction or processing process. The processing of limestone for various purposes such as a mixture of cement, building materials, fertilizer materials, and the pest is carried out to produce pure carbonate elements. The limestone purification process is carried out by releasing carbon elements in the limestone body into the atmosphere. Therefore, if limestone is still preserved, the potential of the karst region in storing this amount of carbon can help in reducing global carbon emissions.

In addition, some karst regions also have the potential to store organic carbon [11], [12]. Karst regions which have a double potential to store carbon both organic and inorganic are karst regions with forest ecosystems cover. Karst regions with forest ecosystems cover are often found in the karst region with a tropical climate called tropical karst [12], [13]. In tropical karst regions, there are still many karst landscapes that are covered by tropical rainforest ecosystems. Tropical rainforest ecosystems are one of ecosystems that has big potential to store organic carbon. In addition, carbon storage in this ecosystem is used as an indicator of achieving efforts to reduce carbon emissions. In previous research, potential value of organic carbon stocks in tropical rainforest ecosystems ranges from 190-221 Pg-C [14] or 40% of all terrestrial carbon stocks above the surface [15], [16], [17]. The ability of tropical karst region to store organic and inorganic carbon can help to achieve carbon emissions reduction targets at the regional, national and global scale.

Tropical karst regions are spread in the country which has tropical climate such as Brazil, South Africa, Malaysia, Indonesia, etc. Karst in Indonesia is spread in all of islands such as Pidie, Baturaja, Sarolangun Karst Region in Sumatera Island; Citatah-Rajamandala, Gombong, Gunungsewu, Kendeng, South Malang Karst Region in Java Island; Sangkulirang-Mangkalihat Karst Region in Kalimantan Island; etc. The total area of karst landscapes that cover Indonesia is 140,000 km2 [18]. One of the tropical karst region in Indonesia which is still dominated by forest ecosystems is Biduk-Biduk Karst Region [19]. The Biduk-Biduk Karst Region is part of the Sangkulirang-Mangkalihat Karst Region. The Biduk-Biduk Karst Region is located at the eastern of Berau Regency, East Kalimantan Province. This area was developed on carbonate rocks in the Lebak, Karangan, and Domaring Formation. The Biduk-Biduk Karst Region is relatively close to the coastal area. The dominant land cover in this karst region is primary and secondary forest ecosystems. Primary forest ecosystems that develop in the Biduk-Biduk Karst Region are included in the Dipterocarp forest with dense of vegetation canopy. Meanwhile, secondary forest ecosystems are developed in forestry concessions that have been abandoned for a long time.

Carbon in the Biduk-Biduk Karst Region is stored in the forest ecosystems and carbonate rock formation. Forest ecosystems as a land cover in the study area can store a number of organic carbon. Organic carbon is stored in aboveground in forest vegetation, belowground in the root zone, and forest litter. Meanwhile, karst ecosystem plays a role in storing a number of inorganic carbon. Inorganic carbon is stored in the carbonate rock as a constituent of the Biduk-Biduk Karst Region. The double potential of Biduk-Biduk Karst Region in storing carbon can help to achieve reducing carbon emissions targets in East Kalimantan and Indonesia by 26% (Presidential Regulation No. 61: 2011) through efforts to increase carbon stocks. Therefore, this area can be used as a supporting area in the REDD + program in Berau District, East Kalimantan Province, Indonesia.

The benefit value produced by the Biduk-Biduk Karst Region in terms of carbon storage needs to be quantified. The direct use value of the Biduk-Biduk Karst Region can be determined by calculating an economic value that can be generated from the region's ability to store carbon. This value can be included in the global carbon trade scheme. The global carbon trade is an incentive or compensation scheme for carbon storage countries from carbon-producing countries because it has safeguarded and preserved these carbon reservoirs. The provision of incentives or compensation is calculated based on the value of carbon stocks that can be produced using standard carbon prices in the global market. The greater potential of carbon storage can be generated the greater an amount of incentives or compensation given. The double potential of the Biduk-Biduk Karst Region to store carbon storage can increase the economic value generated in carbon trade schemes. Therefore, it is necessary to quantify organic and inorganic carbon stocks as well as the economic carbon value that can be generated in the Biduk-Biduk Karst Region. Based on this background, this study aims to (1) identify the potential of carbon stocks value from organic and inorganic in the study area; (2) calculate economic carbon value in the study area.

2. Study Area

This research was conducted in the Biduk-Biduk Karst Region. Biduk-Biduk Karst Region is a part of Sangkulirang-Mangkalihat Karst Region which is located in Berau District, East Kalimantan Province, Indonesia. Biduk-Biduk Karst Region is nearly close to the equator line, so this area has a tropical climate zone. Annual rainfall in the Biduk-Biduk Karst Region is 2521 mm/year with the absence of a dry month for a year (rainfall under 100 mm/month). The average air temperature in the study area is 26.4–27.4oC. Based on the Koppen climate classification, the study area is included in the Af climate with the absence of dry months for a year and relatively warm air temperatures each year.

Biduk-Biduk Karst Region was developed in the area with lithology of coral limestones and crystalline limestones in the Domaring Formation (Tmpd); bioclastic limestones with an interlude of breccia in the Lebak Formation (Toml); and bioclastic limestones with coral limestone inserts in the Karangan Formation (Teok) [20]. Three lithologies were deposited in a shallow-neritic environment in Tertiary times. Each lithology in the Biduk-Biduk Karst Region will form different geomorphological units.

Geomorphological units in the study area were controlled by three main processes, namely solutional, marine, and structural. These processes form three geomorphological units in the Biduk-Biduk Karst Region. Geomorphological units that can be identified are karst plain, terrace marine solutional 1, and terrace marine solutional 2. The Karst Plain has a wavy morphology with a slope ranging from 0–13%. Lithology in the Karst Plain is coral limestones and crystalline limestones in the Domaring Formation (Tmpd). Terrace Marine Solutional 1 has a hilly morphology with a slope ranging from 8–80%. Lithology in the Terrace Marine Solutional 1 is bioclastic limestone with an interlude of coral limestone in the Karangan Formation (Teok). Karst features that can be found in this geomorphological unit are "ceruk", ponors, and limestone pavement. Terrace Marine Solutional 2 was exposed after Terrace Marine Solutional 1. Terrace Marine Solutional 2 has hilly morphology with a slope of 8–55%. Lithology of Terrace Marine Solutional 2 is the intersection of bioclastic limestone with breccia in the Lebak Formation (Toml). Karst features that can be found in this geomorphological unit are clustered conical hills and cockpits.

Biduk- Biduk Karst Region has nine terrestrial ecosystems as a result of interactions among the three components of the abiotic, the biotic, and cultural environments. These ecosystems are spreading with the dominance of tropical forest ecosystems belonging to the Dipterocarpaceae forest type, both primary and secondary. Other ecosystems formed by human activities both crop cultivation and built land area are scattered throughout the Biduk–Biduk coastal area. The distribution and types of terrestrial ecosystems in the Biduk - Biduk Karst Region can be seen in Figure 1.

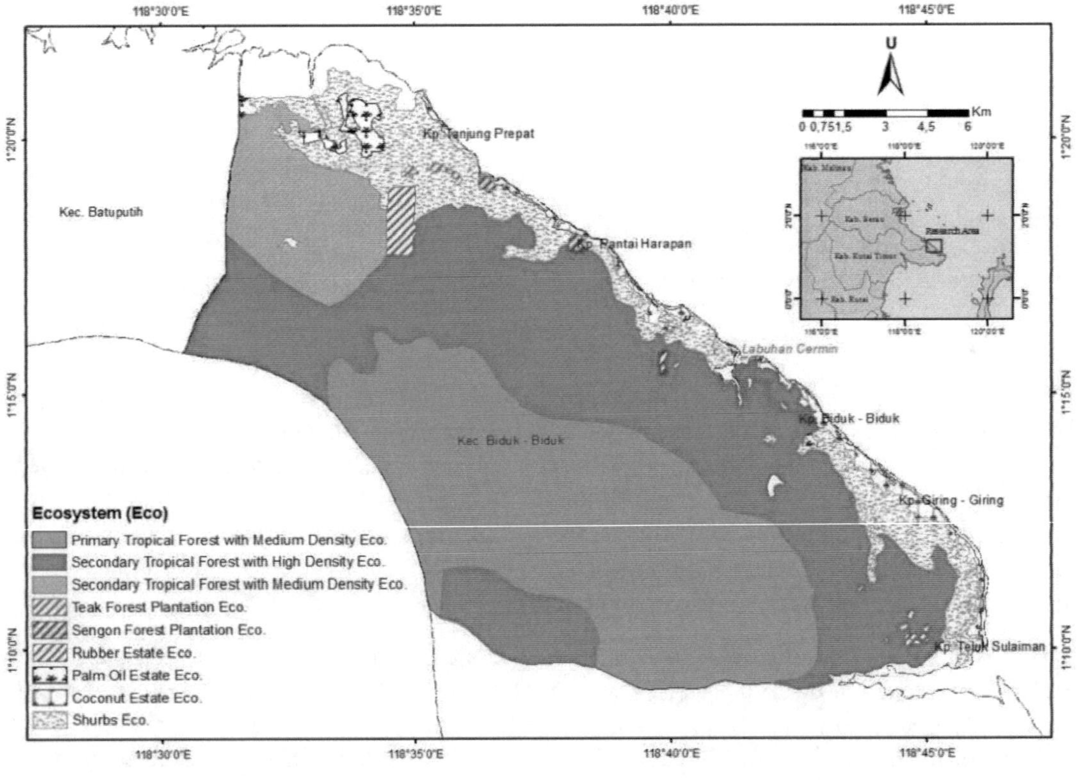

Figure 1. Distribution of Ecosystem in Study Area

3. Method

3.1. Data Collecting

Data collected in this study were organic carbon stocks, NDVI data, limestone volume, CaO content in limestones, and price of carbon. Organic carbon stocks that were measured in this study consists of aboveground carbon in vegetation, belowground carbon in root zone, and litter carbon storage. All of organic carbon stocks data were obtained by making a plot sample measuring 12x12 meters. Organic carbon stocks were measured based on the standard for measuring forest carbon stocks of SNI 7724: 2011. Organic carbon stocks were measured in nine ecosystems at December 2016. NDVI data in each ecosystem were obtained from SPOT 7 image processing in November 2016.

Limestone volume was measured using the cut and fill method. Data used to calculate limestone volume was DEMNAS imagery issued by the Geospatial Information Agency. Limestone volume that was measured in this study was only limestone above the surface ground because the extraction of limestone is only done on limestones above the ground. Limestone volume was calculated in each difference of lithology or rock formations.

CaO content in limestones was obtained from laboratory analysis using X-Ray Fluorescence (XRF) methods. Limestone samples analyzed in the laboratory were limestones in Lebak, Domaring, and Karangan Formation. The difference of depositional environment in each lithological formation can cause differences in the purity of the limestone. Therefore, the inorganic carbon stocks will also vary in each lithological formation.

Price of carbon was obtained by a literature review on previous research and data from government agencies, namely REDD+ Working Group in Berau District. The price of carbon used in this study was $5USD. This value was taken from the lowest compensation value in the carbon trading scheme that has been applied in Indonesia Country.

3.2. Data Analysis

Total organic carbon storage from vegetation, subsurface layer, and litter was analyzed with regression method. Regression was done to create a model of organic carbon stocks determination in vegetation layer based on the NDVI value. NDVI was calculated using the formula number 1. Total organic carbon stocks in the study area can be calculated by summing the value of organic carbon stocks in the nine ecosystems.

$$\text{NDVI} = \frac{(NIR-Red)}{(NIR+Red)} \quad (1)$$

Where NIR is reflectance value of Near-Infrared Band; and Red is reflectance value of Red Band.

Inorganic carbon stocks in the limestone bodies were calculated using the formula number 2.

$$\text{M C} = \left(\frac{Ar\ C}{Mr\ CaCO3}\ x\ \frac{Mr\ CaCO3}{Mr\ CaO}\ \text{x \%CaO x BJ CaCO3 x V CaCO3}\right) \quad (2)$$

Where M C is the mass of carbon in limestone (g); Ar C is the relative atomic mass of carbon (12 g/mol); Mr $CaCO_3$ is the relative atomic mass of $CaCO_3$ (100 g/mol); Mr CaO is the relative atomic mass of CaO (56 g/mol); % CaO is the percentage of CaO content in limestone from laboratory analysis; BJ $CaCO_3$ is the specific weight of CaCO3 (2.71 g/cm^3); and V $CaCO_3$ is the volume of limestone. The total value of carbon stocks potential can be calculated by summing the potential of organic carbon stocks and inorganic carbon stocks.

Journal of Physics: Conference Series **1373** (2019) 012030 doi:10.1088/1742-6596/1373/1/012030

Economic carbon value was calculated using the cost-benefit method. The cost-benefit method is a method to calculate economic carbon value using price of carbon from previous studies or a carbon price standards that have been applied in Indonesia Country. The price of carbon used in this study was $5USD per tonne of carbon. The economic value of carbon in the study area can be calculated using formula number 3.

$$ECV = (OCS + ICS) \times CV \tag{3}$$

Where ECV is a total economic carbon value that can be generated; OCS is the total amount of organic carbon stocks (tons); ICS is the total amount of inorganic carbon stocks (tons); and CV is a standard price of carbon per tones.

4. Results

4.1. Organic Carbon Stocks

Organic carbon stocks are the number of carbon stored in vegetation, litter, and root zone. Organic carbon stocks in Biduk-Biduk Karst Region can be calculated by regression analysis between NDVI values and the organic carbon value in the field measurement. Regression analysis will produce a model that can be used to calculate organic carbon stocks using NDVI data. NDVI is assumed to describe vegetation density in the study area. The NDVI value was obtained by processing imagery that has a red band and near-infrared band.

Modeling results using SPOT 7 show that NDVI values in study area ranging from 0 – 0,904. NDVI with the value of 0 illustrates low vegetation density. This value is spread in residential areas and water bodies like rivers or lake. Whereas, NDVI with the value of 0,904 illustrates high vegetation density both primary and secondary forest ecosystems. The rate of NDVI in the study area has a big value of 0.703. This condition happens because land cover in the study area is still dominated by forest ecosystems, either primary or secondary, which have big NDVI values. The distribution of NDVI values in the study area can be seen in Figure 2.

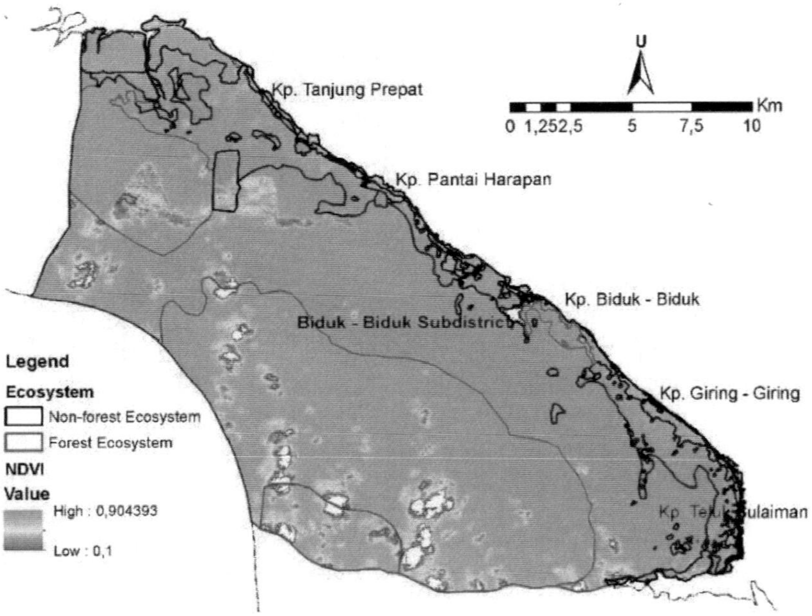

Figure 2. NDVI Map in Biduk-Biduk Karst Region

Based on NDVI values, the value of carbon stocks in aboveground (vegetation), belowground (root zone), and litter can be calculated. This value can be approached using the result of a regression formula between NDVI values and organic carbon stocks in each carbon sink. The regression used in the study area is exponential regression. The regression result is $y = 0.0098e^{13.598x}$. This regression was chosen because it has a greater coefficient of determination and smaller RMSE value than the other regression. RMSE value from this exponential regression is much lower than the standard deviation of aboveground carbon stocks. Spatial distribution of aboveground carbon in the study area has a similar pattern with NDVI value (see Figure 3). The biggest value of aboveground carbon is found in the area with big NDVI values in primary and secondary tropical forest ecosystems.

Based on the spatial distribution of each organic carbon stocks in Figure 3, potential of organic carbon stocks in nine ecosystems can be calculated. The results of organic carbon stocks calculation in nine ecosystems in the study area can be seen in Table 1.

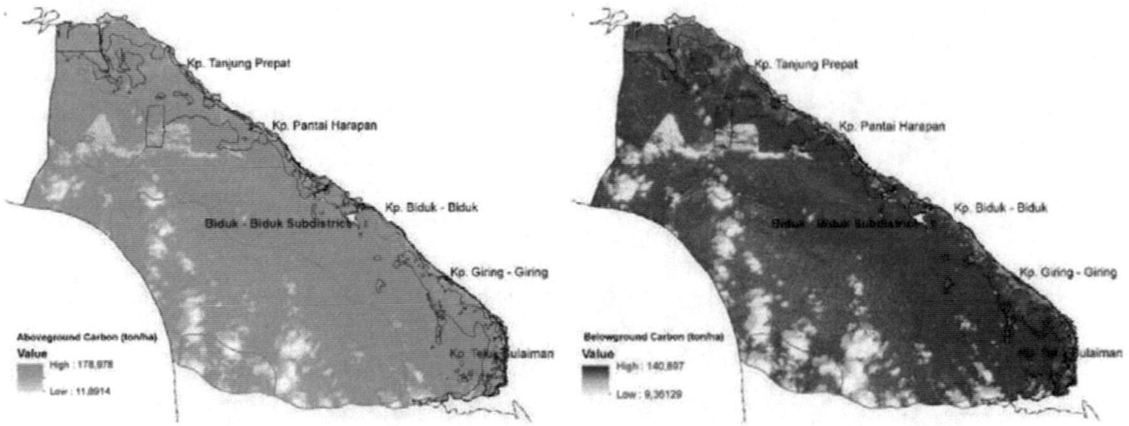

Figure 3. Distribution of Aboveground and Belowground Carbon in Biduk-Biduk Karst Region

Table 1. Organic Carbon Stocks in Biduk-Biduk Karst Region

Ecosystem	Aboveground Carbon Stocks (ton/ha)	Underground Carbon Stocks (ton/ha)	Litter Carbon Stocks (ton/ha)	Total Carbon Stocks (ton/ha)	Area (ha)	Organic Carbon Stocks (ton)
Primary Tropical Forest with Medium Density	132.001	103.916	6.441	242.358	11721.20	2,840,728
Secondary Tropical Forest with High Density	136.997	107.848	6.141	250.986	13868.72	3,480,851
Secondary Tropical Forest with Medium Density	131.976	103.896	2.539	238.412	2701.49	644,068
Teak Forest Plantation	110.438	86.941	4.114	201.492	43.04	8,673
Sengon Forest Plantation	114.393	90.054	4.429	208.877	68.98	14,408
Rubber Estate	111.780	87.997	3.249	203.026	271.16	55,052
Palm Oil Estate	125.573	98.855	1.362	225.790	298.84	67,475
Coconut Estate	93.672	73.742	4.821	172.234	657.93	113,319
Shrubs	82.482	64.932	6.811	154.225	3736.65	576,284
	Total					**7,773,358**

The biggest density value of organic carbon stocks is found in high-density secondary forest ecosystems of 250.9 tons/ha. Meanwhile, the lowest value is in the shrubs ecosystem of 82.4 tons/ha. The average value of organic carbon stocks density in the study area is relatively high of 210.8 tons/ha. This condition happens because forest ecosystems with high organic carbon stocks are spread evenly across all study areas. Secondary tropical forest ecosystem with high vegetation density is dominated in the study area. This secondary forest formed in abandoned forest concessions since the early of the 1990 s. The total value of organic carbon stocks in the Biduk-Biduk Karst Region is 0.007 Pg-C.

4.2. Inorganic Carbon Stocks

Inorganic carbon stocks were calculated by considering volume and carbon content inside limestone bodies ($CaCO_3$). Limestone volume to calculate inorganic carbon stocks was carried out in each difference rock lithology. Each rock lithology is assumed having different values of carbonate purity so the carbon content will also varied. Calculation results of limestone volume in each lithology or rock formation are presented in Table 2. Based on this calculation, the total volume of the Biduk-Biduk Karst Region is 31,822,183.653 m^3 in four lithological formations.

Table 2. Carbonate Rocks Volume in each lithology in Biduk-Biduk Karst Region

Lithology	Volume (m^3)
Domaring Formation	1,455,389,811
Golok Formation	719,749,357
Karangan Formation	9,377,983,970
Lebak Formation	20,329,060,515

Inorganic carbon stocks can be calculated by considering the volume and purity of carbonate content in the limestone. The purity of limestone was analyzed using rock samples in the four lithological formations to obtain CaO values. CaO values was measured using XRF method. The result of CaO calculation in three lithological formation using XRF method can be seen in Table 3.

Table 3. Results of XRF method to identify CaO in each lithology in Biduk-Biduk Karst Region

Lithology	CaO (%)
Domaring Formation	47.6
Golok Formation	40.7
Karangan Formation	46.2
Lebak Formation	50.3

Rock samples with the highest CaO values are found in the Lebak Formation of 50,3%. While the lowest CaO value is found in the Golok Formation. The difference value of CaO happens due to differences in the purity of limestone. Lebak Formation has a high value of CaO because this lithological formation contains of coral limestone. Whereas Golok Formation has lithology of clastic limestone. Depositional environment of coral limestone in shallow sea causing limestone impurity is not much compared than clastic limestone which has depositional environment in the neritic zone of the sea.

A number of inorganic carbon stocks in the study area can be seen in Table 4. The biggest value of inorganic carbon stocks is found in the area with Lebak Formation of 5,93 Pg-C. Inorganic carbon stocks in the Lebak Formation have a big value due to the high volume of limestone with higher purity compared than other rock formations. Based on the calculation results, the total value of inorganic carbon stocks in the Biduk-Biduk Karst Region is 9,03 Pg-C.

Table 4. Inorganic Carbon Stocks in Study Area

Lithology	Volume (m^3)	Limestone Density (ton/m^3)	CaO (%)	$CaCO_3$ (%)	$CaCO_3$ (ton)	Inorganic Carbon Stocks (ton)
Domaring Formation	1,455,389,811	2.71	47.6	85.00	3,352,490,430	402,298,852
Golok Formation	719,749,357	2.71	40.7	72.68	1,417,610,622	170,113,275
Karangan Formation	9,377,983,970	2.71	46.2	82.50	20,966,827,661	2,516,019,319
Lebak Formation	20,329,060,515	2.71	50.3	89.82	49,484,200,464	5,938,104,056
		Total				**9,026,535,502**

4.3. Economic Carbon Value

The important value of the Biduk-Biduk Karst Region as carbon storage can provide economic benefits. The economic value of potential carbon storage can be calculated and included in global carbon trade schemes. The ability to store an amount of carbon can provide benefits in the form of incentive or compensation money from carbon-producing countries. A distinctive feature of the Biduk-Biduk Karst Region in the tropical zone makes the potential of organic carbon stocks in this area similar to other tropical rainforest ecosystems. In addition, the ability of the study area to store some of inorganic carbon in the limestone body can increase the economic carbon value.

The economic carbon value in tropical karst areas, especially in the Biduk-Biduk Karst Region, was calculated from two potential of carbon storage both organic and inorganic carbon stocks. Based on the calculation, the total economic carbon value that can be generated in the study area is $45,181billion USD. The economic value of inorganic carbon stocks can be produced $45,132billion USD or 98% of total economic carbon value in the Biduk-Biduk Karst Region. Meanwhile, the economic value of organic carbon stocks can be produced $38,866 million USD or 2% total economic carbon value in the study area. The result of economic carbon calculation in the Biduk-Biduk Karst Region is presented in Table 5.

Table 5. Economic Carbon Value in Biduk-Biduk Karst Region

	Total Stocks (ton)	**Economic Value per ton (USD)**	**Economic Carbon Value (USD)**
Organic Carbon Stocks	7,773,358	5	38,866,790
Inorganic Carbon Stocks	9,026,535,501	5	45,132,677,506
Total	**9,037,339,666**		**45,171,544,296**

5. Discussion

Tropical karst region especially Biduk-Biduk Karst Region has a great potential to store an amount of carbon. The carbon stored in the tropical karst region is from organic and inorganic. This condition can be seen in the Biduk-Biduk Karst Region which has potential to store organic carbon in the forest ecosystem cover and inorganic carbon in the carbonate rocks. A double potential of carbon stocks in the tropical karst region, especially the Biduk-Biduk Karst Region, make this region has a high carbon stock value.

Based on the measurement results, the value of organic carbon stocks in tropical karst region has a high value as well as other regions in the tropical climate zone. The density of organic carbon stocks in the Biduk-Biduk Karst Region has an average value of 210.8 tons/ha. This value is similar with the value of organic carbon stocks in other tropical climate regions such as Malua Forest, Malaysia [2], Bukit Timah Nature Reserve, Singapore [16], Houzhai Watershed, China [12], and South China Karst [21]. A high density of vegetation cover is the factor which can affect highly organic carbon stocks in tropical karst regions such as Biduk-Biduk Karst Region [16], [3]. The dominance of primary and secondary forest ecosystems in the study area makes carbon stocks in the aboveground (vegetation) have a high value. Vegetation with a large diameter that is still found in many forest ecosystems in the study area can store big amounts of biomass. In addition, the absence of logging activities by forest companies that carried out in the study area in the past of 20 years has made vegetation growth optimally.

The similarity of organic carbon stocks in the study area with forest ecosystems in non-karst regions is contrary with a few previous studies. In previous study, forest ecosystems in the karst region have lower organic carbon stocks compared than forest ecosystems in non-karst regions [22], [11], [12]. This condition is mainly due to that ecosystem productivity in the karst region is limited by the availability of water in the near surface, the thinness of the soil layer, and the limited nutrient content of the soil [23], [13]. This difference finding can be due to the measurement of above-ground carbon in the study area using one allometric formula for various species of vegetation. Therefore, it is very possible that aboveground carbon calculation is overestimated. The measurement of potential carbon stock per vegetation species using different allometric formula needs to be done to improve the validity of organic carbon stocks measurement results.

Organic carbon stocks in the Biduk-Biduk Karst Region are stored in three carbon sinks in the vegetation, root zone, and litter. The highest percentage of organic carbon stocks is contributed by above-ground vegetation of 55%. Vegetation has the highest value because it has potential to store biomass in all parts of the plant such as leaf, root, stem, and branch. Meanwhile, other carbon stocks are only stored in one part such as belowground carbon only stored in plant roots. Several studies conducted in other forest areas such as Malua Forest, Malaysia [2], Bukit Timah Nature Reserve, Singapore [16], Houzhai Watershed, China [12], Kinsangani, Kongo [17], Guizhou Karst, China [11] showed that vegetation carbon stocks have a higher value than belowground and litter carbon stocks.

Besides being able to store organic carbon, tropical karst regions also have potential to store some of carbon in carbonate rocks. Potential of carbon storage in carbonate rocks in the study area has a value of 9.03 Pg-C. This value is greater than organic carbon stocks potential from vegetation of 0.007 Pg-C. The percentage of inorganic carbon stock is 98% from total carbon stocks in the study area. Therefore, it can be concluded that the potential of inorganic stocks is greater than organic carbon stocks. This condition is similar with previous studies explained that potential of carbon storage in rock formations is one of highest carbon stocks on earth below ocean carbon stocks [24]. Previous studies stated that the potential of carbon stocks in rock formations is 10,000 Pg-C, while the potential of carbon stocks in vegetation is only 560 Pg-C [8], [9].

On the other side, inorganic carbon stocks in tropical karst regions are rarely studied. Research on carbon stocks which has been developed is organic carbon storage in vegetation and forest ecosystems [4], [5], [6]. At the global scope, efforts to reduce carbon contained in REDD+ program are still focusing on efforts to conserve organic carbon stocks in forest ecosystems. Various efforts to conserve forest ecosystems have been carried out in various ways to maintain the presence of organic carbon stocks. Whereas inorganic carbon stocks that have greater value are also very vulnerable to exploitation by mining activities. Limestone is to be one of raw materials for making building materials, cement, cosmetics, and other uses. If this is allowed, an amount of carbon released into the atmosphere due to the limestone extraction is higher and higher.

Potential of inorganic carbon stocks in tropical karst region can help increase carbon storage at the global level. The presence of inorganic carbon stocks can increase the economic carbon value in the carbon trade scheme. This condition can be seen in the study area where economic carbon value is $45billion USD. This total value is greater than the economic value of organic carbon ($38 million USD). Therefore, inorganic carbon storage is very important to be taken into carbon trade schemes and carbon reduction efforts through REDD + schemes. Efforts to conserve karst ecosystems as a sink of inorganic carbon also need to be considered, as well as forest ecosystems.

6. Conclusion

Tropical karst areas have a great potential to support efforts in reducing carbon emissions in the atmosphere. The ability of the tropical karst region to store an amount of carbon both organic and inorganic can help to reach the target of reducing global carbon emissions. Organic carbon stocks are stored in forest ecosystems that dominantly cover tropical karst areas. These organic carbon stocks consist of aboveground carbon in vegetation, belowground carbon in the root zone, and litter carbon. The rate of organic carbon density in the study area is 210.8 ton/ha. This potential value of organic carbon storage in tropical karst areas, especially in Biduk-Biduk Karst Region, has a similar value with organic carbon storage in non-karst forest areas. Calculation of organic carbon stocks in the study area is still carried out with an allometric approach assuming a homogeneous forest ecosystem condition. Meanwhile, forest ecosystems in the karst region have a heterogeneous vegetation species. Therefore, calculation of organic carbon stocks using allometric formulas in each vegetation species needs to be done so the calculation results are not overestimated.

The others potential carbon stocks owned by tropical karst areas is inorganic carbon stocks. Inorganic carbon is stored in limestone bodies. The total potential value of inorganic carbon stocks in the study area is 9.03 Pg-C. This value is 98% greater than the potential of organic carbon stocks. This calculation still uses a medium scale of elevation data. Therefore, a more detailed calculation of inorganic carbon stocks needs to be done using a detailed scale of elevation data.

The ability of tropical karst areas to store organic and inorganic carbon can provide higher economic benefits. The tropical karst area, especially the Biduk-Biduk Karst Region has the potential to generate an economic value of $45,171,544,290USD. The biggest contribution to total economic carbon value in the study area came from inorganic carbon stocks of $45,132,677,506USD or 98% of total economic carbon value. Both organic and inorganic carbon stocks in tropical karst areas can be included on carbon trading schemes to provide additional income for Indonesia Country.

Acknowledgments

This research was supported and funded by Karst Research Group Universitas Gadjah Mada and Yayasan Keanekaragaman Hayati (KEHATI) through Tropical Forest Conservation Act (TFCA) Kalimantan Program entitled "Characterization and Revitalization of Important Values in Sangkulirang-Mangkalihat Karst Region, East Kalimantan Province, Indonesia".

References

[1] Saatchi S.S, Harris N.L, Brown S, Lefsky M, and Mitchard E.T.A 2011 *Proceedings of the National Academy of Sciences of the United States of America* **108** 9899-9904.
[2] Saner P, Loh Y.Y., Ong R.C. and Hector A. 2012 *Plos One* **7** 1-11 .
[3] Gara T.W., Murwira A, Dube T, Sibanda M, Rwasoka D.T, Ndaimani H, Chivhenge E., and . Hatendi C. M .2017 *Southern Forests* **79** 31-40.
[4] Rutishauser E, Noor'an F, Laumonier Y, Halperin J, Rufi'ie K, Hergoualc'h L and Verchot 2013 *Forest Ecology and Management* **307** 219–225.
[5] Englhart S, Franke J, Keuck V, and Siegert F 2014 *Land Use and Land Cover Mapping in Europe* **18** 411–427.
[6] Arifanti V.B, Dharmawan I.W.S, and Wicaksono D 2014 *Jurnal Penelitian Sosial dan Ekonomi Kehutanan* **1** 13-31 .
[7] Houghton R.A and Skole D.L, Carbon. In: Turner, B.L., W.C. Clark, R.W. Kates, J.F. Richards, J.T. Matthews, and W.B. Meyer (eds.) 1990 *The Earth as Transformed by Human Action* (New York: Cambridge University Press).
[8] Dixon R.,Winjum J.K., Andrasko J., Lee, P. Schroeder 1994 *Climatic Change* **27** 71-92.
[9] Schlesinger W.H. 1997 An Overview of The Carbon Cycle. In: Lal, R., J.Kimble, E.Levine, and B.A.Stewart, eds. *Soils and Global Change* (Boca Racton: Lewis Publishers).
[10] Haryono E, Danardono, Mulatsih S, Putro S.T, and Adji T.N 2016 *Acta Carsologica* **45**173–185
[11] Liu Y, Liu C, Wang S, Guo K, Yang J, Zhang X, and Li G 2013 *Plos One* **8**1–9.
[12] Liu C, Liu Y, Guo K,Wang S, Liu H, Zhao H, Qiao X, Hou D, and Li S. 2016 *Agriculture,Ecosystems and Environment* **235** 91-100..
[13] Guo K, Liu C.C, Dong M. 2011 *J.Plant Ecology* **35** 991-999.
[14] Prentice I.C, Farquhar, Fasham M.J.R, Goulden M.L, Heimann M, Jaramillo V.J, Kheshgi H.S, Le Quéré C, Scholes R.J, and Wallace D.W.R 2001 *The scientific basis (Contribution of Working Group I to the Third Assessment Report of the Intergovernmental Panel on Climate Change)*, edited by J. T. Houghton et al. (Cambridge: Cambridge University Press).

[15] Pan Y, Birdsey R.A, Fang J, Houghton R, Kauppi P.E, Kurz W.A, Phillips O.L, Shvidenko A, Lewis S.L, Canadell J.G, Ciais P., Jackson R.B, Pacala S.W, McGuire A.D, Piao S, Rautiainen A, Sitch S, and Hayes D. 2011 *Science* **333** 988-993.

[16] Ngo K. M, Turner B.L, Muller-Landau H.C, Davies S.J, Larjavaara M, Hassan N.F.b.N, Lum S. 2013 *Forest Ecology and Management* **296** 81-89 .

[17] Doetterl S, Kearsley E, Bauters M, Hufkens K, Lisingo J, Baert G, Verbeeck H, and Boeckx P, 2015 *Plos One* **10** 1-14.

[18] Haryono E. 2017 *Asian Trans-Disciplinary Karst Conference* 203–207.

[19] Danardono, Haryono E, and Widyastuti E, *E3S Web of Conferences* **76** pp. 1-7.

[20] Djamal B, Sudana D, Soetrisno, Baharuddin, Hasan K. 1995 *Peta Geologi Lembar Tanjung Mangkalihat, Kalimantan* (Bandung: Pusat Penelitian dan Pengembangan Geologi).

[21] Brandt M, Yue Y, Wigneron J.P,Tong X, Tian F, Jepsen M.R, Xiao X, Verger A, Mialon A, Al-Yaari, Wang K, and Fensholt R 2018 *Earth Future* **6** 1017-1028.

[22] Yu W.L, Dong D, and Ni J. 2010 *J. Subtropical Resources Environment* **5** 337-341.

[23] Zhang X.B. and Wang K.L. 2009 *Earth Environment* **5**, 26-30 .

[24] Schumacher B.A, 2002 *Method for The Determination of Total Organic Carbon (TOC) in Soils and Sediments* (Las Vegas: United States Environmental Protection Agency).

Journal of Physics: Conference Series **1373** (2019) 012031 doi:10.1088/1742-6596/1373/1/012031

Earthquake Risk Reduction Study with Mapping an Active Fault at the Southern of East Java

Wien Lestari[1,*], Amien Widodo[1], Dwa Desa Warnana[1], Firman Syaifuddin[1]

[1]Geophysical Engineering Department, ITS Raya Street, Institut Teknologi Sepuluh Nopember, Surabaya, East Java, 60111

[*]Email: wien@geofisika.its.ac.id

Abstract. The southern part of Java Island is located between the two plates of the subduction zone, therefore this region had an active tectonic and volcanic activities. Those made a deformation of the geology around the southern part of Java island. This research was conducted to know the volume of deep geological structures to provide an overview of disaster warning systems in southern East Java region. The research that has been carried out using numerous active geophysical methods such as seismic methods had not been able to indicate internal structures. The needed for innovation used the integration of gravity and magnetotellurics methods. Bouguer anomaly gravity map became the supported data for interpretation. It showed low-density zones (0-80 mGal) between Pacitan-Trenggalek and beneath the Lumajang area which indicates the fault structure. Magnetotellurics (MT) which had a high penetration so that it is able to map deep active geological structures below the surface. The research area covered 8 regencies at the southern of East Java namely Pacitan, Trenggalek, Tulungagung, Blitar, Malang, Lumajang, Jember, and Banyuwangi. Magnetotelluric 1D processing showed indications of faults with orientations tend to be NE-SW between Pacitan and Trenggalek with depth (500 m – 8000 meter). Indications of faults with orientation tend to be NorthEast-SouthWest in Lumajang (100 - 500 meter).

1. Introduction

South East Asia Region was bounded by a Eurasian Plate from north to west and Indian-Australian Plate at the southern (Figure 1.) during the Cenozoic geologic time with some active microplates like Sunda and Banda Plate made up the unique tectonic [1]. Therefore, Indonesian ideal to earthquake and volcanism activity [2][3].

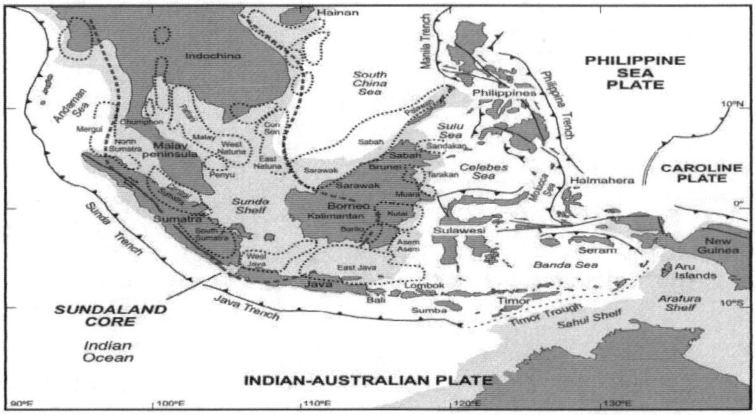

Figure 1. Indonesian Margin and Major Basins

However, many basins in Indonesia are typically back arc basin that is located 100 km from the active volcanic arcs included offshore Northwest and East Java Basin [4]. Sunda-Banda subduction margin built the Quaternary Volcanic Arcs [5]. Current activities of earthquake occured on the southern part of East Java which caused by tectonic activities like Grindulu Fault (NE-SW trend) in Pacitan, Klangon Fault in Madiun was triggered in 2015 and based on Meteorological, Climatological, and Geophysical Agency record data showed the rapid earthquake with less magnitude that occured near Malang and Lumajang since a few last year until now. Figure 2 illustrated the active deformation in Java which accommodated by small-scale structures (km to tens of kilometers) with a wide spread and no dominant structures such as the sumatra fault [6]. Unfortunately, The revision of Indonesia Earthquake source map (2017) did not inform the land fault system which parallel with Opak Fault in Jogjakarta which had a trend tend to NE-SW.

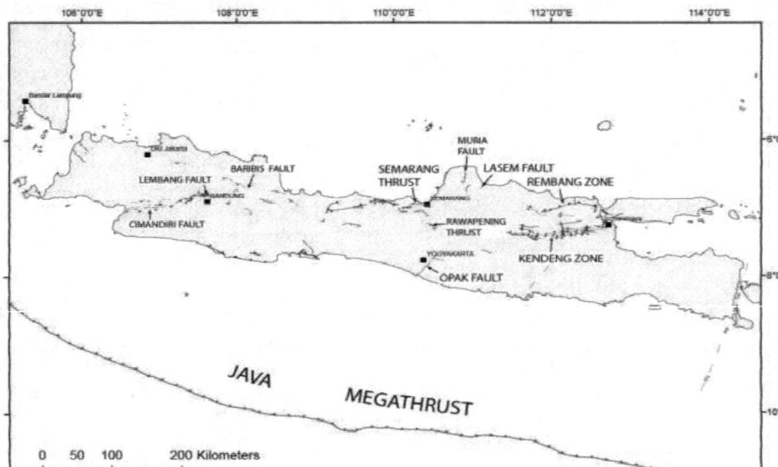

Figure 2. The active fault maps on Java that have been identified and summarized in the National Earthquake Map 2017

This research have the objectives to map the active structural geology which develop at the southern east java particularly the fault at the surface (shallow) and basement (subduction zone) also its associated with the quaternary volcanic arcs. 1D Magnetotelluric-Electromagnetic methods is implemented because of non-destructive methods, high penetration and high resolution in geometry delineation than seismic methods. This method measured the resistivity physical properties of the subsurface which the target had the high resistivity value relatively with its surrounding. Acquisition with the magnetotelluric method are conducted in numerous active fault researches like San Andreas Fault [7];[8], Kalabsha fault in Egypt [9]. Measurement data points are 12 sounding points through numerous districts at the southern East Java (Pacitan, Trenggalek, Tulungagung, Blitar, Malang, Lumajang, Jember, Banyuwangi). Density Bouguer Anomaly from USGS as the supported data for validation of the geological structure. This research is expected to provide comprehensive and more detailed information about the deformation process of the subsurface structure at the Southern of East Java as well as the phenomenon of earthquakes due to active faults seen from the subsurface resistivity distribution and its relation to other geophysical methods. Mapping the active fault became the strong activities to do because the natural hazards aspect must consider for the development of spatial planning in Indonesia.

2. Methodology

This research was conducted using a field approach, modeling the properties of resistivity and magnetism using one of the geophysical methods (Magnetotelluric). The modeling results are followed by an interpretation process that is correlated with geological data or other supporting geophysical methods. Magnetotelluric data acquisition as is illustrated in fig. 3 was carried out for 15 days from 30 June 2018 - 15 July 2018, covering districts located along Southern East Java. (Pacitan, Trenggalek, Tulungagung, Blitar, Malang, Lumajang, Jember and Banyuwangi District). Table 1 shows the retrieval data point.

Table 1. Magnetotelluric Measurement Data

MT POINT	DISTRICT/CITY	VILLAGE	COORDINATE		ELEVATION (m)
			Y	X	
ST01	Pacitan	Ngadirojo	9090313	535347	8
ST03	Trenggalek	Durenan	9100858	588767	108
ST04	Tulungagung	Boyolangu	9101443	596405	52
ST05	Blitar	Maron-Purwokerto	9103696	618442	118
ST06	Blitar City	Pakunden	9105959	626343	163
ST07	Blitar	Tambakan	9109201	643004	317
ST08	Malang	Ngadjum	9102811	670851	325
ST09	Malang	Blayu	9109236	685151	89
ST11	Lumajang	Grobogan	9112046	746186	154
ST13	Jember	Patrang	9100140	800691	132
ST14	Jember	Kemuning Sari	9097408	786431	159
ST15	Banyuwangi	Wonorejo	9122356	209451	68

Figure 3. Magnetotelluric data points along the southern of East Java (yellow pin)

Figure 4 showed the configuration of data points retrieval taken for 8 hours with the radius is 20 meter which is divided into 3 variations of measurement for each station are 30 minutes for High Frequency 16 kHz, 90 minutes for Middle Frequency 4 kHz and 6 hours for Low Frequency 128 Hz.

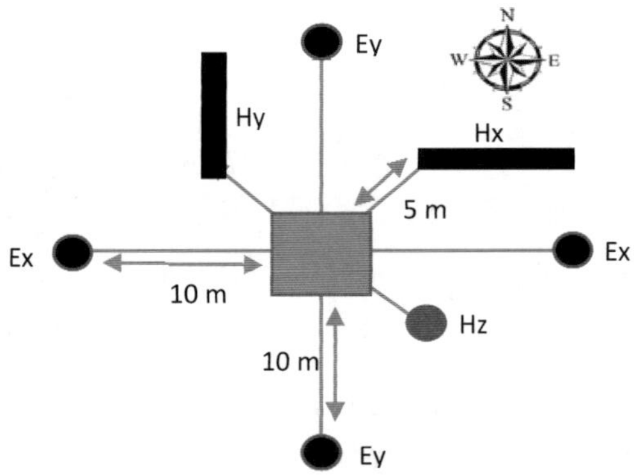

Figure 4. Magnetotelluric Acquisition Configuration

The data is obtained in time series with the format .ats as illustrated in Figure 5. The processing flow of Magnetotelluric data in Figure 6 are done in two steps namely pre-processing and main processing. Further, time series data on magnetic fields and electric fields are transformed into frequency series using the Fast Fourier Transform (FFT) which illustrated in Figure 7.

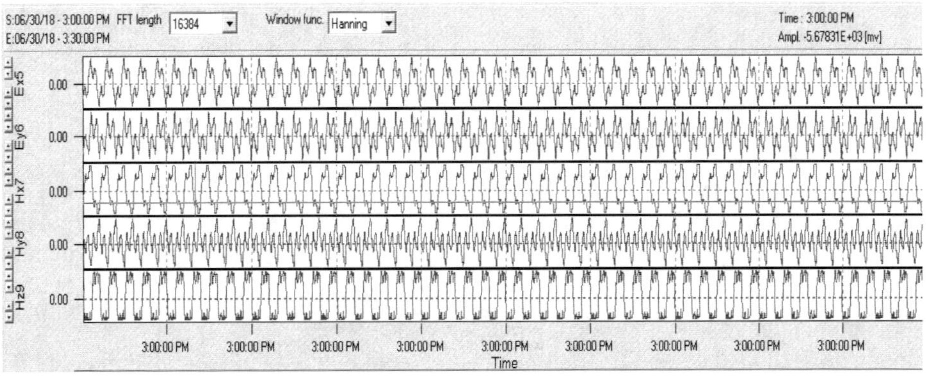

Figure 5. Time Series Data of ST09 Point

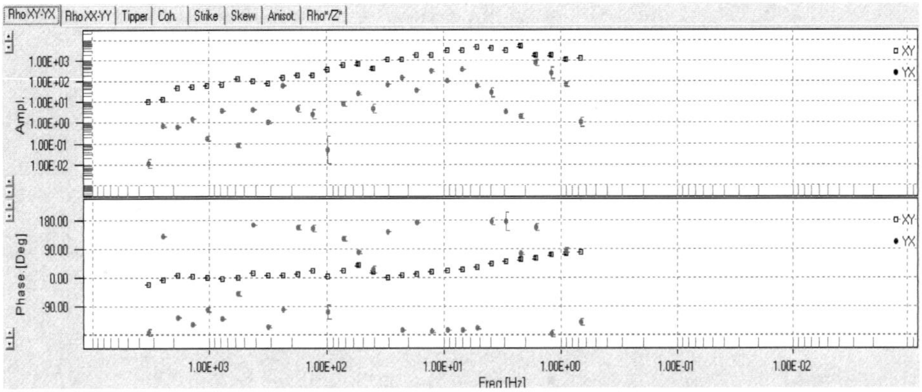

Figure 6. ST 09 Data in Frequency Domain

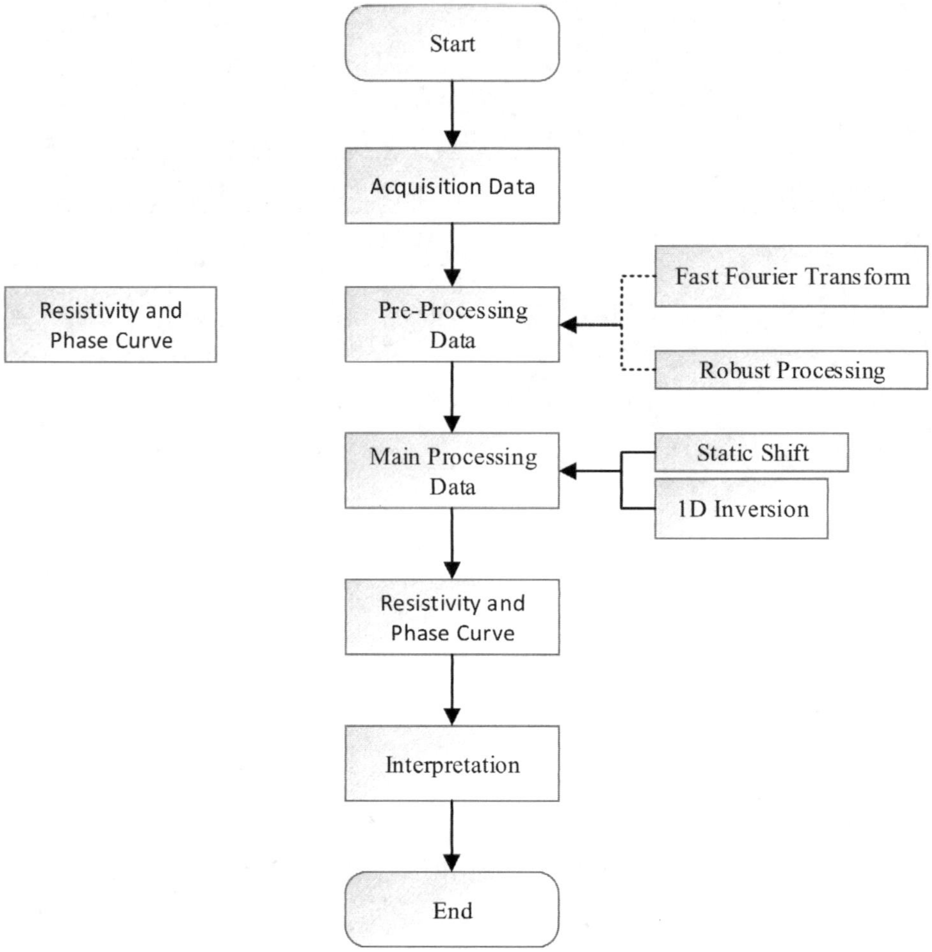

Figure 7. Data Processing Flow

3. Result

This study data was categorized into primary magnetotelluric data and secondary data. Figure 8 described the secondary data in the form of gravity data is taken from the satellite that is used for the integration process when a regional subsurface interpretation. Earthquake data is used with average Magnitude <5 near the study area with a years period of 2004 – 2017 to reinforce the results of the interpretation.

The Bouguer Anomaly distribution map showed an indication of a weak zone at the measurement station between Pacitan (ST01) - Tenggralek District (ST03) and an indication of fault zone by qualitatively looking at the distribution of Bouguer Anomaly data in Lumajang (ST11). Quantitatively, zones that are considered to be potential for weak zone structures had the density value 0-80 mGal. Meanwhile, figure 9 established the apparent resistivity versus period data profile in Transverse Electric (TE) mode as well as Transverse Magnetic (TM) mode (fig.10). The profile indicated a low resistivity value in the area between ST01 (Pacitan) and ST02 (Trenggalek) as an active fault structure.

The cross-section in figure 11 is the results of 1D inversion using Bostic Filter that formed the existence of low resistivity zone (NE-SW) at a depth of 2 km to 8 km above sea level at the area near the Pacitan District and below the Lumajang with a shallow depth of 500-1000 meter (blue) associated with a deep low resistivity zone (green) that predicted as the subduction zone system beneath the Lumajang.

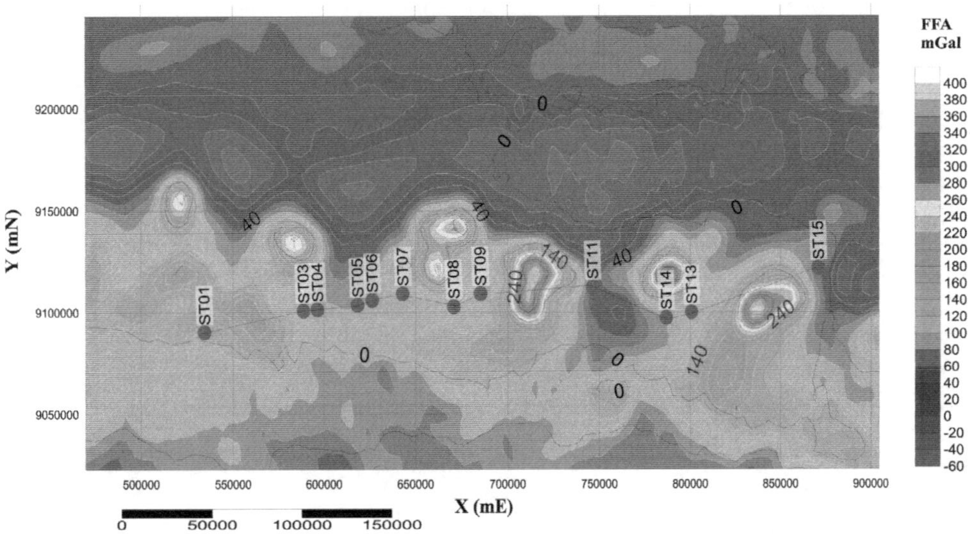

Figure 8. Bouguer Anomaly Gravity Data using Free-Air Correction

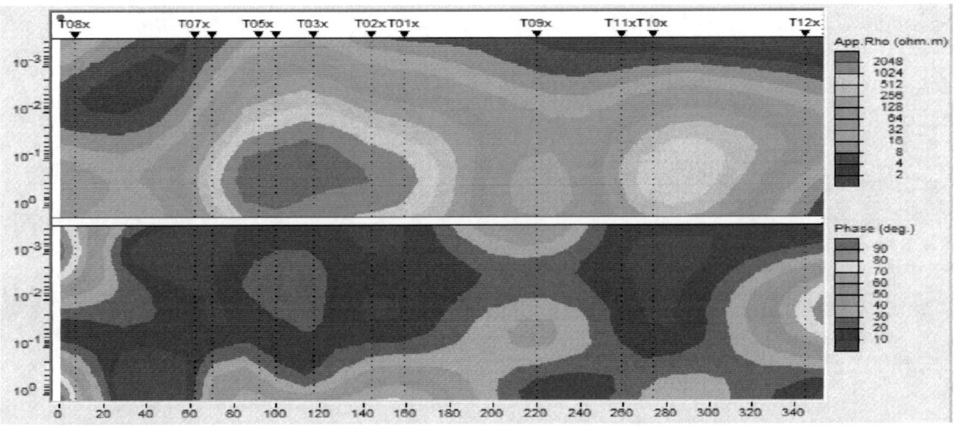

Figure 9. 1D Processing of Magnetotelluric Data in Transverse Electric Mode

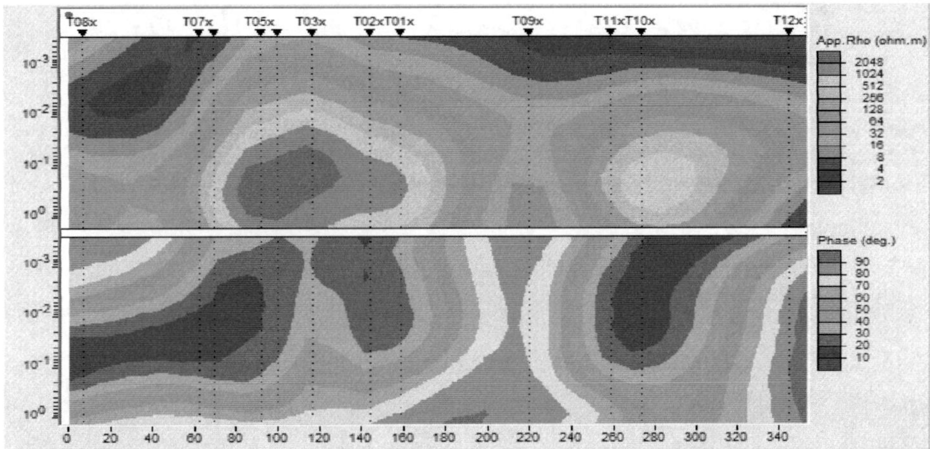

Figure 10. 1D Processing of Magnetotelluric Data in Transverse Magnetic Mode

4. Conclusion
This study aimed to obtain a description of the geological structure such as surface and basement faults oriented relatively parallel with Opak fault NE-SW using Magnetotelluric data. In consequence, it can be concluded from the resulting data are Gravity data method displayed the regional distribution of geological structure which low density predicted as an active fault with orientations tends to be NE-SW near the border area between Pacitan and Trenggralek as well as beneath Lumajang. Processing 1D inversion using The Bostick Method provided quantitative information with differences in resistivity contrast that indicated of faults (low resistivity zone) with orientation tend to be NE-SW between Pacitan and Trenggalek (500-8000 meter) as well as the predicted of fault in Lumajang (100 - 500 meter).

References
[1] Bird P 2003 *Geochemistry Geophys Geosystems G3 An Electronic Journal Of The Earth Science* **4** 1-52.
[2] Hamilton W. R. 1979 *Tectonics of The Indonesia Region* (United States: Geological Survey).
[3] Katili J. A, 1975 *Tectonographysics* **26** 165-188.
[4] Hall R. 1995 *Proceeding Indonesian Petroleum Association* **24** 71-84
[5] Soeriaatmadja R, Maury C. R, Belon H, Pringoprawiro H, Polve M and Priadi B 1994 *Journal of Asian Earth Sciences* **9** 13-27.
[6] Marliyani G. I. 2016 *Neotectonics of Java, Indonesia : Crustal Deformation in The Overriding Plate of an Orthogonal Subduction System* (Doctoral Dissertation, Arizona State University).
[7] Becken M and Ritter O. 2012 *Survey Geophysics* **33** 65-105.
[8] Becken M, Ritter O, Park S. K, Bedrosian P.A, Weckmann U and Weber M. 2008 *Geophys J. Int* **173** 718-732.
[9] Mekkawi M and Saleh A. 2007 *Acta Geodatica et Geophysica Hungarica* **42** 383-397.

Landslide Prone Maps Using Weighting Parameters in Pulung, Ponorogo East Java

D Muriyatmoko [1,*], A Musthafa [1], L Effendi [1], T Harmini [1], and AJ Rozaqi [1]

[1]Department of Informatics Engineering, University of Darussalam Gontor, Siman Street Km. 6, Siman Subdistrict, Ponorogo, East Java, Indonesia 63471

[*]Email: dihin@unida.gontor.ac.id

Abstract. Pulung landslide occurred twice during 2017-2018. The landslide located in the village of Banaran and Wagirkidul, which caused losses. This study aimed to identify the susceptibility level of the landslide in areas of Pulung. This research using four parameters, namely land slope, land use, soil type, and rainfall. This study produces landslide prone maps divided into three levels. It is the level of "very vulnerable" covered area of 3.4% (433.67 ha), level of "vulnerable" covered area of 53.06% (6,767.803 ha), and level of "less vulnerable" covered area of 43.54% (5,553.527 ha). The results are useful for an early warning system in Pulung, especially rainy season. The map presented is also considered as the source for warning landslide system in surrounding Pulung.

1. Introduction

Pulung is included in landslide-prone areas, according to BNPB (National Agency for Disaster Countermeasure) data [1]. April 2017, a landslide occurred in Banaran village [2]. It caused 28 people missing, 32 houses, and five hectares of land destroyed [2]. The landslide is caused by a crack in the upper cliff a few weeks earlier. It causes rainwater to enter the soil and become saturated quickly. So, the weight of the mass and pressure of the pore water increase and the binding capacity decreases. The steep slope, many ginger plants, and lack of firmly rooted plants as slope retaining [2]. On 2018 rainy season, in Banaran were fears for further landslides. As evidenced by the condition of several cracked houses [2]. Another landslide occurred in Wagirkidul. It caused the villagers cannot access the road because almost fifteen-meter had landslides [2].

Previous research has shown that rainfall has an effect on landslides in Ponorogo [3]. Other studies also prove that rainfall, land slope, soil types, and land use also cause landslides in Ponorogo [4]. Three similar studies are using four parameters, namely: land slope, land use, soil type, and rainfall. The first is research in landslide-prone maps in Semarang City produced five classes of vulnerability [5]. The second is research in Ci Tarum river shows that Ci Kondang and Ci Kapundang are the priority in danger of erosion [6]. The third is research in Karang Anyar, Gunung, Semarang City shown that maps can be used as an early warning system [7].

The basic idea behind this research that Ponorogo is the city where the researcher lives. This map design based on four weighting parameters, namely: land slope, land use, soil type, and rainfall. The prone maps are divided into three level areas, namely: less vulnerable, vulnerable, and very vulnerable. This objective research was to determine the susceptibility levels of the landslide in Pulung subdistrict regions.

2. Materials and Methods

2.1 Location and Data Preparation

Pulung is located in Ponorogo City, province of East Java, Indonesia. Pulung also situated between 356 meters to 746 meters above sea level [8]. All data based on Regional Development [8] and Statistic Center of Ponorogo Regency [8]. The data includes Pulung profile, village, land slope, land use, land type, rainfall, etc. Data preparation for draw landslide prone maps is (1) morphology map of Ponorogo with scale 1:25,000 [8], (2) Land type map of Ponorogo with scale 1:100,000 [8], (3) Space pattern map Ponorogo with scale 1:250,000 [8], and (4) Average annual rainfall (mm/years) Pulung, Ponorogo 2017 [8]

2.2 Scoring for Parameters

The parameters choose by UGM (Universitas Gadjah Mada) team, BNPB dan PVMBG when researching in Banaran [2]. It produces four parameters as a cause of landslides, namely land slope, land use, soil type, and rainfall. It is parameter used to this research as a weighting parameter (Table 1-4).

Table 1. The score for land slope

No.	Land slope (°)	Category	Score
1	< 14	Low	1
2	15 - 24	Rather low	2
3	25 - 44	Moderate	3
4	45 - 64	Rather high	4
5	> 65	High	5

Table 2. The score for land use

No.	Land use	Category	Score
1	Secondary dryland forest	Low	1
2	Farm and forest	Rather low	2
3	Settlement	Moderate	3
4	Agriculture, land	Rather high	4
5	Rice fields, grass	High	5

Table 3. The score for land type

No.	Land type	Category	Score
1	Regosol, lithosol, organosol, rendzina	Very sensitive	1
2	Andosol, laterite, grumusol, podzol, and podzolic	Sensitive	2
3	Brown forest land, mediteran land	Moderate sensitive	3
4	Latosol	A little sensitive	4
5	Alluvial, planasol, gray hydro morph, laterit groundwater	Not sensitive	5

Table 4. The score for annual rainfall

No.	Annual rainfall (mm/years)	Category	Score
1	< 500	Low	1
2	500 - 999	Rather low	2
3	1000 - 1999	Moderate	3
4	2000 - 2999	Rather high	4
5	> 3000	High	5

2.3 Weighting Parameters

Weighting parameter based on previous research by Kusratmoko [6] and Purba [5]. Table 5 is the parameter weighting classification.

Table 5. Weighting parameters

No.	Parameter	Weight
1	Land slope (LS)	40%
2	Land use (LU)	30%
3	Land type (LT)	20%
4	Annual rainfall (AR)	10%

The result of four parameter weighting will get the cumulative score shown in the formulation (1):

$$\text{Cumulative score} = (40\% \times \text{LS}) + (30\% \times \text{LU}) + (20\% \times \text{LT}) + (10\% \times \text{AR}) \tag{1}$$

The vulnerability level of a landslide is categorized by three types, shown in Table 6 [5], [6]:

Table 6. Vulnerability level for landslide

No.	Cumulative score	Vulnerability level
1	≤ 2.5	Less vulnerable
2	2.6 - 3.6	vulnerable
3	≥ 3.7	Very vulnerable

2.4 Draw Maps

All map in this research was built by four software. ArcGIS is used to create spatial data, layered map, and fundamental analysis for spatial. Map Server for Windows as a server to for allows a map data to be accessed via the web. PostgreSQL use for manipulates a database. The pmapper framework used as visualization to build the geographic information system via internet. Then the map results from exporting using pmapper plugins.

3. Result and Discussion

Figure 1 is the result of researcher draw maps based on is (1) morphology map of Ponorogo with scale 1:25,000 (2) Land type map of Ponorogo with scale 1:100,000 (3) Space pattern map Ponorogo with scale 1:250,000 [14], and (4) Average annual rainfall (mm/years) Pulung, Ponorogo 2017 [8]. At figure 1 (a) land slope is divided by five (Table 1): (1) high category level is located at the northern of Banaran. (2) rather high level is located at a middle of Banaran, northern of Bekiring and Munggung. (3) A moderate level is located at southern of Banaran, Wagirkidul, middle of Bekiring and Munggung, northern of Wayang and Serag. (4) Rather low level is located at all village except Banaran. (5) low level is located at Sidoharjo and Karangpatihan.

At Figure 1 (b) land use is divided by five (Table 2): (1) high level is rice fields and grass are located spread in all regions except most of the Banaran region. (2) The rather high level is farming and the land is located at all village except Plunturan, Wotan, Patik, Singgahan, Tegalrejo, Bedrug, and Wagirkidul. (3) A middle level is settlement is located at Pomahan, Plunturan, Wotan, Pulungmerdiko, Pulung, Patik, Singgahan, Bedrug, and Wagirkidul. (4) The rather low level is farm and forest are located at all village except Plunturan, Wotan, Patik, Singgahan, Tegalrejo, Bedrug, and Wagirkidul. (5) The low level is dryland forest is located at only the northern of Banaran.

At Figure 1 (c) land type is divided by five (Table 3): (1) The very sensitive level is regosol, lithosol, organosol, rendzina located at southeast of Pulung. (2) A sensitive level is Andosol, laterite, grumusol, podsol, podsolic is located spread in almost all Banaran regions. (3) No region is on a moderate level. (4) a little level is Latasol is located in the middle of Pulung. (5) No region is on the insensitive level. Figure 1 (d) is divided by five (Table 4) and all annual rainfall in Pulung is included in low levels.

Figure 2 shown map of susceptibility landslide level based on cumulative score formulation (1) and using weighting four parameters (Table 5). The vulnerability level based on Table 6. The first parameter is land slope based on Table 1. The second parameter is land use based on Table 2. The third parameter is land type based on Table 3. The fourth parameter is the annual rainfall based on Table 3.

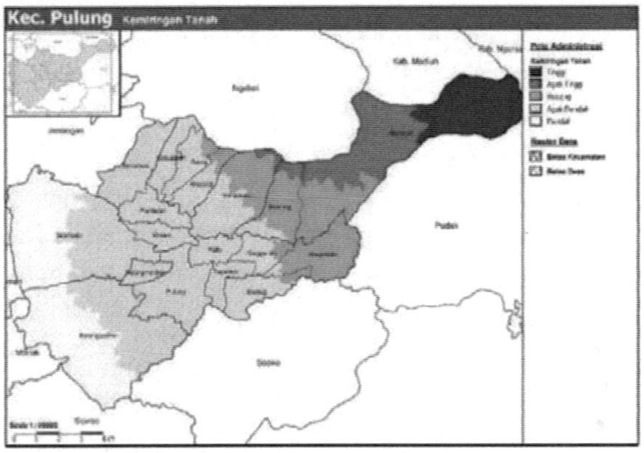

(a) Maps for land slope in Pulung

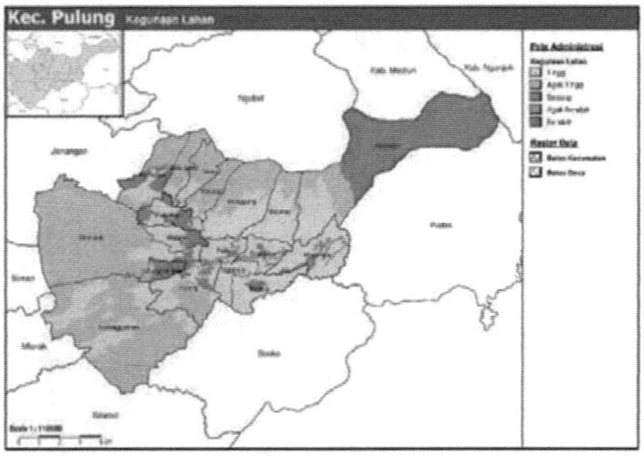

(b) Maps for land use in Pulung

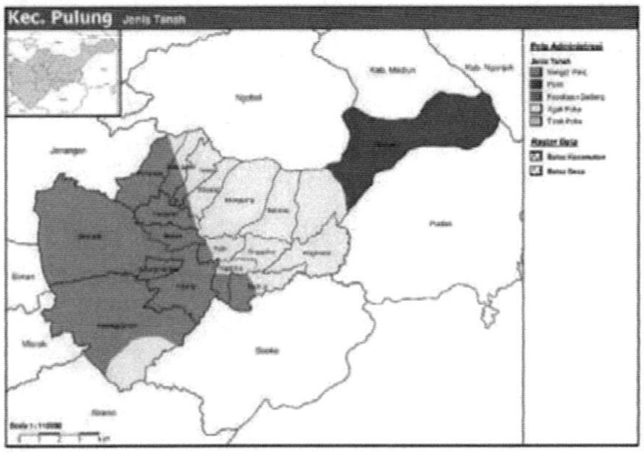

(c) Maps for land type in Pulung

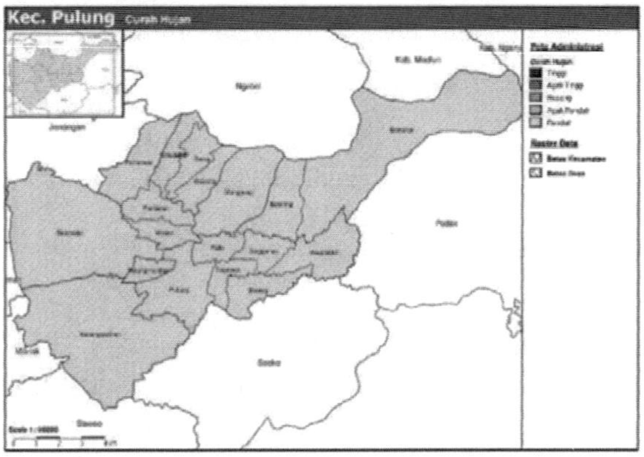

(d) Maps for annual rainfall in Pulung

Figure 1. Result of researcher draw maps with four parameters in Pulung areas

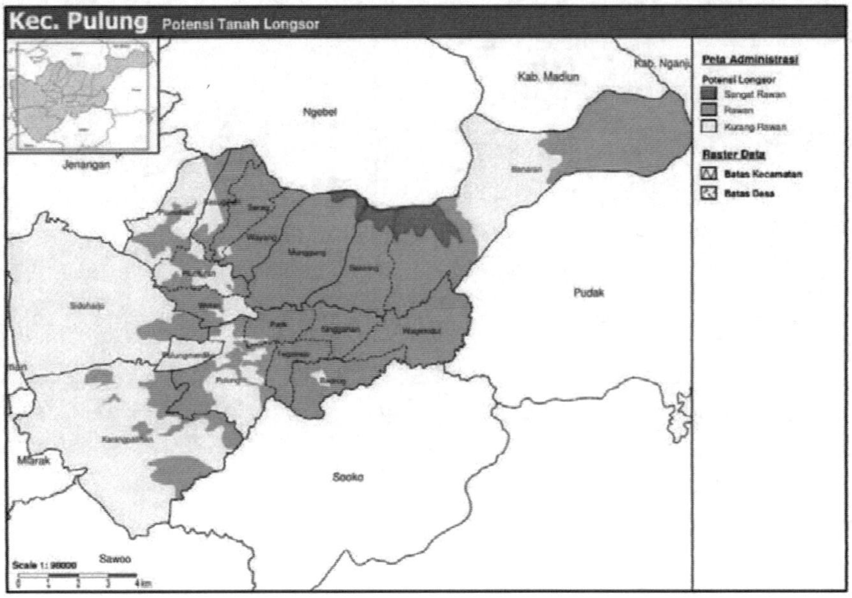

Figure 2. Landslide-prone maps of Pulung areas using weighting four parameters calculate by cumulative score in formula (1)

The research (Figure 2) prove that the village is included very vulnerability levels (red color) of the landslide in western of Banaran, northern of Bekiring, and Munggung. The vulnerability level (orange color) is located in middle and northern of Pulung. The less vulnerability level (low orange color) landslide is located at the southeast of Pulung subdistrict and middle of Banaran village.

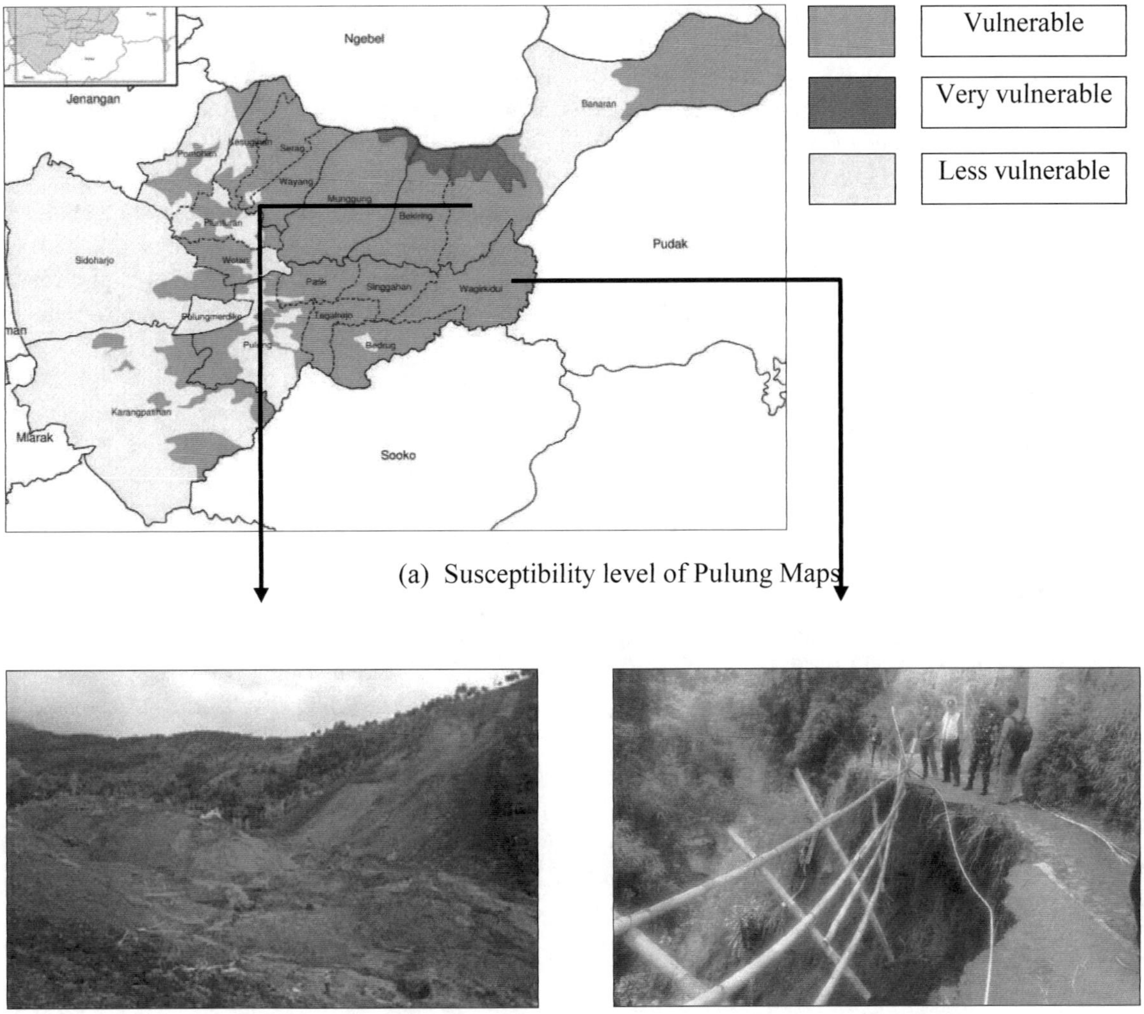

(a) Susceptibility level of Pulung Maps

(b) Location 1. Tangkil hamlet, Banaran village [2]

(c) Location 2. Bangunsari hamlet, Wagirkidul village [6]

Figure 3. The landslide occurred in Banaran and Wagirkidul village

Figure 3. shows the location 1. the landslide happened in the village of Banaran, Pulung subdistrict on April 1, 2017. Among the causes of this landslide was due to the steep slope of the land (40°). The location in Banaran includes as a potential level of "vulnerable". In location 2. The landslide occurred in the village of Wagirkidul, Pulung subdistrict on December 3, 2018. The landslides triggered by heavy rain within four hours. The location in Wagirkidul also includes as a potential level of "vulnerable". This research proves that the percentage accuracy of the weighting parameter in the cumulative score is not yet accurate. So those landslides have occurred in Pulung are not included in "very vulnerable" levels.

4. Conclusion

Based on Landslide prone maps of Pulung areas using weighting four parameters (Figure 2), the susceptibility level was divided into three categories. The first level is "very vulnerable" covered 3.4% (433.67 ha) is located in village of Banaran, Bekiring, and Munggung. The second level is "vulnerable" covered 53.06% (6,767.803 ha) is located in middle and northern of Pulung. The third level is "less vulnerable" covered 43.54% (5,553.527 ha) is located at the southeast of Pulung subdistrict and middle of Banaran village. The prone map shows that landslides occurred in Pulung including the level of "vulnerable" and are not in the level of "very vulnerable". This research proves that the percentage accuracy of the weighting parameter in the cumulative score is not yet accurate. It can be used by the next study to improve the accuracy of the weighting parameter. The advised for villagers, to be alert when the rainy season has arrived.

References

[1] Badan Nasional Penanggulangan Bencana, 2018 *"Data Pantauan Bencana,"* [Online]. Available: http://geospasial.bnpb.go.id/pantauanbencana/data/index.php. [Accessed: 21-Jun-2018].

[2] Ariesta D 2019 *Journal of the Civil Engineering Forum* **5** 149-160.

[3] Muriyatmoko D and Phuspa S.M 2018 *Geosfera Indonesia* **3** 79–89.

[4] Phuspa S M and Muriyatmoko D 2018 *International Conference of Afro-Asian University Forum (AAUF) on The Role of Afro-Asian Universities in Building Civilization* 1063–1082.

[5] Purba J O, Subiyanto S and Sasmito B 2014 *Jurnal Geodesi UNDIP* **3**. 40–52.

[6] Kusratmoko E, Ludiro D, Mataburu I. B, Sobirin and Supriatna, and I. T. L 2002 *Jurnal Geografi* 42-55.

[7] Indrasmoro G. P 2013 *Jurnal GIS Deteksi Rawan Longsor* 1-11.

[8] Badan Pusat Statistik Kabupaten Ponorogo 2018 *http://bappeda.ponorogo.go.id/index.php/data-info/data-statistik (*Accessed: 14-Jan-2019).

The Valuation of Deposit Insurance with Risk Using Fourier Transform

Nur Afnia[1], Mahmud Yunus[1], Endah R. M. Putri[1]

[1]Mathematics Department of Graduate Program Sepuluh Nopember Institute of Technology, Jalan Raya ITS, Keputih, Sukolilo, Surabaya, Jawa Timur 60111, Indonesia

*Email: afnia85@gmail.com, yunusm@matematika.its.ac.id, endahrmp@gmail.com

Abstract. Deposit insurance is a tool to stabilize a banking system. In deposit insurance, the insured risk is the risk of the bank's failure to fulfill its obligations to the customer because of its revoked license. The application of a flat rate deposit insurance system will lead to moral hazard problems, and cause a banking crisis as the banks get involved to risky activities. To anticipate this, it is necessary to design a deposit insurance premium which calculating the different risk levels of each bank. One method that can be used is Fourier transforms. The purpose of this study is to obtain a deposit insurance with risk premium model from the Black-Scholes Model by using Fourier transform. Based on the simulation results, the research found the value of volatility and interest rate is directly proportional to the value of deposit insurance premiums, as well as the value of debt obligations. The high value of deposit insurance premiums is also influenced by the value of dividends, however the value of insurance premiums has changed insignificantly prior to n period, therefore the value of deposit insurance premiums from zero to n period has changed insignificantly.

1. Introduction

In 1980 until 2007 inflation occurred in the economic sector. This provoked the community to get over and overcome the inflation, one of solution is by investing. The main purpose of investment is to have better life in the long run. In other words, they want to have a profit. One of the investments chosen by the community is a bank investment. A bank is chosen based on risk of loss, company reputation, and liquidity. Investment security, company image and easy disbursement are important for an account holder.

Indonesia established a deposit insurance system to maintain stability in the banking sector after the abolition of *Blanket Guarantee* in 2005 [1]. The choice of deposit insurance is better than the implicit protection scheme which is based on four things; first, a deposit insurance system that functions well will produce something faster, smoother and more predictable when the bank is bankrupt. Second, the deposit insurance system provides better protection for small depositors. Third, providing a way to overcome costs through premium. Finally, the deposit insurance system will be more effective to overcome a *Bank Run* [2].

The application of a deposit insurance system with a flat rate will cause *Moral Hazard* problems. The deposit insurance system can also cause a banking crisis as the banks get involved in risky activities [3]. However, most deposit insurance, including in Indonesia, applies it [1]. Economists support the existence of deposit insurance that includes risk as it is considered more fair and efficient than flat premiums [4].

Design determination of deposit insurance can be based on theoretical aspects (option theory) and based on empirical aspects (rating camel). Based on empirical aspects, it refers to the determination made by FDIC. However, FDIC determines premiums on the basis of CAMEL Ratings which the results tend to be underpriced and unfair. Determination of the rate (basis point) is lack of an actual risk calculation [5]. Merton's research was determining deposit insurance premiums using an option theory [6]. However, the absence of future risks consideration on determining premiums. Ronn and Verma anticipated the risks by calculating the decline in the value assets and if the assets value are lower than deposits [4].

Fourier transformation is widely used to obtain the solution of various problems. Jiang used Fourier transformation method to apply a price approach at Asian option [7]. Fourier transform can be used as a tool to simplify partial differential equations in the deposit insurance. In this study, it is expected that a formula will be obtained to evaluate the value of deposit insurance premiums including the risk of Black-Scholes Model by applying Fourier transform.

2. Deposit insurance with risk

Deposit insurance is one of tools to stabilize banking systems [2]. Deposit insurance is a customer protection tool to keep the customers' trust and to ensure a finance stability in a banking system [8]. In deposit insurance, the insured risk is the bank failure in fulfilling its obligations to customers because the license was revoked. The revocation is due to a bank financial health problem and a supervisor tolerance level on the bank condition.

Merton [6] examined " an isomorphic correspondence between loan guarantees and common stock put options, and then to use the well-developed theory of option pricing to derive the formula". Putri [9] described the similarities between the deposit insurance and option mechanism. The stock as the underlying asset of option contract is similar to the bank's asset. The bank and guarantor of an insure deposit are considered as the holder and the writer of deposit insurance.

The mathematical model of deposit insurance proposed by Merton [6] based on a Black-Scholes model and risk adjusted deposit insurance by Ronn and Verma [4]. The dynamic of bank's asset value V is assumed to follow a geometric Brownian motion

$$dV = \mu V\, dt + \sigma V\, dz \tag{1}$$

Where μ is the expected rate return, σ is the volatility and Z is the standart Wiener process.

The risk factor in this deposit insurance is adding dividends. The effect of dividends on an equity is recorded based on the type of issued dividends. When a company issues dividends to its shareholders, the value of the dividends is deducted from their retained earnings. Even if the dividends are issued as additional shares, the value of the shares is deducted. However, a cash dividends result in a reduction in retained earnings directly, stock dividends resulting in the transfer of funds from retained earnings to paid-in capital. While cash dividends reduce an equity, the stock dividends only rearrange the allocation of equity fund. In other words, the value of asset V from a bank will decrease by $V\delta$ so that for n periods the asset value is equal to $V\,(1-\delta)^n$.

A portfolio will consist of a deposit insurance G,V is bank's assets and risk factor $(1-\delta)^n$, dividend per dollar of value of the assets δ and $\Delta = \frac{\partial G}{\partial V}$ unit of bank's asset V. so we have the portfolio as the following,

$$\Pi = G - \Delta V\,(1-\delta)^n \tag{2}$$

If the deposit insurance value $G(V,t)$ depends only on V and t, then by using Ito's lemma based on Eq(1) and Eq(2), the infinitesimal change of the portofolio value implies that

$$d\Pi = \frac{\partial G}{\partial t} + \frac{1}{2}\sigma^2\,(V\,(1-\delta)^n)^2\frac{\partial^2 G}{\partial V^2} \tag{3}$$

Return from Π invested into risky assets can be seen as a growth $r\,\Pi\,dt$ at time dt, with interest rate r and substituting Eq (3), it follows that

$$r\,\Pi\,dt = \left(\frac{\partial G}{\partial t} + \frac{1}{2}\,\sigma^2\,(V\,(1-\delta)^n\,)^2\frac{\partial^2 G}{\partial V^2}\right)dt \tag{4}$$

and Eq (4) implies,

$$\frac{\partial G}{\partial t} + \frac{1}{2}\,\sigma^2\,(V\,(1-\delta)^n\,)^2\frac{\partial^2 G}{\partial V^2} + \frac{\partial G}{\partial V}\,r\,V - r\,G = 0 \tag{5}$$

Eq (5) is called the Black-Scholes model of the deposit insurance with risk.

In the case of asset value V is less than total debt B, assuming that all pre-insurance debt for seniority is similar, the deposit holders will be entitled to the value of their future deposits, or less than total debt.[4]. In other word, they will obtain

$$\min\left\{F_v\,(B_1)\,,\frac{V_T B_1}{B_1 + B_2}\right\}$$

Thus, the maturity value of the deposit insurance is given by

$$\max\left\{0, F_v\,(B_1) - \frac{V_T B_1}{B_1 + B_2}\right\}$$

Where

F_v = denotes the future value operator
V_T = is the terminal value of the bank's assets
B_1 = the face value of the insured deposits
B_2 = the face value of all debt liabilities other than the insured deposits
$B\ = B_1 + B_2$ = face value of total debt liabilities

3. Solution using Fourier transform

In this section we discuss the solution of BlackScholes partial differential equations Eq (5) to find the solution of deposit insurance in BlackScholes model including the risk.

To simplify the calculations and guarantee that eq (5) is integral able and fulfill Fourier domain by following transform;

$$x = \ln\left(\frac{V_T B_1/B}{F_v\,B_1}\right)$$

$$\tau = \frac{\sigma^2}{2}\,(T-t)$$

$$Q\,(x\,,\tau) = \frac{e^{2x}G\,(V,t)}{F_v\,B_1}$$

With $= \frac{2r}{\sigma^2}$, then Eq (5) become:

$$-\frac{\partial Q}{\partial \tau} + k\left(\frac{\partial Q}{\partial x} + 2Q\right) + \left(\frac{\partial^2 Q}{\partial x^2} + 3\frac{\partial Q}{\partial x} + 2Q\right) - k\,Q = 0 \tag{6}$$

In order to fulfill Fourier transformation requirements, the condition of the risk adjusted to deposit insurance is

$$\lim_{|x|\to\infty} Q\,(x,\tau) \;=\; 0$$

and

$$\lim_{|x|\to\infty} Q'\,(x,\tau) \;=\; 0$$

Fourier transform in each equation in Eq(6). So, we result the solution as the following

$$f(\omega,\tau) = e^{(k+2)\tau} f(\omega,0) e^{(-i\omega(k+3)-\omega^2)\tau} \tag{7}$$

After we know that the solution in Fourier space, then we find invers Fourier transform. Previously the characteristic function of X will be seen.
If $X \sim N(\mu,\sigma^2)$ then the characteristic function of X [10] is

$$\chi(\omega) = e^{i\omega\mu - \frac{1}{2}\sigma^2\omega^2}$$

So, the characteristic function of Eq(7) is:

$$\chi(\omega) = e^{(-i\omega(k+3)-\omega^2)\tau}$$

Which mean $-(k+3)\tau$ and varians 2τ . By convolution property in Fourier transform, we get

$$\mathcal{F}\{Q(\omega,\tau)\} = e^{(k+2)\tau}\mathcal{F}\left\{Q(\omega,\tau) * \frac{1}{2\sqrt{\pi\tau}} e^{-\frac{(\omega+(k+3)\tau)^2}{4\tau}}\right\}$$

$$\mathcal{F}^{-1}\{\mathcal{F}Q\{(\omega,\tau)\}\} = e^{(k+2)\tau}\mathcal{F}^{-1}\left\{\mathcal{F}\left\{Q(\omega,\tau) * \frac{1}{2\sqrt{\pi\tau}} e^{-\frac{(\omega+(k+3)\tau)^2}{4\tau}}\right\}\right\}$$

Then

$$Q(\omega,\tau) = e^{(k+2)\tau}\left(Q(\omega,\tau) * \frac{1}{2\sqrt{\pi\tau}} e^{-\frac{(\omega+(k+3)\tau)^2}{4\tau}}\right) \tag{8}$$

Considering $W(\omega,\tau) = \frac{1}{2\sqrt{\pi\tau}} e^{-\frac{(\omega+(k+3)\tau)^2}{4\tau}}$, then Eq (8) can be rewritten as the following

$$Q(\omega,\tau) = e^{(k+2)\tau}\left(Q(\omega,0) * W(\omega,\tau)\right)$$

$$Q(\omega,\tau) = \frac{e^{(k+2)\tau}}{2\sqrt{\pi\tau}}\int_{-\infty}^{\infty} Q(\omega,0)\, e^{-\frac{(\omega-x+(k+3)\tau)^2}{4\tau}}\, d\omega \tag{9}$$

Then substituting the initial condition of deposit insurance with risk $Q(\omega,0) = (e^{-2\omega} - e^{-\omega})^+$ to Eq (9). $(e^{-2\omega} - e^{-\omega})^+$, that will have negative value if and only if $\omega > 0$ then we get the invers Fourier transform is:

$$Q(\omega,\tau) = \frac{e^{(k+2)\tau}}{2\sqrt{\pi\tau}}\int_{-\infty}^{0} e^{-\frac{8\omega\tau + (\omega-x-(k+3)\tau)^2}{4\tau}}\, dw - \frac{e^{(k+2)\tau}}{2\sqrt{\pi\tau}}\int_{-\infty}^{0} e^{\omega}\, e^{-\frac{8\omega\tau+(\omega-x-(k+3)\tau)^2}{4\tau}}\, dw \tag{10}$$

To simplify calculation Eq (10)

$$A_1 = \frac{e^{(k+2)\tau}}{2\sqrt{\pi\tau}}\int_{-\infty}^{0} e^{-\frac{8\omega\tau+(\omega-x-(k+3)\tau)^2}{4\tau}}\, dw$$

$$A_2 = \frac{e^{(k+2)\tau}}{2\sqrt{\pi\tau}}\int_{-\infty}^{0} e^{\omega}\, e^{-\frac{8\omega\tau+(\omega-x-(k+3)\tau)^2}{4\tau}}\, dw$$

To obtain solution of A_1 and A_2 presented with cumulative density function of a standart normal distribution

$$A_1 = e^{-k\tau-2x}\,\Phi\left(\frac{-x-(k-1)\tau}{\sqrt{2\tau}}\right) \tag{11}$$

and

$$A_2 = e^{-x}\,\Phi\left(\frac{-x-(k+1)\tau}{\sqrt{2\tau}}\right) \tag{12}$$

Solution of Eq(9) after invers is

$$Q(x,\tau) = e^{-k\tau-2x}\,\Phi\left(\frac{-x-(k-1)\tau}{\sqrt{2\tau}}\right) - e^{-x}\,\Phi\left(\frac{-x-(k+1)\tau}{\sqrt{2\tau}}\right)$$

Than by reconsidering the initial G form, we will get

$$G(V,t) = B_1\,\Phi(z_2) - \frac{B_1}{B}\,V(1-\delta)^n\,\Phi(z_2) \tag{13}$$

$$z_1 = \frac{\ln\left(\frac{e^{rT}B}{V(1-\delta)^n}\right) - \left(r-\frac{\sigma^2}{2}\right)(T-t)}{\sigma\sqrt{T-t}}$$

$$z_2 = z_1 + \sigma\sqrt{T-t}$$

where $\Phi(\cdot)$ is comulative density functions of normal distribution. Eq (13) is an analytical solution for deposit insurance with risk. Using Fourier transformation, this solution is similar to Ronn's paper.

4. Simulation

Based on Eq (13) in which it's an analytical solution of a deposit insurance including the risk that can be found by numerical simulation using MATLAB. By this simulation, we can see some figures representing premium of deposit insurance that must paid by the bank.

If asset of bank X (trillion in IDR) is $V = 100$, an interest rate $r = 15\%$ and dividend $\delta = 0.00001$, we obtain premium of deposit insurance with risk as the following
For B1= 60 (trillion in IDR)

Tabel 1. Deposit Insurance with Risk $B_1 = 60$

B_1	B_2	G				
		$\sigma^2 = 0.004$	$\sigma^2 = 0.0045$	$\sigma^2 = 0.005$	$\sigma^2 = 0.0055$	$\sigma^2 = 0.006$
60	60	12.4396	12.4397	12.4400	12.4404	12.4411
60	50	8.1288	8.1359	8.1446	8.1549	8.1666
60	40	3.3793	3.4460	3.5110	3.5746	3.6363
60	30	0.4099	0.4757	0.5403	0.6037	0.6658

For 60 (trillion in IDR) there are 3 conditions; face value of total debt liabilities is greater than the bank assets $B > V$ when the face value of all debt liabilities other than B_2 insured deposits are 60 and 50 (trillion in IDR). The face value of total debt liabilities equals to bank $B = V$ assets when the face value of all other insured deposits B_2 debt liabilities is 40 (trillion in IDR), and face value of total debt liabilities smaller than bank assets $B < V$ when the face value of all debt liabilities other than the insured deposits B_2 debt liabilities is 30 (trillion in IDR). The greater the face value of all debt liabilities other than the insured deposits B_2 causes the face value of total debt liabilities B grows bigger, so the value of the deposit insurance premium that must be paid is bigger. Likewise with volatility σ^2, the greater the volatility σ^2 on the face value of total debt liabilities B, the greater the value of deposit insurance premiums.
B1 = 50 (trillion in IDR)

Table 2. Deposit Insurance with Risk $B_1 = 50$

B_1	B_2	G				
		$\sigma^2 = 0.004$	$\sigma^2 = 0.0045$	$\sigma^2 = 0.005$	$\sigma^2 = 0.0055$	$\sigma^2 = 0.006$
50	70	10.3663	10.3664	10.3667	10.3670	10.3676
50	60	6.7740	6.7799	6.7872	6.7957	6.8055
50	50	2.8161	2.8717	2.9258	2.9787	3.0303
50	40	0.3416	0.3964	0.4503	0.5031	0.5549

When $B_1 = 50$ equals to $B_1 = 60$. In this case, there are three conditions; the face value of total debt liabilities is greater than the assets of bank $B > V$ when the face value of all debt liabilities other than the insured deposits B_2 of 70 and 60 (trillion in IDR). The face value of total debt liabilities equals to the assets of bank $B = V$ when the face value of all other than the insured deposits B_2 debt liabilities (trillion in IDR), and the face value of total debt liabilities is smaller than assets bank $B < V$ when the face value of all debt liabilities other than the insured deposits B_2 (trillion in IDR). The greater the face value of all other debt liabilities than the insured deposits B_2 causes the face value of the total debt liabilities B grows bigger, so the value of the deposit insurance premium that must be paid is bigger. Likewise with the volatility value, the greater the value of volatility σ^2 on the face value of the same total debt liabilities B causes the bigger the value of deposit insurance premiums.

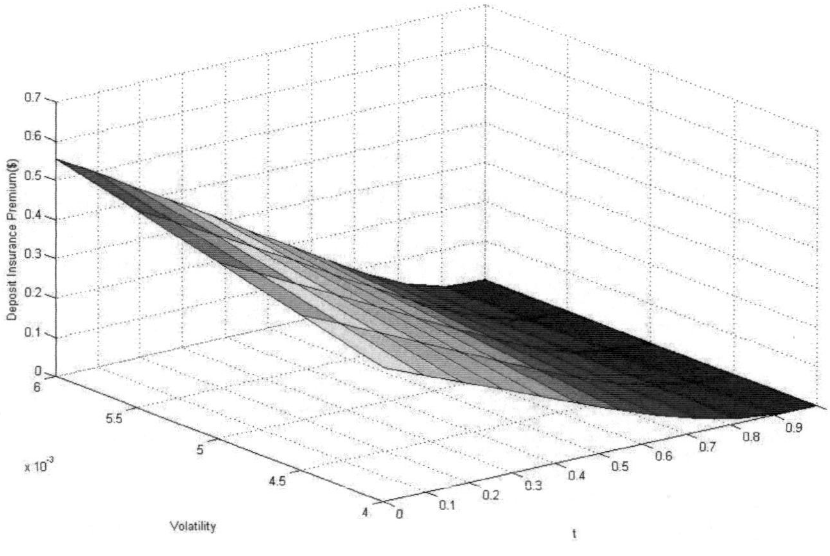

Figure 1. Premium Deposit Insurance with Risk $B_1 = 50$ and $B_2 = 50$

Visualizing the value of deposit insurance premiums including the risks when B1 = 50 and B2 = 50 shown in Figure (1). A debt value of a company influenced by three things, they are: risk rate of return, provisions and limit agreements, and probability that the company can't fulfill the contract partially or fully[11]. These three things are the basis increasing value of deposit insurance premiums. To recognize the effect of interest rates r and volatility on premium values is presented in Figure (2)

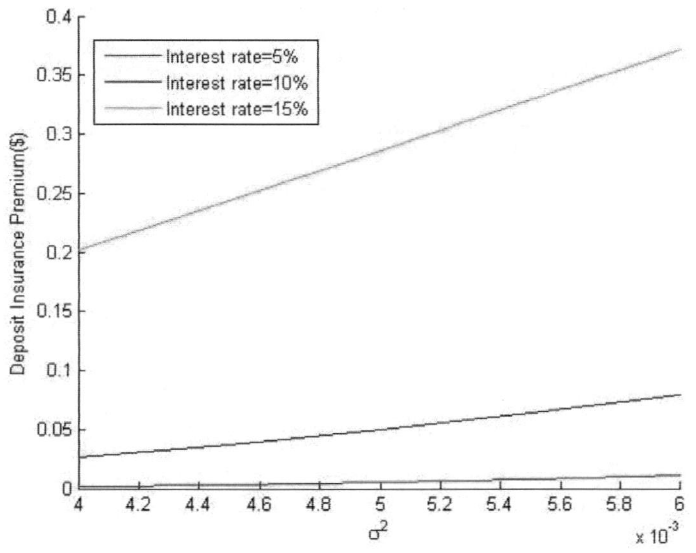

Figure 2. The Effect of Interest Rate and Volatility on Deposit Insurance Premium

Based on figure (2) the interest rate 5%, the increasing of insurance premiums is seen insignificant. For interest rate 10%, the increasing is faster than interest rate 5%. For interest rate 15% causes the increasing of insurance premiums more significant. So, it can be said that by taking the same volatility value, if the interest rate is greater, then the increase in premium value will be sharper. The greater volatility and interest rate cause greater premiums that must be paid off.

Different volatility values show different results. The lowest deposit insurance premium value G is taken at $\sigma^2 = 0.004$ and the highest deposit insurance premium value G is taken at $\sigma^2 = 0.006$. This means that the greater the volatility that is determined, the greater the premium that must be paid.

The effect of dividends on the value of deposit insurance premiums that includes risks shown in Figure (3).

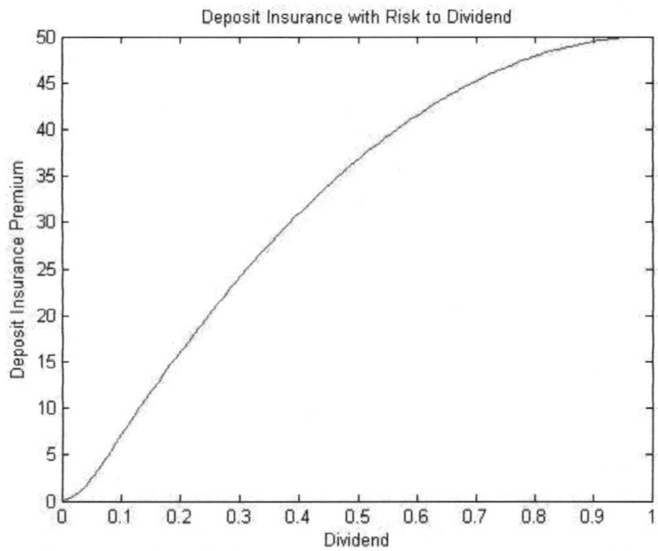

Figure 3. The Effect of Dividend on The Value Deposit Insurance Premium

Journal of Physics: Conference Series **1373** (2019) 012033 doi:10.1088/1742-6596/1373/1/012033

Based on Figure (3) the value of deposit insurance premiums is higher as the value of distributed dividends. The dividend distribution affects the bank assets. The greater distributed dividend, the greater the assets reduction.

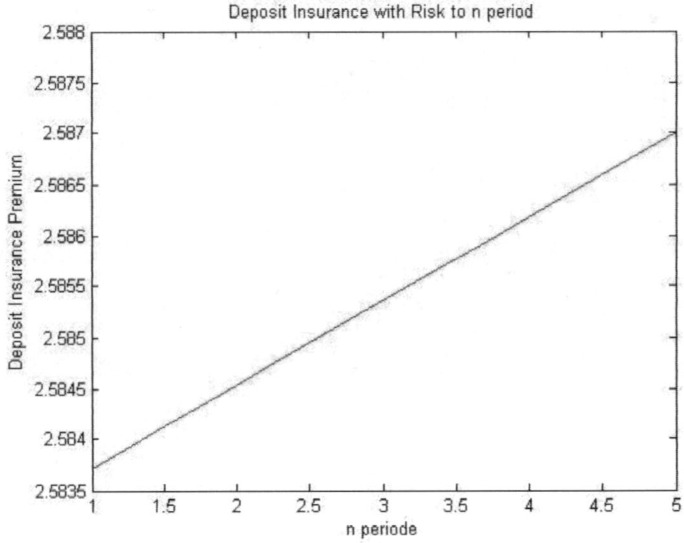

Figure 4. The Effect of n on The Value deposit Insurance Premium

Figure (4) shows the effect on the value of deposit insurance premiums. In this simulation, the value of asset V undergoes small changes up to n-th period, therefore the value of deposit insurance premiums from zero period to n-th period is nearly unchanged.

5. Conclusion

Based on the results and discussion presented in the previous chapter, the following points can be summarized:

1. The model of deposit insurance with risk premium in the form of an equation in second order partial ferential which is influenced by variable V (asset value), t (time), and G (premium value). The influential parameters are σ^2 (volatility), r (interest rate), δ (dividend) and n (time period).
2. The solution of deposit insurance premium including the risk is an analytical solution. The value of deposit insurance premium G is influenced by the amount of nominal value of the insured deposit B_1 minus amount of asset value V. The insured debt value B_1 is influenced by the normal CDF distribution of z_2 which is a combination of z_1 and volatility that is affected by maturity. z_1 is influenced by interest rate r, volatility σ^2, asset value V, total debt value B, dividend δ and maturity date T.
3. Based on the simulation, it can be concluded that:
   - The greater volatility and interest rate, causes greater deposit insurance premium that must be paid.
   - The more debt obligations, the more deposit insurance premiums which must be paid. The value of deposit insurance premiums is higher as the distributed value dividends.
   - The value of deposit insurance premiums has changed marginally up to the *nth* period. Therefore, the value of deposit insurance premiums is nearly seen from zero to period n with no changes.

Acknowledgments
The author would like to thank all those who have supported this research.

References

[1] Kariastanto B 2011 *Munich Personal RePEc Archive* 1-21.
[2] Asnawi KS 2006 *Jurnal Siasat Bisnis* **13** 51-65.
[3] Barth JR, Lee C and Phumiwasana T 2013 *Encyclopedia of Finance* 207-212.
[4] Ronn EI dan Avinash VK 1986 *The Journal of Finance* **41** 871-895.
[5] Asnawi KS 2004 *Jurnal Akuntansi IBil* **11** 51-65.
[6] Merton RC 1977 *Journal of Banking and Finance* **1** 3-11.
[7] Jiang L 2003 *Mathematical Modeling and Method of Option Pricing* (China: Tongji University).
[8] Adlina S and Rubi A 2012 *Bankers Journal Malaysia* **032406** 3-11.
[9] Putri ERM, Tjahjono VR and Utomo DB 2018 *Matematika* 115-128.
[10] Bain LJ 1992 *Introduction to Probability and Mathematical Statistics Second Edition.* Duxbury Press, (Calfornia: An Imprint of Wadsworth Publishing Company).
[11] Merton RC 1973 *Journal of Finance* **29** 449-470.

Forecasting of Vannamei Shrimp Production Based on Weather Factors Using Radial Basis Function Neural Network Approach (Case Study: Lamongan District)

M U Albab[1,*], M I Irawan[1] and D Adzkiya[1]

[1]Department of Mathematics, Institut Teknologi Sepuluh Nopember Surabaya, Sukolilo-Surabaya 60111, Indonesia

[*]Email: mululalbab06@gmail.com

Abstract. One of the economic activity sectors that can be influenced by uncertain weather conditions is aquaculture production. In Lamongan, aquaculture production especially vannamei shrimp is very dependent on the ideal weather conditions. Uncertainty of the weather conditions can cause irregular harvesting to the production of vannamei shrimp. These trend changes can have an impact on the activities of production supply chain, namely the fulfillment of the entry quota of vannamei shrimp production from agents or distributors to exporters of vannamei shrimp to meet market demand. This marketing result can increase the original income of Lamongan area. To find out the development of the trend required a forecasting process and appropriate classification based on past data using artificial neural networks. One structure of artificial neural networks that can be predicting and classifying is a radial basis function neural network (RBFNN). The structure of RBFNN is trained using K-means clustering and gradient descent method. We use average temperature, average humidity and rainfall each month starting from January 2013 until December 2017 as the actual datasets. From those datasets, the training datasets start from January 2013 until December 2016 and the remaining datasets are used as the testing datasets. Built in the Python program, the test results show that our forecasting and classification had an accuracy level with mean absolute percentage error (MAPE) 16.7%.

1. Introduction

Vannamei shrimp production is the commodity of aquaculture in Lamongan regency which has a high selling price in the market and can be harvested in a relatively fast time for 3 months. So it is very profitable for fisheries household from an economic perspective. One of the challenges faced by cultivators of vannamei shrimp is the weather parameters that exixt in the neighborhood of cultivation environment. The current weather parameters have produced uncertain weather conditions that have an impact on human and potentially affect on the economic activities, especially capture fisheries sector [1] and aquaculture [2].

One of the aquaculture activities can be affected by the uncertain weather conditions that is the production of vannamei shrimp that experiences trend of upswings and downswings. These trends will have an impact on interrelated activities in the supply chain of vannamei shrimp conducted by business people, such as agents or distributors and exporters. Hence developing more realistic models to forecast the up and down trends and to classify in the vannamei shrimp production affected by weather conditions is a big challenge for most of the aquaculture business people and professional analysts.

One of the most used methods for forecasting and classification problems is artificial neural networks (ANNs) which are the machine learning method is based on the functional structure of biological neural network. A review of the models and simulations to predict the commercial shrimp growth using the comparison between ANNs of multiple layers perceptron models and nonlinear regression models was performed in [3]. The result showed that ANNs of multiple layers perceptron models provided better descriptions of shrimp growth curves and accuracy than using nonlinear regression models. Irawan, et. al. [4] had applied a forecasting using ANN to obtain predictions on the climatological and hydrological data. The forecasting results are used to determine the minimum crop water requirements in one year and the time of cropping patterns in agriculture to maximize the benefits and minimize crop failure. The application of backpropagation neural network (BPNN) in water environment monitoring system for vannamei shrimp cultivation to study the relationship between the six factors of water environment and vannamei shrimp growth has been discussed in [5]. In contrast, the study comparing BPNN and radial basis function neural network (RBFNN) provided information about speed and accuracy test for small-signal models of active devices. From the computational time and accuracy, RBFNN is better than BPNN as shown in [6].

Sachan [7] had solved a problem of rainfall forecasting using global positioning system (GPS), meteorological data and RBFNN to get the magnitude of rainfall for the next 24 hours. The accuracy of RBFNN was more than 78%. Dubey [8] had modified and implemented RBFNN by using K-means clustering to provide the neural networks set of homogeneous data so the learning time is reduced. By using K-means clustering, the performance of RBFNN increases. Qiao, et. al. had conducted research on the application of RBFNN with K-means clustering algorithm for forecasting the hard landing problem. After the training and testing of the neural network, the results were compared to the results obtained by using support vector machine (SVM) and BPNN. The results obtained by the proposed method were found to be more accurate compared to others.

This paper is organized in 5 sections. Section 2 provides a brief overview of RBFNN and the details of RBFNN based on K-means clustering algorithm. The research methodologies are presented in Section 3. The simulation results obtained using RBFNN with K-means clustering algorithm to solve problems of forecasting and classification are discussed in Section 4. The last section contains the concluding remarks.

2. RBFNN Based on K-Means Clustering Algorithm

This section introduces the basic of RBFNN and its modification using K-means clustering algorithm. The radial basis function for the learning process used a Gaussian function. K-means clustering is used to estimate centers and to calculate the standard width of the Gaussian radial basis function. While gradient descent method is applied to get weights.

2.1 The Radial Basis Function Neural Network

ANNs are a computational method inspired by the behavior of working brains and nervous system. The radial basis function neural network (RBFNN) is a type of ANNs which could be used to solve the problems of supervised learning such as time series prediction, classification and pattern recognition. Keller, et. al. [9] explained that an RBFNN, is an alternative to multilayer perceptrons, which is used to find the function approximation defined by a single weighted sum. While in a multilayer perceptron, the approximation is defined by a nested set of weighted summations. The RBFNN is designed to perform a nonlinear mapping from the input space to the hidden space, followed by a linear mapping from the hidden space to the output space.

In this section, we describe the structure of RBFNN (see Figure 1). Then by using the structure, we explain the relationship between output variables $f(x)$ and input variables x.

The structure of RBFNN, as shown in Figure 1, has three different layers: the input layer, the hidden layer and the output layer. The input layer consists of n source nodes, where n is the dimension of the input vector x. In the hidden layer, each node is activated by a radial basis function ϕ. The output layer consists of a single array of computing unit called an output node $f(x)$. The size of the output layer is not restricted, but commonly it is much smaller than that of the hidden layer to follow the number of length centers. The output layer can be determined by the following equation:

$$f_j(x) = \sum_{i=1}^{l} w_{ij}\phi\left(\left\|x - c_i\right\|\right) \tag{1}$$

where l is the number of centers in the hidden layer, w_{ij} is the connection weights between i^{th} hidden layer and j^{th} output layer node, c_i is the center vector for $i = 1, 2, \ldots, l$ and $\phi\left(\left\|x - c_i\right\|\right)$ is the real-valued radial basis function that depends on the distance between the input vector and the center vector. The distance is usually assumed to be the Euclidean distance $\|\cdot\|$. The performance of RBFNN is greatly dependent on the choice of radial basis function.

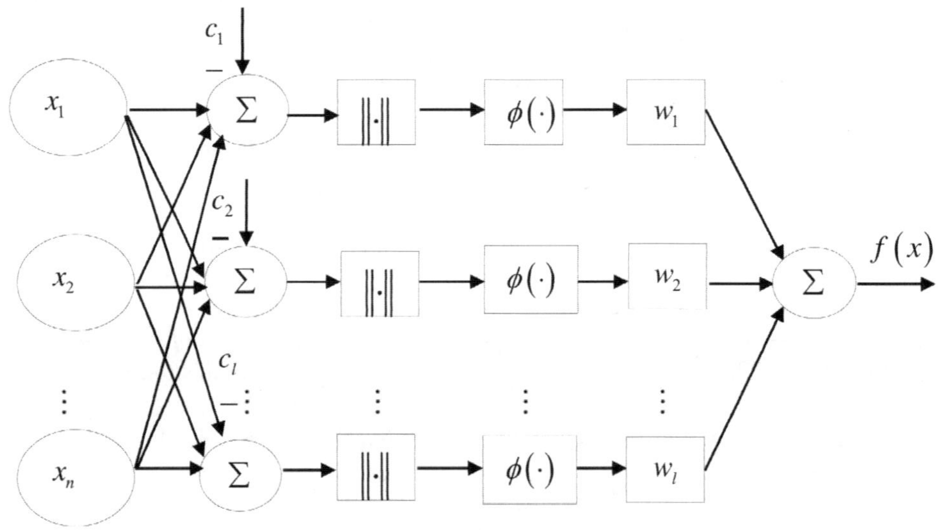

Figure 1. The Structure of RBFNN

In the study, the radial basis function is a Gaussian function which can be expressed by following equation [10]:

$$\phi\left(\left\|x - c_i\right\|\right) = \exp\left(-\frac{1}{2\sigma^2}\left\|x - c_i\right\|^2\right) \tag{2}$$

where σ is the standard width (spread) of the Gaussian radial basis function. Gupta, et. al. [10] explained that a Gaussian function, an unnormalized form of the Gaussian density function, is highly nonlinear and provides good locality for incremental learning. It has been proved that a Gaussian function has many well-defined mathematical features and can be used in the learning and control of nonlinear dynamic systems, and as some powerful schemes for modeling complex input-output mappings.

2.2 The RBFNN Modification Using K-Means Clustering Algorithm

K-means clustering algorithm is one of the clustering algorithms which has been widely used because of its simplicity and ability to produce good result. This algorithm is used in the learning process of RBFNN to group the training datasets into subgroups or clusters and further select the centers of the radial basis function according to the natural measure of the attracting centers in the sense of the

Euclidean distance. Each cluster center is assosiated with one of the hidden Gaussian nodes in the hidden layer. K-means clustering algorithm in this study uses convergent K-means clustering algorithm described by Anderberg and Spath in [10]. It may be used to achieve the goal that the training datasets are finally in the current clusters in the sense of the nearest distance. This task is easily carried out by adding an additional iterative process to the K-means clustering.

The learning process of RBFNN consists of two different stages. First, convergent K-means clustering algorithm was used to determine the centers and the standard width of the Gaussian radial basis function. Second, gradient descent method was applied to obtain the linear weights between the hidden layer and the output layer. The procedure of convergent K-means clustering algorithm is as follow [10]:

1. Select the number l of cluster center c_i for $i = 1,2,\ldots,l$ where l is smaller than the number m of the learning data.
2. Randomly choosing the initial center vectors:

$$c_i = x_i,\ i = 1,2,\ldots,l \tag{3}$$

3. Assign the learning data x_k for $k = l+1, l+2, \ldots m$ to one of the clusters based on the nearest distance criterion using the Euclidean distance:

$$r(x,c) = \sqrt{(x_1 - c_1)^2 + (x_2 - c_2)^2 + \cdots + (x_k - c_i)^2} \tag{4}$$

The learning data x_k belongs to the $i^{*\text{th}}$ cluster if

$$\left\| x_k - c_i^* \right\| = \min_{i^*} \left\| x_k - c_i \right\|,\ 1 \le k \le m \text{ and } 1 \le i \le l \tag{5}$$

4. Recompute the center vectors using the new mean:

$$c_i = \frac{1}{s_i} \sum_{k \in C_i} x_i,\ 1 \le k \le m \text{ and } 1 \le i \le l \tag{6}$$

where s_i is the number of the learning datasets belonging to the ith cluster C_i.

5. Repeat Step 3.
6. If at least one learning data switches to another cluster, then repeat Step 4 and go to Step 5; otherwise, stop the procedure.

Subsequently, let $r_{\max}$ be the maximum distance between the obtained C centers of convergent K-means clustering algorithm. The standard width of all the Gaussian radial basis functions can be expressed as [10]:

$$\sigma = \frac{r_{\max}}{\sqrt{2C}} \tag{7}$$

After determining the update center of the Gaussian radial basis functions and the standard width, the second stage is to apply gradient descent method to identify the linear weights of the output layer. A gradient descent method is based on supervised learning procedure. Assume that the learning

Journal of Physics: Conference Series **1373** (2019) 012034 doi:10.1088/1742-6596/1373/1/012034

process is described by the input-output data pairs $\{x(k), y(k)\}$, then the error signal produced at the output layer at iteration h can be determined by the following equation:

$$e_k(h) = y_k(h) - \sum_{i=1}^{l} w_i(h)\phi(\|x - c_i\|) \tag{8}$$

Then the instantaneous value of the cost function at the output layer can be defined as follows:

$$E(h) = \frac{1}{2}\sum_{k=1}^{m} e_k^2(h) \tag{9}$$

The partial derivative of equation (9) is the gradient

$$\frac{\partial E(h)}{\partial w_i(h)} = -\sum_{k=1}^{m} e_k(h)\phi(\|x - c_i\|) \tag{10}$$

The correction $\Delta w_i(h)$ applied to $w_i(h)$ is defined by the delta learning rule (Widrow-Hoff rule):

$$\Delta w_i(h) = -\eta \frac{\partial E(h)}{\partial w_i(h)} \tag{11}$$

where η is the learning rate parameter. Parameter η must satisfy $0 < \eta \le 1$ [12]. Accordingly, the linear weights parameter between the hidden layer and the output layer can be updated as

$$w_i(h+1) = w_i(h) + \eta\sum_{k=1}^{m} e_k(h)\phi(\|x - c_i\|) \tag{12}$$

for $1 \le i \le l$.

3. Research Methodology

This section introduces the datasets collected and the performance evaluation for the selected RBFNN structure. The collected datasets are devided into the training and testing datasets. The training and testing datasets are time series data in each month. Performance evaluation uses mean absolute percentage error (MAPE) and the value of E (Nash-Sutcliffe model efficient coefficient).

3.1 The Study Datasets

The datasets take the case study in Lamongan district. Astronomically, Lamongan is located between 6°51'54" up to 7°23'6" South latitude, between 112°4'41" up to 112°33'12" East longitude and 12 miles above sea level. Aquaculture in Lamongan especially vannamei shrimp production are classified into several types of culture such as minapadi cultivation, brackish or fresh water pond and fish breeding in paddy fields. Vannamei shrimp production is aquaculture production with the second highest production amount after the production of milkfish for the kind of fish in Lamongan district . Lamongan has two seasons that are rainy and dry season. The high rainy season in Lamongan happens from November until April. In the month of May and June, the rainy season starts decreasing rainfall. While the dry season starts from July until October.

The datasets collected consist of three weather variables which are monthly average temperature, average humidity, rainfall datasets as the input nodes and vannamei shrimp production as the output node from January 2013 until December 2017. Average temperature datasets are obtained from Perak 1 Meteorology Station of Surabaya. Average humidity and rainfall datasets respectively were collected from the satellite data at the website https://www.worldweatheronline.com/lamongan-weather-averages/east-java/id.aspx and the Central Bureau of Statistics joined within the Department of Irrigation Lamongan district. For the output node, vannamei shrimp production is acquired from the Department of Fisheries Lamongan district.

For the five years in the Lamongan district, average number of highest and lowest cases in the study datasets each month is shown in Table 1. The input datasets are normalized to $[0,1]$ using min-max normalization as follows:

$$d_k = \frac{x_k - x_{\min}}{x_{\max} - x_{\min}} \tag{13}$$

where d_k is the normalized result of the kth input node, x_k is the actual value of the kth input node, $x_{\min}$ is the minimum value of the input nodes set and $x_{\max}$ is the maximum value of the input nodes set.

Table 1. Average number of highest and lowest cases in the datasets each month for five years

Datasets	Highest Average	Month	Lowest Average	Month
Average Temperature	29,86 °C	October	27,79 °C	July
Average Humidity	79,4 %	January	60 %	February
Rainfall	294,6 mm	December	9,8 mm	August
Vannamei Shrimp Production	2499,2524 ton	May	196,1686 ton	December

Furthermore, the normalized datasets are divided into training and testing datasets. From each input node or weather variable, 80% samples are January 2013 until December 2016 used as the training datasets and remaining 20% samples are used as the testing datasets. When the input nodes were processed in the RBFNN structure and resulted the approximated value $f(x)$ is normalized so it had to be denormalized using the following equation:

$$d_k^{'} = f_{\min} + y_k\left(f_{\max} - f_{\min}\right) \tag{14}$$

where $d_k^{'}$ is the denormalized result of the kth approximation value, y_k is the value of the kth actual target data, $f_{\min}$ is the minimum value of the approximation datasets and $f_{\max}$ is the maximum value of the approximation datasets.

3.2 Performance Evaluation
The performance evaluation of RBFNN approach method uses MAPE equation and Nash-Sutcliffe model efficient coefficient (E). MAPE and E equation were given as follow:

$$MAPE = \frac{\sum_{t=1}^{N} |Y_t - F_t| / Y_t}{N} 100\% \tag{15}$$

$$E = 1 - \frac{\sum_{t=1}^{N} (Y_t - F_t)^2}{\sum_{t=1}^{N} (Y_t - \overline{Y_0})^2} \tag{16}$$

where Y_t is the value of the actual target data at time t, F_t is the approximation (forecasting) value on the actual target data at time t, N is the number of datasets and $\overline{Y_0}$ is the mean of rainfall over a given period N.

The value of E (Nash-Sutcliffe model efficient coefficient) shows an indication about the ability of forecasting model with actual values. The value of E can vary from $-\infty$ to 1. The closer the value of E to 1, the better the RBFNN structure is selected to be able to forecast the input node especially the rainfall [8]. Meanwhile MAPE shows how big the error value of forecasting within actual datasets. The smaller the value of MAPE, the better the RBFNN structure is selected to be able to forecast.

4. Results and Discussions

This section describes a method to create the best RBFNN structure, the forecasting of each weather as input nodes and the forecasting of vannamei shrimp production. In the creating of the RBFNN structure, RBFNN parameters are obtained by convergent K-means clustering algorithm and gradient descent method. The number of input nodes in the forecasting of weather variables commonly is defined based on the cultivation period of vannamei shrimp conducted by fisheries in Lamongan district. The forecasting result of weather parameters will be used to forecast the production of vannamei shrimp.

4.1 The Best RBFNN Structure

The RBFNN stucture is composed of input nodes, hidden nodes and output node which are respectively in the input layer, hidden layer and output layer. The number of nodes in each layer can be different. In the RBFNN structure, there are the significant RBFNN parameters such as centers c, the standard width σ, radial basis activation function ϕ and update weights w which determine the accuration of approximation results. The process to obtain these parameters that is called the RBFNN training. The parameters are used to obtain the approximation result f in the RBFNN testing. Patterns of RBFNN structure have the number of hidden neurons (centers) different using Gaussian radial basis activation function as shown in Table 2. The update weights of these patterns are obtained by giving the number of epochs 5000 within different standard width and learning rate values. The RBFNN training (see Table 2) was performed starting from datasets January 2013 until December 2016 while the remaining datasets for the RBFNN testing.

Table 2. Simulation results from training and testing RBFNN

Patterns of RBFNN Structure	Number of Centers	Standard width	Learning Rate	Training MAPE	Testing MAPE	Result
3-1-1	2	0.30618	0.01	43.68%	53.64%	Rejected
3-1-1	30	0.16055	0.01	25.01%	35.13%	Rejected
3-1-1	40	0.16101	0.01	19.61%	26.77%	Rejected

Table 2. Simulation results from training and testing RBFNN

Patterns of RBFNN Structure	Number of Centers	Standard width	Learning Rate	Training MAPE	Testing MAPE	Result
3-1-1	47	0.14853	0.01	13.31%	19.1%	Rejected
3-1-1	47	0.14853	0.02	10.94%	18.02%	Rejected
3-1-1	47	0.14853	0.03	9.74%	17.1%	Rejected
3-1-1	**47**	0.14853	**0.04**	**9.44%**	**16.7%**	**Accepted**
3-1-1	47	0.14853	0.05	8.51%	17.89%	Rejected
3-1-1	47	0.14853	0.06	8.12%	16.98%	Rejected
3-1-1	47	0.14853	0.07	7.87%	17.4%	Rejected

During testing on the RBFNN structure for 3 input nodes in input layer, 1 hidden layer with 2 hidden nodes and 1 output layer with standard width value and learning rate respectively 0.30618 and 0.01, the testing MAPE was obtained 53.64% and it was not smallest. Then the number of nodes in hidden layer was determined 30 hidden nodes with standard width value and learning rate respectively 0.16055 and 0.01 and retesting was performed on the RBFNN structure but again the obtained testing MAPE was still big. The number of nodes in hidden layer and learning rate were being changed again and again until the testing MAPE was smallest for 47 hidden nodes in hidden layer and learning rate 0.04. Further the RBFNN structure was tested for bigger learning rate parameter but the testing MAPE was not smallest.

Tabel 2 shows that the best RBFNN structure for training and testing is the pattern of RBFNN structure 3 nodes (average temperature, average humidity, rainfall datasets) taken in input layer, 1 hidden layer with 47 nodes and 1 output layer with standard width value and learning rate respectively 0.14853 and 0.04. This best structure will be used as the structure of RBFNN training to forecast vannamei shrimp production. In this study, the datasets of vannamei shrimp production as the actual-output node was classified in three classes which were shown in Table 3.

Table 3. Classification vannamei shrimp production

Class	Range
1	≤ 650 ton
2	651-999 ton
3	≥ 1000 ton

4.2 Forecasting of Weather Variables

Forecasting of weather variables is used as input nodes in the RBFNN testing to forecast vannamei shrimp production for the next 12 months. In obtaining the new datasets from each weather variabel, the number of nodes in the input layer is 3 input nodes based on the harvesting time conducted by fisheries of vannamei shrimp cultivation Lamongan district each 3 months. The number of nodes in hidden layer is determined 3 centers for average temperature and average humidity variables. While rainfall is determined with 20 centers. The number of nodes in output layer for each weather variabel is 1 output node. Table 4 shows the forecasting results of each weather variable.

Table 4. Forecasting results of weather variables

Month	Actual Average Temperature (°C) 2017	Approximated Average Temperature (°C) 2018	Actual Average Humidity (%) 2017	Approximated Average Humidity (%) 2018	Actual Rainfall (mm) 2017	Approximated Rainfall (mm) 2018
January	28.09	28.89	77	66.95	183	133.47
February	27.94	29.1	78	68.85	259	83.87

Table 4. Forecasting results of weather variables

Month	Actual Average Temperature (°C) 2017	Approximated Average Temperature (°C) 2018	Actual Average Humidity (%) 2017	Approximated Average Humidity (%) 2018	Actual Rainfall (mm) 2017	Approximated Rainfall (mm) 2018
March	28.38	29.07	77	70.76	220	94.03
April	28.61	28.75	77	71.01	173	81.34
May	28.76	28.23	72	72.85	79	24.57
June	28.02	28.26	70	68.28	92	4.4
July	27.73	28.23	65	66.11	26	18.92
August	28.02	29.23	59	65.65	4	105.56
September	29.13	28.37	56	66.14	34	176.16
October	30.39	27.98	57	64.74	106	155
November	28.93	28.03	63	67.45	208	211.99
December	28.55	28.22	71	68.41	249	107.55

Forecasting of new average temperature datasets has performance evaluation with training and testing MAPE respectively 2.32% and 2.79%. For the forecasting of new average humidity datasets, the training and testing MAPE are 9.87% and 8.3%. The forecasting of new rainfall darasets has the value of E for training and testing respectively 0.8201 and -0.33487.

4.3 Forecasting of Vannamei Shrimp Production

Forecasting of vannamei shrimp production is conducted through several stages, namely determining new input datasets using respective forecasting for average temperature, average humidity and rainfall, forecasting training of new datasets, forecasting testing and performance evaluation from the forecasting result. The forecasting training of vannamei shrimp production uses the RBFNN structure that is 3-1-1 with 47 hidden nodes has the testing MAPE was smallest. The plot of training result with RBFNN structure 3-1-1 with 47 hidden nodes and learning rate value 0.04 as shown in Figure 2a. Training MAPE is obtained 9.44%.

The forecasting testing of vannamei shrimp production uses RBFNN parameters which are obtained form the RBFNN structure 3-1-1 with 47 hidden nodes and learning rate value 0.04. The testing result of vannamei shrimp production has the testing MAPE that is obtained 16.7%. The plot of teting result with RBFNN is as shown in Figure 2b.

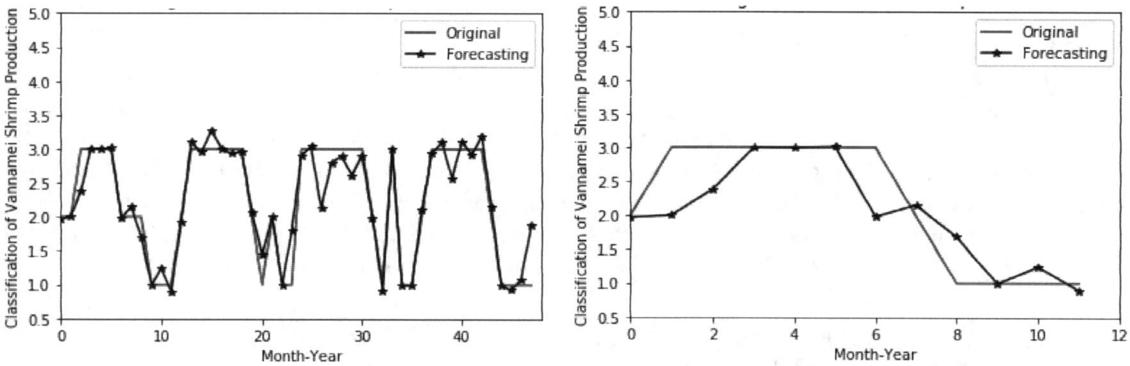

(a) (b)

Figure 2. (a) Training plot of vannamei shrimp production, (b) Testing plot of vannamei shrimp production

Figure 2(b) shows that vannamei shrimp production Lamongan district on September 2018 had increased compared to September 2017. Whereas, vannamei shrimp production Lamongan district on

February, March and July 2018 had decreased compared to these three months 2017. Table 5 shows the forecasting result of vannamei shrimp production.

Table 5. Forecasting result of vannamei shrimp production

Month	Class (2017)	Range (ton)	Output (2018)	Class (2018)	Range (ton)
January	2	651-999	1.8332	2	651-999
February	3	≥1000	2.0162	2	651-999
March	3	≥1000	2.3717	2	651-999
April	3	≥1000	3.0751	3	≥1000
May	3	≥1000	2.9943	3	≥1000
June	3	≥1000	2.9713	3	≥1000
July	3	≥1000	1.9739	2	651-999
August	2	651-999	2.3063	2	651-999
September	1	≤650	1.6803	2	651-999
October	1	≤650	0.9812	1	≤650
November	1	≤650	1.3505	1	≤650
December	1	≤650	1.2128	1	≤650

Table 5 shows vannamei shrimp production is forecasted to produce a total production more than or equal to 1000 ton occuring in months, namely April, May and June. Vannamei shrimp production with a range of 651-999 ton occured in months, namely January, February, March, July, August and September. While the remaining months, vannamei shrimp production decreased that is less than or equal to 650 ton.

5. Conclusions

This study proposes RBFNN with the Gaussian radial basis function using K-means clustering algorithm and gradient descent method to forecast vannamei shrimp production based weather factors (such as average temperature, anerage humidity and rainfall). The following conclusion is drawn (1) The RBFNN structure to forecast average temperature and average humidity is 3 input nodes in input layer, 1 hidden layer with 3 centers or hidden nodes and 1 output layer. While rainfall is determined with 3 input nodes, 1 hidden layer with 20 hidden nodes and 1 output layer. (2) The best RBFNN structure to forecast vannamei shrimp production is 3 input nodes (average temperature, anerage humidity and rainfall variables), 1 hidden layer with 47 hidden nodes and 1 output layer with the standard standard width and learning rate values respectively 0.14853 and 0.04. (3) The accuracy level showed that the error value between the approximation (forecasting) and actual target data, mean absolute percentage error (MAPE) is 16.7%. Based on its MAPE value, the choosen RBFNN structure showed a good performance to forecast vannamei shrimp production.

6. References

[1] Daw T, Adger W N, Brown K and Badjeck M C. 2009 *FAO Fisheries and Aquaculture Technical Paper* **530** 107-150.

[2] De Silva S S and Soto D. 2009 *FAO Fisheries and Aquaculture Technical Paper* **530** 151-212.

[3] Yu R, Leung P S and Bienfang. 2006 *Aquacultural Engineering* **34** 26-32.

[4] Irawan M I, Syaharuddin, Utomo D B and Rukmi A M. 2013 *J. of Theo. and Appl. Infor. Technol.* **58** 657-671.

[5] Shen X, Chen M and Yu J. 2009 *Int. Conf. on Artificial Intelligence and Computational Intelligence* 427-431

[6] Langoni D, Weatherspoon M H, Foo S,Y. and Martinez H,A. 2006 *Engineering Appl. of Artificial Intelligence* **19** 883-890.

[7] Sachan A 2014 *Int. Conf. for Convergence of Technol* (India) (USA:IEEE)

[8] Dubey A D. 2015 *IEEE International Conference on Trends in Automation, Communication and Computing Technologies (IEEE I-TACT)* (India) (USA:IEEE)
[9] Keller J M, Liu D and Fogel D B. 2016 *Fundamentals of Computational Intelligence: Neural Network, Fuzzy Systems and Evolutionary Computation* (New Jersey: John Willy and Sons)
[10] Gupta M M, Jin L and Homma N 2003 *Static and Dynamic Neural Network from Fundamentals to Advanced Theory* (New Jersey: John Willy and Sons)
[11] Fausset L 1994 *Fundamentals of Neural Networks: Architecture, Algorithm and Applications* (New Jersey: Prentice Hall)

Determination of Location and Numbers of Monorail Stops in Surabaya with Max Covering Problem Model

Andika Ardiansyah[1], Mardlijah[1]

[1]Mathematics Department of Graduate Program Sepuluh Nopember Institute of Technology, Jalan Raya ITS, Keputih, Sukolilo, Surabaya, Jawa Timur 60111, Indonesia

*Email: andika.ardiansyah15@gmail.com, mardlijah@matematika.its.ac.id

Abstract. The Surabaya City Government plans to develop Mass Rapid Transit (MRT). The selected MRT is trams and monorails. The MRT procurement is expected to attract public attention to using public transportation so that it can reduce the use of private vehicles. With the reduction of private vehicles, it is expected to reduce congestion and traffic accidents. For the operation of the MRT, supporting facilities are needed, namely the stop for the monorail. So, the author intends to conduct research to determine the location and number of monorail stops. In this study, the determination of the location and number of monorail stops was determined by the Max Covering Problem model when there was a limit in the number of monorail stops established so as to maximize the number of passengers served, which was preceded by calculating the optimal number of monorail stops with the Set Covering Problem model. The results of calculations using the Set Covering Problem model indicate that there are 62 locations for monorail stops in order to serve the point of demand, totaling 151 locations along the route. If it turns out the government has budget constraints in the construction of monorail stops in 62 predetermined locations, then the calculation using the Max Covering Problem model is obtained by choosing 50 locations for monorail stops.

1. Introduction

In this era of globalization, we know that technology is growing. This results in humans being encouraged to do everything quickly and precisely. Because of this, it is necessary to have adequate transportation facilities. The higher human population and human activities result in the need for transportation facilities to increase significantly. As a solution to this problem, the Government plans to develop Mass Rapid Transit (MRT). The selected MRT is trams and monorails.

The City Government intends to allocate monorails on the east-west route, while the trams are allocated to the north-south line [1]. The MRT procurement is expected to attract public attention to using public transportation so that it can reduce the use of private vehicles. With the reduction of private vehicles, it is expected to reduce congestion and traffic accidents. For the operation of the MRT, supporting facilities are needed, namely the stop for the monorail. So, the author intends to conduct research to determine the location and number of monorail stops. The establishment of shelters that are less than optimal can increase transportation problems, this is due to the many people who intend to use this mode but become lazy to use it. This happens because there are difficulties when intending to take advantage of available facilities.

In this study, the determination of the location and number of monorail stops was determined by the Max Covering Problem model when there was a limit in the number of monorail stops established so as to maximize the number of passengers served, which was preceded by calculating the optimal number of monoril stops with the Set Covering Problem. In the first phase the research was examined regarding the data needed to determine the location of the monorail stop. The data includes secondary data about the location of the trip generated from the C-map and Google map. Furthermore, from the data, the location of monorail user requests is determined. From the point of request, the location of the candidate monorail stops is chosen that meets the criteria for building a monorail stop.

Then the distance between the candidate monorail stops and the requested point is measured. The next step is to determine the location and number of monorail stops with the Set Covering Problem model. Furthermore, an analysis of the location of the monorail stop was carried out when the government had limited budget for the construction of monorail stop (consideration of the cost of establishing monorail stop) with the Max Covering Problem model. To simplify the calculation, the problem is done with the help of Lingo 13.0 software. From this calculation, the location and number of monorail stops can serve the maximum number of passengers.

Before this research was conducted, there had been research on determining the location and number of monorail stops which were subsequently used as literature studies. Research has been carried out by determination of location and numbers of tram stops in surabaya with model set covering problem written by Ardiansyah, A. and Mardlijah [2].

2. Monorail stop locations criteria

Criteria for determining the bus stop in this study include:

- The definition of the point of request in this study is the location that is likely to cause a fairly high number of passengers [3].
- The location of the candidate bus stop with a crossing must be more than 50 meters [4] so as not to cause congestion. There are two factors that are taken into consideration in determining stops near the intersection. If the flow of the vehicle turns right, the location of the bus stop is built before the intersection. If the vehicle turns left solid, the stop is built after the intersection.
- Distance of candidate bus stops with hospitals and places of worship must be more than 100 meters, so that the construction of shelters does not disturb the peace of hospitals and places of worship [4].
- The distance of pedestrians to get to the bus stop is a maximum of 400 meters [5].

3. Set covering problem model

The Model Set Covering Problem has the aim of minimizing the number of monorail stop locations but can serve all points of demand. Where the general form of the Set Covering Problem model is as follows [6]:

Objective function:

$$min\, Z = \sum_{j \in J} x_j + \frac{1}{d_j} x_j \tag{1}$$

And constraints function:

$$\sum_{j \in H_i} x_j \geq 1 \ \forall i \in I \tag{2}$$

$$x_j \in \{0,1\} \ \forall j \in J \tag{3}$$

With Z is objective function, I is point of request with index i, J is candidate monorail stop with index j, p_{ij} is distance between point of request i and candidate monorail stop j, P_k is fulfillment distance, d_j is the number of potential demand / passengers at alternative locations j , H_i is $\{j | p_{ij} \leq P_k\}$ which is all candidate monorail stops that can cover point of request i. Where,

$$x_j \begin{cases} 1, & if\ the\ \ alternative\ stops\ \ location\ j\ is\ picked \\ 0, & else \end{cases}$$

Objective (1) to minimize the number of monorail stops. The $\frac{1}{d_j} x_j$ term is included in the model to avoid alternate solutions (ensuring that there will be more candidates for the monorail stop). The constraint function (2) for each point of request can be covered by at least one candidate monorail stops, and , the constraint function (3) to determine the location of the monorail stop chosen or not. The Set Covering Problem model was solved with the help of Lingo 13.0 software.

4. Max covering problem model

The Max Covering Problem model aims to maximize the number of point of requests that are served by limiting the number of location points of service facilities that can serve those points of request [2]. Where the general form of the Max Covering Problem model is as follows [7]:
Objective function:

$$max\ M \ = \textstyle\sum_{i \in I} g_i y_i \tag{4}$$

And constraints function:

$$\textstyle\sum_{j \in H_i} x_j \ \geq 1\ \forall i \in I \tag{5}$$

$$\textstyle\sum_{j \in H_i} x_j - y_i \ \geq 1\ \forall i \in I \tag{6}$$

$$x_j \in \{0,1\}\ \forall j \in J \tag{7}$$

$$y_i \in \{0,1\}\ \forall i \in I \tag{8}$$

With M is objective function, g_i is the number of potential requests / passengers who can served at the point i, q is the number of location limits for monorail stops that can be built, I is point of request with index i, J is candidate monorail stop with index j, p_{ij} is distance between point of request i and candidate monorail stop j, P_k is fulfillment distance, H_i is $\{j | p_{ij} \leq P_k\}$ which is all candidate monorail stops that can cover point of request i. Where:

$$x_j \begin{cases} 1, & if\ the\ \ alternative\ stops\ \ location\ j\ is\ picked \\ 0, & else \end{cases}$$

$$y_i \begin{cases} 1, & if\ the\ alternative\ location\ of\ bus\ stop\ j\ can\ serve\ the\ point\ i\ selected \\ 0, & else \end{cases}$$

In this problem, objective function (1) to maximize total demand that can be fulfilled. Constraint function (2) to determine that the fulfillment of demand at point i is not counted, except in one alternative location that can meet point i, the constraint function (3) to determine that there are restrictions on the number of locations of stops to be built. Constraint function (4) to determine that the location of the location is chosen or not as the location of the establishment of the monorail stop. Constraint function (5) to determine that a point of request i can served by candidate monorail stop j or not. The Max Covering Problem model was solved with the help of Lingo 13.0 software.

5. Simplex method and branch and bround method

The methods that will be used for calculation in Lingo are simplex method and branch and bound method, and can be explained as follows:

- Simplex method is one of the settlement techniques in a linear program that is used as a decision-making technique in problems related to the optimal allocation of resources. [8].
- The Branch and Bound method is a feasible solution approach used in solving linear program problems, especially integers [8], by dividing the feasible solution area into smaller subset, then systematically evaluating the subset to find the best solution. This method consists of node diagrams and branches (branches) as a framework in the process of finding the optimal

solution. In this study the Branch and Bound method is used together with the Simplex method. Examples of node and branch diagrams (branch) are shown in Figure 1.

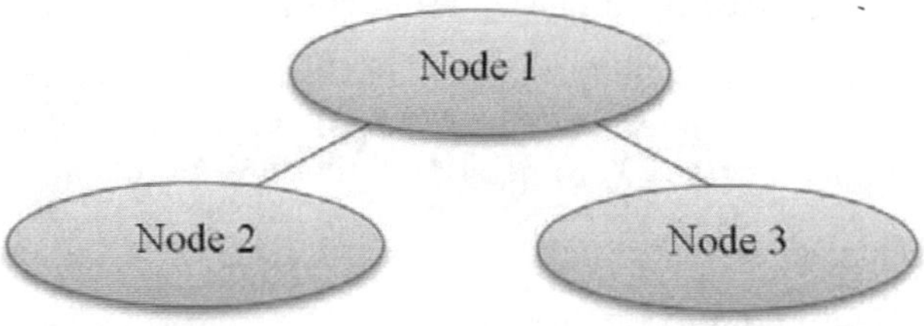

Figure 1. Example Node and Branch Diagram.

6. Defining point of request

In this research, which included the point of request were schools and universities, workplaces (factories and government offices), sports venues, shopping centers and traditional markets, locations of formal housing, tourism objects, and several other places where there might be large movement activities. The point of request includes: East Coast Center M., STAI Luqman A. H. (1), SMP Luqman A. H. (1), SMA Luqman A. H. (1), PPNS (1), PENS (1), Graha ITS (1), ITS (1), H. M. Mansion (1), Oakwood H. and Residences S. (1), Galaxy M. (1), Unair K. C (1), Zoom H. Dharmahusada (1), Maxone H. (1), Sambal Bu Rudy (1), SMKN 5 S./ Halte SMKN 5 S. (1), H. Kassanda (1), Geprek Bensu (1), SMK Berdikari 1 S. (1), Poltekkes Kemenkes S. Jurusan Keperawatan dan Jurusan Kebidanan (1), Unair K. A (1), RS Gigi dan Mulut Unair/ Halte Unair (1), RSUD Dr Soetomo (1), RS Husada Hutama (1), SMAN 4 S. (1), SMPK YBPK 1 GKBW (1), S Gubeng (1), P. S. (1), H. Sahid (1), H. Gubeng (1), Taman Lansia (1), Galaxy H., Sriwijaya Mess, Marvel C. (1), Transmart Ngagel (1), Novotel S. (1), SDN Ngagel 1 (1), S. Wonokromo (1), M. DTC Wonokromo (1), Royal P., RSI S., SMP Khadijah S., SMA Khadijah S., KBS (1), Terminal Joyoboyo (1), Oval H. (1), Pop H. (1), Gereja Kristen Jawi Wetan Jamaah Darmo (1), STIKES William Booth (1), RS William Booth (1), RS RKZ S. (1), S. Town Square (1), Richeese F. Adityawarman (1), Universitas 45 S. (1), Ciwo M. S. (1), Shangri-La H. S. (1), Java Paragon H. (1), SMP Negeri 46 S. (1), Fairfield H. (1), Sumi H. (1), McDonald's M. Sungkono (1), LIPIA S. (1), Ibis Budget S. HR Muhammad H. (1), Vasa H. (1), Grapari (1), Merlion Primary S. (1), SDN P. K. Kendal 1 (1), SMPK Petra 1 (1), Lenmarc M. (1), Mc D' Graha Family (1), SMAK Petra 1 S. (1), P. Graha Family (1), Ranch M. (1), National Hospital (1), Unesa K. Lidah Wetan (1), SMP Lab. Unesa Lidah (1), Citraland (1), SMP Lab. UNESA Lidah (2), Unesa K. Lidah Wetan (2), National Hospital (2), Ranch M. (2), P. Graha Family (2), SMAK Petra 1 S. (2), Mc D' Graha Family (2), Lenmarc M. (2), SMPK Petra 1 (2), SDN P. K. Kendal 1 (2), Merlion Primary S. (2), Grapari (2), Vasa H. (2), Ibis Budget S. H. (2), LIPIA S. (2), Mc D' M. Sungkono (2), Sumi H. (2), Fairfield H. (2), SMPN 46 S. (2), Java Paragon H. (2), Shangri-La H. S. (2), Ciwo M. S. (2), Universitas 45 S. (2), Richeese F. Adityawarman (2), S. Town Square (2), RKZ (2), RS William Booth (2), STIKES William Booth (2), Gereja Kristen Jawi Wetan Jamaah Darmo (2), Pop H. (2), Oval H. (2), Masjid Al Falah, KBS (2), Terminal Joyoboyo (2), M. DTC Wonokromo (2), S. Wonokromo (2), SDN Ngagel 1 (2), Novotel S. (2), Transmart Ngagel (2), Marvel C. (2), RS Siloam, SMA Giki 2 S., G Suites H., H. Santika, Taman Lansia (2), H. Gubeng (2), H. Sahid (2), P. S. (2), S Gubeng (2), SMPK YBPK 1 GKBW (2), SMAN 4 S. (2), RS Husada Hutama (2), RSUD Dr. Soetomo (2), RS Gigi dan Mulut Unair/ Halte Unair (2), Unair K. A (2), Poltekkes Kemenkes S. Jurusan Keperawatan dan Jurusan Kebidanan (2), SMK Berdikari 1 S. (2), Geprek Bensu (2), H. Kassanda (2), SMKN 5 S./ Halte SMKN 5 S. (2), Sambal Bu Rudy (2), Maxone H. (2), Zoom H. Dharmahusada (2), Unair K. C (2), Galaxy M. (2), Oakwood H. and Residences S. (2), H. Mandiri Mansion (2), ITS (2), Graha ITS (2), PENS (2), PPNS (2), SMA Luqman A. H. (2), SMP Luqman A. H. (2), STAI Luqman A. H. (2).

7. Determining candidate stops point that meets the criteria

The point of request to be the candidate bus stop must be in accordance with the criteria discussed earlier. The candidate for the monorail stops are: East Coast Center M. (x_1), PPNS (1) (x_2), Graha ITS (1) (x_3), ITS (1) (x_4), H. Mandiri Mansion (1) (x_5), Galaxy M. (1) (x_6), Unair K. C (1) (x_7), Zoom H. Dharmahusada (1) (x_8), Sambal Bu Rudy (1) (x_9), SMKN 5 S./ Halte SMKN 5 S. (2) (x_10), H. Kassanda (2) (x_11), SMK Berdikari 1 S. (1) (x_12), SMAN 4 S. (1) (x_13), H. Sahid (1) (x_14), H. Gubeng (1) (x_15), Galaxy H. (x_16), Sriwijaya Mess (x_17), Marvel C. (1) (x_18), Transmart Ngagel (1) (x_19), Novotel S. (1) (x_20), SDN Ngagel 1 (1) (x_21), S. Wonokromo (1) (x_22), M. DTC Wonokromo (1) (x_23), SMA Khadijah S. (x_24), Terminal Joyoboyo (1) (x_25), KBS (1) (x_26), Oval H. (1) (x_27), Richeese F. Adityawarman (1) (x_28), Universitas 45 S. (1) (x_29), Ciwo M. S. (1) (x_30), Shangri-La H. S. (1) (x_31), Java Paragon H. (1) (x_32), Sumi H. (1) (x_33), LIPIA S. (1) (x_34), Vasa H. (1) (x_35), SDN P. K. Kendal 1 (1) (x_36), SMPK Petra 1 (1) (x_37), Lenmarc M. (1) (x_38), Mc D' Graha Family (1) (x_39), SMAK Petra 1 S. (1) (x_40), P. Graha Family (1) (x_41), Ranch M. (1) (x_42), Unesa K. Lidah Wetan (1) (x_43), SMP Labschool Unesa Lidah (1) (x_44), Citraland (x_45), SMP Labschool UNESA Lidah (2) (x_46), Unesa K. Lidah Wetan (2) (x_47), Ranch M. (2) (x_48), P. Graha Family (2) (x_49), SMAK Petra 1 S. (2) (x_50), Mc D' Graha Family (2) (x_51), Lenmarc M. (2) (x_52), SMPK Petra 1 (2) (x_53), SDN P. K. Kendal 1 (2) (x_54), Vasa H. (2) (x_55), LIPIA S. (2) (x_56), Sumi H. (2) (x_57), Java Paragon H. (2) (x_58), Shangri-La H. S. (2) (x_59), Ciwo M. S. (2) (x_60), Universitas 45 S. (2) (x_61), Richeese F. Adityawarman (1) (x_62), Oval H. (2) (x_63), KBS (2) (x_64), Terminal Joyoboyo (2) (x_65), M. DTC Wonokromo (2) (x_66), SDN Ngagel 1 (2) (x_67), Novotel S. (2) (x_68), Transmart Ngagel (2) (x_69), Marvel C. (2) (x_70), H. Santika (x_71), Taman Lansia (2) (x_72), H. Gubeng (2) (x_73), H. Sahid (2) (x_74), SMAN 4 S. (2) (x_75), SMK Berdikari 1 S. (2) (x_76), H. Kassanda (2) (x_77), SMKN 5 S./ Halte SMKN 5 S. (2) (x_78), Sambal Bu Rudy (1) (x_79), Zoom H. Dharmahusada (2) (x_80), Unair K. C (1) (x_81), Galaxy M. (2) (x_82), H. Mandiri Mansion (2) (x_83), ITS (2) (x_84), Graha ITS (2) (x_85), PPNS (2) (x_86).

8. Distance measurement between stops candidate and fulfilled demand

Measurements are made in order to find out the location of the request point that can be fulfilled by the location of the candidate monorail stops. Distance measurement results were obtained with the help of C-map and Google Map. This stage is carried out in order to get a distance from one location to another, in order to find out the point of request that can be fulfilled by a candidate monorail stops.

9. Location and numbers of stops determination with set covering problem model

The next process is to determine the location and number of stops using the Set Covering Problem model. Objective function is to minimize the number of monorail stops. Where the shape of the mathematical model is as follows:

$$min\ Z = \sum_{j \in J} x_j + \frac{1}{d_j} x_j$$

where $j = 1, 2, 3, ..., 86$. In this problem, there are 86 candidates monorail stops, with syntax Lingo 13.0 is follows:

```
Min=@sum(haltememenuhisyarat(j):X(j));
```

Where, x_j has 0 and 1 value. So, if the $x_j = 0$, then the candidate monorail stops will not be chosen. If $x_j = 1$, then the candidate will be chosen.

The first constraint function is each point of request can be covered by at least one candidate monorail stops. Where the shape of the mathematical model is as follows:

$$\sum_{j \in H_i} x_j \geq 1\ \forall i \in I$$

Journal of Physics: Conference Series **1373** (2019) 012035 doi:10.1088/1742-6596/1373/1/012035

In this problem, the location of the request points 1, 2, 3, and 4 can only be fulfilled by the candidate bus stop 1, so that the equation model is $x_1 \geq 1$. The request point 5 can only be fulfilled by the candidate stop 2, so the equation model is $x_2 \geq 1$. For request points 6 can be fulfilled by candidates for stops 2 and 3, so the equation model is $x_2 + x_3 \geq 1$≥1. This equation model is used at each request point, with syntax Lingo 13.0 is follows:

```
@for(permintaan(i):@sum(haltememenuhisyarat(j):x(j)*matrix(i,j))>=1);
```

The second constraint function is to determine the location of the monorail stop chosen or not (binary). Where the shape of the mathematical model is as follows:

$$x_j \in \{0,1\} \, \forall j \in J$$

With syntax Lingo 13.0 is follows:

```
@for(haltememenuhisyarat(j):@bin(x(j));
```

Optimization output of Lingo 13.0 result are shown in Figure 2.

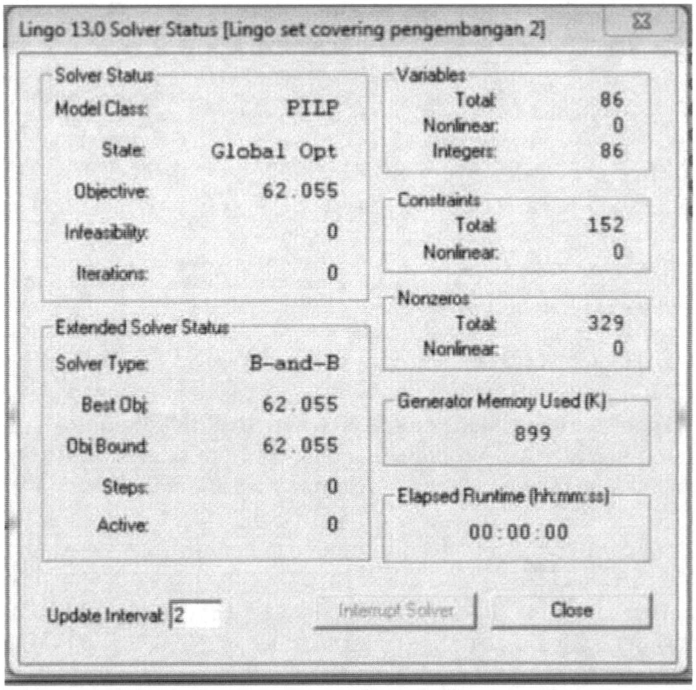

Figure 2. Lingo 13.0 Optimization Output Result (Source: Data Processing, 2019).

From the calculation with the Set Covering Problem model, a monorail stop location is generated. these locations are: East Coast Center M. (x_1), PPNS (1) (x_2), ITS (1) (x_4), H. Mandiri Mansion (1) (x_5), Galaxy M. (1) (x_6), Unair K. C (1) (x_7), Sambal Bu Rudy (1) (x_9), SMKN 5 S./ Halte SMKN 5 S. (2) (x_{10}), SMK Berdikari 1 S. (1) (x_{12}), SMAN 4 S. (1) (x_{13}), H. Sahid (1) (x_{14}), Galaxy H. (x_{16}), Sriwijaya Mess (x_{17}), Marvel C. (1) (x_{18}), Transmart Ngagel (1) (x_{19}), Novotel S. (1) (x_{20}), SDN Ngagel 1 (1) (x_{21}), M. DTC Wonokromo (1) (x_{23}), SMA Khadijah S. (x_{24}), Terminal Joyoboyo (1) (x_{25}), Oval H. (1) (x_{27}), Richeese F. Adityawarman (1) (x_{28}), Ciwo M. S. (1) (x_{30}), Sumi H. (1) (x_{33}), LIPIA S. (1) (x_{34}), Vasa H. (1) (x_{35}), SDN P. K. Kendal 1 (1) (x_{36}), SMPK Petra 1 (1) (x_{37}), Lenmarc M. (1) (x_{38}), Mc D' Graha Family (1) (x_{39}), P. Graha Family (1) (x_{41}), Ranch M. (1) (x_{42}), Citraland (x_{45}), Ranch M. (2) (x_{48}), P. Graha Family (2) (x_{49}), Mc D' Graha Family (2) (x_{51}), Lenmarc M. (2) (x_{52}), SMPK Petra 1 (2) (x_{53}), SDN P. K. Kendal 1 (2) (x_{54}), Vasa H. (2) (x_{55}), LIPIA S. (2) (x_{56}), Sumi H. (2) (x_{57}), Ciwo M. S. (2) (x_{60}), Richeese F. Adityawarman (1) (x_{62}), Oval H. (2) (x_{63}), DTC Wonokromo (2) (x_{66}), SDN Ngagel 1 (2) (x_{67}),

Novotel S. (2) (x_{68}), Transmart Ngagel (2) (x_{69}), Marvel C. (2) (x_{70}), H. Santika (x_{71}), Taman Lansia (2) (x_{72}), H. Gubeng (2) (x_{73}), SMAN 4 S. (2) (x_{75}), SMK Berdikari 1 S. (2) (x_{76}), SMKN 5 S./ Halte SMKN 5 S. (2) (x_{78}), Sambal Bu Rudy (1) (x_{79}), Unair K. C (1) (x_{81}), Galaxy M. (2) (x_{82}), H. Mandiri Mansion (2) (x_{83}), ITS (2) (x_{84}), PPNS (2) (x_{86}).

10. Location and numbers of stops determination with max covering problem model

The Model Max Covering Problem has the aim of to maximize total demand that can be fulfilled. Objective function is to maximizing the number of request points that are served by limiting the number of location points of service facilities that can serve those points of request. Where the shape of the mathematical model is as follows:

$$max\ M = \sum_{i \in I} g_i y_i$$

with i = 1, 2, 3, ..., 151. In this problem, there are 151 locations for request points. With syntax Lingo 13.0 is follows:

```
Max=@sum(Permintaan(i):g(i)*y(i)
```

For each y_i has 0 and 1 value. So, if the $y_i = 0$ then the candidate stops will not be chosen. If $y_i = 1$, then the candidate will be chosen.

Constraint function is a form of constraints that will be allocated optiM.y into the purpose function. There are five constraints function in this research.

The first constraint function is to determine that the fulfillment of demand at point i is not counted, except in one alternative location that can meet point i. Where the shape of the mathematical model is as follows:

$$\sum_{j \in H_i} x_j - y_i \geq 0\ \forall i \in I$$

In this problem, point of request 1 only could be fulfilled by candidate stops 1, do the equation model is $x_1 - y_1 \geq 0$. While point of request 2 only could be fulfilled by candidate stops 1, do the equation model is $x_1 - y_2 \geq 0$. For point of request 3 could be fulfilled by stops candidate 1, so the equation model is $x_1 - y_3 \geq 0$. This equation model used on every point of request. With syntax Lingo 13.0 is follows:

```
@for(permintaan(i):@sum(haltememenuhisyarat(j):x(j)*matrix(i,j))-
y(i)>=0);
```

The second constraint function is to candidate monorail stops at the East Cost Center M., they are always chosen. This location is always chosen because it is the location of the initial departure and final destination of the monorail. Where the shape of the mathematical model is as follows:

$$x_1 = 1$$

With syntax Lingo 13.0 is follows:

```
init :
 x1 = 1;
endinit;
```

The third constraint function is to determine that there are restrictions on the number of locations of stops to be built. Where the shape of the mathematical model is as follows:

$$\sum_{j \in J} x_j = q$$

With syntax Lingo 13.0 is follows:

```
@sum(Haltememenuhisyarat(j):x(j))= q;
```

In this analysis the value of q is determined as 50.

The fourth constraint function is to determine that the location of the location is chosen or not as the location of the establishment of the monorail stop (binary). Where the shape of the mathematical model is as follows:

$$x_j \in \{0,1\} \; \forall j \in J$$

With syntax Lingo 13.0 is follows: `@for(haltememenuhisyarat(j):@bin(x(j));`

The fifth constraint function is to determine that a point of request i can served by candidate monorail stop j or not (binary). Where the shape of the mathematical model is as follows:

$$y_i \in \{0,1\} \; \forall i \in I$$

With syntax Lingo 13.0 is follows:

```
@for(Permintaan(i):@bin(y(i)))
```

Optimization output of Lingo 13.0 result are shown in Figure 3.

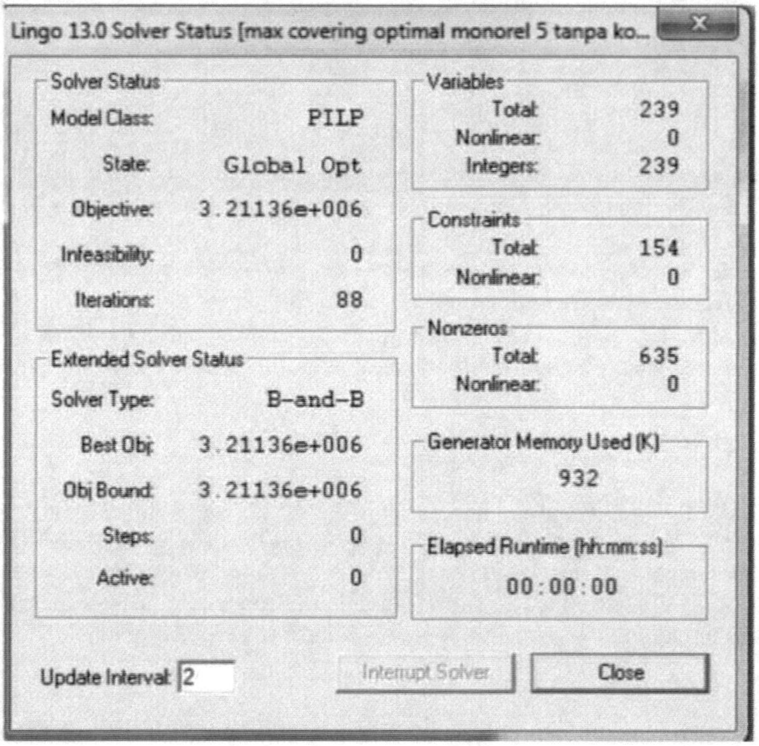

Figure 3. Lingo 13.0 Optimization Output Result (Source: Data Processing, 2019).

From the calculation with the Max Covering Problem model, a monorail stop location is generated. these locations are: East Coast Center M. (1), PPNS (1), ITS (1), Galaxy M. (1), Unair K. C (1), SMKN 5 S./ Halte SMKN 5 S. (2), SMK Berdikari 1 S. (1), SMAN 4 S. (1), H. Sahid (1), Marvel C. (1), Transmart Ngagel (1), S. Wonokromo (1), SMA Khadijah S., Terminal Joyoboyo (1), Oval H. (1), Richeese F. Adityawarman (1), Ciwo M. S. (1), Sumi H. (1), Vasa H. (1), SDN P. K. Kendal 1 (1), SMPK Petra 1 (1), Lenmarc M. (1), Mc D' Graha Family (1), SMAK Petra 1 S. (1), Ranch M. (1), Citraland, Unesa K. Lidah Wetan (2), P. Graha Family (2), Mc D' Graha Family (2), Lenmarc M. (2), SMPK Petra 1 (2), SDN P. K. Kendal 1 (2), Vasa H. (2), Sumi H. (2), Ciwo M. S. (2), Richeese F. Adityawarman (1), Oval H. (2), M. DTC Wonokromo (2), SDN Ngagel 1 (2), Transmart Ngagel (2), Marvel C. (2), H. Santika , H. Gubeng (2), SMAN 4 S. (2), SMK Berdikari 1 S. (2), H. Kassanda (2), Unair K. C (1), Galaxy M. (2), ITS (2), PPNS (2).

11. Simulation with limits on the number of different monorel stop

The location of the bus stops selected from the calculation using the Set model Covering Problems as many as 65 stops. So this analysis will be tried simulation by providing input variations limiting the number of stops 20, 25, 30, 35, 40 and 45. The simulation results show that:

Table 1. Simulation results by providing variations in input limits on the number of stops.

Number of Monorail Stop	Underserved passengers	Percentage
20	2453747	68,96%
25	2845517	79,97%
30	3105258	87,27%
35	3279672	92,17%
40	3377472	94,92%
45	3438532	96,64%
50	3482354	97,87%

12. Conclusion

From calculations using the Max Covering Problem model, it can be concluded that the number of monorails chosen to be built when the government has a limited budget for the construction of shelters with a limit of 50 stops are East Coast Centre M. (1), PPNS (1), ITS (1), Galaxy Mall (1), Unair Kampus C (1), SMKN 5 Surabaya/ Halte SMKN 5 Surabaya (2), SMK Berdikari 1 Surabaya (1), SMAN 4 Surabaya (1), Hotel Sahid (1), Marvel City (1), Transmart Ngagel (1), Stasiun Wonokromo (1), SMA Khadijah Surabaya, Terminal Joyoboyo (1), Oval Hotel (1), Richeese Factory Adityawarman (1), Ciwo Mall Surabaya (1), Sumi H. (1), Vasa Hotel (1), SDN Pradah Kali Kendal 1 (1), SMPK Petra 1 (1), Lenmarc Mall (1), Mc Donalds Graha Family (1), SMAK Petra 1 Surabaya (1), Ranch Market (1), Citraland, Unesa Kampus Lidah Wetan (2), Plaza Graha Family (2), Mc Donalds Graha Family (2), Lenmarc Mall (2), SMPK Petra 1 Surabaya (2), SDN Pradah Kali Kendal 1 (2), Vasa Hotel (2), Sumi H. (2), Ciwo Mall Surabaya (2), Richeese Factory Adityawarman (1), Oval Hotel (2), Mall DTC Wonokromo (2), SDN Ngagel 1 (2), Transmart Ngagel (2), Marvel City (2), Hotel Santika , Hotel Gubeng (2), SMAN 4 Surabaya (2), SMK Berdikari 1 Surabaya (2), Hotel Kassanda (2), Unair Kampus C (1), Galaxy Mall (2), ITS (2), PPNS (2).

References

[1] Dephub, "Fenomena Pembangunan MRT di Kota-Kota Besar", accessed from <URL: http://m.dephub.go.id> on 20 September 2018

[2] Ardiansyah, A. and Mardlijah 2017 *International Conference on Advanced Mechatronics, Intelligent Manufacture, and Industrial Automation (ICAMIMIA),* (Indonesia/Surabaya), pp. 337-341.

[3] Rahmawati M 2009 Penentuan Jumlah dan Lokasi Halte Rute I Monorail Rapid Transit (BRT) di Surakarta dengan Model Set Covering Problem thesis UNS Surakarta.

[4] Departemen Perhubungan 1996 Pedoman Teknis Perekayasaan Tempat Perhentian Kendaraan Penumpang Umum Jakarta: Direktorat Jenderal Perhubungan Darat.

[5] Kementerian Pekerjaan Umum 2014 Pedoman Perencanaan, Penyediaan, dan Pemanfaatan Prasarana dan Sarana Jaringan Pejalan Kaki di Kawasan Perkotaan Jakarta: Menteri Pekerjaan Umum.

[6] Buzna L, Ekel P, Mohan C, and Wang M 2013 *Hindawi* **2013** 1-10.

[7] Cordeaua J F, Furinib F, and Ljubi I. 2018 *European Journal of Operational Research*, **275** 882-896

[8] Pratiwi E W, 2012 Optimasi Penyusunan Timetable Angkutan Umum Menggunakan Metode Branch and Bound dan Implementasinya pada Monorail Transjakarta thesis Universitas Indonesia.

Indonesian community welfare levels clustering using the fuzzy subtractive clustering (FCM) method

P Laras Sakti Cahyaningrum[1], Nursyiva Irsalinda[2]
[1,2]Department of Mathematics, University Ahmad Dahlan, Yogyakarta, Indonesia

*Email: [1] putrilarassakti16@gmail.com

Abstract. Clustering is a technique used to classify objects or cases into groups based on their similarity, called clusters or groups. Objects in each group tend to resemble each other and differ greatly (not the same) with objects from other clusters. Public welfare is a condition of fulfilling the material, spiritual and social needs of citizens in order to be able to live properly. Fuzzy Subtractive Clustering (FSC) method is a clustering algorithm that can form the number and centroid of clusters in accordance with data conditions. This study aims to determine the FSC results in grouping the level of welfare of the Indonesian people in 2017. The testing results of the cluster validity index show 2 values of Partition Entropy and Classification Entropy forming into 2 clusters that have the best value, indicating that the provincial group has a high welfare level and the provincial group has a low welfare level.

1. Introduction

Clustering is a technique used to classify objects or cases into relatively homogeneous groups, called clusters or groups. Objects in each group tend to resemble each other and differ greatly (not the same) with objects from other clusters [1].

There are many methods that can be used in grouping, one of which is most often used is Fuzzy C-Means (FCM). Fuzzy C-Means method is a supervised clustering algorithm, because in FCM we need to know in advance the number of clusters to be formed [2]. The FCM method has a weakness, it requires the number of groups and the matrix of group membership predetermined [3]. Meanwhile, if the number of groups is not known beforehand, there are alternative grouping methods that can be used, namely the Subtractive clustering (SC) method. SC gets more consistent results compared to FCM [4].

Fuzzy subtractive clustering is an unsupervised clustering algorithm that can form the number and centroid of clusters that match the data conditions [5], [6], [7]. The basic concept of subtractive clustering is to determine the areas in a variable that have a high density of the surrounding points. Fuzzy subtractive clustering method produces very good estimation model accuracy in modelling non-linear systems [8], [9]. The community welfare is a condition of fulfilling the material, spiritual and social needs of citizens to live properly and be able to develop themselves, so that they can carry out their social functions. The level of community welfare is caused by many factors including economic growth factors, education factors, health factors and so forth. Equitable distribution of people's welfare needs to be accelerated by taking into account employment growth and absorption, people's access to employment opportunities in parallel must be followed by efforts to improve their skills in order to compete in the free market [10],[11].

Therefore, researcher will use the Fuzzy subtractive clustering method to classify the level of welfare of Indonesian people. Thus the purpose of this study is to determine the results of the application of fuzzy subtractive clustering method to welfare level of Indonesian.

2. Problem characteristics and assumption

The data source in this study is secondary data. The data used in this study are the percentage of education level, average income level, percentage of good nutritional status, percentage of national road length and average floor area per capita. Data was obtained through the official website of the Central Statistics Agency (BPS), the Ministry of Public Works and Public Housing, and the Ministry of Health of the Republic of Indonesia

The steps of the FSC algorithm in this study are as follow:

1. Input the data to be clustered x_{ij} , with $i = 1,2, \ldots, n \; and \; j = 1,2, \ldots, m$, where n is amount of data and m is the amount of data attributes.
2. Set the parameter values, i.e r (radius each data attributes), q (squash factor), Accept ratio, Reject ratio, $XMin$ (minimum data), $XMax$ (maximum data).
3. Normalize each data using the S-curve membership function using Equation (1) as follows:

$$\mu(x) = \begin{cases} 0, & x \leq a \\ 2\left(\frac{x-a}{b-a}\right)^2, & a < x < \frac{a+b}{2} \\ 1 - 2\left(\frac{x-b}{b-a}\right)^2, & \frac{a+b}{2} \leq x < b \\ 1, & x \geq b \end{cases} \quad (1)$$

with x are elements of the decision matrix (research data) and a and b are intervals of the growth S-curve membership function.

4. Determine the initial potential of each data point (D_i; $i = 1,2,\ldots,n$). The steps are as follows:
   a. Calculate the distance of each data for T_j with formula:

$$Dist_{ij}(x_i) = \left(\frac{T_j - x_{ij_{norm}}}{r}\right) \quad (2)$$

   where $T_j = x_{ij_{norm}}$; $j = 1,2,\ldots,m$. To calculate the $Dist_{ij}(x_1)$, value T_j used is $x_{1j_{norm}}$ with value T_j not changing until $i = n$. And so on until $Dist_{ij}(x_n)$.

   b. Determine the initial potential of each data point using equation (3) for $m = 1$ and equation (4) for $m > 1$.

$$D_i = \sum_{i=1}^{n} e^{-4(Dist_{ij}(x_i)^2)} \quad (3)$$

$$D_i = \sum_{i=1}^{n} e^{-4(\sum_{j=1}^{m} Dist_{ij}(x_i)^2)} \quad (4)$$

5. Search the highest potential value:.
   $M = \max[D_i | i = 1,2,\dots,n]$; for first iteration. (5)
   $Z = \max[D_i | i = 1,2,\dots,n]$; for the next iteration. (6)
6. Calculate the value of prospective cluster centroid ratio by using Equation (7) as follows:
   $$Rasio = \frac{Z}{M} \quad (7)$$

   Particularly in the first iteration $Z = M$.
7. Checking the eligibility of prospective cluster centroids into cluster centroid, there are 3 conditions that can occur as follows:
   a. If Ratio > Accept ratio the prospective cluster centroid can be accepted as a new cluster centroid
   b. If Reject ratio < Ratio ≤ Accept ratio, then checking the credibility of the prospective cluster centroid can be accepted as a new cluster centroid, if its existence is far enough from the existing cluster centroid.
   The procedure in this situation is as follows:
   For example : $Md = -1$;
   For $k = 1$ to $k = p; p =$ the amount of $cluster$.
   $$Sd_k = \sum_{j=1}^{m} \left(\frac{V_j - C_{kj}}{r}\right)^2 \quad (8)$$

   If $(Md < 0)$ or $(Sd < Md)$, then $Md = Sd_l$, if $(Sd_l > Md)$, then Md not changing.
   $$Mds = \sqrt{Md};$$

   with $M\ ds$ is the closest distance from the prospective cluster data centroid to the cluster centroid. If $(Rasio + Mds) \geq 1$, prospective cluster centroid is accepted as a new cluster centroid. If$(Rasio + Mds) < 1$, prospective cluster centroids are not accepted and will not be reconsidered as new cluster centroid and the potential data set is zero. If Ratio ≤ Reject ratio, then there are no more data points to be considered to be candidates for the cluster centroid, the iteration is stopped.
8. If a prospective cluster centroid can be accepted as a new cluster centroid, a reduction in the potential of the data around the first cluster centroid is performed with Equation (9) as follows:

$$D_i^t = D_i^{t-1} - D_{C_{kj}} \quad (9)$$

To calculate value $D_{C_{lj}}$using Equation (10).

$$D_{C_{kj}} = M \times e^{-4\left[\sum_{j=1}^{m}(S_{ij})^2\right]} \quad (10)$$

$$S_{ij} = \frac{C_{kj} - x_{ij_{norm}}}{r*q} \quad (11)$$

9. Returns the cluster centroid from the normalized form to the original form using the Equation (12).

$$C_{kj_{denorm}} = C_{kj} * \left(X_{max_j} - X_{min_j}\right) + X_{min_j} \tag{12}$$

10. Calculate the sigma cluster value of each data attribute by using the Equation (13).

$$\sigma_j = \frac{r*\left(X_{max_j} - X_{min_j}\right)}{\sqrt{8}} \tag{13}$$

11. Calculate the degree value of membership using the gauss function with Equation (14).

$$\mu_{ki} = e^{-\sum_{j=1}^{m} \frac{\left(x_{ij} - c_{kj}\right)^2}{2\sigma_j^2}} \tag{14}$$

2.1 Validity Index

1. ***Partition Coefficient* (PC)**

$$PC = \frac{1}{N}\left(\sum_{i=1}^{N}\sum_{j=1}^{K}\mu_{ij}^2\right) \tag{15}$$

with N is the number of research objects, K is the number of clusters, and μ_{ij} is the value of membership of the i-th object with the centroid of the j-th group.

2. ***Partition Entropy* (PE)**

$$PE = -\frac{1}{N}\left(\sum_{i=1}^{N}\sum_{j=1}^{K}\mu_{ij}\ {}^{2}log\ \mu_{ij}\right) \tag{16}$$

3. ***Coefficient Entropy* (CE)**

$$CE = -\frac{1}{N}\left(\sum_{i=1}^{N}\sum_{j=1}^{K}\mu_{ij}\log(\mu_{ij})\right) \tag{17}$$

with N is the amount of research objects, K is the amount of clusters, and μ_{ij} is the value of membership of the i-th object with the centroid of the j group.

3. Result and Discussion

Inputting Data

Data is made by a matrix with size 34 ×5, The size of this matrix shows that the sample amounts are 34 provinces and there are 5 variables used in the analysis, namely the percentage of Education Level, Average Income Level, Percentage of Good Nutrition Status, Percentage of National Road Length and Average Floor Area Per Capita.

Determining Parameter Values

a. In testing the effect of the radius's value in this study there were 5 values of the radius used, namely 1.35, 1.2, 0.95, 0.92 and 0.85. Researchers took 5 values of this radius because based on experiments using laboratory of math software if taking a value of radius> 1.35 will produce a very small group whereas taking a value of radius <0.85 produces a very large group.

Journal of Physics: Conference Series **1373** (2019) 012036 doi:10.1088/1742-6596/1373/1/012036

b. Squash factor used in this study was 1.25. Squash factor of 1.25 shows that the radius of the data points around the centroid of the cluster to be measured decreases in potential data is 1.25.
c. Accept ratio used in this study is 0.5, indicating that the lower boundary value of the data points that are prospective cluster centroids to be allowed to become cluster centroid is 0.5.
d. The rejection ratio used in this study is 0.15, indicating that the upper boundary value from the data points that become prospective cluster centroids is not allowed to be the cluster centroid is 0.15.
e. Minimum Data: V_{1min} =5.385; V_{2min} =2010062; V_{3min} =71.7; V_{4min} =0.54; V_{5min} =13.54. The minimum data on the first attribute is 5,385 indicates, showing the lowest level of education, the minimum data on the second attribute is 2010062, showing the lowest average income, the minimum data on the third attribute of 71.7 shows the lowest level of good nutrition, the minimum data on the fourth attribute of 0.54 shows the lowest national road length presentation and the minimum data on the fifth attribute of 13.54 shows the lowest average floor area.
f. Maximum data: V_{1max} =11.755; V_{2max}=4089123; V_{3max}=91.4; V_{4max}=8.3658; V_{5max} =34.13. The maximum data on the first attribute of 11,755 shows the highest level of education, the maximum data on the second attribute of 4089123 shows the highest level of income, the maximum data on the third attribute of 91.4 shows the highest level of good nutrition, the maximum data on the fourth attribute of 8.3658 shows the long presentation the highest national road and the maximum data on the fifth attribute of 34.13 shows the highest average floor area.

Normalizing the data

In normalizing data the researcher uses the growth S-curve membership function in the equation (1).

Finding Cluster Centroid Values

Analysis with equations r = 1.35 produces 2 clusters, r = 1.2 produces 3 clusters, r = 0.95 produces 4 clusters, r = 0.92 produces 5 clusters and r = 0.85 produces 6 clusters. The results of the cluster centroid values obtained are as follows:

$$C_{r=1.35} = \begin{bmatrix} 0.3230 & 0.1211 & 0.5251 & 0.0092 & 0.3554 \\ 0.9946 & 0.2338 & 0.1044 & 0.1581 & 0.3958 \end{bmatrix}$$

$$C_{r=1.2} = \begin{bmatrix} 0.2407 & 0.0501 & 0.5051 & 0.1270 & 0.3579 \\ 0.9946 & 0.2338 & 0.1044 & 0.1581 & 0.3958 \\ 0.5855 & 1.0000 & 0.8497 & 0.7486 & 0.3050 \end{bmatrix}$$

$$C_{r=0.95} = \begin{bmatrix} 0.2407 & 0.0501 & 0.5051 & 0.1270 & 0.3579 \\ 0.9946 & 0.2338 & 0.1044 & 0.1581 & 0.3958 \\ 0.0580 & 0.9413 & 0.3811 & 0.4072 & 0.2863 \\ 0.5754 & 0.5414 & 0.7687 & 1.0000 & 0.6213 \end{bmatrix}$$

$$C_{r=0.92} = \begin{bmatrix} 0.2407 & 0.0501 & 0.5051 & 0.1270 & 0.3579 \\ 0.9946 & 0.2338 & 0.1044 & 0.1581 & 0.3958 \\ 0.0580 & 0.9413 & 0.3811 & 0.4072 & 0.2863 \\ 0.5754 & 0.5414 & 0.7687 & 1.0000 & 0.6213 \\ 0.9983 & 0.0885 & 0.8384 & 0.1626 & 0.2775 \end{bmatrix}$$

$$C_{r=0.85} = \begin{bmatrix} 0.2407 & 0.0501 & 0.5051 & 0.1270 & 0.3579 \\ 0.7847 & 0.0671 & 0.0249 & 0.1486 & 0.3329 \\ 0.0580 & 0.9413 & 0.3811 & 0.4072 & 0.2863 \\ 0.9122 & 0.1982 & 0.5918 & 0.1413 & 0.3604 \\ 0.0004 & 0 & 0.6364 & 0.2351 & 0.9844 \\ 0.5754 & 0.5414 & 0.7687 & 1.0000 & 0.6213 \end{bmatrix}$$

The number of rows in the C matrix is the number of clusters formed. The first row shows the centroid of the first cluster, the second line shows the centroid of the second cluster, and so on. Whereas the first column shows the Education Level value, the second column shows the Income Level value, the third column shows the Good Nutrition value, the fourth column shows the National Road Length value and the fifth column shows the Floor Area value.

Returning the Cluster Centroid from the Normalized Form to the Original Form

In returning the cluster centroid from the normalized form to the original form we use equation (12).

Table 1. The Cluster Centroid Value by $r = 1.35$

Cluster	Education	Income	Good Nutrition	PJN	Floor Area	Cluster Average	Cluster Rating
1	7.4426	2261753.95	82.04	0.61	20.86	452372.981	2
2	9.0504	3135754.06	86.84	8.37	26.33	627176.9306	1

Table 2. Cluster Centroid Value by $r = 1.2$

Cluster	Education	Income	Good Nutrition	PJN	Floor Area	Cluster Average	Cluster Rating
1	7.595	2339093	81.6	2.512	22.25	422863.505	3
2	11.425	2720928	76.2	2.740	22.7	499257.0158	2
3	17.71	4089123	86	5.591	21.58	817849.3543	1

Table 3. Cluster Centroid Value by $r = 0.95$

Cluster	Education	Income	Good Nutrition	PJN	Floor Area	Cluster Average	Cluster Rating
1	7.595	2339093	81.6	2.512	22.25	422863.505	4
2	11.425	2720928	76.2	2.740	22.7	499257.0158	3
3	6.47	3732904	80.3	4.071	21.33	793432.9007	1
4	8.82	3093605	84.7	8.366	25.17	627176.9306	2

Table 4. Cluster Centroid Value by $r = 0.92$

Cluster	Education	Income	Good Nutrition	PJN	Floor Area	Cluster Average	Cluster Rating
1	7.595	2339093	81.6	2.512	22.25	422863.505	5
2	11.425	2720928	76.2	2.740	22.7	499257.0158	3
3	6.47	3732904	80.3	4.071	21.33	793432.9007	1
4	8.82	3093605	84.7	8.366	25.17	627176.9306	2
5	11.57	2447512	85.8	2.771	21.21	438853.7075	4

Table 5. Cluster Centroid Value by r= 0.85

Cluster	Education	Income	Good Nutrition	PJN	Floor Area	Cluster Average	Cluster Rating
1	7.595	2339093	81.6	2.512	22.25	422863.505	5
2	9.665	2390924	73.9	2.673	21.94	429941.2914	4
3	6.47	3732904	80.3	4.071	21.33	793432.9007	1
4	10.42	2664486	82.5	2.620	22.28	484432.7932	3
5	5.48	2010062	83	3.223	32.31	402037.5625	6
6	8.82	3093605	84.7	8.366	25.17	627176.9306	2

Calculate the sigma cluster

Calculate the sigma cluster using Equation 13 and the results obtained as in Table 6 below:

Table 6. Sigma Cluster Values

r	*Sigma Cluster*				
	Education	Income	Good Nutrition	PJN	Floor Area
1.35	3.04	992329.74	9.40	3.74	9.83
1.2	2.70	882070.88	8.36	3.32	8.74
0.95	2.13	698306.11	6.62	2.63	6.92
0.92	2.07	676254.34	6.41	2.56	6.70
0.85	1.91	624800.21	5.92	2.35	6.19

Calculate the Membership Degree

In calculating the degree of membership using the Gauss function in equation (14), the degree of membership of each data can be found in each cluster.

Determine the Location of Clusters

The location of clusters in each province is determined based on the value of the membership's degree. The highest degree of membership indicates a high tendency for a province to join the cluster.

Calculating Cluster Validity

The Partition Coefficient (PC) is calculated using Equation (15) and produces the following values:

Table 7. Index of PC Value Validity Results of the FSC Grouping

Cluster Amount	PC Value
2	0.3252
3	0.3963
4	0.2982
5	0.3073
6	0.3316

From Table 7 the highest PC value is 0.3963 so it shows the formation of 3 clusters is better than the others. Partition Entropy (PE) is calculated using Equation (16) and produces values in Table 7:

Table 8. Index of PE Value Validity Results of the FSC Grouping

Cluster Amount	PE Value
2	0.7502
3	0.9093
4	0.8444
5	0.9606
6	1.0670

From Table 8 the smallest PE value is 0.7502, so it shows the formation of 2 clusters is better than the others. Classification Entropy (CE) is calculated using equation (17) and produces the following values:

Table 9. Index of CE Value Validity Results of the FSC Grouping

Cluster Value	CE Rating
2	0 .5200
3	0.6303
4	0.5853
5	0.6659
6	0.7396

from Table 9 the smallest CE value obtained is 0.5200, so it shows the formation of 2 clusters is better than the others.

Based on Table 6 it can be seen that table 8 has the formation of 2 clusters for better grouping at the level of Indonesian welfare if compared to 3,4,5 or 6 clusters. From the validity index it can be seen that 2 clusters have 2 levels of validity that are better than other clusters with the best PE value of 0.7502 and the CE validity index obtained the best value of 0.5200. The formation of 3 clusters has only one good validity index namely the PC value of 0.3963.

4. Conclusion

It can be concluded that Community Welfare in Indonesia can be grouped into 2 clusters. The first cluster for the second group level will consist of the provinces of Aceh, North Sumatra, West Sumatra, Riau, Jambi, South Sumatra, Bengkulu, Lampung, Kepulauan Bangka Belitung, West Java, Central Java, East Java, West Nusa Tenggara, East Nusa Tenggara, West Kalimantan, Central Kalimantan, South Kalimantan, East Kalimantan, North Kalimantan, Central Sulawesi, South Sulawesi, Southeast Sulawesi, Gorontalo, West Sulawesi, Maluku, North Maluku, West Papua and Papua. Whereas the second Cluster for the 1st group level will consist of Kepulauan Riau province, DKI Jakarta, DI Yogyakarta, Banten, Bali and North Sulawesi.

References

[1] Supranto. J 2004 *Analisis Multivariat Arti dan Inferensi* (Jakarta: Rieka)

[2] Kusumadewi. S and Purnomo. H 2010 *Aplikasi logika Fuzzy untuk Pendukung Keputusan* (Yogyakarta: Graha Ilmu)

[3] Le. T and Altman. T 2011 *Proc. International Conference on Information and knowledge Enginering Las Vegas USA* **1** 144-150

[4] Bataineh. K M, Naji. M and Saqer. M 2011 *Jordan Journal of Mechanical and Industrial Engineering* **5** 335-343

[5] Bezdek. J 1981 *Pattern Recognition with Fuzzy Objective Function Algorithm* (New York: Plenum)

[6] Chamzini. A Y, Razani. M and Yakhchali. S H 2013 *Automation in Construction* **35** 113

[7] Chiu. S L 1994 *Fuzzy Model Identification Based on Cluster Estimation Rockwell Science Center* (California)

[8] Haqiqi. B N and Kurniawan. R 2015 *Statistika* **8**

[9] Hossen. J, Rahman. A, Sayeed. S, Samsuddin. K, and Rokhani. F 2011 *Australian Journal of Basic and Applied Sciences* **5** 674-681

[10] Kusumadewi. S and Hartiati. S 2006 *Neuro Fuzzy Integrasi Sistem Fuzzy dan Jaringan Syaraf* (Yogyakarta: Graha Ilmu)

[11] Kusumadewi. S and Purnomo. H 2004 *Aplikasi Logika Fuzzy Untuk Pendukung Keputusan* (Yogyakarta: Graha Ilmu)

Mushroom Images Identification Using Orde 1 Statistics Feature Extraction with Artificial Neural Network Classification Technique

Abdul Fadlil[1,*], Rusydi Umar[2] and Sapriani Gustina[2]

[1] Department of Electrical Engineering Universitas Ahmad Dahlan Yogyakarta, Indonesia
[2] Department of Informatics Technology Universitas Ahmad Dahlan Yogyakarta, Indonesia

*Email: fadlil@mti.uad.ac.id

Abstract. There are many kinds of mushrooms difficult to identified manually. Because of that, a certain system that can be used to identify mushrooms is needed. One feature that artificial intelligence has is image identification. One image that can be identified mushroom image. Mushroom image identification can contribute to artificial intelligence technology development. Computer-based mushroom image identification can be done by conducting a segmentation process that converts the original image to a grayscale image. The mushroom image pattern characteristics are selected and separated using a feature extraction process. Mushrooms feature extraction conducted by using orde 1 statistics. Feature extraction results are classified using the Artificial Neural Network method with the Backpropagation Algorithm. Classification process carried out by training and testing with neurons variations 5, 10, 15 and 20, while hidden layers are 0.1, 0.3, 0.5, 0.7, and 0.9 with 10,000 times iteration. 30 images that are consist of 15 images for training data and 15 images for test data. From research can be seen that mushroom image identification using orde 1 statistics features extraction with artificial neuron network has the best result with 93% accuracy on neuron 20. Mushroom's image identification system that is developed can be implemented in other applications.

1. Introduction

Progress and application of pattern recognition have included in various fields, one of them is the agriculture area. Agricultural and plantation products industry is growing very rapidly. Post-harvest events are closely related to the quality of the product produced which ultimately determines the selling price that can be accepted by the farmers [1].

The development of computer technology continues to increase and has been able to recognize many characteristics of the image with its pattern. Pattern recognition is something that is easily done by a human but not so for computers. Someone can easily recognize the type of fungus based on its shape or color. However, it is a difficult task for computers that are not yet equipped with intelligent systems. The similarities and differences between types of fungi can be ideas to be developed into a new study [2] [3].

There are many kinds of Feature Extraction methods, which one is the feature extraction process with Orde 1 Statistics which has several parameter values namely Mean, Variance, Skewness, Kurtosis, and

Journal of Physics: Conference Series **1373** (2019) 012037 doi:10.1088/1742-6596/1373/1/012037

Entropy [4]. The use of technology is so great that computers can work by mimicking the workings of the human brain by utilizing artificial neural network methods [5].
Classification technique used by Artificial Neural Networks with Backpropagation Algorithm. Artificial Neural Networks trains the network to get a balance between the ability of the network to recognize patterns used during training and the ability to respond correctly to input patterns similar to the patterns used during training [6]. As for this study, the type of fungus that will be identified is Kancing Mushroom, Kuping Mushroom, Merang Mushroom, Lingzhi Kerang Mushroom, and Tiram Merah Mushroom.

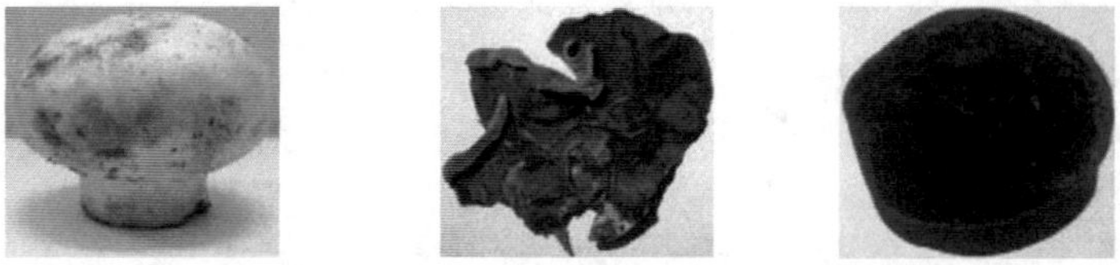

a) Kancing Mushroom b) Kuping Mushroom c) Merang Mushroom

d) Lingzhi Kerang Mushroom e) Tiram Merah Mushroom

Figure 1. Image of mushroom Identification

Figure 1 shows the camera being a module that serves to capture the characteristics of the mushroom image. The captured image data is stored in the file format * .jpg. Some examples of images used in this study as shown in Figure 1.
Of the types of mushroom that will be tested for each of the 30 mushroom images, 15 data will be used as training data and 15 data will be used as test data as Table 1.

Table 1. Source of mushroom image data

Mushroom Image	Amount of data	Training Data	Test Data	Format
Kancing Mushroom	30	15	15	*.jpg
Kuping Mushroom	30	15	15	*.jpg
Merang Mushroom	30	15	15	*.jpg
Lingzhi Kerang Mushroom	30	15	15	*.jpg
Tiram Merah Mushroom	30	15	15	*.jpg

2. Methods

The stages of the process in the system developed to include the process of image acquisition, image processing to obtain image features, and the classification process with Backpropagation neural networks consisting of training and testing [7].

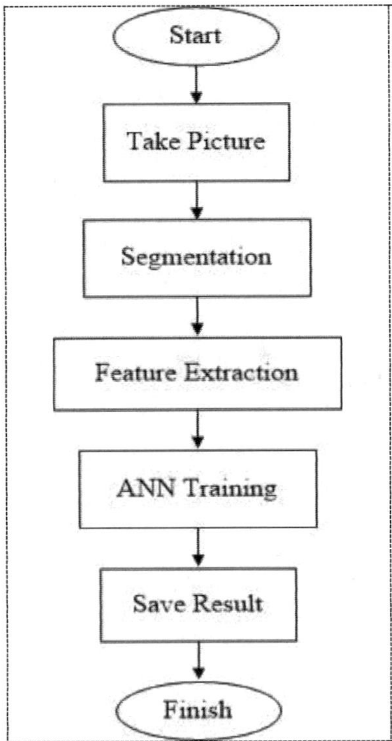

Figure 2. Shows the Training Process stages. Flowchart of Artificial Neural Networks (ANN) Training begins with taking a picture then the segmentation process from the original image is changed to grayscale, then performs feature extraction and training of ANN with neurons, rate, and iteration that has been varied than the weight values will be stored in .mat file which will be tested later.

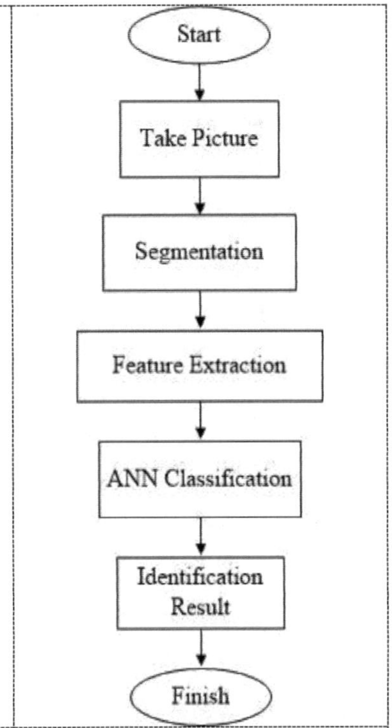

Figure 3. Shows the ANN recognition process stages. It starts with taking a picture and then followed with the segmentation process. The original image is changed to grayscale. Then it performs feature extraction and classification of Artificial Neural Networks by selecting the weight that has been stored in the form of a file.

2.1. Orde 1 Statistics Feature Extraction

Orde 1 Statistics Feature Extraction is a feature retrieval method based on the characteristics of the image histogram. The histogram shows the probability of the occurrence of the value of the gray degree of pixels in an image. From the values in the resulting histogram, we can calculate several orde 1 characteristic parameters, including mean, skewness, variance, kurtosis, and entropy [4].

a. *Mean* (μ)

$$\mu = \sum_{n} fn\, P(fn) \qquad (1)$$

Formula (1) used to calculate the Mean value. It indicates the size of the dispersion of an image where fn is a gray intensity value and p (fn) shows the value of the histogram (probability of occurrence of that intensity in the image)

b. *Variance* (σ2)

$$\sigma^2 = \sum_n (fn - \mu)^2 P(fn) \quad (2)$$

Formula (2) shows the variation of elements on the histogram of an image

c. *Skewness* (α3)

$$\alpha_3 = \frac{1}{\sigma^3} \sum_n (fn - \mu)^3 P(fn) \quad (3)$$

Formula (3) used to calculate Skewness and it indicates the degree of slackness of the relative histogram curve of an image.

d. *Kurtosis* (α4)

$$\alpha^4 = \frac{1}{\sigma^4} \sum_n (fn - \mu)^4 P(fn) - 3 \quad (4)$$

Formula (4) shows the level of the relative curve of the histogram curve of an image

e. *Entropy* (H)

$$H = -\sum_n P(fn).^2 \log P(fn) \quad (5)$$

Formula (5) shows the size of form irregularities in an image

2.2. Artificial Neural Network

Artificial Neural Networks can store knowledge obtained from the results of training. This ability is similar to the function of the human brain so that artificial neural network systems can be used in an area that requires human intelligence. In carrying out the learning process, Artificial Neural Network can modify its behavior by the circumstances of its environment and can regulate itself to produce a response that is consistent with a series of inputs. Some Artificial Neural Networks have the ability to abstract the essence of a series of inputs. For several years generally, the application of Artificial Neural Networks centered on three main areas, namely data analysis, pattern recognition, and control functions. Artificial neural networks have excellent abilities in pattern recognition techniques [8]. Backpropagation architecture consists of one or more input units plus one bias unit, one hidden screen consisting of one more unit plus one bias unit, and one or more output units such as Figure 4.

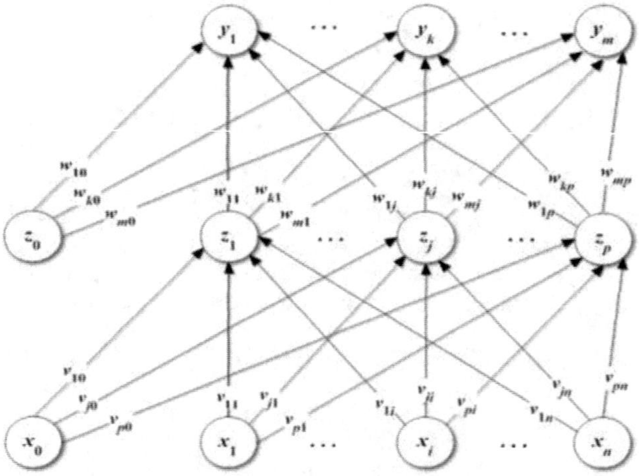

Figure 4. Backpropagation Architecture

Journal of Physics: Conference Series **1373** (2019) 012037 doi:10.1088/1742-6596/1373/1/012037

Figure 4 Illustration of backpropagation architecture with n input (x_1, x_2, ..., x_i, ..., x_n) plus one unit of bias (x_0),a hidden screen consisting of p units (z_1, ..., z_j, ..., z_p) plus one unit of bias (z_0),and the units of output (y_1, y_2, ..., y_k, ..., y_m). v_{ji} is the line weight of the input unit x_i to the hidden screen unit z_j. v_{j0} is the line weight that connects the bias in the input unit to the hidden screen unit z_j. w_{kj} the weight of the hidden screen unit z_j to the output unit y_k. w_{k0} is the weight of the bias on the hidden screen to the output unit y_k.

2.3. Backpropagation Algorithm

The Backpropagation training /network learning algorithm steps are shown as follows:

1. Present the pattern on the input screen (Xi), give the target for each pattern (ti), and specify the number of hidden screen neurons and the rate of training
2. Initialize the initial weight values (WO), on the hidden screen and output screen
3. For each training data, do step 4 through step 11
   Advanced Propagation
4. Hidden screen

$$z\_net_j = v_{jo} + \sum_{i=1}^{n} X_i V_{ji} \qquad (6)$$

$$z_j = f(z\_net_j) \rightarrow \text{sigmoid activation function} \qquad (7)$$

with (j= 1,2,…p)

5. Output screen

$$y\_net_k = W_{ko} + \sum_{j=1}^{p} Z_j W_{kj} \qquad (8)$$

$$y_j = f(y\_net_j) \rightarrow \text{sigmoid activation function} \qquad (9)$$

with (k= 1,2,…m)

Reverse Propagation

$$\delta_k = (t_k - y_k) f(y_{net_k}) = (t_k - y_k) y_k (1 - y_k) \qquad (10)$$

$$\delta\_net_j = \sum_{k=1}^{m} \delta_k \, W_{kj} \tag{11}$$

$$\delta_k = \delta_{net_j} f\left(y_{net_j}\right) = \delta\_net_j \, z_j \left(1 - Z_j\right) \tag{12}$$

6. Calculate all changes in weight

$$\Delta W_{kj} = \alpha \delta_k Z_J \quad k = 1,2,\ldots\ldots, m; j=0,1,\ldots.p \tag{13}$$

$$\Delta V_{kj} = \alpha \delta_J X_J \quad j = 1,2,\ldots\ldots, p; i=0,1,\ldots.p \tag{14}$$

Change in line weight leading to output unit:

$$W_{kj}(new) = W_{kj}(old) + \Delta V_{kj} \; k= 1, 2, \ldots m ; j=0,1,\ldots.,p \tag{15}$$

Change the line weight to the hidden unit:

$$V_{Jj}(new) = V_{Jj}(old) + \Delta V_{Jj} \; J= 1, 2, \ldots m ; i=0,1, \ldots,p \tag{16}$$

The results of the network learning process obtained the final weight value in the last iteration or iteration when the error value was reached, this weight value will be used in testing the test image shown as follows:

1. Present the input pattern that will be tested (Xi)
2. Use the weight of the training process to do advanced propagation
3. Advanced Propagation
4. Hidden Screen

$$(j= 1,2,\ldots p) \tag{17}$$

$$z\_net_j = V_{Jo} + \sum_{i=1}^{n} X_i \, V_{ji} \tag{18}$$

$$z_j = f(z\_net_j) \rightarrow \text{Sigmoid activation function} \tag{19}$$

5. Output screen

$$(k=1,2,\ldots.m) \tag{20}$$

$$y\_net_k = W_{ko} + \sum_{j=1}^{p} Z_j \, W_{kj} \tag{21}$$

$$y_j = f(y\_net_j) \rightarrow \text{Sigmoid activation function} \tag{22}$$

6. Output testing process

2.4. Accuracy

Decision making based on feature extraction and classification is the result of a system decision to identify mushroom images to measure the performance of a mushroom image identification system. The percentage accuracy of system performance can be calculated by Formula (23).

$$\text{Accuracy}(\%) = \frac{ricognized\ amount}{N} \times 100 \tag{23}$$

N is the total number of images tested.

3. Result

The best training process results using the weight variations in the number of hidden screen neurons are shown in Table 2.

Table 2. Training Results with variations in neurons

Learning Rate	Value Error Accuracy Training	Accuracy (%)
0.1	0.04	80
0.3	0.08	100
0.5	0.09	100
0.7	0.17	40
0.9	0.12	46

Table 2 the results of the training with variations in neurons were obtained at the learning rate of 0.1 Final Training Error Value 0.04 with 80% accuracy. Results with a learning rate of 0.3 Final Training Error Value 0.08 with 100% accuracy. Results with a learning rate of 0.3 Final Training Error Value 0.09 with 100% accuracy. Results with learning rates 0.7 Final Training Error Value 0.17 with 40% accuracy. Results with learning rates 0.9 Final Training Error Value 0.12 with 46% accuracy.

Table 3. Best Image Recognition Results with 93% accuracy

Input /Output	Kancing	Kuping	Merang	Lingzhi Kerang	Tiram Merah
Kancing	14	0	0	0	1
Kuping	2	13	0	0	0
Merang	0	0	15	0	0
Lingzhi Kerang	0	1	0	14	0
Tiram Merah	1	0	0	0	14

To calculate accuracy using the formula (25) to obtain accuracy (%) = 70/75 × 100 = 93% Table 3 Test results with 30 images that consist of 15 images as standard images and 15 images as test images. In the results above, it can be seen that for 14 Mushroom Button, the image is identified as Button Mushroom and 1 Image is known as Tiram Merah, for Kuping mushroom 13 the image is recognized as Kuping mushroom, 2 images are recognized as Button mushrooms, for Merang mushroom 15 images are recognized as Merang mushrooms, for Lingzhi Shellfish mushrooms 1 image of mushrooms is recognized as Kuping mushrooms and 14 images are known as Lingzhi Shellfish mushrooms, for Red Oyster Mushrooms 1 image is identified as Button Mushroom and 14 images are known as Oyster Mushrooms. Accuracy values obtained from the accuracy formula. The number of results identified is divided by the total of all images multiplied by 100 so that the accuracy of this test is obtained by 93%.

Research on identification of this mushroom has been done using different methods namely, by using Orde 1 Statistics Extraction Method and Distance Classification, the data used are 30 data, of which 3 data are training data and 27 data are experimental data. 4 distance classifications are used, namely the Euclidean Distance Classification method, giving an accuracy of 78%, while the Manhattan Distance Classification method produces an accuracy value of 82%. The experimental results carried out on the One Minus Correlation Coefficient Distance Classification method provide an accuracy rate of 57%, and the Minkowski Distance Classification method produces an accuracy value of 65%. From these results, it can be seen that the highest accuracy value is from the Manhattan Distance Classification Method and the Minkowski Distance Classification method produces the lowest accuracy value [9].

4. Conclusion
One thing that can be concluded in this study is that the accuracy of the system to identifying and recognizing the image of the fungus is strongly influenced by internal network parameters, namely the rate of learning, the number of hidden screen neurons, and Iterations. The highest percentage of accuracy reaches 93% it means the result is good, and for further research can be tried without changing to grayscale first.

5. References
[1] Li L, Zhang Q, and Huang D 2014 *Journal MDPI* **14** 20078-20111.
[2] Permadi Y and Murito 2015 *Jurnal Informatika Universitas Ahmad Dahlan* **9** 1028
[3] Fadlil, A 2015 *Jurnal Spektrum Industri* **10** 19-29.
[4] Nurhayati and Dwi O 2015 *Jurnal Sistem Komputer* **5** 79-82.
[5] Santoso D Y, and Suhartono 2017 *Jurnal Khazanah Informatika* **3** 1-5
[6] Feri W and Harjoko A 2017 *Jurnal Khazanah Informatika* **3** 100-104
[7] Lestari D. Y 2017 *Jurnal ISD* **2** 40-46.
[8] Sulistiyasni dan Winarko 2014 *Journal of Mathematics and Natural Science* **24** 215
[9] Gustina, Sapriani, Fadli A, and Umar R 2017 *Techno Com* **16** 378-386.

Journal of Physics: Conference Series **1373** (2019) 012038 doi:10.1088/1742-6596/1373/1/012038

Optimal Control Of Dynamic IS-LM Bussiness Cycle Model With Two Time Delay

Airin Nur Hidayati[1,*], Erna Apriliani[1] and I Gst Ngr Rai Usadha[1]

[1]Department of Mathematics, Institut Teknologi Sepuluh Nopember,Sukolilo-Surabaya 60111, Indonesia

[*]Email: airinnurhidayati@gmail.com

Abstract. One of the business cycle model in the dynamics economy is the IS-LM business cycle model with time delay. This model talks about stability in the micro-economic system. Meanwhile, the time delay in the IS-LM business cycle model involve a change in stability at the equilibrium point so that a bifurcation is occurs. In this study, analysis of stability and optimal control on the IS-LM business cycle model with time delay. Based on simulation with numerical computation, show that there is a change in the stability when the delay value was given exceed the critical delay value. The stability change occur when the delay value arose a pure eigen value so that there was a limit cycle that show a Hopf bifurcation. Furthermore, optimal control in the IS-LM business cycle model given when the system changes to be unstable, i.e. when the delay value passed the critical delay value. Variable control use in the interest rate function. While the objective function maximize the total money supply from the country, the optimal solution is obtained by using the Pontryagin Maximum Principle. The results of this simulations with numerical computation show that maximizing the rate of income, the rate of interest rates, and the rate of capital stock lead to the stability point at thirtieth time.

1. Introduction

Dynamic systems can be developed in various fields of life, such as engineering [1], biology [2], economics [3], robotic [4], encryption [5] and other social sciences [6]. Generally dynamic systems can be modeled in the form of differential equations. In general there are two kinds of differential equations, namely ordinary differential equations and partial differential equations. Ordinary differential equations are differential equations which only have one independent variable and contain ordinary derivatives, while partial differential equations are differential equations which have more than one independent variable and contain partial derivatives[7].

At present, many delay differential equations are developed or commonly called time delays. One model that contains a delay differential equation in the economic field is the IS-LM business cycle (Investment Saving-Liquidity Money) model. This model involves the investment function (I), the deposit function (S), the money demand function (L) and consists of three non-independent variables that represent income, interest rates, and capital stock. In the IS-LM business cycle model, with two delays, the value of delay is given assuming that investment depends on income when the investment is first made and in the stock of capital when the time to invest ends [8], [9],[10], [11], [12].

The IS-LM business cycle model has been developed in many researches. One of them is the research conducted by Cai [13]. This study examines the IS-LM business cycle model with an additional lag time in capital accumulation, which is assumed that the invested profits and capital growth are due to past investment decisions. Furthermore, Zhou and Li examined the IS-LM business cycle model by assuming that investment depends on past income and accumulated capital stock, and also on different preparation periods. Zhou and Li used the Hopf bifurcation theorem to show that the delay in the process of capital accumulation can lead to loss of balance or stability of profits and cycles in dynamic macroeconomics [14].

Whereas in 2016, research conducted by Zhou and Li was continued by Rosmely by investigating the occurrence of oscillations in time delay parameters in the IS-LM business cycle model with two time delays. The results show that when the Hopf bifurcation occurs, graph in the solution field shows a constant oscillation movement from the level of income, interest rates and capital stock. If the value of the delay time given is less than the critical value of delay, then the system has a controlled solution towards a balanced condition. Then when the value of the delay time is greater than the critical value of delay, then the system solution will continue to fluctuate, causing unstable system conditions [15].

In this paper, we discuss the optimal stability and control analysis in the IS-LM business cycle model that contains time delay. Stability analysis on the surface is to determine the effect of the value of delay given to the system [16], [17]. Then, the control is given to the interest rate variable. Given the control provided, it is expected to help reduce interest rates when interest rates begin to increases.

2. The IS-LM Bussiness Cycle Model With Two Time Delays

Research by Zhou and Li (2009) studied the IS-LM business cycle model of time delay in the equation of capital accumulation, which is assumed that investment depends on past income and capital accumulation, and also on different preparation periods. The IS-LM Zhou and Li business cycle models are given by the following equation:

$$\begin{cases} \dot{Y}(t) = \alpha\left[I_1\big(Y(t), r(t)\big) + \beta_1 K(t) - S\big(Y(t), r(t)\big)\right] \\ \dot{r}(t) = \beta\left[L\big(Y(t), r(t)\big) - \bar{M}\right] \\ \dot{K}(t) = I_1\big(Y(t-\tau_1), r(t)\big) - (\delta - \beta_1)K(t-\tau_2) \end{cases}$$

In 2005, De Casare and Sportelli formulated functions I, S and L in the following forms [18]:

$$\begin{cases} I_1\big(Y(t), r(t)\big) = A\dfrac{[Y(t)]^a}{[r(t)]^b} \\ S\big(Y(t), r(t)\big) = s_1[Y(t)]^a[r(t)]^b \\ L\big(Y(t), r(t)\big) = L_1\big(Y(t)\big) + L_2\big(r(t)\big) = gY(t) + \dfrac{h}{r(t) - \hat{r_1}} \end{cases}$$

Thus the IS-LM Zhou and Li (2009) business cycle models and De Casare and Sportelli (2005) can be stated as follows:

$$\begin{cases} \dot{Y}(t) = \alpha\left[A\frac{[Y(t)]^a}{[r(t)]^b} + \beta_1 K(t) - s_1[Y(t)]^a[r(t)]^b\right] \\ \dot{r}(t) = \beta\left[gY(t) + \frac{h}{r(t)-\hat{r_1}} - \bar{M}\right] \\ \dot{K}(t) = A\frac{[Y(t-\tau_1)]^a}{[r(t)]^b} - (\delta - \beta_1)K(t-\tau_2) \end{cases} \tag{1}$$

with $\dot{Y}(t)$ is the rate of income that is affected by the function of income, function of interest rates, function of capital stock, and parameters in goods market index (α), technology productivity (A), level of decline in investment in capital stock (β_1), level growth of savings against income (s_1). Whereas $\dot{r}(t)$ is the interest rate that is influenced by the income function, interest rate function and the money market index parameter (β), the amount of money demand for income (g), the amount of money demand from the interest rate (h), the lowest fixed rate of the interest rate $(\widehat{r_1})$, the money supply constant $(\overline{M})$. Similary, $\dot{K}(t)$ is the rate of capital stock affected by the function of income, function of interest rates, function of capital stock, and parameters of technological productivity, capital depreciation constant (δ) rate of decline in investment in capital stock.

3. Analysis Of The Stability Of The IS-LM Bussiness Cycle Model With Two Time Delays

Stability analysis is obtained by determining the equilibrium point or the critical point, linearizing the model if the model used is nonlinear and determining the characteristic value of the model. The IS-LM business cycle model in Equation (1) contains a delay differential equation, so to obtain the equilibrium point, the case is considered without delay $(\boldsymbol{\tau_1 = \tau_2 = 0})$. The equilibrium point is obtained from $\dot{\boldsymbol{Y}}(\boldsymbol{t}) = \dot{\boldsymbol{r}}(\boldsymbol{t}) = \dot{\boldsymbol{K}}(\boldsymbol{t}) = \boldsymbol{0}$ so that the equilibrium pointis obtained

$$\boldsymbol{E_1} = \left(\boldsymbol{0}, \frac{\boldsymbol{h+\overline{M}\widehat{r_1}}}{\boldsymbol{\overline{M}}}, \boldsymbol{0}\right) \text{ and } \boldsymbol{E_2} = \left(\frac{\boldsymbol{1}}{\boldsymbol{g}}\left(\boldsymbol{\overline{M}} - \frac{\boldsymbol{h}}{\boldsymbol{r(t)-\widehat{r_1}}}\right), \left(\frac{\boldsymbol{A\delta}}{\boldsymbol{s_1(\delta-\beta_1)}}\right)^{\frac{\boldsymbol{1}}{\boldsymbol{2b}}}, \frac{\boldsymbol{A[Y(t)]^a}}{\boldsymbol{(\delta-\beta_1)[r(t)]^b}}\right).$$

Furthermore, The IS-LM business cycle model is a non-linear model, for it to be linearized or to obtain the characteristic values of the model. Equation (1) contains a delay parameter, so that the linearization of the IS-LM business cycle model is as follows:

$$\begin{pmatrix}\dot{Y}\\ \dot{r}\\ \dot{K}\end{pmatrix} = J_1\begin{pmatrix}Y(t)\\ r(t)\\ K(t)\end{pmatrix} + J_2\begin{pmatrix}Y(t-\tau_1)\\ r(t-\tau_1)\\ K(t-\tau_1)\end{pmatrix} + J_3\begin{pmatrix}Y(t-\tau_2)\\ r(t-\tau_2)\\ K(t-\tau_2)\end{pmatrix}$$

where the matrix J_1, J_2 and $J_3$are the Jacobi matrix as follows:

$$J_1 = \begin{pmatrix}\alpha d_1 & \alpha d_2 & -\alpha\beta_1\\ \beta g & \beta d_3 & 0\\ 0 & d_4 & 0\end{pmatrix}, J_2 = \begin{pmatrix}0 & 0 & 0\\ 0 & 0 & 0\\ d_5 & 0 & 0\end{pmatrix}, J_3 = \begin{pmatrix}0 & 0 & 0\\ 0 & 0 & 0\\ 0 & 0 & -\delta+\beta\end{pmatrix}$$

with

$$d_1 = \frac{Aa[Y(t)]^{a-1}}{[r(t)]^b} - as_1[Y(t)]^{a-1}[r(t)]^b$$

$$d_2 = -\frac{bA[Y(t)]^a}{[r(t)]^{b+1}} - bs_1[Y(t)]^a[r(t)]^{b-1}$$

$$d_3 = -\frac{h}{(r(t)-\widehat{r_1})^2}$$

$$d_4 = -\frac{Ab[Y(t-\tau_1)]^a}{[r(t)]^{b+1}}$$

$$d_5 = \frac{Aa[Y(t-\tau_1)]^{a-1}}{[r(t)]^b}$$

Furthermore, the characteristic equation of the IS-LM business cycle model will be determined by completing

$$\det\left(\lambda I - J_1 - J_2e^{-\lambda\tau_1} - J_3e^{-\lambda\tau_2}\right) = 0$$

So that the following results are obtained:

$$\lambda^3 + p_2\lambda^2 + p_1\lambda + p_0 + (q_1\lambda + q_0)e^{-\lambda\tau_1} + (r_2\lambda^2 + r_1\lambda + r_0)e^{-\lambda\tau_2} = 0 \quad (2)$$

Equation (2) contains $e^{-\lambda\tau_1}$ and $e^{-\lambda\tau_2}$ so that it cannot be solved exactly. For that, a numerical approach is needed to solve it.

Analysis of the stability of the IS-LM business cycle model can be seen from the characteristics of the model. If the characteristic value is less than zero then the model is stable. Stability analysis is carried out in several cases, i.e:

1. Case 1. When $\tau_1 = \tau_2 = 0$ (without delay)
   In the IS-LM business cycle model without delay, the following characteristics are obtained:
   $$\lambda^3 + (p_2 + r_2)\lambda^2 + (p_1 + q_1 + r_1)\lambda + (p_0 + q_0 + r_0) = 0$$
   - For fixed points$E_1 = \left(0, \frac{h + \bar{M}\widehat{r_1}}{\bar{M}}, 0\right)$obtained eigenvalue
   $$\lambda_1 = 0,\ \lambda_2 = -\frac{\bar{M}^2\beta}{h},\ \lambda_3 = -\delta + \beta_1$$
   - For fixed points$E_2 = \left(\frac{1}{g}\left(\bar{M} - \frac{h}{r(t) - \widehat{r_1}}\right), \left(\frac{A\delta}{s_1(\delta - \beta_1)}\right)^{\frac{1}{2b}}, \frac{A[Y(t)]^a}{(\delta - \beta_1)[r(t)]^b}\right)$

   obtained the following characteristic equation:
   $$\lambda^3 + (p_2 + r_2)\lambda^2 + (p_1 + q_1 + r_1)\lambda + (p_0 + q_0 + r_0) = 0$$

   Based on Routh-Hurwitz criteria for the third degree characteristic equation, so that the local asymptotic stable E_2 equilibrium point must meet the following criteria:
   $$p_2 + r_2 > 0,\ \ p_0 + q_0 + r_0 > 0$$
   and
   $$(p_2 + r_2)(p_1 + q_1 + r_1) - (p_0 + q_0 + r_0) > 0$$

2. Case 2. When $\tau_1 \neq 0$ and $\tau_2 = 0$
   In the IS-LM business cycle model at the time $\tau_1 \neq 0$ and $\tau_2 = 0$, obtained the following characteristic equation:
   $$\lambda^3 + (p_2 + r_2)\lambda^2 + (p_1 + r_1)\lambda + p_0 + r_0 + (q_1\lambda + q_0)e^{-\lambda\tau_1} = 0 \quad (3)$$

   Let$\lambda = i\omega,\ \omega > 0,\ \omega \in R$, then Equation (2) can be written in the form of:
   $$(i\omega)^3 + (p_2 + r_2)(i\omega)^2 + (p_1 + r_1)(i\omega) + p_0 + r_0 + (q_1(i\omega) + q_0)\left(\cos(\omega\tau_1) - i sin(\omega\tau_1)\right) = 0 \quad (4)$$

   So that the real part is obtained in Equation (4)
   $$-(p_2 + r_2)\omega^2 + (p_0 + r_0) + q_1\omega\sin(\omega\tau_1) + q_0\cos(\omega\tau_1) = 0$$
   $$(p_2 + r_2)\omega^2 - (p_0 + r_0) = q_1\omega\sin(\omega\tau_1) + q_0\cos(\omega\tau_1) \quad (5)$$

   while the imaginary part is
   $$-\omega^3 + (p_1 + r_1)\omega + q_1\omega\cos(\omega\tau_1) - q_0\sin(\omega\tau_1) = 0$$
   $$\omega^3 - (p_1 + r_1)\omega = q_1\omega\cos(\omega\tau_1) - q_0\sin(\omega\tau_1) \quad (6)$$

   Next, Equation (5) and Equation (6) can be simplified to become

$$\omega^6 + (p_2^2 + 2p_2r_2 + r_2^2 - 2(p_1 + r_1))\omega^4 + ((p_1 + r_1)^2 - 2(p_2 + r_2)(p_0 + r_0) - q_1^2)\omega^2 + ((p_0 + r_0)^2 - q_0^2) = 0$$

Or it can be rewritten to

$$\omega^6 + G\omega^4 + H\omega^2 + I = 0 \qquad (7)$$

with

$$G = p_2^2 + 2p_2r_2 + r_2^2 - 2(p_1 + r_1)$$
$$H = (p_1 + r_1)^2 - 2(p_2 + r_2)(p_0 + r_0) - q_1^2$$
$$I = (p_0 + r_0)^2 - q_0^2$$

For example $\omega^2 = z$, then

$$z^3 + Gz^2 + H + I = 0 \qquad (8)$$

Suppose Equation (8) has positive roots z_1, z_2 and z_3, then the roots of Equation (7) are

$$\omega_1 = \sqrt{z_1},\ \omega_2 = \sqrt{z_2} \text{ dan } \omega_3 = \sqrt{z_3}$$

Review Equation (5) and Equation (6)

$$(p_2 + r_2)\omega^2 - (p_0 + r_0) - q_1\omega\sin(\omega\tau_1) = q_0\cos(\omega\tau_1)$$

so that it is obtained

$$\cos(\omega\tau_1) = \frac{(p_2+r_2)\omega^2-(p_0+r_0)-q_1\omega\sin(\omega\tau_1)}{q_0} \qquad (9)$$

Next, substitute Equation (9) in Equation (6)

$$\omega^3 - (p_1 + r_1)\omega = q_1\omega\left(\frac{(p_2 + r_2)\omega^2 - (p_0 + r_0) - q_1\omega\sin(\omega\tau_1)}{q_0}\right) - q_0\sin(\omega\tau_1)$$

Thus obtained

$$\sin(\omega\tau_1) = \frac{q_1\omega^3(p_2+r_2)-q_1\omega(p_0+r_0)-(\omega^3-(p_1+r_1)\omega)q_0}{(q_1\omega)^2-q_0^2} \qquad (10)$$

Next, substitute Equation (9) in Equation (8)

$$\cos(\omega\tau_1) = \frac{(p_2 + r_2)\omega^2 - (p_0 + r_0) - q_1\omega\left(\frac{q_1\omega^3(p_2+r_2)-q_1\omega(p_0+r_0)-(\omega^3-(p_1+r_1)\omega)q_0}{(q_1\omega)^2-q_0^2}\right)}{q_0}$$

$$\cos(\omega\tau_1) = \frac{q_0\omega^2(p_2 + r_2) - q_0(p_0 + r_0) + q_1\omega(\omega^3 - (p_1 + r_1)\omega)}{(q_1\omega_j)^2 - q_0^2}$$

Thus obtained

$$\tau_1 = \frac{1}{\omega}\cos^{-1}\frac{q_0\omega^2(p_2 + r_2) - q_0(p_0 + r_0) + q_1\omega(\omega^3 - (p_1 + r_1)\omega)}{(q_1\omega_j)^2 - q_0^2}$$

Changes in characteristic values occur because τ_1 increases, so a bounded value is needed called a delay critical value or denoted by $\tau_{1_{bif}}$ which is

$$\tau_{1,i}^j = \frac{1}{\omega_j}\, arc\cos\frac{-(p_2 + r_2)q_0\omega_j^2 + (p_0 + r_0)q_0 + \left(\omega_j^2 - (p_1 + r_1)\omega_j\right)q_1\omega_j}{(q_1\omega_j)^2 - q_0^2} + \frac{2i\pi}{\omega_j}$$

With $i = 0,1,2,\ldots$and $j = 1,2,3$.

Because $\tau_{1,i}^{j}$ has an infinite value, the critical value of delay is $\tau_{1_{bif}} = \min_{i=0,1,2,\dots \text{ and } j=1,2,3} \tau_{1,i}^{j}$ and ω_j are selected from the three positive roots of Equation (3).

To prove the occurrence of the Hopf bifurcation at $\tau_1 = \tau_{1_{bif}}$, it is necessary to point out that $Re(\frac{d}{d\tau_1}(\lambda(\tau_1)) \neq 0$.

Suppose Equation (3) can be rewritten with

$$G(\lambda(\tau_1), \tau_1) = \lambda^3 + A\lambda^2 + B\lambda + C + D\lambda e^{-\lambda\tau_1} + Ee^{-\lambda\tau_1}$$

Then, differential $G(\lambda(\tau_1), \tau_1)$ produces

$$dG = 0$$
$$\frac{\partial G}{\partial \lambda} d\lambda + \frac{\partial G}{\partial \tau_1} d\tau_1 = 0$$
$$\frac{d\lambda}{d\tau_1} = -\frac{\frac{\partial G}{\partial \tau_1}}{\frac{\partial G}{\partial \lambda}}$$
$$= \frac{(D\lambda^2 + E\lambda)e^{-\lambda\tau_1}}{3\lambda^3 + 2A\lambda + B + (D - \tau_1 D\lambda - \tau_1 E)e^{-\lambda\tau_1}}$$

Next, the derivative of G is calculated at the time $\tau_1 = \tau_{1bif}$ dan $\lambda = i\omega_{bif}$

$$\frac{d}{d\tau_1}(\lambda(\tau_1)) = \frac{\left(D\left(i\omega_{bif}\right)^2 + E\left(i\omega_{bif}\right)\right)e^{-i\omega_{bif}\tau_1}}{3\left(i\omega_{bif}\right)^2 + 2A\left(i\omega_{bif}\right) + B + \left(D - i\tau_{1_{bif}} D\omega_{bif} - \tau_{1_{bif}} E\right)e^{-i\omega_{bif}\tau_1}}$$

Substitute $e^{-i\omega_{bif}\tau_1} = \cos(\omega_{bif}\tau_{1bif}) - i\sin(\omega_{bif}\tau_{1bif})$, and with simplification obtained

$$\begin{aligned} Re(\frac{d}{d\tau_1}(\lambda(\tau_1)) &= 3\omega_{bif}^4(D\omega_{bif}\cos(\omega_{bif}\tau_{1bif}) - E\sin(\omega_{bif}\tau_{1bif})) \\ &- B\omega_{bif}(D\omega_{bif}\cos(\omega_{bif}\tau_{1bif}) - E\sin(\omega_{bif}\tau_{1bif})) \\ &-D^2\omega_{bif}^2 + 2A\omega_{bif}^2(D\omega_{bif}\sin(\omega_{bif}\tau_{1bif}) + E\cos(\omega_{bif}\tau_{1bif})) \end{aligned}$$

4. Optimal Control of the IS-LM Business Cycle Model with Two Delays

The problem of optimal control of the IS-LM business cycle model in Equation (2) is to balance or stabilize the supply of money and demand for money in the community. This is expected to stabilize the economy of a country. In this study, what is used as a controller is the supply of money $\overline{M}$.

Optimal control of the IS-LM business cycle model to get the optimal shape so as to minimize the objective function with the $\overline{M}$ control are:

$$J(\overline{M}) = \int_0^T (r(t) + c_1\overline{M}^2)\, dt$$

Journal of Physics: Conference Series **1373** (2019) 012038 doi:10.1088/1742-6596/1373/1/012038

Completion of Optimal Control:

1. Determine the Hamiltonian function

$$\mathcal{H} = r(t) + \bar{M}^2 + \lambda_1\left(\alpha\left[A\frac{[Y(t)]^a}{[r(t)]^b} + \beta_1 K(t) - s_1[Y(t)]^a[r(t)]^b\right]\right)$$
$$+\lambda_2\left(\beta\left[gY(t) + \frac{h}{r(t)-\hat{r_1}} - \bar{M}\right]\right) + \lambda_3\left(A\frac{[Y(t-\tau_1)]^a}{[r(t)]^b} - (\delta - \beta_1)K(t-\tau_2)\right)$$

2. Stationary conditions

$$\frac{\partial\mathcal{H}}{\partial\bar{M}} = 0 \text{ then } \bar{M}^*(t) = \frac{\beta}{2c_1}\lambda_2$$

So that it is obtained:

$$\mathcal{H}^* = r(t) + \left(\frac{\beta}{2c_1}\lambda_2\right)^2 + \lambda_1\left(\alpha\left[A\frac{[Y(t)]^a}{[r(t)]^b} + \beta_1 K(t) - s_1[Y(t)]^a[r(t)]^b\right]\right)$$
$$+\lambda_2\left(\beta\left[gY(t) + \frac{h}{r(t)-\hat{r_1}} - \frac{\beta}{2c_1}\lambda_2\right]\right) + \lambda_3\left(A\frac{[Y(t-\tau_1)]^a}{[r(t)]^b} - (\delta - \beta_1)K(t-\tau_2)\right)$$

3. Equation State

$$\dot{Y}^*(t) = \frac{\partial\mathcal{H}^*}{\partial\lambda_1}$$
$$= \alpha\left[A\frac{[Y(t)]^a}{[r(t)]^b} + \beta_1 K(t) - s_1[Y(t)]^a[r(t)]^b\right]$$
$$\dot{r}^*(t) = \frac{\partial\mathcal{H}^*}{\partial\lambda_2}$$
$$= \beta\left[gY(t) + \frac{h}{r(t)-\hat{r_1}} - \bar{M}\right]$$
$$\dot{K}^*(t) = \frac{\partial\mathcal{H}^*}{\partial\lambda_3}$$
$$= A\frac{[Y(t-\tau_1)]^a}{[r(t)]^b} - (\delta - \beta_1)K(t-\tau_2)$$

4. Costate equation

$$\dot{\lambda_1}^*(t) = -\frac{\partial\mathcal{H}^*}{\partial Y^*(t)} - \frac{\partial\mathcal{H}^*}{\partial Y^*(t-\tau_1)}$$
$$= -\lambda_1\left(\alpha a A\frac{[Y^*(t)]^{a-1}}{[r^*(t)]^b} - \alpha a s_1[Y^*(t)]^{a-1}[r^*(t)]^b\right) - \lambda_2\beta g - \lambda_3 a A\frac{[Y^*(t-\tau_1)]^{a-1}}{[r^*(t)]^b}$$
$$\dot{\lambda_2}^*(t) = -\frac{\partial\mathcal{H}^*}{\partial r^*(t)}$$
$$= -1 - \lambda_1\left(-\alpha b A\frac{[Y^*(t)]^a}{[r^*(t)]^{b+1}} - b s_1[Y^*(t)]^a[r^*(t)]^{b-1}\right) - \lambda_2\beta\frac{h}{(r(t)-\hat{r_1})^2}$$
$$-\lambda_3 b A\frac{[Y^*(t-\tau_1)]^a}{[r^*(t)]^{b+1}}$$
$$\dot{\lambda_3}^*(t) = -\frac{\partial\mathcal{H}^*}{\partial K^*(t)} - \frac{\partial\mathcal{H}^*}{\partial K^*(t-\tau_2)}$$
$$= -\lambda_1\alpha\beta_1 + \lambda_3(\delta - \beta_1)$$

5. Numerical Simulation

In the next section a simulation will be performed to show changes in system stability and optimal control in the IS-LM business cycle model. Model simulations are shown in graphical form, by substituting parameter values in Tabel 1.

Tabel 1. Parameter values and initial conditions for the IS-LM business cycle model

Parameter	description	values
A	technology productivity	0.1
α	index in the goods market	1
β	index in the money market	1
a	coefficient of adjustment in the goods market	1.03
b	coefficient of adjustment on the money market	0.6
β_1	the level of decline in investment in the capital stock	-0.1
δ	capital depreciation constant	0.1
s_1	growth rate of deposits against income	0.08
g	amount of money demand for income	0.05
h	the amount of money demand for interest rates	0.01
$\widehat{r_1}$	lowest fixed rate of interest rates	0.001
$\bar{M}$	money supply constant	0.1

5.1 Stability on IS-LM Business Cycle Model

Analysis of the stability of the IS-LM business cycle model can be seen from the characteristics of the model. If the characteristic value is less than zero then the model is stable. Stability analysis is carried out in several cases, i.e: Case 1. When $\tau_1 = \tau_2 = 0$ (without delay) and *Case 2: for* $\tau_1 \neq 0$ *and* $\tau_2 = 0$

5.1.1 Case 1. When $\tau_1 = \tau_2 = 0$ (without delay)

Based on the parameter values in Tabel 1, obtained a fixed point $E_1 = (0, 0.101, 0)$ and $E_2 = (1.7037, 0.6759, 1.0949)$. For fixed points $E_1 = (0, 0.101, 0)$, the eigenvalue is $\lambda_i = (-0.1, 0, -0.2)$ with $i = 1, 2, 3$. Because there is an eigen value of 0, then at a fixed point E_1does not oscillate. Whereas the fixed point $E_2 = (1.7037, 0.6759, 1.0949)$is asymptotically stable because it meets the Routh-Hurwitz criteria. Then simulations are performed for different s_1 parameter values, where the parameter s_1 is a parameter of the growth rate of deposits against income. Selected parameters $s_1 = 0{,}08$ and $s_1 = 0{,}01$ with initial values $Y = 1.5, r = 0.5$, and $K = 1$.

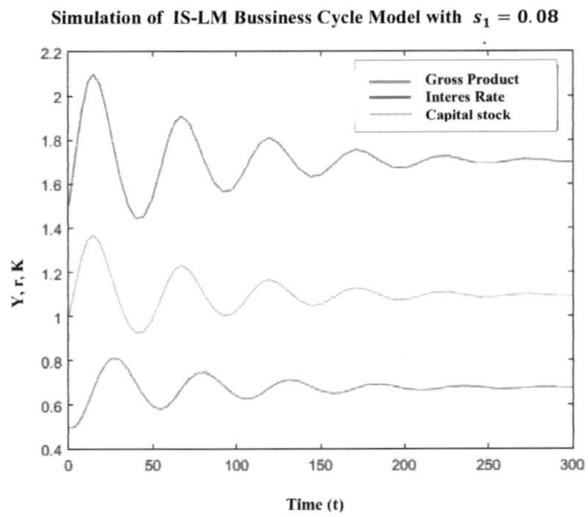

Figure 1. The graph dynamic of IS-LM Bussiness Cycle Model with $s_1 = 0.08$

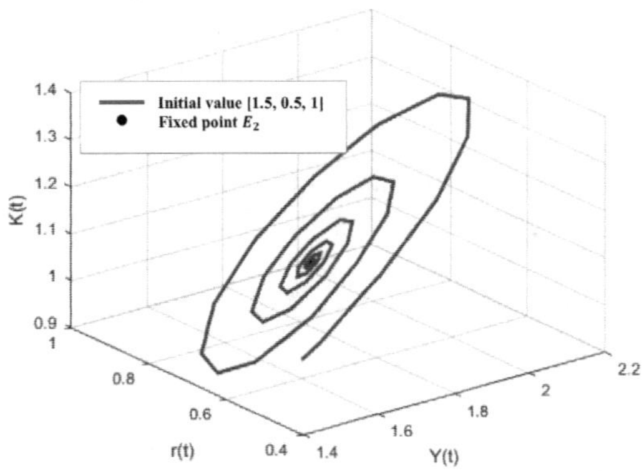

Figure 2. The phase field IS-LM Bussiness Cycle Model with $s_1 = 0.08$

In Figure 1 it can be seen that when the parameter value $s_1 = 0.08$ the graph solution goes to the fixed point E_2 so that the system is stable.
In Figure 2 it can be seen that when the parameter value $s_1 = 0.08$ the phase field solution goes to the fixed point E_2 that the system is stable.

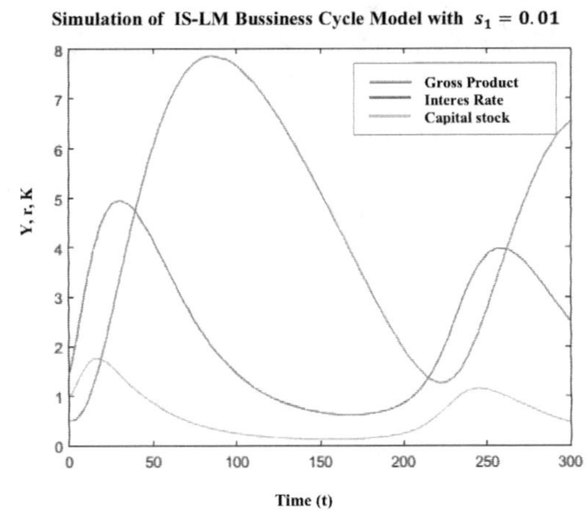

Figure 3. The graph dynamic of IS-LM Bussiness Cycle Model with $s_1 = 0.01$

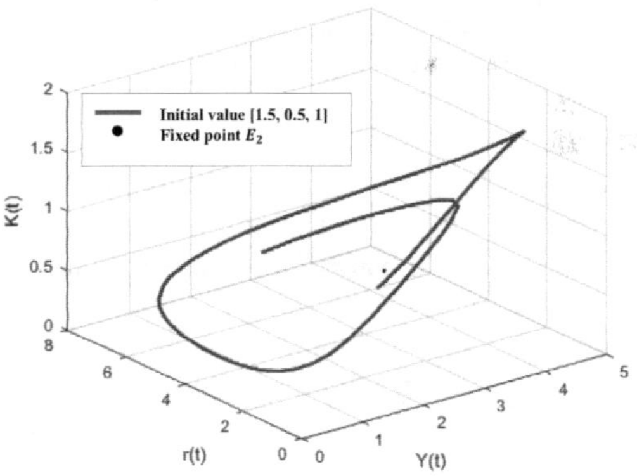

Figure 4. The phase field IS-LM Bussiness Cycle Model with $s_1 = 0.01$

In Figure 3 it can be seen that when the parameter value $s_1 = 0.01$, the graph solution has oscillation and is further away from the fixed point E_2 so that the system is unstable.

In Figure 4 it can be seen that when the parameter value $s_1 = 0.01$, the phase field is further away from the fixed point E_2 so that the system is unstable.

Thus, the simulation results above show that at fixed point E_2 there is a bifurcation with a change in the value of the parameter s_1.

5.1.2 Case 2: for $\tau_1 \neq 0$ and $\tau_2 = 0$
Based on the parameter values in Tabel.1, obtained a fixed point $E_2 = (1.7037, 0.6759, 1.0949)$ and a delay critical value $\tau_1 = 2.7076$. Parameter values meet the Routh-Hurwitz criteria so that at a fixed point $E_2$is stable. Simulations were carried out at $\tau_1 = 1$ and $\tau_1 = 3.8$ with initial values $Y = 1.5, r = 0.5$ and $K = 1$.

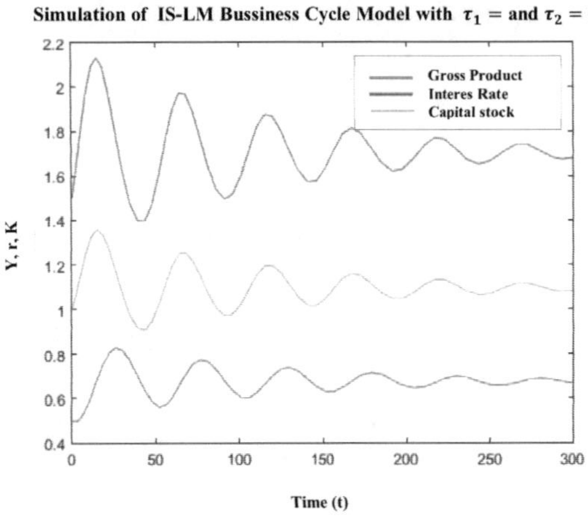

Figure 5. The graph dynamic of IS-LM Bussiness Cycle Model with $\tau_1 = 1$ and $\tau_2 = 0$

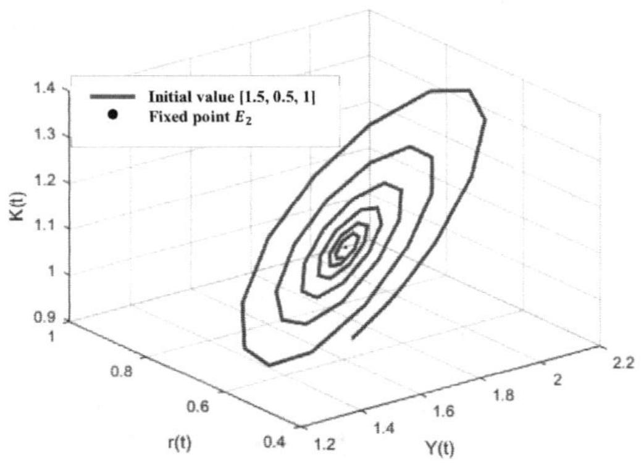

Figure 6. The phase field IS-LM Bussiness Cycle Model with $\tau_1 = 1$ and $\tau_2 = 0$

In Figure 5, it can be seen that when the delay time is $\tau_1 = 1$, it is explained that the time delay on income causes the rate of income, interest rate, and rate of capital stock to decrease, consequently the system will move towards E_2 so that the system is stable.

In Figure 6, it can be seen that when the delay time is equal $\tau_1 = 1$, the phase field solution goes to the fixed point E_2 that the system is stable.

Journal of Physics: Conference Series **1373** (2019) 012038 doi:10.1088/1742-6596/1373/1/012038

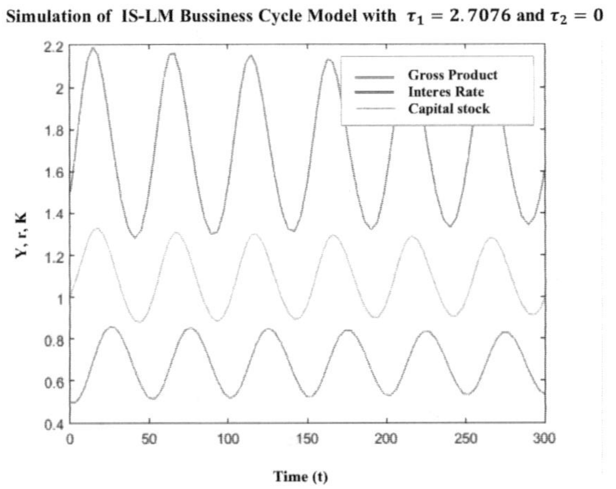

Figure 7. The graph dynamic of IS-LM Bussiness Cycle Model with $\tau_1 = 2.7076$ and $\tau_2 = 0$

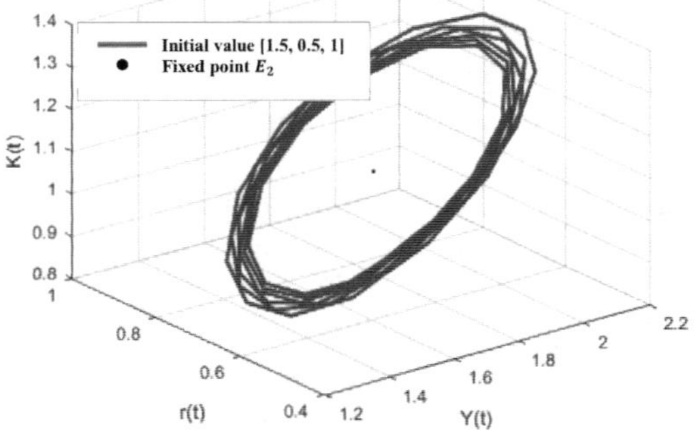

Figure 8. The phase field IS-LM Bussiness Cycle Model with $\tau_1 = 2.7076$ and $\tau_2 = 0$

In Figure 7, can be seen when the delay time $\tau_1 = 2.7076$ system shows a contant oscillation movement. The rate of income, interest rates, and the rate of capital stock do not seem to move towards a certain value. This is because there is an eigenvalue into a pure imaginary form, so that the stability of the system changes. Eigenvalues in the form of pure imaginary indicate a limit cycle. This limit cycle illustrates that when the delay value is 1 = 2: 7076, the system will not go to a certain value but only oscillates over time which will only change when the time value of the delay gets bigger.

In Figure 8, it can be seen when $\tau_1 = 2.7076$ phase fields form a constant spiral, so the system experiences oscillation.

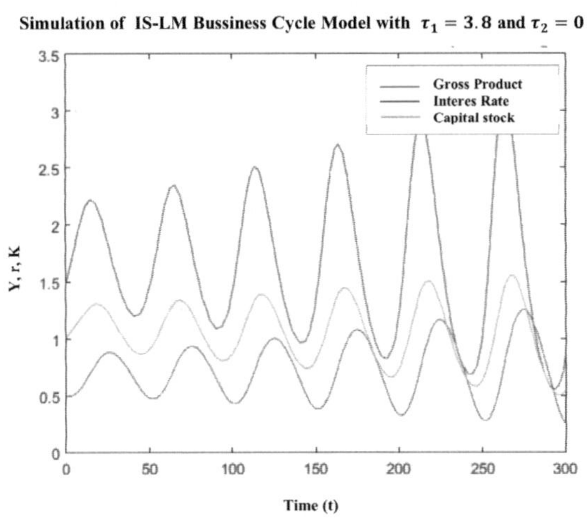

Figure 9. The graph dynamic of IS-LM Bussiness Cycle Model with $\tau_1 = 3.8$ and $\tau_2 = 0$

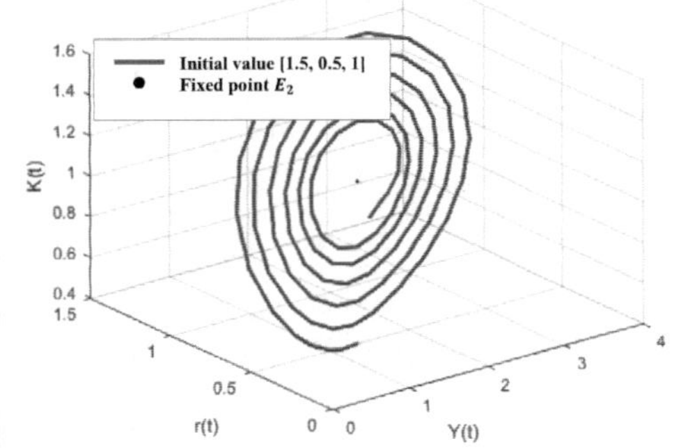

Figure 10. The phase field IS-LM Bussiness Cycle Model with $\tau_1 = 3.8$ and $\tau_2 = 0$

Journal of Physics: Conference Series **1373** (2019) 012038 doi:10.1088/1742-6596/1373/1/012038

In Figure 9, it can be seen that when $\tau_1 = 3.8$ shows a very large change in oscillation, especially in the income variable, as a result the income rate, interest rate, and the rate of capital stock will increase because the delay time will increase. to the point of stability.

In Figure 10, it can be seen that when $\tau_1 = 3.8$ phase fields form a spiral that gets bigger and does not lead to E_2 equilibrium point, so the system is unstable.

The simulation results show that there is a stability change when $\tau_1 = 3.8$ and $\tau_1 = 3.8$ so that according to the theorem occurs bifurcation. Stability changes occur when the value $\tau_1 = 2.7076$ is when the eigenvalues in the form of pure imaginary arise so that there is a limit cycle that reveals that there is a Hopf bifurcation. Hopf bifurcation at the equilibrium point describes the existence of a boundary value called the delay critical point value. The income rate, interest rate, and capital stock rate have a stable system solution when given a value of delay that is less than the value of the delay critical point $0 < \tau_1 < 2.7076$. If the value of the delay given is more than the value of the delay critical point ($\tau_1 > 2.7076$), the system solution shows the rate of income, the interest rate, and the rate of capital stock will continue to fluctuate, causing unstable economic conditions.

5.2 Optimal Control of the IS-LM Business Cycle Model

In this section numerical simulations optimal control in the IS-LM business cycle model is performed. To solve the state equation, the forward of 4th order Runge Kutta method is used, while the completion of the co-state equation is used as the backward of 4th order runge kutta method.
In this simulation, parameter values are based on Table 1 with a value of $\tau_1 = 3.5$ and $\tau_2 = 6.5$. The simulation for optimal control $\bar{M}$ in the interest rate equation can be seen in the following figure.

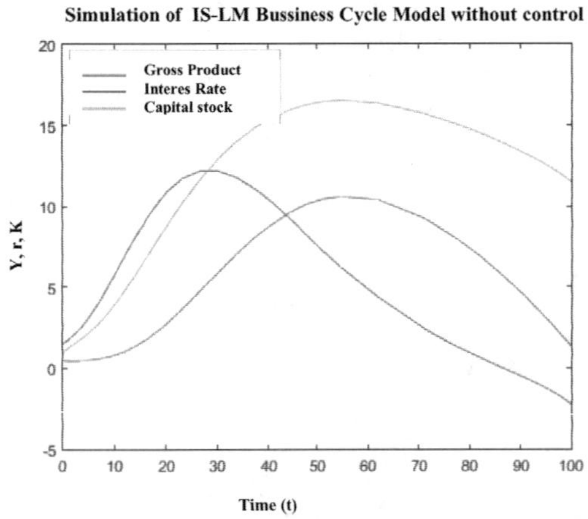

Figure 11. The graph dynamic of IS-LM Bussiness Cycle Model with $\tau_1 = 3.5$ and $\tau_2 = 6.5$ without control

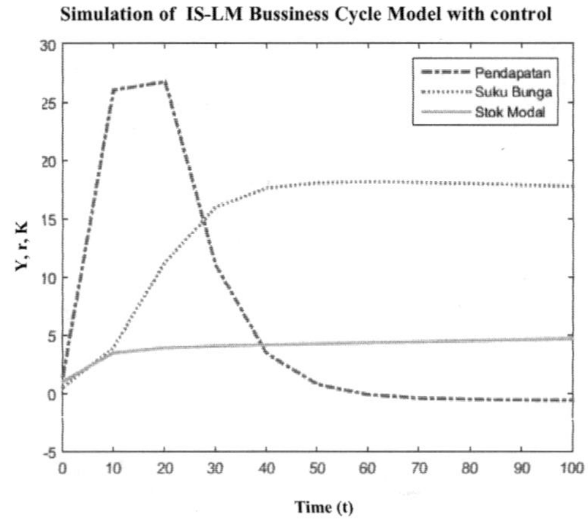

Figure 12. The phase field IS-LM Bussiness Cycle Model with $\tau_1 = 3.5$ and $\tau_2 = 6.5$ with control

In Figure 11, it can be seen that when the values $\tau_1 = 3.5$ and $\tau_2 = 6.5$ pass the critical delay value, the changes in oscillations are very large, especially in the capital stock variable. As a result, the rate of income, interest rates, and the rate of capital stock will increase because the delay time is getting bigger, so the system is not unstable. Furthermore optimal control is carried out so that the system becomes stable.

In Figure 12, it can be seen that when giving money supply control to the interest rate variable, the income rate decreases and starts towards the stability point in the 50th time. While the interest rate starts towards the stability point at the 40th time. Likewise with the rate of capital stock starting towards the stability point at the 30th time. Thus it can be concluded that with the control given, the rate of income, the interest rate and the rate of capital stock will go to the point of stability.

6. Conclusion

Based on the discussion above it can be concluded that based on the analysis of the stability of the IS-LM business cycle model The simulation results show that there is a stability change when the delay value given exceeds the critical value of delay. Stability changes occur when the delay value given arises a pure eigenvalue in the form of pure imaginary so that there is a limit cycle that reveals that there is a Hopf bifurcation. While based on optimal control provided by maximizing the total money supply, the rate of income, interest rate, and the rate of capital stock decrease. As a result, the system will lead to stability.

7. References

[1] Sambas A , Vaidyanathan S, Mamat M, Mohamed M A and Sanjaya W S M 2018 *International Journal of Electrical and Computer Engineering* **8** 4951-4958.
[2] Vaidyanathan S, Feki M, Sambas A and Lien C H 2018 *International Journal of Simulation and Process Modelling* **13** 419-432.
[3] Vaidyanathan S, Sambas A, Kacar S and Cavusoglu U 2019 *Nonlinear Engineering* **8** 193-205.
[4] Vaidyanathan S, Sambas A, Mamat M and Sanjaya W M 2017 *Archives of Control Sciences* **27** 541-554.
[5] Mobayen S, Vaidyanathan S, Sambas , Kaçar S and Çavuşoğlu Ü 2019 *Iranian Journal of Science and Technology, Transactions of Electrical Engineering* **43**, 1-12.
[6] Vespignani A 2012 *Nature physics* **8**, 32.
[7] Tastrawati N K T 2012 *Mathematical Journal* **2** 1.
[8] Neamţu M, Opriş D and Chilarescu C 2007 *Chaos, Solitons & Fractals* **34**, 519-530.
[9] Kaddar A and Alaoui H T 2008 *Electronic Journal of Differential Equations* **134**, 1-9.
[10] Sportelli M, De Cesare L and Binetti M T 2014 *Applied Mathematics and Computation* **243,** 728-739.
[11] Ma J and Ren W 2016 *International Journal of Bifurcation and Chaos* **26**, 1650181.
[12] Liu X, Cai W, Lu J and Wang Y 2015 *Communications in Nonlinear Science and Numerical Simulation* **25,** 149-161.
[13] Cai J 1986 *Electronic Journal of Differential Equations* **15**, 1-6.
[14] Zou L and Li Y 2009 Journal of Computational and Applied Mathematics **228** 182-187.
[15] Rosmely, Nugrahani E H and Sianturi P 2016 *Mathematical Journal* **15** 1-10.
[16] Kaddar A and Alaoui H T 2008 *Applied Mathematical Sciences* **2** 1529-1539.
[17] Neri U and Venturi B 2007 *International Review of Economics* **54** 53-65.
[18] De Cesare L and Sportelli M 2005 *Chaos, Solutions and Fractals* **25** 233-244.

Journal of Physics: Conference Series **1373** (2019) 012039 doi:10.1088/1742-6596/1373/1/012039

Application of Intelligent Search Algorithms in Proportional-Integral-Derivative Control of Direct-Current Motor System

Alfian Ma'arif [1*], Hanamirza Nabila[2], Iswanto[3], Oyas Wahyunggoro[4]

[1] Department of Electrical Engineering, Universitas Ahmad Dahlan, Yogyakarta, Indonesia
[2] Department of Electrical Engineering, Universitas Islam Indonesia, Yogyakarta, Indonesia
[3] Department of Electrical Engineering, Universitas Muhammadiyah Yogyakarta, Yogyakarta, Indonesia
[4] Department of Electrical Engineering and Information Technology, Universitas Gadjah Mada, Yogyakarta, Indonesia

*Email: alfianmaarif@ee.uad.ac.id

Abstract. Proportional-Integral-Derivative (PID) control is one of the famous controllers applied in industrial and robotics world. It is quite easy to understand and to be implemented. It is known that, with a proper combination of its gain parameters, a system with good performance can be obtained. This parameter tuning can be tricky since the number of the combination is almost infinite. Hence, methods to tune the parameter gains of proportional, integral and derivative is needed. Some optimization methods that can be used in tuning parameters are Genetic Algorithm (GA), Particle Swarm Optimization (PSO) and Differential Evolution (DE). The paper will analyse the performance of those methods by applying them in PID Controller to control a Direct Current (DC) motor system by comparing the iteration time, the number of parameters, and augmented system performance. The comparison is done on MATLAB simulation by using the same computer. It results as DE as the best among them with the least number of parameters and best DC motor system performances.

1. Introduction

PID Controller was first published officially by Minorsky in the early 19th century with his theory of three-term control. It is the most used controller in industrial and robotics world until now [1] [2]. Some advantages of PID Controller are easy to understand, easy to be implemented, and able to provide a good system's stability [3]. Many systems are able to be stabilized, from non-linear to linear system.

Nevertheless, it is not problem-free. Two major problems in PID Controller are its inability to control high order system and trivial task in tuning its parameter. Its ability to stabilize is limited to second order system or less. This can be solved by doing modification in its integrator part.

The tuning parameter of PID is important because it can guarantee the controller efficiency [4]. The most frequently used method in designing a PID controller's parameter is a trial and error method. It is practically done by adding random value in its parameter then tested on system or simulation. If the parameter value was 'right', it can provide good system's performance by chance. However, it is not the best result. Besides, it is ineffective and does not have a standard procedure.

There are many methods that are able to function as tune the parameter such as Ziegler Nichols [5], Coefficient Diagram Method (CDM) [6] [7], Fuzzy tuning [8], Neural network and Intelligent Search Algorithm [9]. Unfortunately, few methods above have limitations and weaknesses. Ziegler Nichols method is only suitable for first order system. CDM has inflexibility issue in its system model representation preference, which it has to be in state space. Meanwhile, Fuzzy is quite difficult to be designed, especially for a system with no previous experimental data.

Intelligent Search Algorithm, or Heuristic method, is another method to tune parameter by doing a repeated iteration to obtain the best parameter value. It copies natural phenomenon so that it has a structured procedure to solve complex problems which cannot be solved by an analytical method. Its weakness is it has a great number of iterations which requires longer iteration time. Besides, the obtained result somehow not always consistent if the searching process is retaken.

Due to the many innovations in technology nowadays, those weaknesses can be overcome. High-speed data process in modern computers helps to decrease the number of iterations and computing time. Hence, Intelligent Search Algorithm is the most potent method to optimize parameter tuning in PID Controller.

The algorithm itself has many variations due to which natural phenomenon in the world. Some of them are Genetic Algorithm (GA) [10] [11], Particle Swarm Optimization (PSO) [12], and Differential Evolution (DE) [13] [14]. They are the most commonly used algorithm in Intelligent Search Algorithm. To test the algorithm, it is applied in the PID Controller of DC Motor System. The performance of the system along with a number of iterations and iteration time will be observed and analysed. DC Motor is chosen due to its commonly-used in industrial [13] and robotics world.

The structure of this paper is written as follows. The first section is started with modelling of DC motor, and a brief explanation about PID controller. The next section discusses the Genetic Algorithm, Particle Swarm Optimization, Differential Evolution and its objective functions. Result and discussion will be described next and it will be closed with conclusions.

2. Model of Direct Current (DC) Motor

The model diagram of DC motor system that is used in the research is shown in **Figure 1**. Based on **Figure 1**, V_{DC} is the power source $(Volt)$, i is current (A), R is resistance (Ω), L is inductance (H), f_k is the friction force, J is inertia moment, T is torque, and ω is angular speed (rad/s).

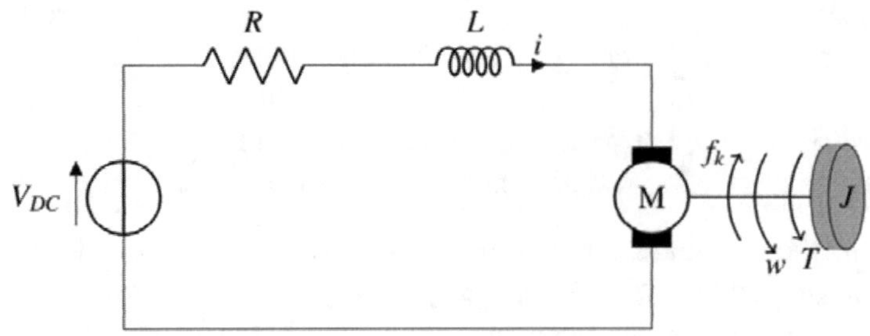

Figure 1. Model Diagram of DC Motor System

Model of DC motor system is derived from the electrical and mechanical point of view analysis. The equation from the electrical point of view is based on the Kirchhoff voltage law that the total voltage in the closed circuit is zero, which is written as:

$$V_L + V_R + V_{Emf} = v(t) \tag{1}$$

$$L\frac{di(t)}{dt} + Ri(t) + K_e\omega(t) = v(t) \tag{2}$$

where K_e is Back-Emf (Electro-mechanical force) constant (Vs/rad). Then, the equation (2) is transformed with Laplace Transformation so that it becomes

$$I(s)(Ls + R) = V(s) - K_e\omega(s) \tag{3}$$

Meanwhile, when the system is seen from a mechanical point of view, the Newton second law of rotation can be applied

$$T - f_k = J\dot{\omega}(t)$$

$$K_t i(t) - K_b\omega(t) = J\frac{\omega(t)}{dt}, \tag{4}$$

where K_t is torque constant, K_b is friction constant, $\dot{\omega}$ is angular acceleration. The equation (4) is also transformed into Laplace as stated,

$$Js\omega(s) + K_b\omega(s) = I(s)K_t$$

$$I(s) = (Js + b)\frac{\omega}{K_t} \tag{5}$$

By substituting (3) to (5), the transfer function of the DC motor system can be obtained as

$$\omega(s) = \frac{K_t}{(Ls + R)(Js + K_b) + K_t K_e} \tag{6}$$

3. Proportional-Integral-Derivative Controller

PID Control [15] equation in time domain is

$$u = K_p e(t) + K_i \int_{t=0}^{t} e(t)\, dt + K_d \frac{e(t)}{dt}, \tag{7}$$

where e is the error between the set point and feedback value, u is the control signal. The K_p is the proportional gain, K_i is the integral gain, and K_d is the derivative gain. The K_p, K_i and K_d later will be tuned by using the optimization method.

The block diagram of the proposed system shown in **Figure 2**.

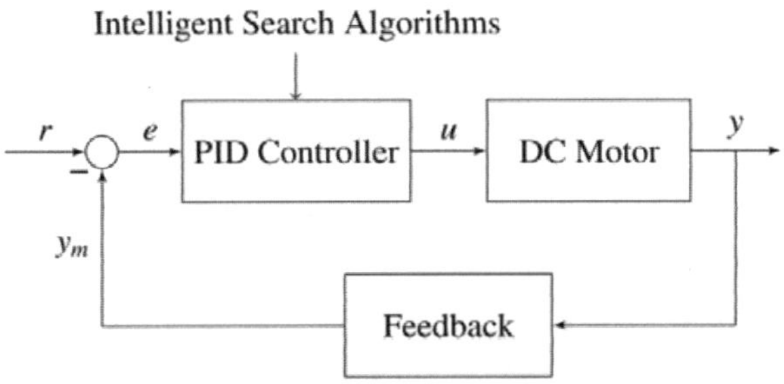

Figure 2. The Diagram Block of Proposed Controller

Based on **Figure 2**, r is the set point or speed reference, y_m is the feedback value (speed value from the sensor). The value of K_p, K_i, and K_d are determined as the output from the tuning process which uses intelligent search algorithms (GA, PSO, DE).

4. Objective Function

In every optimization process, the optimization problem needs to be identified first in the objective function. It represents problems that will be solved that guides the searching process. It can be the minimum function or maximum function based on conditions.

In this design, the function itself also represents the fitness value. The control objective of the controller is to achieve minimum steady state error, minimum rise time, minimum settling time, and also minimum overshoot. The objective function of this design can be stated as follows,

$$Min(f) = (w_1 * sse) + (w_2 * t_s) + (w_3 * t_r) + (w_4 * O_s), \tag{8}$$

where, w_i is weight, sse is steady state error, t_r is rise time, t_s is settling time and O_s is overshoot.

When compared in the iteration of the optimization process, fitness value should be compared in maximum function. Therefore, it is logical that the solution with bigger fitness value is a better solution. In order to make this happen, the obtained fitness value is converted to maximum function by applying

$$Max(f) = \frac{1}{Min(f)}. \quad (9)$$

5. Genetic Algorithm (GA)

GA was firstly declared by Holland in 1962. It copies the natural selection process in population which is important to genetic evolution. Some important terms used in GA are selection, crossover, mutation, and elitism. Selection is a process to pick individuals which will be used as parents in the reproduction process. Crossover is a process which chromosomes of individual parents are exchanged with each other to inherit traits. The mutation is a process where traits are changed. Elitism is a replacement of individual with less-quality to the new one in a population.

Procedures of the genetic algorithm can be explained briefly as follows,

1. Initialize population and decode an individual;
2. Evaluate the fitness value of the individual. It is normally formulated by its objective function;
3. Do selection to determine new parents in the reproduction process;
4. Do reproduction with selected parents to create a new individual. This step requires two operators: crossover and mutation;
5. Evaluate the fitness value of the new individual in the population. If it is better than another individual, replace the old individual with this new one (do elitism).

There are many kinds of selection operators. There are roulette wheel selection, tournament selection, and rank-based selection. The most popular among them all is the roulette-wheel selection. In this kind of selection, a random number r is generated and the chosen parents are individuals which have closest relative fitness value to r.

Two most popular crossover operators are a 1-point crossover and n-point crossover. 1-point crossover is the simplest where a point in chromosome will separate parent's chromosomes into two segments and the new individual will be made by exchanging them. It can be improved by adding more points which are then called n-point crossover.

6. Particle Swarm Optimization (PSO)

PSO was first introduced by Kennedy and Eberhart in 1995 which made based on social and communication behaviour of bird's flock. In PSO, a large number of particles move spread in multidimensional problem space. Every individual is assumed as a vector position and represents a potential solution of optimization problems.

Unlike other intelligent search algorithms which use evolution operator to manipulate individuals, PSO uses dynamic velocity which may change based on the individual's experience during moving in the swarm. Therefore, every individual is pulled to average stochastic weight from a previous best local point near its flock.

Procedures of Particle Swarm Optimization (PSO) can be explained briefly as follows,

1. Initialize particles and generate velocity randomly;
2. Evaluate particles by comparing their fitness values. This will be the benchmark to determine the best local point ($Pbest$) and best global point ($Gbest$);
3. Compare $Pbest$ of before and after iteration. If the new fitness value of a particle is better than previous $Pbest$ then it will be the new $Pbest$;
4. Get $Gbest$ by finding the highest fitness value of $Pbest$;
5. Update the velocity and position of the particle.

$$V_i^{kg+1} = w(t) + V_i^{kg} + c_1 r_1 (p_i^{kg} - X_i^{kg}) + c_2 r_2 (BestS_i^{kg} - X_i^{kg}) \tag{10}$$

$$X_i^{kg+1} = X_i^{kg} + V_i^{kg} \tag{11}$$

where $g = 1,2, \dots, G$, $i = 1,2, \dots, Size$, r_1 and r_2 is random values between $0 - 1$, c_1 is a local learning factor, c_2 is a global learning factor. Commonly, c_2 is bigger than c_1, or can be written as $c_2 > c_1 > 0$.

7. Differential Evolution (DE)

DE was first proposed by Storn and Price in 1995. Although it may seem that DE was created based on natural phenomenon, actually it was rather based on geometry argument. It has mutation and crossover operator in the searching process like in GA yet the new solution was created by selecting three vectors as 'parents' or strategic vectors. One of them is used as a base vector while two others are compared to find their vector difference to be added to the base vector.

This competition strategy, together with encoding and simple mutation differential, the complexity of the DE algorithm can be reduced. Main advantages of DE can be summarized in three points: less parameter, not easily trapped in local solution, and fast to converge.

Standard DE Algorithm needs at least 4 steps,

1. Initialize population generation,

$$P_{ij} = rand_{p1j}(0,1)(x_{ij}^U + x_{ij}^L) + x_{ij}^L \tag{12}$$

   where i is the number of individuals (population), j is the number of variables, $rand_{p1j}(0,1)$ is the random value from 0 to 1, x_{ij}^U is the upper limit, x_{ij}^L is the lower limit;

2. Do mutation by selecting three random individuals in the population,

$$h_{ij}(t+1) = x_{p1j}(t) + F\left(x_{p2j}(t) - x_{p3j}(t)\right), \tag{13}$$

while the function for no local solution is,

$$h_{ij}(t+1) = x_{bj}(t) + F\left(x_{p2j}(t) - x_{p3j}(t)\right), \tag{14}$$

where, x_{bj} is the best individuals. F is the mutation factor that selected between $0-2$. Generally, it can be set from 0.3 to 0.6. It will make fast convergence and the process may not trap in the local solution;

3. Do crossover to create a new vector,

$$v_i(t+1) = \begin{cases} h_{ij}(t+1), & rand_{ij} \leq CR \\ P_{ij}(t), & rand_{ij} > CR \end{cases} \tag{15}$$

where CR is the crossover factor in the range $[0,1]$. Generally, it can be set from 0.6 to0.9;

4. Evaluate vector by comparing it with its parents. If it is better than its parents, then do the replacement,

$$P_i(t+1) = \begin{cases} v_i(t+1), & f(v_{ij}(t+1), \dots, v_{in}(t+1)) < f(x_{x1}(t), \dots, x_{in}(t)) \\ x_{ij}, & f(v_{ij}(t+1), \dots, v_{in}(t+1)) \geq f(x_{x1}(t), \dots, x_{in}(t)) \end{cases} \tag{16}$$

8. Numerical Simulation and Discussion

The method will be simulated in Simulink Matlab. The chosen simulation stop time is 5 seconds. The following are specifications of the computer used to do the computation: Windows 10 Home Basic, i5 8th generation core and a solid-state drive. The transfer function model of DC motor system is,

$$\omega = \frac{0.01}{0.005s^2 + 0.06s + 0.001} V, \tag{17}$$

where ω is the angular velocity which is stated in rad/s (radian per second). In order to convert it to rotation per minute (rpm), the obtained result should be multiplied with $60/(2 * \pi)$ or $30/\pi$.

In every test taken, data are compared by some criteria. Those are the number of parameter variables, number of generation, number of iteration, iteration time, the obtained result (K_p, K_i, K_d), the system's steady state error (sse), system's rise time (t_r), system's settling time t_s, and the system's overshoot (O_s). The result of the experiment can be shown on **Figure 3**, **Table 1**, and

Table 2.

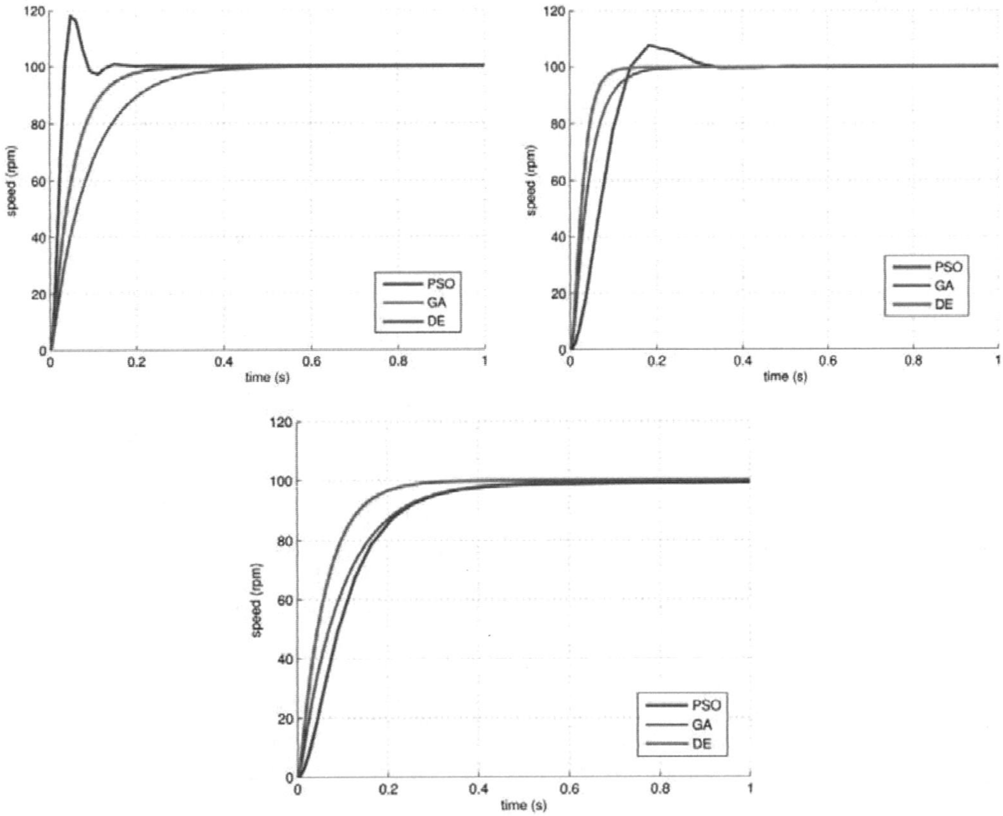

Figure 3. Best Result in First, Second, and Third Experiment

Table 1. Compared Parameters

	Variables	Population	Iteration	Time (s)
PSO	11	10	10	48.854
GA	9	10	10	10.753
DE	7	10	10	33.998

Table 2. Parameters of PID Control and Performances System

Method	K_p	K_i	K_d	sse	t_r	t_s	O_s
			First Experiment				
PSO	20	12.6999	0.22419	-0.0005334	0.0252	0.1178	17.9839
GA	18.1836	1.0156	1.5625	0.0571	0.1937	0.3516	0
DE	20	1.702	0.906	0.0092	0.1108	0.2040	0

			Second Experiment				
PSO	19.6734	16.732	0.59466	0.0000053151	0.0975	0.2848	7.7253
GA	15.625	1.875	0.625	0.0296	0.0869	0.1621	0
DE	20	2.57	0.0235	0.53017	0.0527	0.0970	0
			Third Experiment				
PSO	17.5011	7.3813	1.2912	0.1644	0.2069	0.4322	0
GA	12.5	0.41016	1.1523	0.4554	0.2124	0.4043	0
DE	18.6362	1.6704	1.0618	0.0184	0.1271	0.2321	0

Overall, GA and DE often give good system's performance. PSO tends to give poor result such as oscillated or overshoot response. PSO's best result is obtained through many experiments. As seen in **Figure 3**, PSO gives overshoot response in both, first and second experiment. However, in the third experiment, PSO gives a response with no overshoot but slower response than other algorithms. GA and DE are more excellent than PSO in terms of providing good system performance.

Speaking of iteration time in **Table 1**, PSO needs longer time to find solutions rather than DE and GA with the same number of generation. GA takes the shortest iteration time while DE is less fast (slower) than GA but quite faster than PSO. In term of the number of variables which must be tuned, DE has the least variables among them all. Otherwise, PSO has the most variables.

More details of system performance can be seen in

Table 2. Neglecting the occurrence of overshoot, system with best steady state error is achieved through PSO tuning process in fist experiment and second experiment. However, best steady state error performance with no overshoot in response, achieved by DE in the first experiment with $sse = 0.0092$.

Aggressively of the system can be observed from the rise time performances. More aggressive system commonly has faster or less rise time performances. Overall, PSO provides the fastest rise time with 0.0252 seconds in the first experiment. However, DE gives fastest rise time among all algorithms in the second experiment and third experiment. Thus, indicates that DE tends to give best aggressively of the system.

In the terms of fast or slow response, all algorithms tend to give fast response which can be seen in the obtained rise time and settling time which is almost under 1 seconds. Again, PSO provides the best settling time response in first experiment but DE is better in second and third experiment. Hence, in case of giving best settling time performance, DE is the best. Meanwhile, from the obtained K_p, K_i, and K_d, consistency of tuning process result can be observed.

As in every optimization problem, the objective function roles to guide the searching process. Therefore, the objective function needs to be formulated correctly so that the searching process can lead to the best solutions. Based on experimental data taken so far, the previous objective function can be modified to find the best K_p, K_i, and K_d. It would be better if some conditions are added. For example, proportional values or K_p should be greater than K_i and K_d. Also, K_i should be greater than K_d. Range values should be saturated to make a better result.

9. Conclusions

In the paper, parameters of PID Controller such as proportional, integral and derivative gains, have tuned by Genetic Algorithm (GA), Particle Swarm Optimization (PSO) and Differential Evolution (DE). Based on the simulation, Differential Evolution gives the best result with some advantages than the others. Differential Evolution provides the least number of parameter variables and fasts to converge. The best value of PID Controller for controlling the speed of DC motor is proportional gain $K_p = 20$, integral gain $K_i = 2.57$, and derivative gain $K_d = 0.0235$ with steady-state error $= 0.53017$, rise time $= 0.0527$, settling time $= 0.097$, overshoot $= 0$.

The future work of the research is as follow. The paper only compares three intelligent search method. Thus, it needs to compare with other methods such as bat algorithm. To make the result useful, ten methods are needed. The key to the optimization problem is in the objective function. The modification in the objective function will obtain the best and consistent solution.

10. References

[1] Gunawan S.A, Yuwono Y.C.H, Pratama G.N.P, Cahyadi A.I and Winduratna B. 2018 *4th International Conference on Science and Technology (ICST)* (Indonesia) 1–6.

[2] Sirisantisamrid K, Wongvanich N, Gulpanich S, and Tammarugwattana N, 2018 *Proceedings of the International MultiConference of Engineers and Computer Scientists* (Hong Kong) 14-16.

[3] El-Deen A.T, Mahmoud A.A.H, and El-Sawi A.R. 2015 *Int. Rev. Autom. Control* **8** 81–82.

[4] Vishal V, Kumar V, Rana K.P.S and Mishra P. 2014 *IEEE International Advance Computing Conference (IACC)* 1342–1347.

[5] Ogata K 2010 *Modern Control Engineering*. (Boston: Prentice Hall).

[6] Ma'arif A,Cahyadi A. I., Wahyunggoro O, and Herianto. 2017 *3rd International Conference on Science and Technology - Computer (ICST)* 22–27.

[7] Ma'arif A, Cahyadi A.I, and Wahyunggoro O. 2018 *Int. J. Adv. Sci. Eng. Inf. Technol.*, **8** 930.

[8] Hasanjani R.A, Javadi S, and Nadooshan R.S. 2015 *Trans. Inst. Meas. Control* **37** 164–176.

[9] Liu J. 2017 *Intelligent Control Design and MATLAB Simulation*. (Singapore: Springer) p. 33-56.

[10] Wati D.A.R and Hidayat R. 2013 *International Conference on Robotics, Biomimetics, Intelligent Computational Systems* (Indonesia) 30–34.

[11] Kumar P, Chatterjee S, Shah D, Saha U. K, and Chatterjee S. 2017 *Cogent Engineering.* **4** 1-20.

[12] Wati D.A.R 2013 *Proceeding of IEEE International Conference on Computational Intelligence and Cybernetics (CYBERNETICSCOM)* (Indonesia) 40–44.

[13] Idir A, Kidouche M, Bensafia Y, Khettab K, and Tadjer S. A 2018 *International Journal of Intelligent Engineering and Systems* **20** 241-249.

[14] Jigang H, Hui F, and Jie W 2019 *Automatika* **60** 135–148.

[15] Franklin G.F, Powell J, and Emami A, 2015 *Feedback Control of Dynamic Systems, Global Edition* (New Jersey : Pearson Education Limited).

Identification of Cyber Bullying by using Clustering Methods on Social Media Twitter

Nina Meliana[1,*], Sunardi[1], Abdul Fadlil[1,*]

[1]Magister of Information Technology Program, Universitas Ahmad Dahlan, Jl.Prof. Dr. Soepomo S.H., Janturan, Yogyakarta 55164.

Corresponding Email: nmeliana@uad.ac.id, fadlil@mti.uad.ac.id

Abstract. In today's era social media is very important for some people, because it is the nature of social media that can make people addicted to using social media, this has been widely proven in terms of medical, social interaction has been reduced due to social media, things related to the physical to be reduced, for example going out, and playing outdoors for children, for adults social media can be positive and negative, positive if used to offer old friends who don't meet for long, but negative things can be used to crime or things that are not good, what else will soon be the presidential election, social media becomes a place or means of defamation of one party with another, the initial goal of social media is to build good relations Actually, there is a sentence that says "Yout thumb your tiger" with comments that are not good on social media. If we take us to the criminal section, and our words on social media can be legally justified, one example is intimidation, intimidation is one of the ITE Law articles, intimidation will be lifted from Twitter's social media, with Twitter we can see examples how much of this intimidation. In retrieving data on social media there are several methods used in retrieving data, the method used is the clustering method or data grouping method. The study used the Naive Bayes and Decision Tree J48 classification method. The results to be obtained in this study are the Naive Bayes method which obtained an average value of the overall scenario of 92% success rate and 8% not detected and the result of Decision Tree J48 method with an accuracy value of 100%. The most widely used type of cyberbullying is psychology related bullying or related to one's ignorance or nature

1. Introduction

Twitter is one of the fastest growing social networking sites nowadays where users can interact with other users from their computers or mobile devices anywhere and anytime. After its first introduction in July 2006, the number of Twitter users has increased very rapidly. As of September 2018, an estimated number of Twitter users registered around 326 million users. Twitter's own social media users consist of various groups of users who can interact with friends, family and colleagues. Twitter gives its users access to send a short message consisting of a maximum of 140 characters (called a tweet). The Tweet itself can consist of a text and photo message. It is through this tweet that Twitter users can interact more closely with other Twitter users by sending about what they are thinking, what is being done, about the events that have just happened, about the latest news and other things they have experienced. In 2018, the number of tweets posted almost reached 500 million tweets / day. In this paper, the author conducts research on data mining on social media, which aims to get the results and its correlation to cyber-bullying identification.

In the paper discusses the phenomena of fake news that occur in America, the spread of fake news through social media where social media has become the land to spread false news and in the discussion fake news has become an influential tool in US elections in 2016. False news is increasingly rising. Thus handling is needed in eradicating false news. One of them is using the "Fake News" detection tool where the tool can detect the truth of a news, but psychological factors are also needed so that people can be more careful in making or spreading false news. Overall the need to develop and apply methods in detecting false news is important and must be further developed so that the news becomes clear and ethical for the community [1]. Based on the research conclusions, it can be seen clearly that hoax news is a common enemy that must be anticipated. Public awareness of media literacy must be increased so that it can counteract the dangers of hoax news.

The hope of this researcher is that more research on this in more depth will have a significant influence on the wider environment or community [2]. In comparison to an automated approach to verifying information online with multimedia content can be presented with 3 different methods. The first method is the MCG-ICT method. The second method is the CERTH-UNITN method. The third method is the TweetCred method. There are several ways to overcome challenging verification, to measure verification by evaluating these methods using cross validation, and reporting scores per case. The MCG-ICT method has the highest precision score with a very low false positive rate [3]. With the support of social media, SPL can provide contributions, in the form of presenting an interface disaster report that is currently happening, and can also provide information such as evacuation points in the disaster area and the safe point of disaster victims. With social media support, SPL and describe information in real time about the location of the disaster that is happening [4].

In making news or information must be sourced from various speakers or informants and also compare experiments with each other so that the information produced does not favor one experiment. In the future it is expected that there will be an explanation and detection for the hoax news language style. People or news recipients can compare the truth of the news if they get more explanation about the characteristics of hoax news [5]. Fake news affects various fields such as the economy, law, environment and also one's mentality. The impact of fake news can be more dangerous than the impact of a fraud case. Fake news can also affect a person's level of trust in others. To reduce the existence of false news, it would be better if a group of people held a direct appointment rather than having to receive information in the mass media that was still doubtful. Because it could be that what was conveyed by Party A would be changed by the party B and received by Party C. Therefore, it is better for the A B C party to hold discussions directly [6].

According to this paper the news of hoaxes is the most serious threat that must be addressed immediately, because it can divide society. To be able to overcome it, there needs to be cooperation between the government and the community in order to reduce or eliminate the hoax news. One way we can do it is not to be easy to believe in news that is not necessarily the truth. [7]. According to this paper the netizens must be smart in avoiding the hoax news so that the news is not widespread. [8]. There is also a harmonic method leading to the smallest difference in classification starting from two different basic truths. But this harmonic method has two disadvantages. First, this is based on a very specific graph algorithm that cannot be extended to additional features in the same way as other machine-learning algorithms. Second, adjusting harmonious methods to provide good results is a very time-consuming process. [9] The goal in social media marketing is sincerity as the key to being able to communicate useful.

Disbelief in social media is due to the many fake news or hoaxes that are spread on social media that are not useful, so we as social media users must use social media as wisely and well as possible. [10]. So a comparative study for an automated approach to verifying information online with multimedia content has three methods, namely: the MCG-ICT method, using the highest score with some unique cases and very low false positive rates. So the accuracy is certainThe CERTH-UNITN method, produces consistently high results, and especially on events with lots of tweets. TweetCred method, From the method above to measure the accuracy of verification using cross validation and

Journal of Physics: Conference Series **1373** (2019) 012040 doi:10.1088/1742-6596/1373/1/012040

each case report itself. [11]. So Hoax news is very dangerous and will make it complicated with the emergence of debates. And will face legal problems if the news is declared Hoax [12]. Of the 13 million tweets, 15% of the stories were wrong, in the attempt to detect the initial hoax of social media, they were: 1. Early detection within minutes of the first report or first tweet; 2. Compile benchmark datasets 3. Use the word embeddings model for the entire dataset [13]. The journal above contains the perception of hoax news in various social media in the view of law and science. Then the youth also have responsibility and have high digital skills to protect the public from hoax news on social media. [14].

2. Experimental Method

Web pages clustering methods can be categorized into four types based on features employed: link-based, content-based, URL-based and structure-based. Link-based methods state that links are considered to be topically similar if they share similar parent nodes. Content-based approaches categorize web pages according to the words and sentences contained. These methods argue that pages with similar functions share similar topic features. Vector space model is commonly used to represent documents. However, content-based approaches are limited due to vagueness of topical differences and diffculties in understanding cross-language websites. URL-based methods have an advantage of clustering pages without downloading less. However, dealing with duplication and redirection problems is a non-trivial task. Once the patterns of URLs are decided, regular expression can be used to cluster and classify web pages. Structure-based approaches are more general ways to cluster web pages. Some previous research estimates the similarity of pairwise web pages through their tree structures. However, it is computationally expensive to do pairwise page comparison and not suitable for web-scale applications.

Another idea is representing web pages with vector space models. Two pages are considered to be structurally similar if they have common links placed in the same location. Minimum Description Length Principle (MDL) is adopted in combining small clusters. Further exploration on this idea has been done by exploiting Xpaths in recording positions of specific links. The method, known as UXCluster, is claimed to be more accurate and efficient than the method produced by pairwise distance. Another alternative is to cluster web pages represented by repetitive region patterns extracted by web wrappers. A drawback of most of structure-based clustering is that it requires heuristic threshold to decide whether to form a new cluster or to merge into an existing cluster [14].

We consider two common crawling Algoritma, crawling for user created content (crawling-for-UCC) and crawling for a target page type (crawling-for-target) to show how a structure-oriented sitemap supports different social media crawling strategies.

1. Crawling for UCC. Typically when crawling social media, some types of pages are considered less interesting or useful. For example, near-duplicates and non-user-created pages are less valuable than pages shared by users.
2. Crawling for target. Often it is desirable to focus crawling on pages that match the page type of an example page provided by some other process (a target page type) [15].

The clustering algorithm method in data mining can be used to find data clusters naturally derived from data extracted or examined using formulas from data mining.

2.1. Naive Bayes Classifier

Naïve Bayes Classifier is a classification method rooted in the Bayes theorem. Naïve Bayes algorithm is a classification method using probability and statistical methods. This method uses the Bayes theorem, which was discovered by Thomas Bayes in the 18th century.
The stages of the Naive Bayes algorithm process are:

a. Calculate the number of classes / labels.
b. Count the number of cases per class.

c. Multiply all class variables.
d. Compare results per class

Bayes Theorem Equation

$$P(C|X) = \frac{P(X|C)P(c)}{P(x)} \quad (1)$$

where,
x = Data that is unknown to the class
c = The data hypothesis is a specific class
P (c | x) = Probability of a hypothesis based on conditions (posteriori probability)
P (c) = Probability of hypothesis (prior probability)
P (x | c) = Probability based on conditions in the hypothesis
P (x) = Probability c

The above formula explains that the chance of the entry of certain characteristic samples in class C (Posterior) is the chance of the emergence of class C (before the entry of the sample, often called prior), multiplied by the probability of the emergence of sample characteristic characteristics in class C (also called likelihood), divided by opportunities appearance of sample characteristics globally (also called evidence). Therefore, the formula above can also be written as follows:

$$posterior = \frac{prior\ x\ likelihood}{evidence} \quad (2)$$

Evidence values are always fixed for each class in one sample. The value of the posterior will be compared with the posterior value of the other classes to determine to what class a sample will be classified. Further elaboration of the Bayes formula is done by describing (c | x1, ..., xn) using the multiplication rule as follows:

$$\begin{aligned} &P(C|X1,...,Xn)=P(C)P(X1,...,Xn|C) \\ &= P(C)P(X1|c)(X2,...,Xn|C,X1) \\ &= P(C)P(X1|c)P(X2|C,X1)(X3,...,Xn|C,X1,X2) \\ &= P(C)P(X1|c)\ P(X2|C,X1)P(X3|C,X1,X2) \ldots\ P(Xn|C,X1,X2,...,Xn\text{-}1) \end{aligned} \quad (3)$$

It can be seen that the results of the translation cause more and more complex factor factors that affect the probability value, which is almost impossible to analyze one by one. As a result, the calculation becomes difficult to do. This is where the assumption of very high independence (naive) is used, that each of the instructions is free from each other. With these assumptions, a similarity applies as follows:

$$P(c|X1,\ldots,X_n) = P(C)\prod_{i=1}^{n} P(Xi|C)$$

$$P(c|X) = P(x_1|c)P(x_2|c)\ldots P(x_n|c)P(c) \quad (4)$$

The above equation is a model of the Naive Bayes Theorem which will then be used in the classification process. For classification with continuous data, the Gauss Density formula is used:

$$P = (X_i = x_i|Y_i = y_i) = \frac{1}{\sqrt{2\pi\sigma ij}}e^{-\frac{(x_i-u_{ij})2}{2\sigma^2 ij}} \quad (5)$$

where,

P : Opportunity
Xi : Attribute to i
xi : Value of attribute to i
Y : Class sought
yj : Y sub-class searched
u : Mean, declares the average of all attributes

o : Standard deviation, expresses the variance of all attributes

$$\mu = \frac{1}{n}\sum_{i=1}^{n} x_i$$

Mean Formula

$$\sigma = \left[\frac{1}{n-1}\sum_{i=1}^{n}(x_i - \mu)^n\right] 0{,}5$$

Standard Deviation

2.2. *Decision Tree J48 Classifier*

This test uses classification using the Decision Tree J48 method. Decision Tree J48 is an implementation of the C4.5 algorithm that produces Decision Tree used in machine learning as a standard algorithm. Decision Tree is one of the classification algorithms in data mining data.

From the results of testing using the Decision Tree J48 algorithm that has been done through WEKA on the dataset, the results of the evaluation of the accuracy of classification of the Decision Tree J48 algorithm have an accuracy of 93.60% and error rate of 6.40%, as shown in the Table 1.

Table 1. Learning Dataset Table

No	Word bully	Type of bullying	cyberbuly	ID
1	anjing;	animals;	negatif;	non-bully;
2	babi;	animals;	negatif;	non-bully;
3	cebong;	animals;	positif;	bully;
4	cebong;	animals;	positif;	bully;
5	buta;	Someone's inability;	negatif;	non-bully;
6	buta;	Someone's inability;	negatif;	non-bully;
....	....	....	....	....
34	....	....	....	....

The table shows training data on category attributes to determine the word including bullying or non bullying.

$$\boldsymbol{Entropy(S) = \textstyle\sum_{s-1}^{n} - pi * log2\ pi} \quad (6)$$

S : Set of cases
k : Number of partitions S
P_i : The probability obtained from the number (Yes / No) is divided by the total cases

Information Gain calculations for each formula attribute are:

$$Gain\ (A) = Entropi\ (S) - \sum_{i=1}^{k} \frac{|S_i|}{|S|} \times Entropi(S_i) \quad (7)$$

2.3. Stage of Data Mining

The stages carried out in the data mining process starts from the selection of data from the source data to the target data, the preprocessing stage to improve data quality, transformation, data mining and the stages of interpretation and evaluation that produce output in the form of new knowledge that is expected to contribute better. To choose an attribute as root, it is based on the highest Gain value of the existing attributes, as shown in the Figure 1.

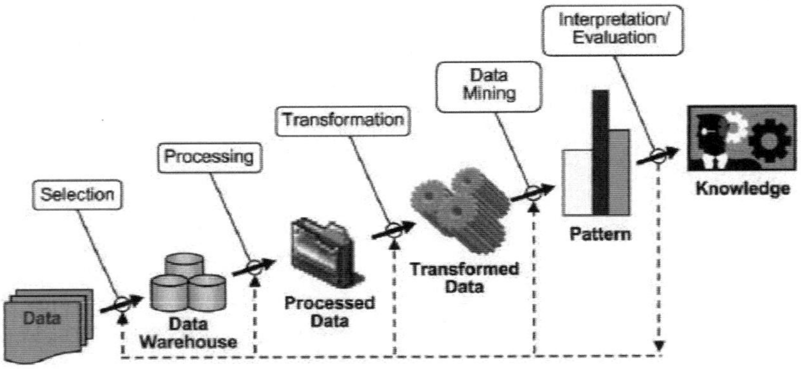

Figure 1. Stage Of Data Mining

1. *Data selection*
   Selection of data from a set of operational data needs to be carried out before the information excavation stage in KDD starts. The selection data used for the data mining process is stored in a file, separate from the operational database.
2. *Pre-processing / cleaning*
   Before the data mining process can be carried out, it is necessary to do a cleaning process on the data that is the focus of KDD. The cleaning process includes, among other things, removing duplicate data, checking inconsistent data, and correcting errors in data.
3. *Transformation*
   Coding is a transformation process on selected data, so that the data is suitable for the data mining process. The coding process in KDD is a creative process and is very dependent on the type or pattern of information to be searched in the database.
4. *Data mining*
   Data mining is the process of finding patterns or interesting information in selected data using a particular technique or method. Techniques, methods, or algorithms in data mining vary greatly. The choice of the right method or algorithm depends on the overall purpose and process of the KDD.
5. *Interpretation / evalution*
   The pattern of information generated from the data mining process needs to be displayed in a form that is easily understood by interested parties. This stage is part of the KDD process called interpretation. This stage includes examining whether the pattern or information found is contrary to the facts or hypotheses that existed before.

3. Results And Discussion

3.1. R Sistem Using Data Mining To Get Data From Twitter

R is a programming language and software system that is specifically and specifically designed to do everything related to computational statistical data that is so complicated. This programming language was first developed in 1993 by two statisticians who wanted to simplify statistical calculations, namely Ross Ihaka and Robert Gentleman at Auckland University, New Zealand. R programming language continues to grow rapidly along with the increasingly popular terminology "Big Data" and the increasing need for a company to be a data scientist to process and analyze data in the company as a basis for policy making and automation of business processes into data-driven. Programming languages like Python and R have become the main choice for researchers and users as well as practitioners in the field of data science to process and analyze data both for research and business purposes. Therefore, for someone just starting out in the field of data science, R is a programming language that is highly recommended for mastering.

3.2. Investigation Simulation

3.2.1. Collection

The collection phase of data is the stage of taking data (Tweet) from Twitter using the Rstudio program by utilizing the Twitter API (Application Programing Interface) as shown in the Figure 2.

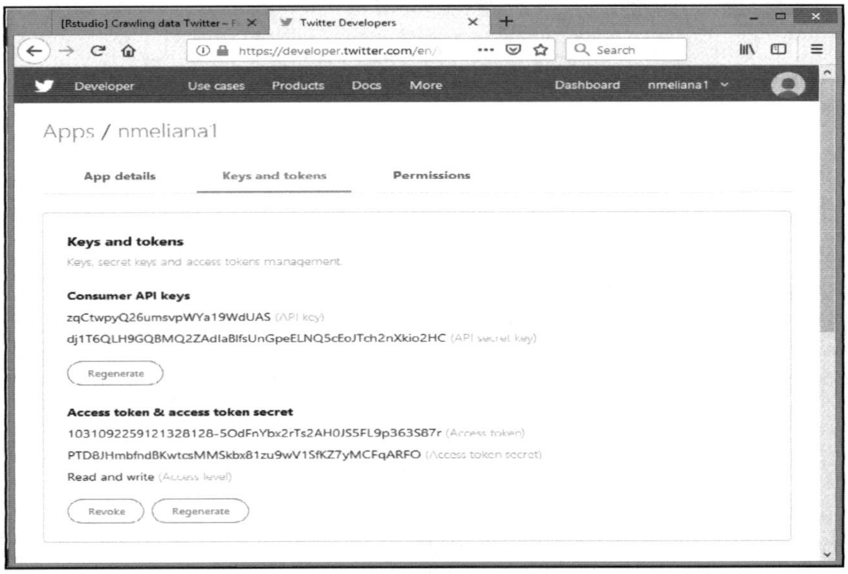

Figure 2. Twitter Developer API page

After getting the Twitter API key the next step is crawling or collection of data using Rstudio software. Previously installed necessary libraries such as library twitteR, ROAuth library, wordcloud library, and tm library, as shown in the Figure 3.

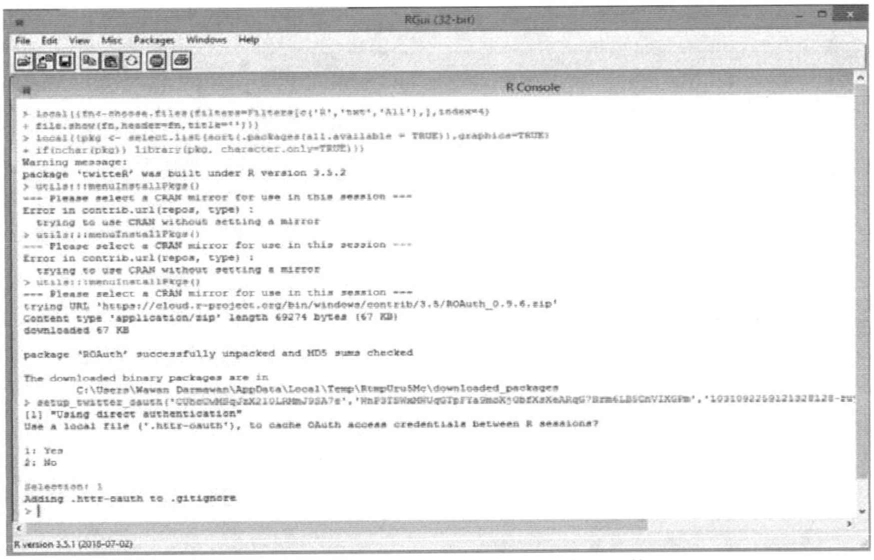

Figure 3. The console page on Rstudio

3.3. Examination

The next stage of the digital forensic investigation process is Examination, at this stage the data sources taken in the collection phase are processed from various data into more regular data because not all words reflect the meaning or content contained in a document. For this reason, an initial process is called preprocessing text.

Data obtained were 250 data, then cleansing data from irrelevant tweets to 167 relevant data. The data is then done manually labeling (tagging) based on predetermined categories. Based on these data can be calculated the number of types of categories. The dataset obtained will be divided into two parts, namely training data by 70% and testing data by 30%, as shown in Table 2.

Table 2. Classification Table

No.	Type of Cyberbullying	Bullying	Non Bullying
1	Psychology	52	3
2	Animal	43	42
3	General	8	4
4	Behavior	16	9
5	Someone's inability	48	25
	Sum	167	83
	Percentage	70%	30%

3.4. Analysis

The data obtained will then be implemented using the Naive Bayes classification method and Decision Tree J48. The classification process using Naive Bayes is done by preparing data that has been labeled as a reference or reference in the classification process.

3.4.1. Naive Bayes Mehtods

In the WEKA application, 80% percentage splits are filled. This means that 80% of the data becomes training data, the remaining 20% becomes testing data. The data validation uses 10 validations. The results get a success rate of 92% and an error rate of 8%, as shown in Figure 4.

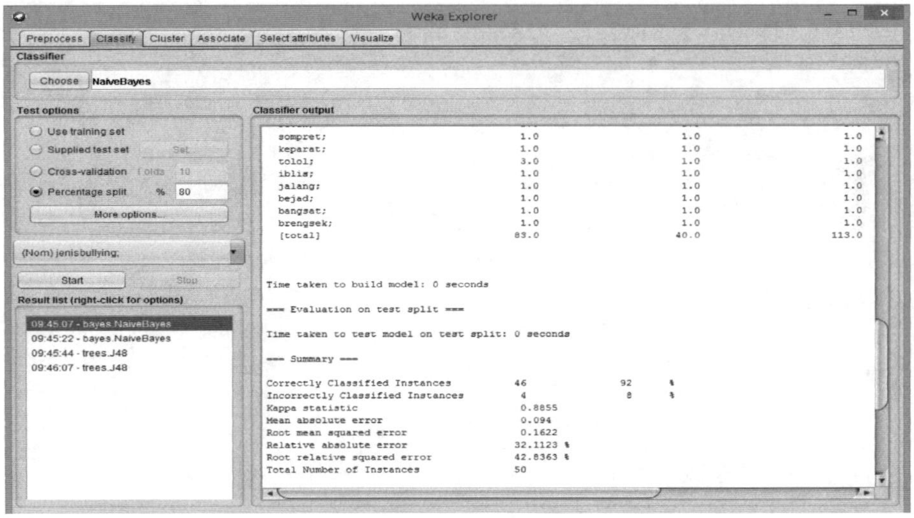

Figure 4. Testing the Naïve Bayes Method with a Percentage of 80%

The opportunity value obtained is:
P(Bully Type = P | wordbully= goblok) = 22 / 167 = 0.13
P(Bully Type = P | wordbully= gila) = 20 / 167 = 0.11
P(Bully Type = P | wordbully= idiot) = 6 / 167 = 0.03
P(Bully Type = P | wordbully= tolol) = 2 / 167 = 0.011
P(Bully Type = P | wordbully= saraf) = 2 / 167 = 0.011
P(Bully Type = P | wordbully= cebong) = 14 / 167 = 0.08
P(Bully Type = P | wordbully= kampret) = 8 / 167 = 0.04
P(Bully Type = P | wordbully= monyet) = 8 / 167 = 0.04
P(Bully Type = P | wordbully= babi) = 6 / 167 = 0.035
P(Bully Type = P | wordbully= anjing) = 5 / 167 = 0.029
P(Bully Type = P | wordbully= kunyuk) = 2 / 167 = 0.011
P(Bully Type = P | wordbully= sialan) = 3 / 167 = 0.017
P(Bully Type = P | wordbully= udik) = 2 / 167 = 0.011
P(Bully Type = P | wordbully= jancuk) = 2 / 167 = 0.011
P(Bully Type = P | wordbully= jelek) = 1 / 167 = 0.005
P(Bully Type = P | wordbully= item) = 0 / 167 = 0
P(Bully Type = P | wordbully= kampungan) = 0 / 167 = 0
P(Bully Type = P | wordbully= tai) = 0 / 167 = 0
P(Bully Type = P | wordbully= bangsat) = 4 / 167 = 0.02
P(Bully Type = P | wordbully= bajingan) = 2 / 167 = 0.011
P(Bully Type = P | wordbully= setan) = 2 / 167 = 0.011
P(Bully Type = P | wordbully= keparat) = 2 / 167 = 0.011
P(Bully Type = P | wordbully= jalang) = 2 / 167 = 0.011
P(Bully Type = P | wordbully= bejad) = 1 / 167 = 0.005
P(Bully Type = P | wordbully= iblis) = 1 / 167 = 0.005
P(Bully Type = P | wordbully= berengsek) = 1 / 167 = 0.005
P(Bully Type = P | wordbully= sompret) = 1 / 167 = 0.005
P(Bully Type = P | wordbully= lgbt) = 0 / 167 = 0
P(Bully Type = P | wordbully= banci) = 0 / 167 = 0
P(Bully Type = P | wordbully= jahanam) = 0 / 167 = 0
P(Bully Type = P | wordbully= biadab) = 0 / 167 = 0
P(Bully Type = P | wordbully= gembel) = 22 / 167 = 0.13

P(Bully Type = P | wordbully= budek) = 20 / 167 = 0.11
P(Bully Type = P | wordbully= kampungan) = 6 / 167 = 0.03

P(psychology) = 0.292
P(animals) = 0.230
P(general) = 0.044
P(behavior) = 0.084
P(Someone's inability) = 0.27

Cyberbullying(Positive) = P(psychology) × P(animals) × P(general) × P(behavior) × P(Someone's inability)
Cyberbullying(Positive) = (0.292) × (0.230) × (0.044) × (0.084) × (0.27) = **0.925**
Cyberbullying(Negative) = 1 – 0.925= **0.075**

Table 3. Probability Table (manual calculation)

Cyberbullying	Probability	Percentage
Positive	0.929263476	92 %
Negative	0.070736524	8 %

3.4.2. Decision Tree J48 Mehtods

From the test results using the Decision Tree J48 algorithm that has been done through WEKA on the dataset, the results of the evaluation of the accuracy of the classification of the Decision Tree J48 algorithm have an accuracy of 93.60% and an error rate of 6.40%. In the WEKA application, a percentage of 50% was filled. This means that 50% of the data becomes training data, the remaining 50% becomes testing data. Then for data validation use 10 times validation. The results get a success rate of 93.60% and an error rate of 6.40% and can be explained in the following Figure 5.

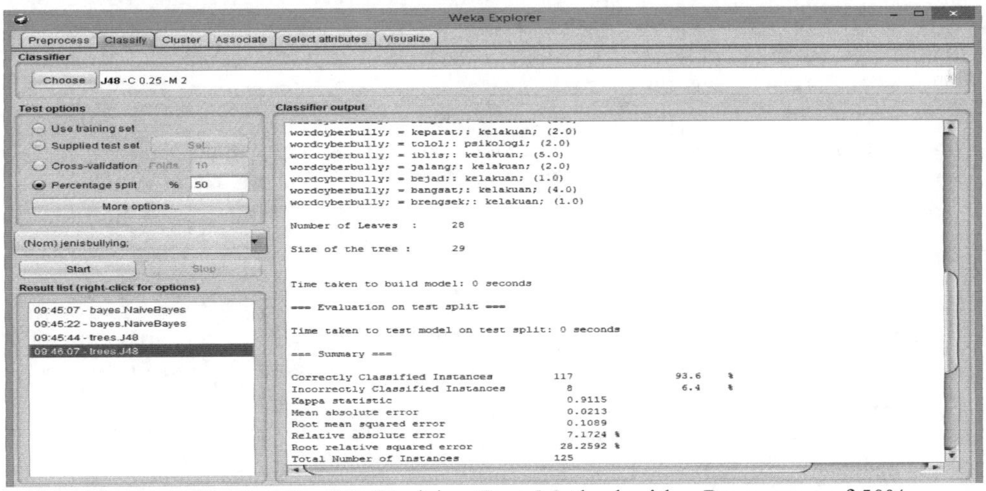

Figure 5. Testing the J48 Decision Tree Method with a Percentage of 50%

In the second test, a percentage of 80% was filled. This means that 80% of the data becomes training data, the remaining 20% becomes testing data. Then for data validation use 10 times validation. The results get a success rate of 100% and an error rate of 0%.

3.5. Reporting
At this stage is the reporting phase of data collection until testing the data using the Naïve Bayes classification method and Decision Tree J48. From the classification that has been done, Naive Bayes has an accuracy rate of 92%, while the Decision Tree is 100%. Both methods have different levels of accuracy, but are still relatively good because they are still above 80%. If calculated based on this level of accuracy, both methods are quite appropriate for the classification of existing datasets, as shown in Table 4.

Table 4. Classification (Reporting)

Classification	Data Training		Data Prediction			
			Naïve Bayes		Decision Tree	
	Positive	Negative	Positive	Negative	Positive	Negative
Cyberbullying	63%	37%	92%	8%	100%	0%
Psychology	31%	2%				
Animal	26%	25%				
General	5%	2%	86%	14%	93.60%	6.40%
Behavior	10%	5%				
Someone's inability	29%	15%				

4. Conclusions
From this work, it can be concluded that (1) Stage of Detecting cyberbullying on Twitter can be done with several techniques. Starting from the Twitter API registration, lowering data with input access tokens from the Twitter API, storing data in the database, doing preprocessing analysis and data cleaning, classifying using WEKA and doing weighting and validation and classification with Naïve Bayes method and Decision Tree J48 Method. (2) Based on the accuracy obtained by each method in the detection of Cyberbullying on Twitter for each scenario, the method that produces the best accuracy is obtained by the Decision Tree J48 method with an accuracy value of 100% success rate, and for the Naive Bayes method the average value of the overall scenario is 92% success rate and 8% not detected. (3) The results of the classification can be seen that the highest type of cyberbullying on Twitter social media is the type of psychology or bullying related to ignorance / trait that is equal to 31%.

References
[1] J. Meinert, M. Mirbabaie, S. Dungs, A. Aker "Is It Really Fake? – Towards An Understanding Of Fake News In Social Media Communication", University Of Duisburg-Essen, Duisburg, Germany 2018.
[2] D. S. Adhiarso, P. Utari, S. Hastjarjo, "The Influence Of News Construction And Netizen Response To The Hoax News In Online Media", Universitas Sebelas Maret Surakarta 2018.
[3] C Boididou, S. E Middleton, Z. J. S. Papadopoulos, D. D. Nguyen, G. Boato, Y. Komppatsiaris "Verifying Information With Multimedia On Twitter" 2018.
[4] C Slamet "Social Media-Based Identifier For Natural Disaster", Et Al Iop Conf. Ser Mater. Sci. Eng. 288 012039, 2018.
[5] O. Ray, " A Survey On Natural Language Processing For Fake News Detection", Natural Sciences I College Of Arts And Sciences The University Of Tokyo 2018.
[6] A. Yap, Ph, " The Information War In The Digital Society: A Conceptual Framework For A Comprehensive Solution To Fake News", Academy Of Social Science 2018.

[7] M. K. Gunawan, A Wijaya, Salma, A. H. Idrus "Handling Of Hoax Messages From The Legal Perspective: A Comparative Study Between Indonesia And Singapore" 2018.
[8] D. S. Adhiarso, P. Utari, S. Hastjarjo "The Influence Of News Construction And Netizen Response To The Hoax News In Online Media" 2018.
[9] L. D. Alfaro, "Reputation Systems For News On Twitter: A Large-Scale Study" 6 Maret 2018 – 8 Januari 2018.
[10] T. Pesonen. "The Effects Of Fake News On Consumer Trust In Social Media Marketing" April 2018.
[11] C. Boididou, S. E. Middleton, "Verifying Information With Multimedia Content On Twitter", 2018.
[12] A. Yap, Ph.D,"The Information War In The Digital Society: A Conceptual Framework For A Comprehensive Solution To Fake News", 2018.
[13] A.Zubiaga. "Learning Class-Specific Word Representations For Early Detection Of Hoaxes In Social Media", 5-6, 2018.
[14] A. Shensa, MA, J. E. Sidani, MPH, PhD, M. A. Dew, PhD, C. G. E. Viera, MD, PhD, B. A. Primack, MD, PhD. " Social Media Use and Depression and Anxiety Symptoms: A Cluster Analysis", 5.101.220.175 on: Mon, 30 Jul 2018.
[15] K. Xu, K. Yingkai Gao, J. Callan. " A Structure-Oriented Unsupervised Crawling Strategy for Social Media Sites", arXiv:1804.02734v1, 8 Apr 2018.

Fuzzy Features for Facial Shape Classification on Panoramic Dental Image

Nur Nafi'iyah[1,*], Chastin Fatichah [2,*]

[1]Department of Informatics, Faculty of Informatic Technology, Universitas Islam Lamongan,
[2]Institut Teknologi Sepuluh Nopember, Surabaya, Indonesia

*Email: mynaff26@gmail.com, chastine@cs.its.ac.id

Abstract. Research on human face shape identification can assist forensic teams in reconstructing an unidentified victim's facial features. Human face shape identification using panoramic dental imaging is suitable for use by forensic teams in identifying a large number of victims due to the teeth's ability to withstand heat of up to 1,000°C. This study proposes an application for face shapes classification on panoramic dental image using fuzzy features and decision tree method. It will be used to assist forensic scientists in reconstructing unidentified people identifying numerous victims from their facial features. There are three classifications of human face shapes, namely oval, tapered, and square. The steps in this study are digitizing panoramic dental images into files; segmenting the upper jaw's incisor teeth; then extracting the features by area, parameter, width, length, width-to-length ratio, area-to-parameter ratio, center_x and center_y. Fuzzy theory is used to convert numeric features into category features, while decision tree method will be used for training features. The experimental results show that the proposed method obtain accuracy 67% of 42 panoramic dental image.

1. Introduction

Forensic identification has been growing rapidly with a variety of science fields being implemented to solve an algorithm in order to facilitate in identifying a large number of victims of natural disasters or mass casualties. Forensic odontology identifies individuals based on their teeth. The reason for using forensic odontology is the teeth's ability to withstand fires and heat.

Identifying individuals with panoramic dental imaging can be used as an alternative to determine a person's sex, age, or forensic identification. Forensic identification is used to identify victims of crime or mass casualties. An advantage of using panoramic dental imaging is the tooth's resistance to heat and can withstand temperatures of up to 1,000°C. Using panoramic dental imaging on mass victims can be done more efficiently compared to individual DNA testing.

Damaged bodies in accidents or disasters is relatively difficult to identify. The most accurate methods of identifying bodies in such cases are DNA testing and reviewing dental records. DNA tests take a relatively long time to produce results, making dental records the better option. Having dental radiography records makes the person much more easily identifiable.

Identifying teeth can uncover age, sex, and can be used to reconstruct facial shape. The purpose of facial reconstruction using dental imagery is to identify victims when their corpses are not intact or damaged, and when the face is unrecognizable. Identification of human facial features is beneficial to reconstruct faces as well as identifying the shape of the face (oval, tapered, or round). This research is related to classifying human facial shape on dental panoramic image, where the face can be classified as either tapered, square or oval. This study proposes an application for face shapes classification on panoramic dental image using fuzzy features and decision tree method. Fuzzy theory is used to convert numeric features into category features, while decision tree method will be used for training features. It will be used to assist forensic scientists in reconstructing unidentified people identifying numerous victims from their facial features.

Decision tree is a tree where each internal node indicates an attribute test, each branch represents the test results, and the leaf nodes show the classes or class distribution [1]. Decision tree is a structured tree consisting of a set of attributes to be tested with the purpose of forecasting their results. It is one of the most popular tools in classification because the results can be understood in the form of decision rules. In determining which attribute must be tested first, we must look at the attribute with the highest gain. Decision tree attempts to find the appropriate attribute to complete and define the class. In other words, the uppermost attribute of the Decision tree is the most influential attribute in determining the predicted outcome. Therefore, the Decision Tree modelling process is an interactive process to assess the most influential attribute to branch off from the tree-like structure. The branching is also commonly referred to as the split point [2]. This process can be divided into two steps: calculating the magnitude of the influence of each attribute and choosing the most appropriate attribute to split from the tree structure. For the first step, there are several matrices that can be used to calculate the magnitude of influence such as entropy, information gain [2].

ID3 algorithm is a model used to generate Decision Trees. The input of ID3 algorithm is a database consisting of several variables, referred to as attributes. Each input in the database presents an object of a domain called an independent variable. An attribute is designed to classify an object called a dependent variable [2].

The algorithm process of ID3 [3] is: Input: Training samples, samples; Output: Decision Tree; Method: Create node N; (1) If samples are all of the same class, C then return N as a leaf node labeled with the class C; (2) If atribut-list is empty then Return N as a leaf node labeled with the most common class in samples; select test-atribut, atribut among attribute-list with the highest information gain; (3) Label node N with test-atribut; (4) For each known value a, of test-atribut: Grow a branch from node N for the condition test atribut ai; Let Si be the set of samples in samples for which test-atribut = ai; If Si is empty then Attach a leaf labeled with the most common class in samples; Else attach the node returned by Generate_decision_tree (Si,atribut-list-test-atribut).

Fuzzy logic was first developed by Prof. Lotfi A. Zadeh, a researcher from the University of California [4] who introduced the theory of Fuzzy and Fuzzy logic as a way of overcoming inaccuracy and uncertainty. Each member in the Fuzzy set has a membership value degree that determines whether a potential member can fit into a Fuzzy set. A Fuzzy set is a development of the classical logic that models everything using boolean terms (Yes and No, 0 and 1).

Research [5] proposed local fuzzy thresholding based on measuring fuzzy similarity in interactive segmentation of panoramic dental imagery. The proposed method [5] consists of three main stages:

region splitting to identify a local region, user marking to obtain the initial seed background and object, and measuring fuzzy similarity is each local region to obtain a threshold value. The results of the panoramic dental imagery test succeeded in segmentation with an average misclassification error (ME) of only 5.47 percent.

Forensic radiology is a part of forensic medicine, used to study the identification of humans through postmortem radiology images of various body parts such as the spine, skull, and teeth. Identification is done through comparing postmortem (PM) and antemortem (AM) records.

Research [6] diagnosed acute leukemia with the use of a Fuzzy Decision Tree. Fuzzy was used to process the input on a 0 to 1 scale. The results are used as a dataset Decision Tree. The process was for studying, which yielded a tree and rule, where that rule was used to diagnose the disease.

The Fuzzy method is used to classify teeth based on periapical radiography [7], where each analyzed tooth was based on a set of criteria, such as area-to-perimeter and width-to-height ratios. The tooth classification process used for this study is periapical radiography imaging. The proposed method of classification [7] is implemented as a submodule system in identifying individuals from the teeth of the lower jaw.

Identification is dental forensics is the application of all disciplines of dentistry related to an investigation in obtaining postmortem data, useful in determining authenticity and identity of the victim and suspect for the sake of law in a judicial process and upholding the truth [8]. The Naïve Bayes algorithm, Artificial Neural Network, and ID3 are all appropriate algorithms for classification. Evidence of accuracy of these logarithms can be found in [7,9]. The ID3 algorithm does an initial study, it then generates a Tree and Rule. We can prove a classification of an object using the resulted Tree and Rule.

Research [9] developed and established a system to identify the sex of an individual based on panoramic dental pictures. Research [9] used 20 panoramic photos of human teeth which was then digitized and used as input. Next, the pictures went through pre-processing, an enhancing of the pictures in order to give maximal yields. The feature extraction stage takes the most significant features of an image, such as the length of the mandibula, the length of the arch, and other attributes. The results of this stage were processed using backpropagation algorithm. The purpose of the testing phase was the determine accuracy of the system.

In study [10] computerized human face identification required high accuracy due to the actual 3D form of a human face and a computers limitation to 2D. Previous research also developed an application to restore a face as realistically as possible [10].

Research [11] discusses the problem of the 3D face rendering approach of occluding contour, which is the limits between the facial area and background. The purpose of [11] was to study the relationship between a 2D and 3D face shape using linear regression. [11] found the matrix approach in linear regression was able to estimate a 3D face shape from a 2D shape trained from the selected contour. The approach of [11] offered a differentiating face shape against a 2D and 3D picture.

Research [12] made an offline system used to identify a 3D face shape. The identification process of a face shape helped with reconstruction at an appropriately accurate level. Identification was based on robust estimation of albedo for illumination.

Study [13] focused on picture identification based on features. Preparation of feature were based on color, texture and shape. The color feature used a histogram picture, the texture feature used Gray

Level Co-occurrence Matrix (GLCM), and the shape feature used canny edge detection. Color, texture, and shape extraction techniques resulted in 18 usable features in the Clustering process of pictures in the following study. Testing methods included Purity and Precision of a picture against a group of practice pictures.

Edge detection [14] is finding a part of an image experiencing intensely drastic changes. There are two ways to find the parts, namely using the first instance (magnitude of intensity is greater than the defined threshold) and using the second instance (color intensity has zero crossing). In research [14], using Canny Edge Detection commonly uses the following algorithms: Smoothing to reduce the effects of noise on edge detection; Calculating the potential image gradient; Non-maximal suppression of the image gradient to precisely localize the edge; and Hysteresis thresholding for final classification. The next step was to calculate the edge direction histogram using five bin directions, namely 0°, 45°, 90°, 135°, and 180° with three similar pixels.

The feature used to identify the leaf type included shape, color, and texture. Not all features needed to be used to compute the extraction, but some features needed to be selected based on their significance of influence in the retrieval system of the leaf image [15]. Correlation-based Feature Selection (CFS) was used to select features based on their correlations between one another, increasing the performance of the retrieval system of the leaf image. Feature selection types used included CFS, CFS with Genetic Search (GS), and chi square. Analysis of correlations was done through a combination of selected features and counting the similarities with the retrieval system. Lp norm, manhattan, euclidean, cosine, and mahalanobis proximities were used. The results of [15] show that the retrieval value was highest when using CFS with measuring mahalanobis proximity. There were six types of features used, namely: slimness, the ratio between the length and width of the leaf; form factor/roundess, the area and circumference of the leaf; rectangularity, drawing similarities in shape with the leaf and a box shape; narrow factor, the ratio between diameter and length; the roving ratio, the ratio between the circumference (P) and diameter (D); the perimeter ratio, the ratio between length and width; and the width (Wp).

The purpose of this study is to classify human face shape using panoramic dental imagery with ID3 algorithm. This research is expected to help identify individuals based on their facial features, and prove the accuracy of ID3 algorithm in classifying face shape.

This research serves to help forensic teams in identifying victims or people from the shape of their faces. This application can help save time in identifying unidentified people when compared to DNA testing.

This article was organized into introduction, research methodology, results and discussion, and conclusion.

2. Research Methodology

The diagram system of this research can be seen in Figure 1. The steps for the Fuzzy and ID3 algorithms for this research are:

a) Panoramic dental images are enhanced using median filter
b) Thresholding using Iterative Adaptive Thresholding
c) Segmenting images of incisor teeth
d) Take features of panoramic incisor images

e) Feature value normalized using Fuzzy. Fuzzy graphic used in a membership set is triangular. The limits for the triangle are taken from the lowest of each feature, and the cap is the highest of each feature, and average of each feature
f) Studying Decision Tree
g) Testing Rule

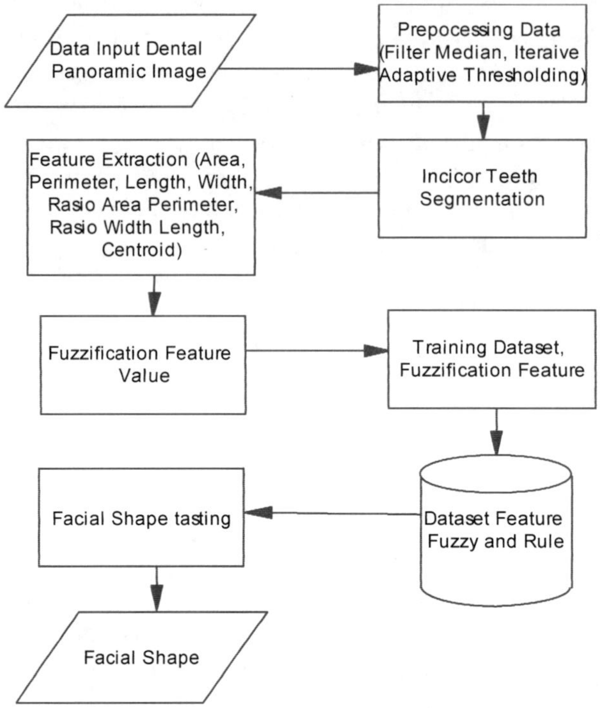

Figure 1. The diagram system of facial shape classification on panoramic dental image

Data used in this includes 26 panoramic images of teeth. Segmentation was done on the images to view the top incisor teeth, resulting in 52 incisors teeth. The panoramic images show the upper and lower jaws, and the only incisors used in this study are the ones located on the upper and lower jaw. The input data of the panoramic dental images are shown in Table 1.

Table 1. Panoramic Dental Images

No	Panoramic Dental Images
1	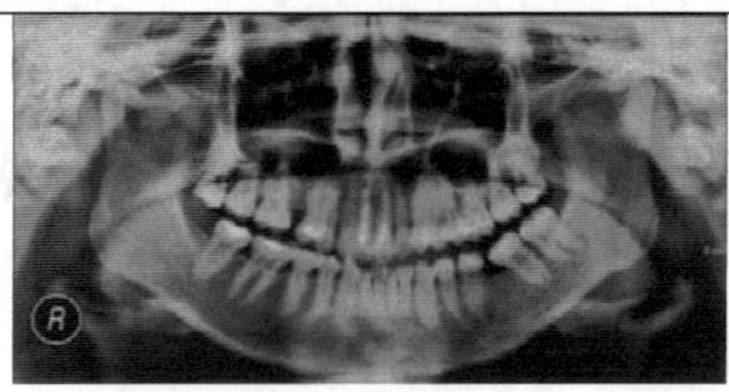

2	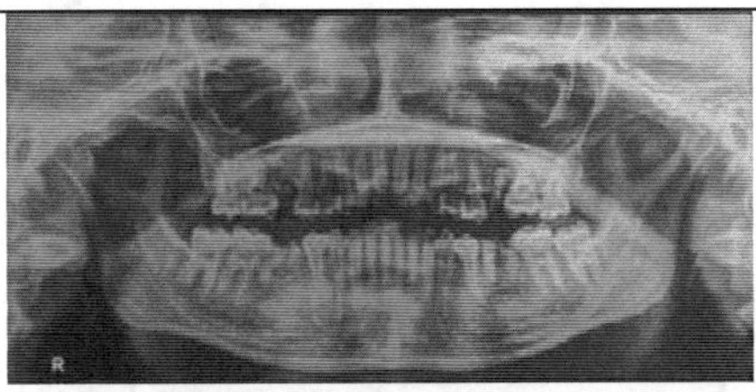
3	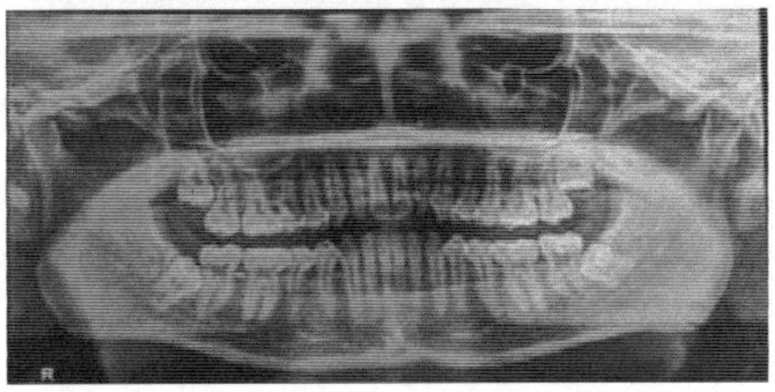

This research enhances the images and extracts the features. The fuzzification of the features are saved in the database. This study takes the panoramic dental data. After it is taken in the form of a panoramic radiographic scan, it is enhanced and segmented to take the incisors of the upper jaw. They are put in a median filter for enhancement. The median filter is done with iterative adaptive thresholding. The area, perimeter, length, width, area-to-perimeter ratio, width-to-length ratio, center_x, and center_y are taken. It is then fuzzified, with the fuzzification value entered into the dataset for ID3 training to generate a tree and rule. The tree and rule are then used to test the system.

The extracted features-starting from the area, perimeter, length, width, length, width-to-length ratio, area-to-perimeter ratio, center_x, and center_y-are normalized using fuzzification. The Fuzzy used is a triangular graphic shown in Table 2. The limits of the triangle are: a = lower limit taken from each feature; b = highest value taken from the average of each feature; and c = maximum value of each feature.

Table 2. Fuzzy Feature Extraction

Membership Features	Graphic
Area [3500 11750 20000]	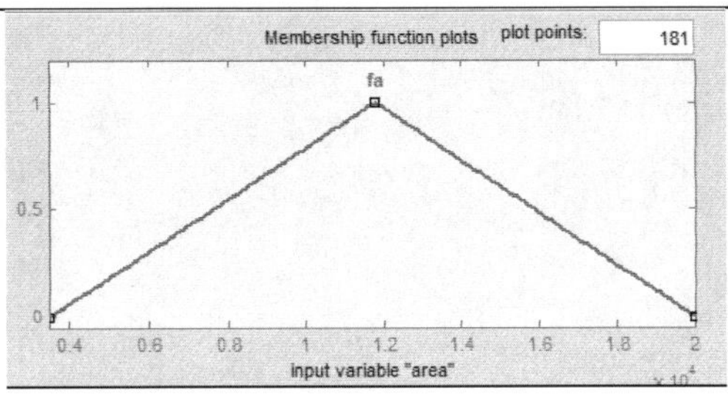

Journal of Physics: Conference Series **1373** (2019) 012041 doi:10.1088/1742-6596/1373/1/012041

Perimeter [90 150 200]

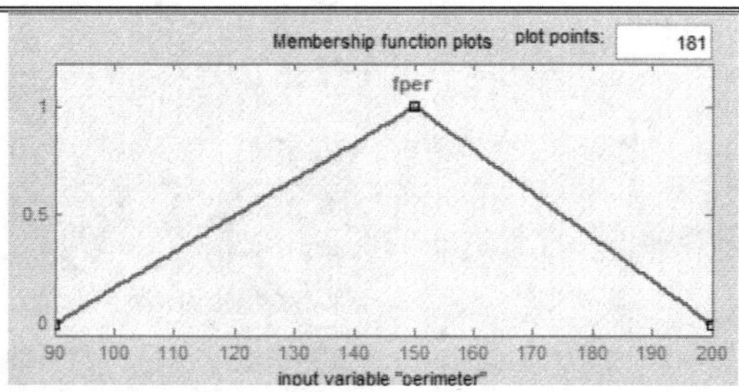

Width [100 150 200]

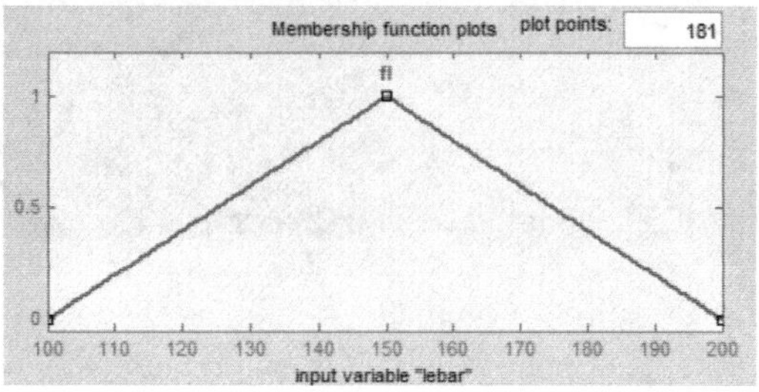

Length [50 130 200]

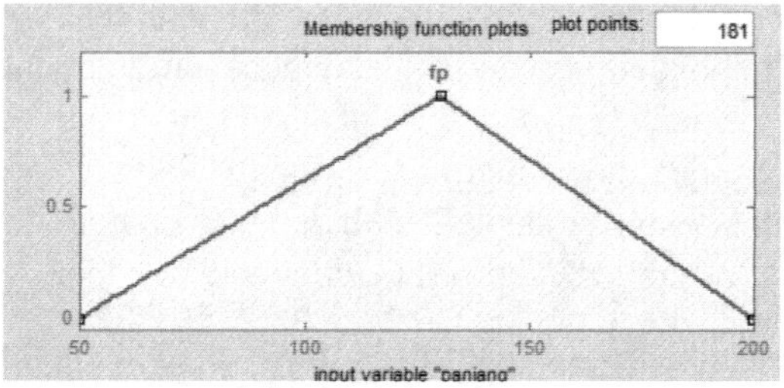

W/D Ratio [1 2 3]

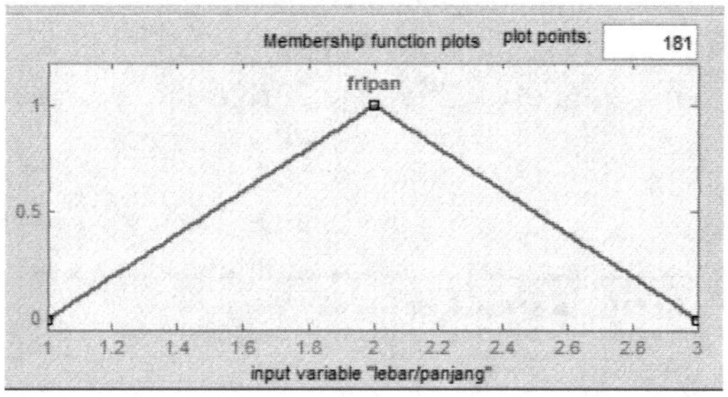

A/P Ratio [30 80 120]

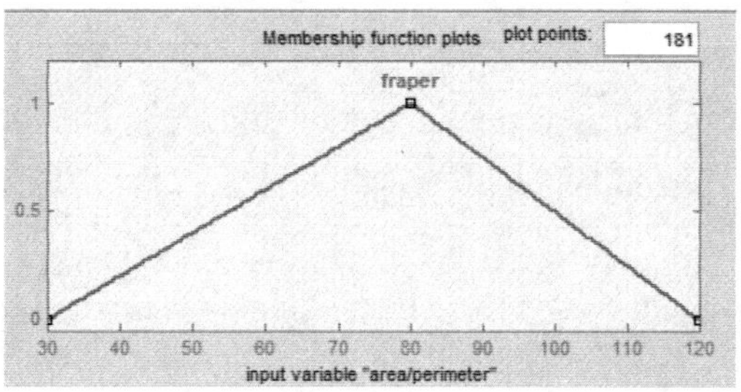

Center_x [30 60 90]

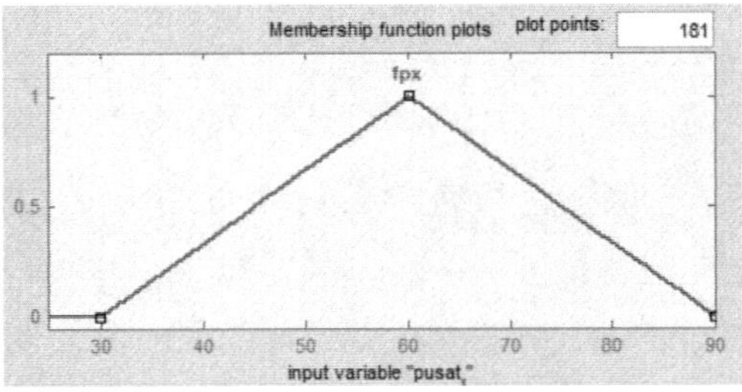

Center_y [35 65 90]

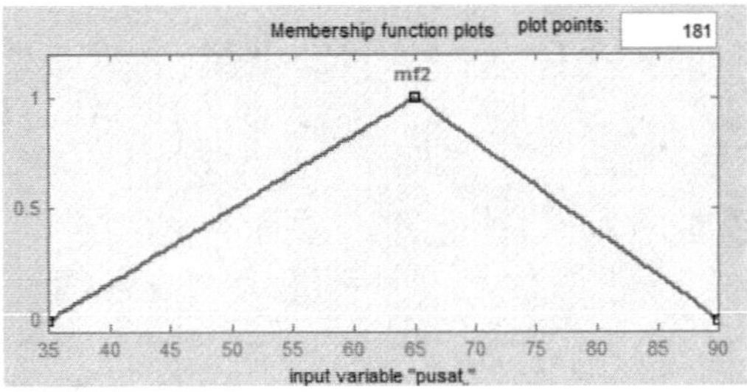

Table 3 shows the initial data of the feature extraction of the panoramic dental image of the incisor teeth. Fuzzification is shown in Table 4. The data is saved to the database and the CSV file is converted to a descriptive value shown in Table 5.

Table 3. Incisor Feature Values

A	P	W	L	W/L	A/P	C_x	C_y	Class
8090	172	173	81	2.14	47.03	38.12	71.01	Tapered
6916	156	157	65	2.42	44.33	35.18	68.57	Tapered
8404	158	159	94	1.69	53.19	48.54	69.92	Square
6706	149	150	79	1.90	45.01	37.86	75.12	Square
14854	176	177	128	1.38	84.40	65.24	79.43	Square
15175	184	185	120	1.54	82.47	61.52	84.12	Square
10115	164	165	108	1.53	61.68	56.23	69.30	Oval
9496	156	157	96	1.64	60.87	44.98	67.79	Oval
8995	155	156	87	1.79	58.03	40.77	75.51	Tapered

Table 4. Results of Fuzzy Extraction

A	P	W	L	W/L	A/P	C_x	C_y	Class
0.56	0.56	0.54	0.39	0.86	0.34	0.27	0.76	Tapered
0.41	0.88	0.86	0.19	0.58	0.29	0.17	0.86	Tapered

0.59	0.84	0.82	0.55	0.69	0.46	0.62	0.80	Square
0.39	0.98	0.83	0.36	0.90	0.30	0.26	0.60	Square
0.62	0.48	0.46	0.98	0.38	0.89	0.83	0.42	Square
0.58	0.32	0.30	0.88	0.54	0.94	0.95	0.24	Square
0.80	0.72	0.70	0.73	0.53	0.63	0.87	0.83	Oval
0.73	0.88	0.86	0.58	0.64	0.62	0.50	0.89	Oval
0.67	0.90	0.88	0.46	0.79	0.56	0.36	0.58	Tapered

Table 5. Feature Category Values

A	P	W	L	W/L	A/P	C_x	C_y	Class
Wide	Wide	Wide	Narrow	Wide	Narrow	Narrow	Wide	Tapered
Narrow	Wide	Wide	Narrow	Wide	Narrow	Narrow	Wide	Tapered
Wide	Wide	Wide	Wide	Wide	Narrow	Wide	Wide	Square
Narrow	Wide	Wide	Narrow	Wide	Narrow	Narrow	Wide	Square
Wide	Wide	Wide	Wide	Narrow	Narrow	Wide	Narrow	Square
Wide	Narrow	Narrow	Wide	Wide	Wide	Wide	Narrow	Square
Wide	Narrow	Narrow	Wide	Wide	Wide	Wide	Wide	Oval
Wide	Wide	Wide	Wide	Wide	Wide	Narrow	Wide	Oval

3. Results and Discussion

The results of each stage of this research are as follows:

a) Readings of the panoramic dental image is shown in Figure 2.(a)
b) Panoramic dental images are segmented, taking only the upper and lower incisors. This is shown in Figure 2.(b)
c) Figure 2.(c) is median filtered, shown in Figure 5. Binary process was done using iterative adaptive thresholding.

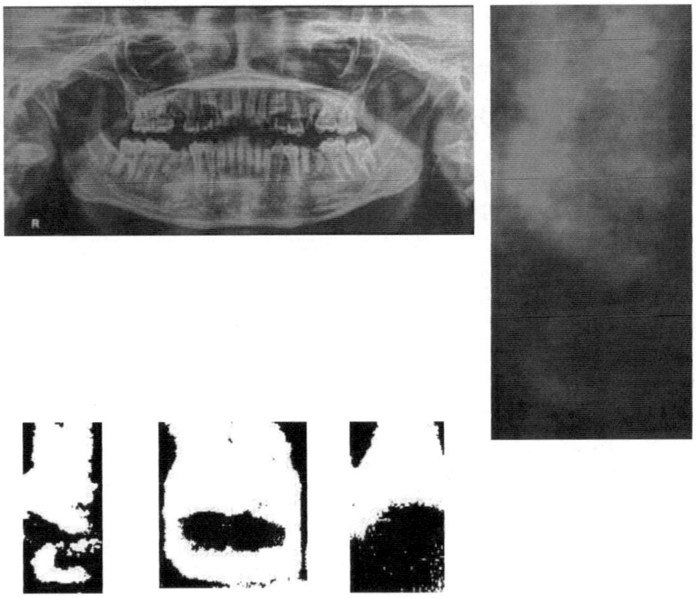

Tapered Square Oval

Figure 2.(a) Original panoramic dental image, (b) Upper Incisors after segmentation, and (c) Binarization of panoramic incisor image

After binarization of the left and right incisors of the upper and lower jaw, the features taken are: area, perimeter, width, length, width-to-length ratio, area-to-perimeter ratio, center_x, and center_y. The features were put into a database and normalized using fuzzy and studying ID3 algorithm. Data used to analyze the ID3 algorithm to classify facial shape consisted of 42 rows. The initial data in the form of numbers can be seen in Table 3. Afterwards, this was normalized using fuzzy on a scale of 0 to 1 as seen in Table 4 and convert to category as seen in Tabel 5. The normalized data shown in the triangular format can be seen in Table 2. The data was then saved to the database. The researchers used two method models to analyze the ID3 algorithm: the first had nine columns, which included area, perimeter, width, length, width-to-length ratio, area-to-perimeter ratio, center_x, center_y, and facial shape; the model had using ID3 algorithm.

The Tree for the model using ID3 algorithm. It resulted in the following rules:

[1] if ratio_w/l=wide and length=wide and ratio_are/per=wide and perimeter=wide and centroid_x=wide then face=oval

[2] if ratio_w/l=wide and length=wide and ratio_are/per=wide and perimeter=wide and centroid_x=narrow then face=oval

[3] if ratio_w/l=wide and length=wide and ratio_are/per=wide and perimeter=narrow then face=square

[4] if ratio_w/l=wide and length=wide and ratio_are/per=narrow then face=square

[5] if ratio_w/l=wide and length=narrow and centroid_x=wide then face=oval

[6] if ratio_w/l=wide and length=narrow and centroid_x=narrow and area=wide then face=tapered

[7] if ratio_w/l=wide and length=narrow and centroid_x=narrow and area=narrow then face=tapered

[8] if ratio_w/l=narrow and area=wide and ratio_are/per=wide and perimeter=wide then face=oval

[9] if ratio_w/l=narrow and area=wide and ratio_are/per=wide and perimeter=narrow then face=square

[10] if ratio_w/l=narrow and area=wide and ratio_are/per=narrow then face=square

[11] if ratio_w/l=narrow and area=narrow then face=oval

The Rule of ID3 model resulted in eleven rules with an accuracy of 67 %, as seen in Figure 3.(a). The accuracy of both models ID3 algorithm is shown in Figure 3.(b), there are nine columns used or variables. Figure 3.(b) shows the variables used for model in Weka with nine column. The results for facial shape classification shows 28 correct rows of data from the dataset of 42 using ID3 algorithm. Testing of classifications can be seen in Figure 3.(c). The results of the facial shape classification showed that 28 rows of data were correct out of the 42 in the dataset, meaning an accuracy of 67 % using ID3 algorithm.

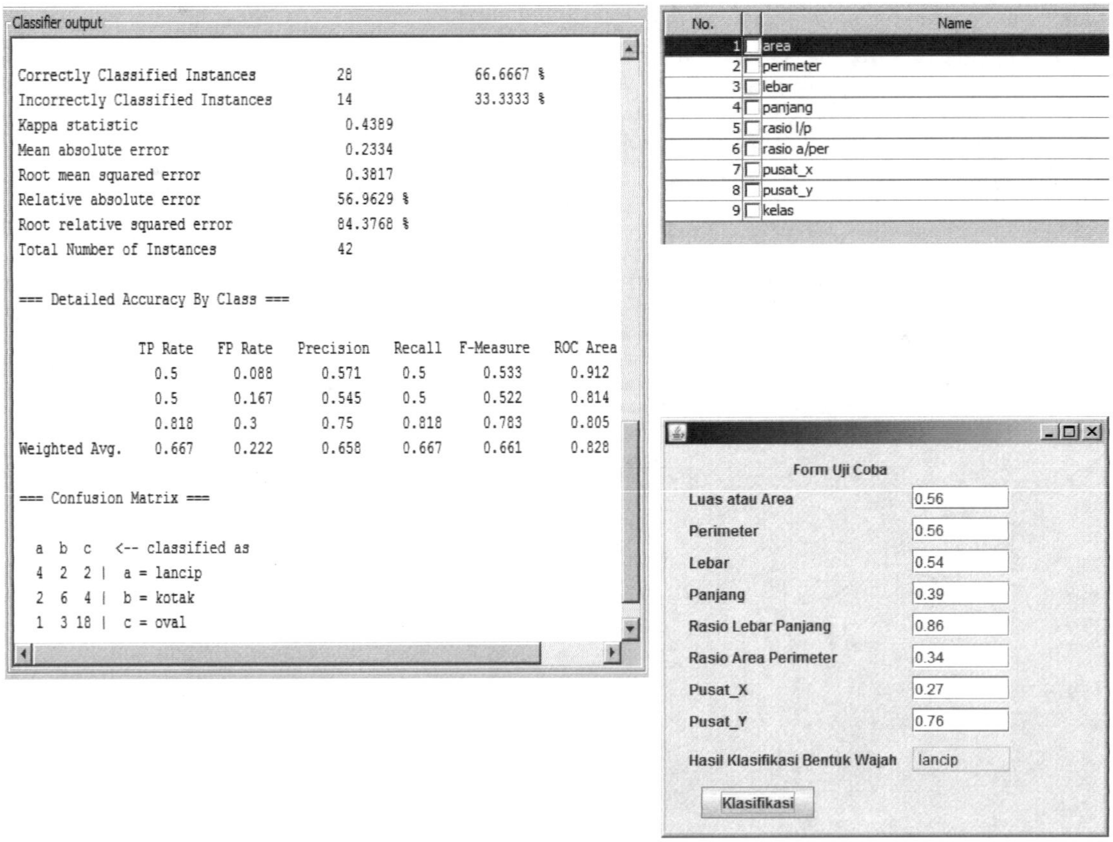

Figure 3.(a) Accuracy of ID3 in Weka, (b) Variables for Model in Weka, and (c) Testing of Classifications of Facial Shapes

4. Conclusions

This study proposes an application for face shapes classification on panoramic dental image using fuzzy features and decision tree method. Fuzzy theory is used to convert numeric features into category features, while decision tree method will be used for training features. From the experimental results, it can be concluded that the ID3 algorithm resulted in an accuracy level of 67 %.

References

[1] Han J and Kamber M 2012 *Handbook of Data Mining: Concepts and Technques 3rd ed* (Amsterdam: ScienceDirect).
[2] Giwani RA, Mudjihartono P and Pranowo 2011 *Jurnal Buana Informatika* **2** 102-108.
[3] Frank VB 2003 *Classification Trees: C4.5* (Belgium :University Libre de Bruxelles).
[4] Zadeh LA 1965 *Inf. Control* **8** 338-353.
[5] Gunawan W and Arifin AZ 2017 *Jurnal Infotel* **9** 40-47.
[6] Amaliah B, Yuniarti A, Nugroho AS and Arifin AZ 2011 *Jurnal Ilmiah Kursor* 6 123-130.
[7] Fatichah C, Tangel ML, Yan F, Betancourt JP, Widyanto MR, Dong F and Hirota K 2015 *International Journal of Control, Automation and Systems* **13** 1-11.
[8] Tangel ML, Fatichah C, Yan F, Betancourt JP, Widyanto MR, Dong F and Hirota K 2013 *IEEE*

304-309.
[9] Lukman D 2006 *Ilmu Kedokteran Gigi Forensik,* (Jakarta: Sagung Seto).
[10] Nafi'iyah N and Wardhani R 2016 *Seminar Hasil Penelitian dan Pengabdian Masyarakat* (Jember: Politeknik Negeri Jember) p 120-125.
[11] Zhang J, Sun R, Zhan X and Hong C 2015 *Elsevier* 5-14.
[12] Escobedo DS, Castelan M and Smith WAP 2016 *Computer Vision and Image Understanding* **142** 111-124.
[13] Hu JF, Zheng WS, Xie X and Lai J. 2017 *Pattern Recognition* **68** 272-285.
[14] Sugiartha and Agung IGR 2016 *Seminar Nasional Teknologi Informasi dan Multimedia* (Yogyakarta: STMIK AMIKOM) p 431-436.
[15] Sari YA, Dewi RK and Fatichah C 2014 *JUTI* **12** 1-8.

An Application of Stochastic Dominance with Truncated Normal Disribution on LQ45 Index

Iga Rahmawati[1], Isnandar Slamet[1,*] and Etik Zukhronah[1]
[1]Department of Statistics, Universitas Sebelas Maret, Surakarta, Indonesia, 57126

*Email: isnandarslamet@staff.uns.ac.id

Abstract. Return is the profit obtained by investors from the capital invested in a company. The movement of stock prices in an exchange affect the condition of stock returns. Stochastic dominance with truncated normal distribution is an analysis method of stock dominance with parametric approach where the stock returns is bounded between two values. The purpose of this research is to obtain the optimum stocks listed in the LQ45 index using stochastic dominance with truncated normal distribution. The assumption of a truncated normal distribution is derived from the normal distribution of monthly stock returns that symmetrically truncated on two sides. The dominance of stocks then obtained by comparing the truncated cumulative distribution function of returns based on first order stochastic dominance (FSD) and second order stochastic dominance (SSD). Based on the analysis, it can be concluded that there are 17 stock combinations that dominated by FSD and 14 stock combinations that dominated by SSD.

1. Introduction

Investment is the activity of infuse a number of funds with the aim of gaining profits in the future [1]. Invest in stocks is one of investment types that can be made in the capital market. The LQ45 is one of stock market index in the Indonesia Stock Exchange or IDX, which one indicator of stock prices movements, consists 45 combined stocks that have high liquidity and large market capitalization.

Fluctuations of stock prices in the exchange make the return face uncertainty. In other word, stocks are classified as risky assets because of the possibility of loss that can be happen. Stochastic dominance (SD) is a method that suitable to use in uncertain condition, such as choosing the best stock based on diverse investor preferences. The fact is investors prefer strong stocks such as from companies listed in LQ45 index. The domination of stock that obtained by SD method can be used to form optimal portfolio as well as Multi Index Models [2]. Corresponding to calendar anomaly on IDX, which has effect to the condition of stock returns, the SD method give information that the returns on Wednesday and Friday are more optimum than the other days, also the return in the second week is the most optimum in a respective month [3].

The SD method provides a sequence of stocks that investors can choose according to their preferences [4]. In some cases, SD method need to be derive to give more alternative decision analysis. The almost stochastic dominance (ASD) can be used when the SD criterias have negative values or there is violation area based on SD criteria [5]. Moreover, with almost marginal condition stochastic dominance (AMCSD), the derivative method of SD, we also can choose efficient set of stocks and obtain an efficient portfolio [6].

The SD method shows the domination of a stock over other stock stochastically by comparing the cumulative distribution of the both stocks based on the utility function [7]. There are three criterias representing the investor preferences for the desired stock, that are first order stochastic dominance (FSD), second order stochastic dominance (SSD), and third order stochastic dominance (TSD).

Although, the application of the SD method can be carried out without any specific assumptions, but sometimes a parametric approach will give more relevant results of dominance [8]. A parametric approach assuming a truncated normal distribution can be applied to the SD method [9]. Stock returns commonly have a range that is not so large and the stock price that forms a return is always positive. Therefore, the assumption of a truncated normal distribution is more appropriate than a normal distribution.

2. Research Method

This research uses monthly stock closing price data from stocks that listed in the LQ45 index over the period of August 2014 to July 2018 [10]. Those data are used to determine the amount of return for each stock based on the equation

$$R_{t(i)} = \frac{P_{t(i)} - P_{t-1(i)}}{P_{t-1(i)}}, \tag{1}$$

where $R_{t(i)}$ is the value of stock returns i in the period t, $P_{t(i)}$ is the closing price of stock i in the period t, and $P_{t-1(i)}$ is the closing price of stock i in the period $t-1$ [11]. Furthermore, stock samples are selected by the Jarque Bera normality test on stock returns formed. Based on the mean value and standard deviation of the return it can be determined that two truncation points will truncate the normal return distribution symmetrically on the lower and upper tail. Finally, the analysis of stock dominance with the SD method with a truncated normal distribution can be done by the criteria of FSD and SSD.

3. Stochastic Dominance with Truncated Normal Distribution

Stochastic dominance with a truncated normal distribution is a method of selecting risky investment assets with a parametric approach that compares the two functions of the cumulative distribution of returns [7]. The assumption of a truncated normal distribution is derived by bounding the normal distribution of returns at two points defined as

$$A_i = \mu_i - \delta\sigma_i \text{ and } B_i = \mu_i + \delta\sigma_i, \tag{2}$$

where A_i and B_i are the lower and upper truncated points respectively, μ_i is the mean of return, σ_i is the standard deviation of the return, and δ is a constant that shows the distance from the mean to the point of each truncation point.

Truncation in the normal distribution can be carried out symmetrically on two tails or non symmetrically that is only on one tail. The dominance rules for symmetrical truncation and non symmetrical truncation are the same. If F^* is the truncated cumulative distribution of stock I and G^* is the truncated cumulative distribution of stock II, then the dominance of stock I to stock II is given by rules based on the relationship of mean and standard deviation of both stocks [9].

There is a relationship between the mean and standard deviation of stock I and stock II, namely $\mu_{\mathrm{I}} > \mu_{\mathrm{II}}$ and $\sigma_{\mathrm{I}} > \sigma_{\mathrm{II}}$, so that by defining the truncation point of the upper tail of both stocks as B_{I} and B_{II} it can be clearly seen that $B_{\mathrm{I}} > B_{\mathrm{II}}$. The relationship of the lower tail of the truncation point for both stocks can be identified $A_{\mathrm{I}} \gtreqless A_{\mathrm{II}}$ if and only if $(\mu_{\mathrm{I}} - \mu_{\mathrm{II}})/(\sigma_{\mathrm{I}} - \sigma_{\mathrm{II}}) \gtreqless \delta$. Figure 1 shows that the cumulative distribution of stock I (F) will has a thicker left tail than the cumulative distribution of stock II (G) because $\sigma_{\mathrm{I}} > \sigma_{\mathrm{II}}$ where F and G intersect at $x_0 = [(\mu_{\mathrm{I}}.\sigma_{\mathrm{II}}) - (\mu_{\mathrm{II}}.\sigma_{\mathrm{I}})]/\sigma_{\mathrm{II}} - \sigma_{\mathrm{I}}$.

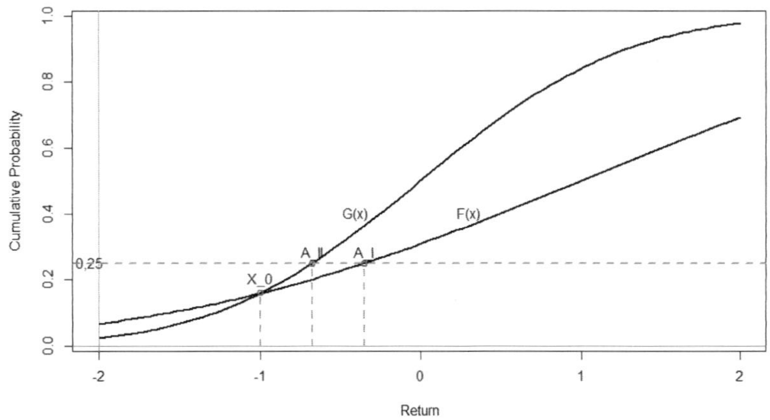

Figure 1. The cumulative normal distribution for case $\mu_{\mathrm{I}} > \mu_{\mathrm{II}}$ and $\sigma_{\mathrm{I}} > \sigma_{\mathrm{II}}$

Based on Fig. 1 it is known if $A_{\mathrm{I}} > A_{\mathrm{II}}$ then $A_{\mathrm{I}} > A_{\mathrm{II}} > x_0$, so that F^* dominates G^* by FSD if and only if

$$\frac{(\mu_{\mathrm{I}}-\mu_{\mathrm{II}})}{(\sigma_{\mathrm{I}}-\sigma_{\mathrm{II}})} > \delta. \tag{3}$$

The case where the mean of stock I is greater than the mean of stock II which is $\mu_{\mathrm{I}} > \mu_{\mathrm{II}}$ and the standard deviation of stock I is greater than the standar deviation sof stock II which is $\sigma_{\mathrm{I}} < \sigma_{\mathrm{II}}$, so this will have implications for $A_{\mathrm{I}} > A_{\mathrm{II}}$. The relation between B_{I} and B_{II} can be identified $B_{\mathrm{I}} \gtreqless B_{\mathrm{II}}$ if and only if $(\mu_{\mathrm{I}} - \mu_{\mathrm{II}})/(\sigma_{\mathrm{II}} - \sigma_{\mathrm{I}}) \gtreqless \delta$. Figure 2 shows that F intersects G from below at $x_0 = [(\mu_{\mathrm{I}}.\sigma_{\mathrm{II}}) - (\mu_{\mathrm{II}}.\sigma_{\mathrm{I}})]/\sigma_{\mathrm{II}} - \sigma_{\mathrm{I}}$.

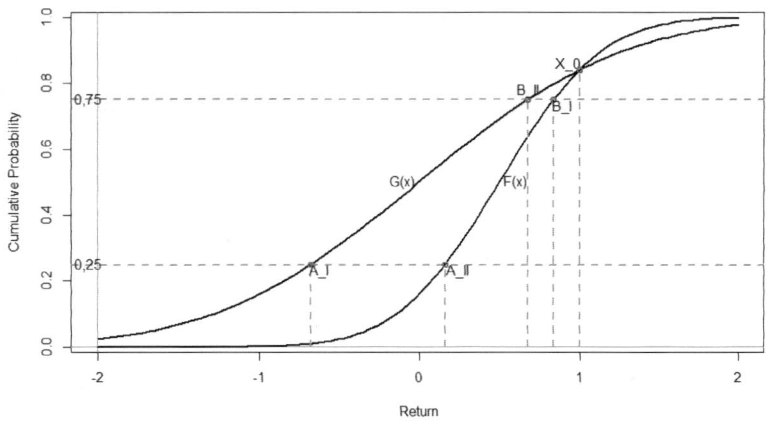

Figure 2. The cumulative normal distribution for case $\mu_{\mathrm{I}} > \mu_{\mathrm{II}}$ and $\sigma_{\mathrm{I}} < \sigma_{\mathrm{II}}$

Based on Fig. 2 it is known if $B_{\mathrm{I}} > B_{\mathrm{II}}$ then $x_0 > B_{\mathrm{I}} > B_{\mathrm{II}}$, so that F^* dominates G^* by FSD if and only if

$$\frac{(\mu_{\mathrm{I}}-\mu_{\mathrm{II}})}{(\sigma_{\mathrm{II}}-\sigma_{\mathrm{I}})} > \delta. \tag{4}$$

If the dominance of stock I over stock II by FSD with inequality (4) is met, then stock I will dominate stock II by SSD automatically. This property is derived from the dominance rule on stochastic dominance with a normal distribution where F will dominate G by SSD based on the mean variance rule, i.e. $\mu_{\mathrm{I}} > \mu_{\mathrm{II}}$ and $\sigma_{\mathrm{I}} < \sigma_{\mathrm{II}}$ [7].

4. Result and Discussion

There are 31 stocks included in the calculation of the LQ45 index over the period of August 2014 to July 2018 [10]. Then, the monthly stock closing price data is used to calculate the return of 31 stocks by equation (1). Jarque Bera normality test on stock return data formed by the null hypothesis which is stock returns are normally distributed and the significance level used is 0,05. From the Jarque Bera normality test, a number of 22 stocks with normal distribution were obtained.

Table 1. The list of stocks which have normally distributed returns.

Number	Stock	Number	Stock	Number	Stock	Number	Stock
1	ADHI	7	BMRI	13	JSMR	19	SCMA
2	AKRA	8	BSDE	14	KLBF	20	SMGR
3	ASII	9	GGRM	15	LPKR	21	TLKM
4	BBCA	10	INCO	16	LPPF	22	UNTR
5	BBRI	11	INDF	17	PTPP		
6	BBTN	12	INTP	18	PWON		

Using a significance level (α) of 0,05, the truncation points for each tail corresponds to 2σ from the mean. The illustration of the normal distribution of INTP stock returns which are truncated symmetrically based on equation (2) represents in Fig. 3.

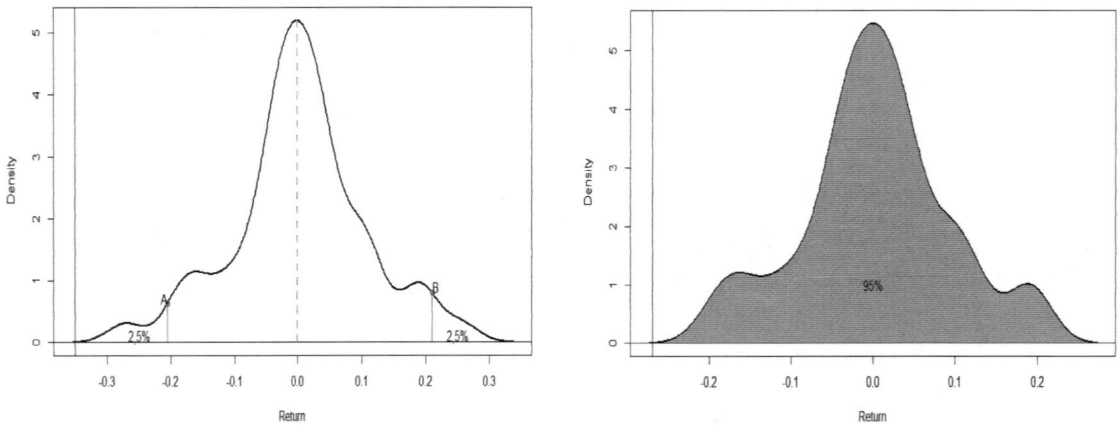

(a) (b)

Figure 3. (a) The density of normal distribution (b) The density of truncated normal distribution

All stocks are then combined and identified to choose the rule of dominance inequality (3) or (4) to be used. From the analysis for stock combinations with the case of $\mu_{\text{I}} > \mu_{\text{II}}$ and $\sigma_{\text{I}} > \sigma_{\text{II}}$, there are 17 stock combinations which have dominance by FSD.

Table 2. The results of stock dominance by FSD.

Stock	Domination	Total
ADHI	LPKR	1
BBRI	AKRA, INDF, JSMR, SCMA	4
BBTN	LPPF	1
GGRM	KLBF, TLKM	2
INTP	ADHI, LPKR	2
PTPP	LPKR	1
PWON	LPPF	1
TLKM	KLBF	1
UNTR	AKRA, BSDE, JSMR, SCMA	4

Based on Table 2, we can see the stock ranking based on the number of domination. The first rank is BBRI and UNTR which both dominates four stocks and three of them are the same stocks namely AKRA, JSMR, and SCMA. The difference is, BBRI also dominates INDF while UNTR dominates BSDE. The second rank are GGRM and INTP which are GGRM dominates KLBF and TLKM while INTP dominates ADHI and LPKR. Furthermore, there are ADHI and PTPP which both dominates LPKR, BBTN and also PWON which dominates LPPF. Finally the last rank is TLKM that dominates KLBF.

The domination analysis for stock combinations with the case of $\mu_{\mathrm{I}} > \mu_{\mathrm{II}}$ and $\sigma_{\mathrm{I}} < \sigma_{\mathrm{II}}$ give 14 stock combinations that have dominance in FSD and SSD. The results of the analysis are given in Table 3.

Table 3. The results of stock dominance by FSD and SSD.

Stock	Domination	Total
ASII	SCMA	1
BBRI	BSDE, LPPF, SMGR, UNTR	4
BBTN	ADHI, INTP, LPKR, PTPP	4
BMRI	KLBF	1
INDF	SCMA	1
PWON	LPKR	1
UNTR	LPPF, SMGR	2

Based on Table 3, BBRI and BBTN were ranked first with both dominating four stocks. Where BBRI dominates BSDE, LPPF, SMGR, and UNTR, while BBTN dominates ADHI, INTP, LPKR, and PTPP. Furthermore, in the second rank there were UNTR which had two dominance, namely towards LPPF and SMGR. In the next rank, there are ASII and INDF which both have dominance over SCMA, BMRI which dominates KLBF, and PWON which dominates LPKR.

5. Conclusion

Based on the discussion, it can be concluded that from 12 stocks that have dominance, there are 17 combinations of stocks that dominate by FSD and 14 stocks combinations that dominate by SSD. Also based on the total number of dominance both FSD and SSD, it can be obtained the ranking represent the optimum stocks, namely BBRI with 8 dominance, UNTR with 6 dominance, BBTN with 5 dominance, INTP, GGRM, PTPP, and PWON with 2 dominance, ADHI, ASII , BMRI, INDF, and TLKM with 1 dominance.

References

[1] Tandelilin E. 2010 *Portofolio dan Investasi: Teori dan Aplikasi* edisi pertama (Yogyakarta: Kanisisus)
[2] Slamet I, Arifan S N and Sutrima. 2018 *J. Phys.: Conf. Ser.* **1028** 012217
[3] Slamet I, Laela N and Pangadi. 2016 *AIP: Conference Proceedings* **1746** 20044
[4] Hadar J and Russell W R. 1969 *American Economic Review* **59** 25-34
[5] Slamet I, Wibowo A A and Roswitha M. 2017 *J. Phys.: Conf. Ser.* **943** 012048
[6] Slamet I, Carissa S M P and Pratiwi H. 2017 *J. Phys.: Conf. Ser.* **893** 012035
[7] Levy H 2016 *Stochastic Dominance: Investment Decision Making under Uncertain* 3rd ed (Springer International Publishing)
[8] Kuosmanen T. 2004 *Management Science* **50** 1390-1406
[9] Levy H. 1982 *J. Finance.* **37** 1299-1303
[10] Bursa Efek Indonesia, https://www.idx.co.id/data-pasar/data-saham/indeks-saham/ (Mei, 2018)
[11] Brigham E F and Houston J F. 2007 *Fundamental of Financial Management* (USA: Thomson Rights)

Journal of Physics: Conference Series **1373** (2019) 012043 doi:10.1088/1742-6596/1373/1/012043

Stability Analysis on the Spread of the Ebola Virus Considering Changes in Human Behavior and Climate (Case Study: Democratic Republic of the Congo)

Sitti Rosnafian. S[1], Mardlijah[1], and Hariyanto[1]

[1]Mathematics Department of Graduate Program Sepuluh Nopember Institute of Technology, Jalan Raya ITS, Keputih, Sukolilo, Surabaya, Jawa Timur 60111, Indonesia

*Email: rosnafian@gmail.com, mardlijah@matematika.its.ac.id, hariyantomath@gmail.com.

Abstract. In this paper construction of the SEIR (Susceptible-Exposed-Infected-Recovery) type mathematical model consider account the influence of changes in human behavior and climate in the Democratic Republic of the Congo (DR Congo) in the form of differential equations to describe the spread of the Ebola virus. The relationship between climate and behavior change is represented in the form of parameters (α). The constructed system is then validated with existence and uniqueness. Furthermore, an analysis of the spread of the virus was carried out with stability analysis. Based on the results of stability analysis, the system around the endemic equilibrium point is unstable. In the final stage a simulation was carried out to obtain a visual picture of the spread of the Ebola virus in DR Congo.

1. Introduction

The Ebola virus is a type of zoonotic virus that can be fatal if not treated immediately. The Ebola virus first appeared in 1976. Infection with the Ebola virus is caused by a person's mucous membrane coming into contact with goods or the environment that is contaminated with bodily fluids from an infected person or damaged skin. Based on data from WHO the mortality rate caused by the Ebola virus is quite high, on average 50% and in the past it can reach the range of 25% -90% [1].

Ebola outbreaks have reappeared in the DR Congo in May 2018, this outbreak is the 9th Ebola outbreak in the past four decades [2]. The location of the emergence of the Ebola virus is in remote areas with difficult road access, causing people to have very little knowledge regarding the Ebola virus. The outbreak of the Ebola virus in the DR Congo is allegedly due to the influence of changes in behavior of people in DR Congo that are influenced by the climate.

Research on stability analysis on the spread of disease has been examined by Nihaya et al with the SEITR multi-path epidemic model [3]. Then research about transmission of Ebola virus in Guinea using the $SEIT$ type model has been researched by Li et al [4], model of the spread of the Ebola virus using SLIR type models with two control parameters to control the spread of Ebola virus in Nigeria by Ndanusa et al [5], and then developed by constructing a mathematical model of the impact of behavior change on the spread of the Ebola virus by considering the settings of medical care and different types of burial by Conrad et al [6].

In contrast to previous research, in this study using a *SEIR* type mathematical model that refers to [4] but it was developed by adding new parameters namely α in the susceptible population and exposed population. The new parameter describes the level of change in human behavior that is influenced by climate.

2. Main Result

2.1. *Effects of Climate on Behavior Change, in Transmission Spread of the Ebola Virus*

DR Congo is a country located in Central Africa and has a tropical climate that tends to have high temperatures, so the outbreak of the Ebola virus can spread easily in the country. Climate change that often occurs in the DR Congo is extreme temperatures which can cause a prolonged dry season and cause drought. Drought can cause difficulties in obtaining food ingredients because of the large number of dead plants. Besides that, climate change can also cause increased pest attacks and facilitate the spread of disease. The DR Congo people who have difficulty obtaining food ingredients for a long time are at risk of starvation. the risk of starvation can encourage people to behave deviant.

The deviant behavior of the DR Congo community based on information from WHO is that the DR Congo people still eat fruit that has been eaten by animals (bats), the DR Congo people still consume wild animals with low meat maturity, the DR Congo tradition in traditional burial ceremonies has direct contact between mourners and corpses, and DR Congo people still rely on traditional medicine. The deviant behavior above can cause an increase in transmission of the spread of the Ebola virus if the animal, person or corpse is infected with the Ebola virus.

Based on the explanation above, there is a correlation between climate and behavior change which can be represented by parameter α. The value of α in this paper uses proportion numbers so that the value of α is in interval [0,1] or $0 < \alpha < 1$. The value of alpha in this paper uses propostional numbers of changes in human behavior compared to the number of population, so that we obtain the value of alpha is 0.7 from the journal [7].

2.2. *Model Construction*

In this paper the object observed is the spread of the Ebola virus in the DR Congo. The mathematical model was constructed by referring to the model of the spread of the Ebola virus that had been previously studied [4]. In contrast to the previous study [4], in this study using a SEIR type mathematical model which was developed by adding new parameters namely α. The parameter α is placed in the susceptible population and the population is exposed because of the community in that population has a considerable opportunity to change behavior.

The mathematical model in [4] that has been developed by adding the α parameter is as follows :

$$\begin{aligned}
\frac{dS}{dt} &= \Lambda - \beta_{10} IS - \beta_{20} ES - \mu S + \gamma_0 R + \alpha S \\
\frac{dE}{dt} &= \beta_{10} IS + \beta_{20} ES - \varepsilon_0 E - (\mu + \mu_{10})E + \alpha E \\
\frac{dI}{dt} &= \varepsilon_0 E - \nu_0 I - (\mu + \mu_{20})I \\
\frac{dR}{dt} &= \nu_0 \mathrm{I} - \gamma_0 \mathrm{R} - (\mu + \mu_{30})R
\end{aligned} \tag{1}$$

with initial conditions,
$S(0) = S_0, E(0) = E_0, I(0) = I_0, R = R_0$

The total population at time t is represented by $N(t) = S(t) + E(t) + I(t) + R(t)$. With $S(t)$ is the susceptible population at time t, $E(t)$ is the exposed population at time t, $I(t)$ is the infected population at time t, $R(t)$ is the recovery population / free from the Ebola virus at time t, Λ rate natural birth, β_{10} rate of Ebola virus transmission in population I and S, β_{20} rate of Ebola virus transmission in population E and S, μ rate natural mortality, α rate of change in human behavior affected by climate, ε_0 transition rate from population E to population I, v_0 rate transition from population I to population R, γ_0 rate transition from population R back to population S, and $\mu_{10}, \mu_{20}, \mu_{30}$ is the rate of death due to disease or virus.

2.3. *Existence and Uniques Solution (Validation Model)*

Analysis of existence and uniques solution can be seen by looking for the value of the *lipschitz constant* $k(t)$ that satisfies:

$$\|f(x^1,t) - f(x^2,t)\| \leq k(t)\|x^1 - x^2\|$$

Assuming that

$$\|f(x^1,t) - f(x^2,t)\| = \left\|\begin{matrix} f(S^1,t) - f(S^2,t) \\ f(E^1,t) - f(E^2,t) \\ f(I^1,t) - f(I^2,t) \\ f(R^1,t) - f(R^2,t) \end{matrix}\right\| = \left\|\begin{matrix} a_1 \\ a_2 \\ a_3 \\ a_4 \end{matrix}\right\|$$

$\|f(x^1,t) - f(x^2,t)\| \leq k(t)\|a_i\|$,with $\|a_i\| = maks\left\|\sum_{i=1}^{4}|a_i|\right\|$

Based on the results of simplifying the system with model reduction, the results are :

$$\left\|\begin{matrix} (-\lambda_1 - \lambda_2 - \mu + \lambda_3 + \alpha)(S^1 - S^2) \\ (\lambda_4 + \lambda_5 - \varepsilon_0 - (\mu + \mu_{10}) + \alpha)(E^1 - E^2) \\ (\lambda_6 - v_0 - (\mu + \mu_{20}))(I^1 - I^2) \\ (\lambda_7 - \gamma_0 - (\mu + \mu_{30}))(R^1 - R^2) \end{matrix}\right\| \leq \max_i\{|(-\lambda_1 - \lambda_2 - \mu + \lambda_3 + \alpha)|, |(\lambda_4 + \lambda_5 - \varepsilon_0 - (\mu + \mu_{10}) + \alpha)|, |(\lambda_6 - v_0 - (\mu + \mu_{20}))|, |(\lambda_7 - \gamma_0 - (\mu + \mu_{30}))|\} \left\|\begin{matrix} (S^1 - S^2) \\ (E^1 - E^2) \\ (I^1 - I^2) \\ (R^1 - R^2) \end{matrix}\right\|$$

$$\left\|\begin{matrix} (-\lambda_1 - \lambda_2 - \mu + \lambda_3 + \alpha)(S^1 - S^2) \\ (\lambda_4 + \lambda_5 - \varepsilon_0 - (\mu + \mu_{10}) + \alpha)(E^1 - E^2) \\ (\lambda_6 - v_0 - (\mu + \mu_{20}))(I^1 - I^2) \\ (\lambda_7 - \gamma_0 - (\mu + \mu_{30}))(R^1 - R^2) \end{matrix}\right\| \leq k(t) \left\|\begin{matrix} (S^1 - S^2) \\ (E^1 - E^2) \\ (I^1 - I^2) \\ (R^1 - R^2) \end{matrix}\right\|$$

so it can be obtained

$$k(t) = \left\{|(\lambda_3 + \alpha)_{max} - (\lambda_1 + \lambda_2 + \mu)_{min}|, \left|(\lambda_4 + \lambda_5 + \alpha)_{max} - \left(\varepsilon_0 + (\mu + \mu_{10})\right)_{min}\right|, \left|(\lambda_6)_{max} - \left(v_0 + (\mu + \mu_{20})\right)_{min}\right|, \left|(\lambda_7)_{max} - \left(\gamma_0 + (\mu + \mu_{30})\right)_{min}\right|\right\}$$

It is assumed that the infected population can spread the Ebola virus widely in DR Congo. Therefore, observations are only made to the infected population. The mathematical model of the spread of the Ebola virus in this study has a single and complete solution at the time $k(t) =$

$\left|(\lambda_6)_{max} + \left(\nu_0 + (\mu + \mu_{20})\right)_{min}\right|$. Because the model in this study has a solution and uniqueness, the constructed model is valid.

2.4 Equilibrium Point

The equilibrium point is obtained from $\frac{dS}{dt} = 0, \frac{dE}{dt} = 0, \frac{dI}{dt} = 0, \frac{dR}{dt} = 0$.

By using a substitution method, an equilibrium point for the spread of Ebola virus in the DR Congo is obtained as follows :

$$S^* = \frac{\Lambda(-B - B\varepsilon_0 A - B\mu A - B\mu_{10}A + B\alpha A + \gamma_0\nu_0\varepsilon_0)}{B\alpha - B\mu - B\varepsilon_0 A\mu + B\varepsilon_0 A\alpha - BA\mu^2 + B\mu A\alpha - B\mu_{10}A\mu + B\mu_{10}A\alpha + B\alpha A\mu - BA\alpha^2 + \gamma_0\nu_0\varepsilon_0\mu - \gamma_0\nu_0\varepsilon_0\alpha}$$

$$E^* = \frac{A(-\Lambda + \mu S^* - \alpha S^*)}{B}$$

$$I^* = \frac{\varepsilon_0 A(-\Lambda + \mu S^* - \alpha S^*)}{B\left(\nu_0 + (\mu + \mu_{20})\right)}$$

$$R^* = \frac{\nu_0\varepsilon_0(-\Lambda + \mu S^* - \alpha S^*)}{B}$$

with

$$A = \left(\gamma_0 + (\mu + \mu_{30})\right)\left(\nu_0 + (\mu + \mu_{20})\right),$$

$$B = A(-\varepsilon_0 - (\mu + \mu_{10}) + \alpha) + \gamma_0\nu_0\varepsilon_0.$$

2.5 Stability Analysis

The mathematical model in equation (1) is linearized. The results of linearization are then written into the form of the *Jacobian* matrix.

So that the *Jacobian* matrix is obtained from the system of differential equations as follows

$$J = \begin{bmatrix} -\beta_{10}I - \beta_{20}E - \mu + \alpha & -\beta_{20}S & -\beta_{10}S & \gamma_0 \\ \beta_{10}I + \beta_{20}E & \beta_{20}S - \varepsilon_0 - (\mu + \mu_{10}) & \beta_{10}S & 0 \\ 0 & \varepsilon_0 & -\nu_0 - (\mu + \mu_{20}) & 0 \\ 0 & 0 & \nu_0 & -\gamma_0 - (\mu + \mu_{30}) \end{bmatrix}_{[S^*,E^*,I^*,R^*]}$$

Assuming that

$$a_{11} = -\beta_{10}I^* - \beta_{20}E^* - \mu + \alpha$$
$$a_{12} = -\beta_{20}S^*$$
$$a_{13} = -\beta_{10}S^*$$
$$a_{14} = \gamma_0$$
$$a_{21} = \beta_{10}I^* + \beta_{20}E^*$$
$$a_{22} = \beta_{20}S^* - \varepsilon_0 - (\mu + \mu_{10})$$
$$a_{23} = \beta_{10}S^*$$
$$a_{32} = \varepsilon_0$$
$$a_{33} = -\nu_0 - (\mu + \mu_{20})$$
$$a_{43} = \nu_0$$
$$a_{44} = -\gamma_0 - (\mu + \mu_{30})$$

So obtained

$$|J-\lambda I| = \begin{vmatrix} \lambda - a_{11} & a_{12} & a_{13} & a_{14} \\ a_{21} & \lambda - a_{22} & a_{23} & 0 \\ 0 & a_{32} & \lambda - a_{33} & 0 \\ 0 & 0 & a_{43} & \lambda - a_{44} \end{vmatrix}$$

So we can get the characteristic equation
$|J-\lambda I| = a_4\lambda^4 + a_3\lambda^3 + a_2\lambda^2 + a_1\lambda + a_0$
with assumption
$a_4 = -1$
$a_3 = a_{11} + a_{22} + a_{33} + a_{44}$
$a_2 = -a_{11}a_{22} - a_{22}a_{33} - a_{11}a_{33} - a_{22}a_{44} - a_{11}a_{33} - a_{33}a_{44} + a_{23}a_{32} + a_{12}a_{21}$
$a_1 = a_{11}a_{22}a_{33} + a_{11}a_{22}a_{44} + a_{22}a_{33}a_{44} + a_{11}a_{33}a_{44} - a_{23}a_{32}a_{44} - a_{11}a_{23}a_{32} - a_{12}a_{21}a_{44} - a_{12}a_{21}a_{33} - a_{13}a_{21}a_{32}$
$a_0 = -a_{11}a_{22}a_{33}a_{44} - a_{11}a_{23}a_{32}a_{44} + a_{12}a_{21}a_{33}a_{44} + a_{13}a_{21}a_{32}a_{44} + a_{14}a_{21}a_{32}a_{43}$

To find out the sign of the coefficient a_i use the following *routh-hurwitz* criteria table:

Table 1. Table Routh-Hurwitz

Root Characteristics	Coefficients			
λ^4	a_0	a_2	a_4	a_6
λ^3	a_1	a_3	a_5	a_7
λ^2	b_1	b_2	b_3	b_4
λ^1	c_1	c_2	c_3	c_4
λ^0	d_1	d_1	d_3	d_4

The system around the endemic equilibrium point is said to be stable if the signs from the first column $(a_0, a_1, b_1, c_1, d_1)$ in the *routh-hurwitz* table have the same sign that is positive.

Table 2. Parameter Value Used

Parameter	Value	Source
Λ	1,159	[4]
β_{10}	0,055	[5]
β_{20}	0,394	[5]
μ	0,0097	[4]
μ_{10}	0,5	[4]
μ_{20}	0,6647	[4]
μ_{30}	0,8	[4]
ε_0	0,0596	[4]

Parameter	Value	Source
v_0	0,8613	[4]
γ_0	0,2995	[4]
α	0,70519	[7]

By entering the parameter values found in table 2, the values obtained :
$a_0 = 0.67182$
$a_1 = -1.07500$
$b_1 = 0.56263$
$c_1 = 0.36345$
$d_1 = -1$

Because a_1 end d_1 are negative, the system around the endemic equilibrium point is unstable. Thus, the spread of the Ebola virus is categorized as high at time t and the population transition rate E to I is quite large.

2.6 Numerical Simulation
In this section a simulation is done using matlab software with the aim to see the effect of α on the spread of the Ebola virus in the DR Congo.

The data used in this paper are data that spread the Ebola virus that occurred in DR Congo in 1976,1977, 1995, 2007, 2008-2009, 2012, 2014, 2017 and may – july 2018. Related information from WHO and journals ([1],[2]) obtained data on the total population of DR Congo is 77,267,000; $S = 77{,}265{,}875$; $E = 5{,}625$; $I = 1{,}125$; $R = 285$.

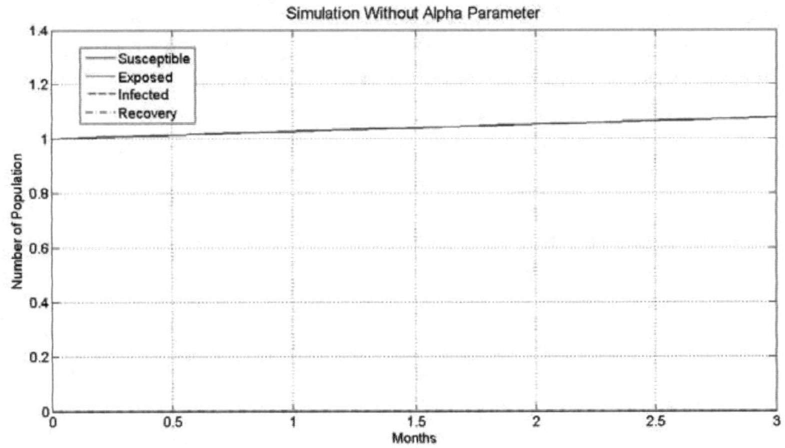

Figure 1. Graph simulation the rate of spread the ebola virus without α parameter

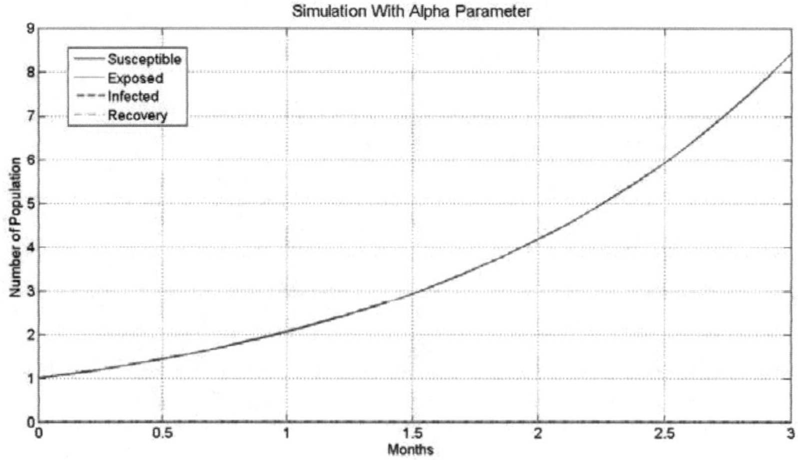

Figure 2. Graph simulation the rate of spread the ebola virus with α parameter

Figure 1 shows the spread of the Ebola virus without the influence of changes in human behavior or the same as the initial model in [4]. While Figure 2 shows the spread of the Ebola virus due to the influence of changes in human behavior (α).

The population value S without the parameter α is 1.078 when $t = 3$ months. After entering the parameter α, the population value S has increased by 87.17% so that it becomes 8.4 when $t = 3$ months. This shows that changes in human behavior influenced by climate can increase the population of people who are susceptible to infection with the Ebola virus.

The value of population E without the α parameter is 1.76×10^{-5} when $t = 3$ months. After entering the parameter α, the population value S has increased by 99.36% so that it becomes 2.76×10^{-3} when $t = 3$ months. This shows that changes in human behavior influenced by climate can increase the population of people who are susceptible to infection with the Ebola virus. This shows that changes in human behavior can also increase the number of people showing signs of being infected with the Ebola virus.

3. Conclusion

From the results of the analysis carried out in the epidemic model of the spread of the SEIR type Ebola virus, the following conclusions were drawn :

1) The results of the analysis of the mathematical models that have been constructed show that the system of equations has a single and a solution. With the *Lipschitz Constant* value $k(t) = \left|(\lambda_6)_{max} - \left(v_0 + (\mu + \mu_{20})\right)_{min}\right|$ which shows that the maximum condition of the spread of the Ebola virus occurs when more population E moves to population I. While the minimum conditions for the spread of the Ebola virus in the DR Congo occur when the mortality rate increases in population I and more people recover from Ebola virus infection.
2) From the stability analysis, it was concluded that the system in this study was unstable around the endemic equilibrium point (S^*, E^*, I^*, R^*). This shows that the Ebola virus still infects the DR Congo community with a transition from population E to a fairly high population I with a parameter indicator value $\varepsilon_0 = 0{,}4$.
3) Simulation results with parameter values $\Lambda = 0.03602$, $\beta_{10} = 0.055$, $\beta_{20} = 0.394$, $\mu = 0.0097$,$\mu_{10} = 0.5$, $\mu_{20} = 0.5061$, $\mu_{30} = 0.8$, $\varepsilon_0 = 0.4$, $v_0 = 0.5140$, $\gamma_0 = 0.2$, and $\alpha =$

0.70519 which shows that deviant behavior can increase populations S and E by an average of 93.26%.

References

[1] World Health Organization 2018a *Risk Communication and Community Engagement (RCCE) Considerations: Ebola Response in the Democratic Republic of the Congo*, WHO Health Emergencies Programme Switzerland, p. 4-19.

[2] World Health Organization 2017 *WHO Regional Director for Africa, Dr Matshidiso Moeti travels to Kinshasa to discuss Ebola outbreak response*, http://www.who.int/news-room/detail/13-05-2017-who-regional-director-for-africa-dr-matshidiso-moeti-travels-to-kinshasa-to-discuss-ebola-outbreak-response, accessed 27-Juny-2018.

[3] Nihaya A C, Dewi, Haryanto, and Mardlijah 2018 ICAS *AIP Conference Proceedings* (*Indonesia*) Vol 2014 (AIP Publishing / America) p. 1-10.

[4] Li Z, Teng Z, Feng X., Li Y, and Zhang H 2015 *Hindawi* **2015** 1-11.

[5] Ndanusa A., Abdulrahman S, end Abdulmalik A. 2015 *Int J Human Manage Sci*, **3** 144-148.

[6] Conrad J R, Xue L, Dewar L, and Hyman J M 2016 *Mathematical and Statistical Modeling for Emerging and Re-emerging Infectious Diseases* Springer International Publishing (Switzerland).

[7] World Health Organization 2018b *Outbreak Of Ebola Virus Desease In The Democratic Republic Of The Congo*, April-May, 2018 : An Epidemiological Study", Elselvier Ltd., http://dx.doi.org/10.1016/S0140-6736(18)31387-4. ScienceDirect.

Journal of Physics: Conference Series **1373** (2019) 012044 doi:10.1088/1742-6596/1373/1/012044

Design of Automatic Braking System Using Interval Type-2 Fuzzy Logic System

Dwi Setiady[1] ,Oyas Wahyunggoro[1], and Prapto Nugroho[1]

[1]Department of Electrical Engineering and Information Technology Universitas Gadjah Mada ,Yogyakarta,Indonesia.

Email: dwi.setiady@mail.ugm.ac.id, oyas@ugm.ac.id, tatok@ugm.ac.id

Abstract. The car braking system is a very important part of car safety. The effectiveness of Conventional braking system is strongly influenced by the ability and the driver's experience. It is very vulnerable to accidents if the driver fails to operate the braking system. Because of that, an automation of the braking system is needed that can decrease the accident's impact. The Automatic braking systems require the reasoning to manage information before making a decision that brakes or slows down. There is uncertainty in determining membership function of distance and velocity. It is difficult to determine an exact membership function. In this paper, the system use Interval type-2 fuzzy logic control that can work well in an unstructured environment and have the ability to overcome the uncertainty of information. This system will be compared with the braking system using T1FLS. The result is that the pressure on the brake pads is gradually increased. At a distance of 34 m, the pressure is at maximum value. The changes produced by the automatic braking system that uses IT2FLS increase the average pressure of 4,167% while in T1FLS system, a change in growth greater pressure by an average of 5%. In IT2FLS braking system, braking is gradually increasing smoother than T1FLS.

1. Introduction

It is undeniable that the current level of traffic accidents is still very high. Based on data compiled by the Indonesian Police, the number of traffic accident in Indonesia for the last quarter tends to increase, which is about 26,000 cases with a death toll of about 7,000 people. Based on these data, about 30% of these accidents are car accidents [1].

The braking system is one of important thing that supporting the driver's safety. The car's braking system is usually operated manually with the brake pedal. If the driver fails to break, there will be a collision or even accidents that cause damage or victim.

In recent years, a lot of research have been developed for an automatic car braking systems [2][3][4][5]. Some studies rely on mathematical modeling but the fact is that in the case of the braking system, driver behavior is strongly influenced by the experience, not the appropriate mathematical modeling [2][5]. Because of that, the system need a model that can apply human reasoning based on experience in a linguistic form that does not need to use complex mathematical models. Fuzzy logic systems are very good at applying human thinking based on experience to handle non-linear systems such as braking systems [2].

Fuzzy was first developed by Lotfi Zadeh and does not yet have the ability to handle uncertainty [6]. In this braking system, there is uncertainty information in determining a safe distance at a certain speed when braking. Safe distance is only determined according to the estimates and experiences have been obtained. conventional fuzzy is less effective in dealing with uncertainty [7].

Over the past decade, many researchers have developed Type-2 Fuzzy Logic System (T2FLS) which is an improvement from conventional fuzzy or Type-1 Fuzzy Logic System [8][9][10][11]. Type 2 has a better ability to overcome uncertainty. Compared to T1FLS, T2FLS has more parameters in representing Fuzzy set elements [7]. T2FLS has 3-dimensional functions and an uncertainty range called Footprint-Of-Uncertainty offering additional degrees of freedom to deal with various uncertainties.

Interval Type-2 Fuzzy Set (IT2FLS) is a type of type-2 Fuzzy Set that is widely used because it has lower computational complexity than general T2FS. IT2FLS has easy-to-manage computing that makes IT2FLS easy to implement [12].

This paper will discuss the braking system that applies IT2FLS methods for managing and analyzing the conditions in performing automatic braking. The purpose of this study is to test IT2FLS modeling capabilities in handling non-linear systems such as braking systems.

2. Braking System Design

2.1. Car Braking System

In this paper, the automatic braking system uses IT2FLS modeling to apply brakes when the distance between the car and obstacles is close. The output is the pressure required by the canvas brake when clamping the disk plate in Figure 1.

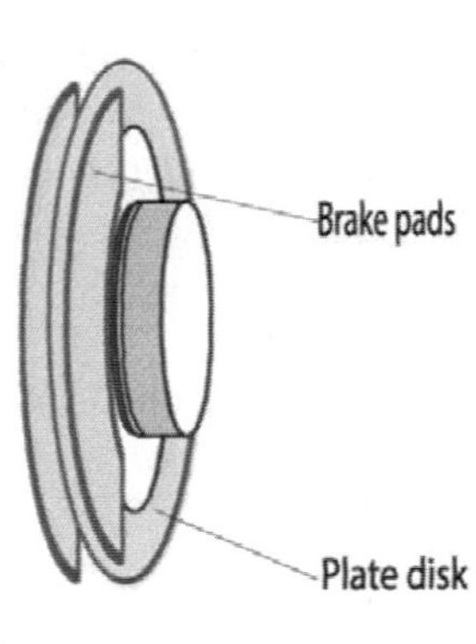

Figure 1. Illustration of the braking system mechanism

It is assumed that the pressure is 0 at minimum and 150 at the maximum. This system will be implemented using the fuzzy type-2 interval Matlab Simulink toolbox [13]. In this car's braking system, there are several parameters that have been ignored, namely engine dynamics, friction, and gravity.

2.2. System Modeling

System modeling is generally needed in the development of a system. Figure 2 shows the system process in general. Starting from the input process, the process of fuzzification, up to the results.

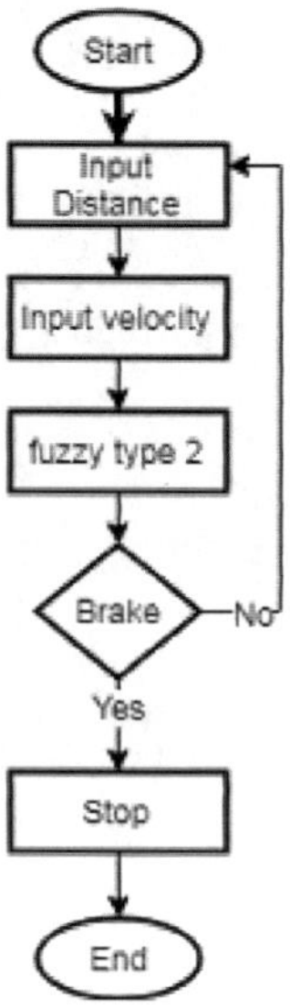

Figure 2. Block diagram Automatic braking system using IT2FLS

2.3. *System planning*

The method used is Grid Partition [14]. This partitioning method is often used in designing fuzzy controllers that have few variables as input to controllers. This partitioning strategy requires only a small number of MF for each input. Fuzzy models with 2 inputs and 3 MF in each input will produce 9 Fuzzy if-then rules. Rule modeling uses simple modeling, namely Takagi Sugeno Kang (TSK) model. The rule model is written in Equation (1) [15].

$$R^n\text{: If } \chi_1 \text{ is } \tilde{X}_1 \text{ and } \chi_p \text{ is } \tilde{X}_p \text{ then y is Y} \quad (1)$$

With (p = 1, 2, …, Np) is IT2 FS for IT2 FLS. whereas $Y^n=[\underline{y}^n,\bar{y}^n]$ is a simple interval. This model is used $\underline{y}^n = \bar{y}^n$ In Table 1, the attribute is attached. Distance variables have 3 membership functions namely near, medium and far. Speed variable which has 3 membership functions, namely slow, medium, and fast.

Table 1. Data set attribute

Attribute	Unit	Range
Distance	Near(1), Medium(2), Far(3)	0-70(1),35-105(2),70-150(3)
Velocity	Slow(1), Medium(2), Fast(3)	0-45(1),30-70(2),50-100(3)

2.4. Design the Membership Function

Design the Membership Function After generating the rule base, the thing to do is to determine the membership function as the basis of fuzzification and defuzzification. The first thing to do is to design a membership function from T1 and become a T1 MF baseline. Furthermore, based on the MF baseline, the IT2 FOU was determined to be the presentation of randomness and uncertainty of MF data.

2.5. Designing the basis of Fuzzy Type-1 membership functions

In designing the type 1 membership function using a triangular model and trapezium with each variable has 3 parameters. Distance variable consists of near, medium, and far. Velocity variables consist of slow, medium, and fast.

2.6. Designing Footprint of Uncertainty (FOU)

This research using type 1 membership function as the basis for designing the FOU described in the previous stage. Next, determine the two fields of FOU, namely LMF and UMF. It is assumed that there is no doubt at the center point of type 1 membership function. Illustration of FOU design is shown in Figure 3.

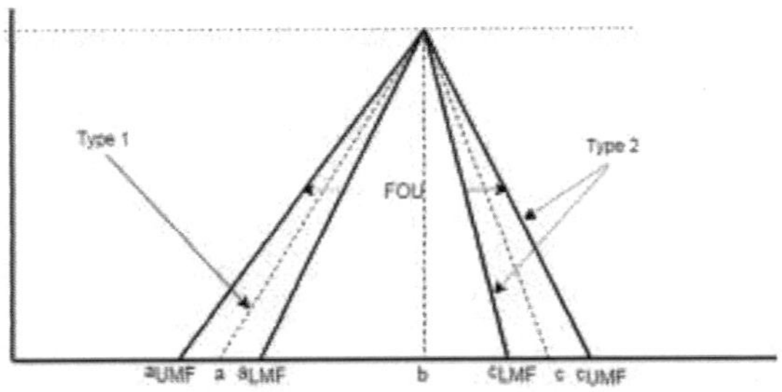

Figure 3. Design of IT2MF FOU

a_{UMF} is the left point of the UMF determined by Equation (2), ba_{UMF} is the center point of UMF with Equation (3), c_{UMF} is the right point of UMF with Equation (4), a_{LMF} is the LMF left point with Equation (5), bc_{LMF} is the LMF center point with Equation (6), c_{LMF} is the center point of LMF with Equation (7), a is the left point T1 MF, b is the center point T1 MF and c is the right point T1 MF. The results of the FOU design are shown in Figure 4 and Figure 5.

$$aUMF = a - 5 \tag{2}$$

$$baUMF = b \tag{3}$$

$$cUMF = c + 5 \tag{4}$$

$$aLMF = a + 5 \tag{5}$$

$$bcLMF = b \tag{6}$$

$$cLMF = c - 5 \tag{7}$$

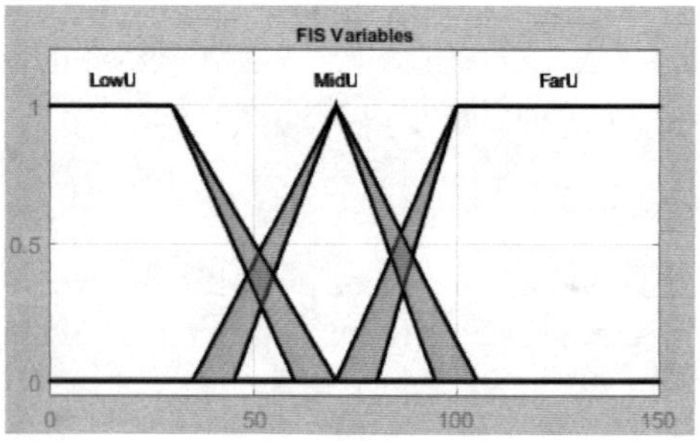

(a)

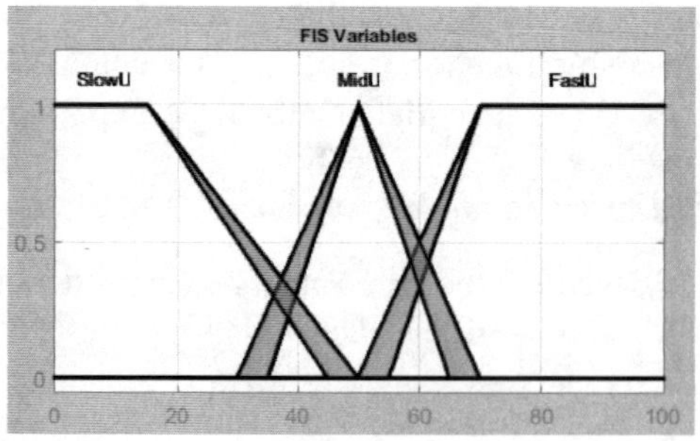

(b)

Figure 4. (a) Distance membership functions at IT2FLS, (b) Velocity membership functions at IT2FLS

2.7. Interval type-2 fuzzy logic design

2.7.1. Fuzzifier

The input is a vector n crisp element $x = (x_1, x_2, \ldots x_n)$ T in the discourse universe $x_1 \times x_2 \times \ldots \times x_n$ mapped into a type-2 Fuzzy set.

2.7.2. Rule Base

The composition of the rules in set type-2 has no difference with that of type-1 shown in figure 6. The l-rule for n-input, $x_1 \in X_1, \ldots, x_n \in X_n$, and m-output, $y_1 \in Y_1, \ldots, y_m \in Y_m$, from type-2 Fuzzy logic systems are given in Equation (8).

$$\text{Rp: IF x1 is } \tilde{F}_1^p \text{ and } xn \text{ is } \tilde{F}_n^p \text{ THEN y is } \tilde{Y}1 \text{ p=1,2,...N} \tag{8}$$

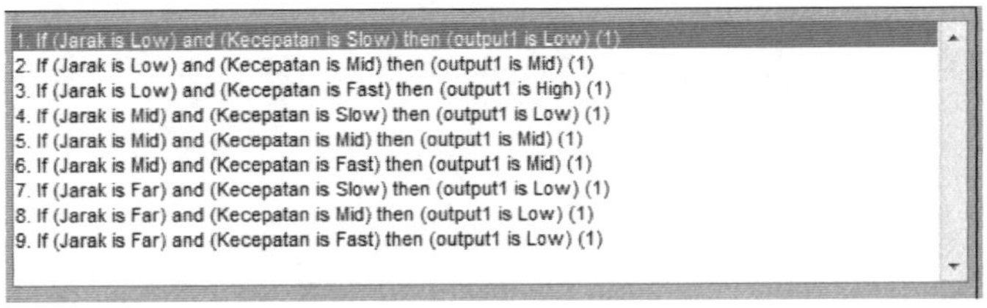

Figure 5. The Base Rule

2.7.3. Inference

The Inference is needed to combine rules and do a mapping from input to output. This is important for calculating the join ∪ (union), slices ∩ (intersection) the sup-star composition of type 2 relations. In this FLS, the type-2 Fuzzy set and product t-norm intervals are used, so the input and antecedent results are in the firing set $\prod_{n=1}^{p} \mu_{\tilde{F}_{n n}} (x_n \equiv F^p(x'))$ is the type 2 interval set, as seen in Equation (9).

$$F^p(x') = \left[\underline{f}^p(x'), \overline{f}^p(x')\right] \equiv \left[\underline{f}^p, \overline{f}^p\right] \tag{9}$$

where

$$\underline{f}^p(x') = \underline{\mu}_{\tilde{F}_1^1}(x1') \times \ldots \times \underline{\mu}_{\tilde{F}_n^p}(xn') \tag{10}$$

and

$$\overline{f}^p(x') = \overline{\mu}_{\tilde{F}_1^1}(x1') \times \ldots \times \overline{\mu}_{\tilde{F}_n^p}(xn') \tag{11}$$

2.7.4. Type Reducer

The type reducer produces a type 1 Fuzzy set which will then become a defuzzification input to get the crisp value. For this research, center of sets (cos) type reduction which Y_{cos} is used is stated in Equation (12).

$$Y_{cos}(x) = \underset{y^n \in Y^n}{\bigcup f^n \in f^n(x')} \frac{\sum_{n=1}^{p} f^n y^n}{\sum_{n=1}^{p} f^n} = [y_l, y_r] \tag{12}$$

with

$$y_l = \min_{k\in[1,N-1]} \frac{\sum_{n=1}^{k} \overline{f}^n \underline{y}^n + \sum_{n=k+1}^{N} \underline{f}^n \underline{y}^n}{\sum_{n=1}^{k} \overline{f}^n + \sum_{n=k+1}^{N} \underline{f}^n} \equiv \frac{\sum_{n=1}^{L} \overline{f}^n \underline{y}^n + \sum_{n=L+1}^{k} \underline{f}^n \underline{y}^n}{\sum_{n=1}^{L} \overline{f}^n + \sum_{n=L+1}^{k} \underline{f}^n} \tag{13}$$

$$y_r = \max_{k\in[1,N-1]} \frac{\sum_{n=1}^{k} \underline{f}^n \overline{y}^n + \sum_{n=k+1}^{N} \overline{f}^n \overline{y}^n}{\sum_{n=1}^{k} \underline{f}^n + \sum_{n=k+1}^{N} \overline{f}^n} \equiv \frac{\sum_{n=1}^{R} \underline{f}^n \overline{y}^n + \sum_{n=R+1}^{N} \overline{f}^n \overline{y}^n}{\sum_{n=1}^{R} \underline{f}^n + \sum_{n=R+1}^{N} \overline{f}^n} \tag{14}$$

Where L and R switch points are determined by Equation (15) and (16).

$$\underline{y}^L \leq y_l \leq \underline{y}^{L+1} \tag{15}$$

$$\overline{y}^R \leq y_r \leq \overline{y}^{R+1} \tag{16}$$

y_l and y_r Where L and R switches are determined by Equation (15) and (16) [16] (Karnik-Mendel).

2.7.5. *Defuzzifier*

After passing the type reducer, Y_{cos} interval value will be obtained. To defuzzify this value, only the average values of y_l and y_r are needed. The equation used is Equation (17).

$$y(x) = \frac{y_l + y_r}{2} \tag{17}$$

3. Results and Discussion

This system is implemented using Matlab. Testing is done by setting the initial speed of a car. Then place objects at a certain distance right in front of the car. This test is carried out by testing a system that applies to brake using the IT2FLS system. The car speed value is 100 km/hr. The distance to the object is 100 m. The test illustration is shown in Figure 7.

For the evaluation of the test refers to the formula for a safe distance determined by the National Police Corps, which was 3 seconds. If the speed of the car was at 100km / h, the safety distance to the object or the obstacle in front of it was 83,3 m. Based on the above calculations, braking must started at a distance of 83,3 m from the obstacle because it had passed the safe distance.

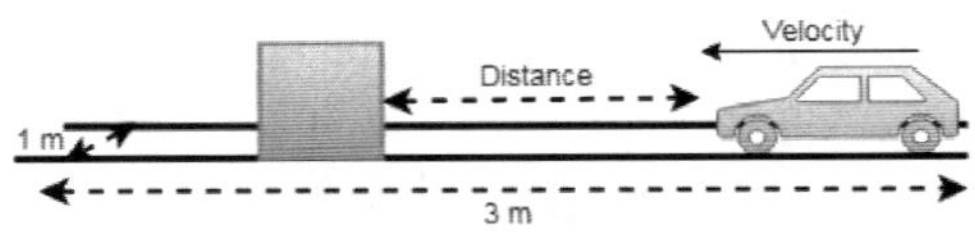

Figure 6. Illustration of Braking System Testing

Table 2. Result of The IT2FLS Bracking System

No	*Distance*	Preasure of Bracking System	
		IT2FLS (KM)	*T1FLS*
1	34	150	150
2	36	148.7	150
3	40	143.3	150
4	44	137.8	137
.	.	.	.
.	.	.	.
.	.	.	.
15	88	32.4	30
16	96	19.6	20
17	98	6.81	5
18	100	4.68	0

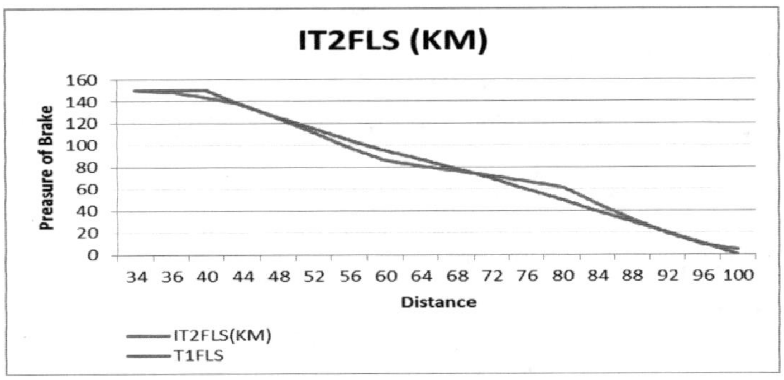

Figure 7. Result test for an automatic braking system for distance

The IT2FLS braking system would be compared to the braking system using T1FLS based on previous research [2][5]. In testing the automatic braking system, the pressure of the IT2FLS system starts at a distance of 100 m with a pressure value of 4.68. In the T1FLS, braking starts at 98 m from the obstacle with a pressure value of 5. The pressure on the canvas increases slowly along the distance approaching. The value of IT2FLS brake pressure at a distance of 80 m is 61.6. While the T1FLS is exactly 50. At a distance of 34 m, the pressure on the IT2FLS system brake canvas is at a maximum value of 150. The pressure on the T1FLS system is at a maximum value of 40 m shown in table 2.

System testing used the same input value. Based on Figure 7, a system that uses T1FLS shows good results. This system of braking gradually increases. In systems that use T2FLS, braking is done gradually increasing smoother than T1FLS. Changes generated by the automatic braking system that uses IT2FLS increase the pressure by an average of 4.167% while the system uses T1FLS, the change in pressure increases is greater by an average of 5%.

4. CONCLUSION

In this paper, we discuss the automatic braking system using IT2FLS which has two inputs and one output. Input in the form of distance and speed while the output is in the form of pressure on the brake pads. The results show a pressure In braking system using IT2FLS smoothly increase with an average value of 4.167 %. Whereas the braking system using T1FLS has a bigger changes pressure with an average value of 5%. For further research, it can be improved by adding input and adding rules so that the system's capabilities are better.

References

[1] Korlantas-irsms.info. (2018). KORLANTAS POLRI - *Accident Reports by Polda.* [online] Available at: http://www.korlantas irsms.info/graph/accidentTypeTable [Accessed 1 Nov. 2018].

[2] Hirulkar S, Damle M, Rathee V and Hardas B 2014 *International Conference on Electronic Systems, Signal Processing, and Computing Technologies* (India) (USA: IEEE) p 413-418.

[3] Jin P, Kim J and Kim J 2015 *12th International Conference on Ubiquitous Robots and Ambient Intelligence* (URAI).

[4] Chen X, Zhang J and Liu Y 2016 *Mathematical Problems in Engineering* 1-12.

[5] Rizianiza I and Djafar A 2017 *4th International Conference on Electric Vehicular Technology* (ICEVT).

[6] Zadeh L 1975 *Information Sciences* **8** 199-249.

[7] Naik K and Gupta C 2017 *4th International Conference on Signal Processing, Computing and Control (ISPCC).*

[8] Castillo O 2012 *Type-2 Fuzzy Logic in Intelligent Control Applications* vol 272 (Heidelberg: Springer Berlin Heidelberg).

[9] Karnik, N. and Mendel, J. (n.d.). Introduction to type-2 fuzzy logic systems. 1998 *IEEE International Conference on Fuzzy Systems Proceedings.* IEEE World Congress on Computational Intelligence (Cat. No.98CH36228).

[10] Castillo O and Melin P 2012 Recent advances in interval type-2 fuzzy systems (Heidelberg: Springer).

[11] Saxena V, Yadala N, Chourasia R and Rhee F 2017 *IEEE International Conference on Fuzzy Systems (Italy)* (USA: IEEE).

[12] Wu D and Mendel JM 2014 *IEEE International Conference on Fuzzy Systems* (China) (USA: IEEE)

[13] Taskin A and Kumbasar T 2015 *IEEE Symposium Series on Computational Intelligence* (South Africa) (USA: IEEE).

[14] Castillo O, Melin P, Kacprzyk J and Pedrycz W 2007 *IEEE International Conference on Granular Computing* (GRC 2007) (USA: IEEE).

[15] Sajiah A, Setiawan N and Wahyunggoro O 2016 *Communications in Science and Technology* **1** 51-55.

[16] Wu D 2010 *University of Southern California* 1-13.

Journal of Physics: Conference Series **1373** (2019) 012045 doi:10.1088/1742-6596/1373/1/012045

Indonesian Financial Crisis Prediction Using Combined Volatility and Markov Regime Switching Model Based on Indicators of Real Interest Rate on Deposits and Lending Interest Rate/Deposit Interest Rate

Sugiyanto[1,*], and Aisyah Zahrotul Hidayah [2]

[1] Department of Statistics, Faculty of Mathematics and Natural Science, Universitas Sebelas Maret, Jalan Ir. Sutami 36 Kentingan, Jebres, Surakarta, Central Java, Indonesia

[2] Department of Statistics, Faculty of Mathematics and Natural Science, Universitas Sebelas Maret, Jalan Ir. Sutami 36 Kentingan, Jebres, Surakarta, Central Java, Indonesia

Email: [1,*] sugiyanto61@staff.uns.ac.id

Abstract. The financial crisis that occurred in mid-1997 made it aware of the importance of building an early detection system because the crisis had a severe impact on the Indonesian economy. The crisis occurred due to several macroeconomic indicators experiencing very high fluctuations and the existence of changes in the structure (regime). The combined volatility and Markov regime switching models are very suitable to explain the crisis. Indicators of real interest rates on deposits and lending interest rate/deposit interest rates from 1990 to 2018 were used to build the combined model. The result showed that the MRS-ARCH (2,1) model for real interest rate on deposits indicator and the MRS-GARCH (3,1,1) model for lending interest rate/deposit interest rate indicator could explain the crisis. Markov regime switching predictions indicate that in 2019 there are no signs of a financial crisis in Indonesia.

1. Introduction

The financial crisis that occurred in mid-1997 began when the value of the Thai currency fell sharply. Disruptions to the financial system in economic rules can cause a currency crisis. Currency crises can be detected through banking performance. Seeing the effects of the financial crisis on the decline in economic activity made it aware of the importance of building an early detection system because the crisis had a severe impact on the Indonesian economy. Indicators of real deposit interest rates and the ratio of interest rates on loans and deposits represent performance in the banking sector that has fluctuations and can change conditions. Changes in fluctuations can be explained using the volatility model. While changing conditions can be explained using the Markov switching regime model.

The autoregressive conditional heteroscedasticity (ARCH) model was introduced by Engle [1], to model the variance of inflation data in the UK from 1958 to 1977 and obtained a more realistic prediction of variance. Bollerslev [2] introduced generalized autoregressive conditional heteroscedasticity (GARCH) to model US GNP data from 1948 to 1983.Hamilton [3] introduced the Markov switching (MS) model as a time series data model that contained changes in conditions.

Hamilton and Susmel [4] combined the MS and ARCH models to produce the Markov switching autoregressive conditional heteroscedasticity (SWARCH) model. They applied the SWARCH model on U.S gross national product (GNP) data from 1952 to 1984; the model can explain the state changes that occur in the U.S GNP data economic variables. Chang [5] used the SWARCH model to identify stock market and exchange rate volatility in Korea and the global financial crisis. Sugiyanto [6] through output real, domestic credit per gross domestic product (GDP), and ICI indicators detect the financial crisis using a combination of volatility and Markov switching models. Sugiyanto [7] also explained the financial crisis using the MS-GARCH model through banking indicators. This research discusses currency crises that might occur in Indonesia through real interest rate on deposits and lending interest rate/deposit interest rate indicators using a combination of Markov switching and volatility models.

2. Material and methods

2.1. Material

The model used for stationary time series data is the autoregressive AR (p) model which is written as follows

$$r_t = \phi_0 + \phi_1 r_{t-1} + \cdots + \phi_p r_{t-p} + \varepsilon_t, \tag{1}$$

where r_t is the log return at time t, ε_t is the residual model AR(p), and $\phi_0, \phi_1, \dots, \phi_p$ is the parameter model AR(p). The autoregressive conditional heteroscedasticity ARCH(m) model is an alternative to modeling non-constant variance that can be written as [1]

$$a_t = \sigma_t \epsilon_t, \tag{2}$$

where $\epsilon_t \sim N(0,1)$ and $a_t | \psi_{t-1} \sim N(0, \sigma_t^2)$,

$$\sigma_t^2 = \alpha_0 + \alpha_1 a_{t-1}^2 + \cdots + \alpha_m a_{t-m}^2 = \alpha_0 + \textstyle\sum_{i=0}^{m} \alpha_i a_{t-i}^2, \tag{3}$$

$\alpha_0 > 0$, $\alpha_i \geq 0$, for $i > 0$, $\sigma_t^2 = E(a_t^2 | \psi_{t-1})$ is a conditional variance residual at time t, and ψ_t is all information sets up to time t. The GARCH(m,s) model is a generalization of the ARCH model by involving past variances that can be written as [8]

$$\sigma_t^2 = \alpha_0 + \alpha_1 a_{t-1}^2 + \cdots + \alpha_s a_{t-s}^2 + \beta_1 \sigma_{t-1}^2 + \cdots + \beta_m \sigma_{t-m}^2 = \alpha_0 + \sum_{i=0}^{s} \alpha_i a_{t-i}^2 + \sum_{j=0}^{m} \beta_j \sigma_{t-j}^2. \tag{4}$$

Changing conditions can be seen using the Markov regime switching. The MRS-ARCH model can be written as [4]

$$r_t = \mu_{s_t} + a_t, \ a_t = \sigma_t \epsilon_t, \tag{5}$$

$$\sigma_{t,s_t}^2 = \alpha_{0,s_t} + \textstyle\sum_{i=1}^{m} \alpha_{i,s_t} a_{t-i}^2, \tag{6}$$

Equation (5) and (6) are MRS-ARCH with the k regime and order m. Furthermore, the MRS-GARCH can be written as [9]

$$\sigma_{t,s_t}^2 = \alpha_{0,s_t} + \textstyle\sum_{i=1}^{m} \alpha_{i,s_t} a_{t-i}^2 + \sum_{j=1}^{s} \beta_{j,s_t} \sigma_{t-j}^2, \tag{7}$$

For r_t is a vector of observed variable and s_t denote an unobserved random variable satisfy the first-order Markov chain that can take on the value 1,2,…, T. The variable $\mathrm{s_t}$ is regarded as the state or regime that the process is in date t and s_t governs that a first-order Markov chain governs parameters of the conditional distribution of the unobserved random variable with constant transition probability given by

$$\mathrm{P[s_t = j | s_{t-1} = i] = p_{ij}}, \textstyle\sum_{\mathrm{j=1}}^{\mathrm{T}} \mathrm{p_{ij}} = 1, \text{ for i, j} = 1,2, \dots, \mathrm{T}. \tag{8}$$

In matrix notation, P can be defined by

$$P = \begin{pmatrix} p_{11} & p_{21} & \cdots & p_{T1} \\ p_{12} & p_{22} & \cdots & p_{T2} \\ \vdots & \vdots & \ddots & \vdots \\ p_{1T} & p_{2T} & \cdots & p_{TT} \end{pmatrix}. \quad (9)$$

Based on Kim and Nelso [10], smoothed probability value $(\Pr(S_t = i|\psi_T)), t = 1,2, \ldots, T$, t = 1,2, ..., T can be written as follows,

$$\Pr(S_t = i|\psi_T) = \sum_{s=1}^{T} \Pr(S_{t+1} = s|\psi_T) \Pr(S_t = i|S_{t+1} = T, \psi_T), \quad (10)$$

Based on Sopipan et al [11], smoothed probability value at time $T + 1$ can be predicted with

$$\Pr(S_{t+1} = i|\psi_T) = p_{1i}\, P(S_t = 1|\psi_T) + p_{2i}\, P(S_t = 2|\psi_T) + \cdots + p_{ji}\, P(S_t = j|\psi_T). \quad (11)$$

Where p_{ji} shows the elements of the P_{jxj} transition matrix. Short-term crisis signals can be indicated by the number of predictive of smoothed probability values.

2.2. Method

Monthly data on real interest rate on deposits and lending interest rate/deposit interest rate from January 1990 to October 2018 were taken from International Financial Statistics (IFS). With data January 1990 to December 2017 as training data and data from January 2018 to October 2018 as testing data. The steps in this study are as follows.

1. Stationary test using ADF (augmented Dickey Fuller) test. If the data is not stationary then transformation data using log return.
2. Create an ARMA model and perform diagnostic tests on the best ARMA models.
3. Check volatility clustering using DTW (Dynamic Time Warping) distance and test heteroscedasticity ARMA residual uses the Lagrange multiplier test.
4. If there is heteroscedasticity on the residual ARMA model, then identify the volatility model. Next is a diagnostic test (autocorrelation test, heteroscedasticity and normality).
5. Perform a combination of volatility models and the Markov regime switching model.
6. Calculate the smoothed probability value from data real interest rate on deposits and lending interest rate/deposit interest rate.
7. If the smoothed probability value of of the real interest rate on deposits more than 0.89 then a financial crisis is predicted and more than 0.921 for lending interest rate/deposit interest rate.
8. Predict the financial crisis for the next year.

3. Result and discussions

Stationary data can be seen from the ADF test. Based on the ADF test, the probability value is more than 0.05, which means that the data is not stationary so the log return transformation is needed. Based on the ADF test the probability value is less than 0.05 which means that the data is stasionary. After the data is stationary, ACF and PACF tests are carried out to determine the ARMA model used. Based on the AIC value, the best ARMA(2,0) model for real interest rate on deposits and ARMA(1,0) model for lending interest rate/deposit interest rate. ARMA(2,0) model and ARMA(1,0) model can be written as follows

$$r_t = -0.42556 r_{t-1} - 0.20071 r_{t-2}$$
$$r_t = 0.49647 r_{t-1}$$

where r_t is the log return of real interest rate on deposits and lending interest rate/deposit interest rate at time t, and ε_t is the residue at time t

The residual ARMA model is leptokurtic, the leptokurtic curve indicates the existence of different range values, so clustering is necessary. The best clustering analysis for time series data is cluster analysis using DTW (Dynamic Time Warping) distance. Based on the result of clustering, the number of clusters that are suitable for real interest rate on deposits and lending interest rate/deposit interest rate are respectively 2 and 3.

Futhermore, heteroscedasticity test of the residual ARMA(2,0) model and ARMA(1,0) model. Based on Lagrange multiplier test, obtained a probability value less than 0.05. this means that the residual ARMA model contains heteroscedasticity. Volatility models are used to handle heterocedasticity. The volatility model for real interest rate on deposits is ARCH(1), which can be written as follows

$$\sigma_t^2 = 0.003404 + 2.812066 a_{t-1}^2$$

and volatility model for lending interest rate/deposit interest rate is GARCH(1,1), which can be written as follows

$$\sigma_t^2 = 0.00009936 + 0.2307 a_{t-1}^2 + 0.6676 \sigma_{t-1}^2.$$

where σ_t^2 is the variance at time *t*, and a_t is the residue at time *t*-1.
Next, a diagnostic test of the ARCH(1) model and GARCH(1,1) model. Based on Ljung-Box test, the probability value is more than 0.05, which means that the residual volatility model does not contain autocorrelation. From the Lagrange multiplier test, the probability value is more than 0.05, which means that the residual volatility model is homogeneous. The probability value of the Kolmogorov-Smirnov test is more than 0.05, which means that the residual volatility model is normally distributed.
Regime is a change in conditions that occur in the Markov regime switching. Regime Regime is assumed to follow the first order markov chain with the transition probability p_{ij} with $i, j = 1,2$ for indicator real interest rate on deposits and $i, j = 1,2,3$ for indicator lending interest rate/deposit interest rate. Thus, the Markov regime switching autoregressive conditional heteroscedasticity MRS-ARCH(2,1) for indicator real interest rate on deposits is an ARCH(1) model which has 2 regimes, with regime 1 having low volatility configuration and regime 2 having high volatility configuration. The MRS-GARCH(3,1,1) model for indicator lending interest rate/deposit interest rate is a GARCH(1,1) model that has 3 regimes, with 1,2,3 consecutive regimes describing low, similar and high volatility. A regime has the possibility to survive in the same regime or move to another regime the next time. The probability of a regime change can be seen in the form of a transition probability matrix. Transition probability matrix 2 regime for real interest rate on deposits is as follows.

$$P = \begin{pmatrix} 0.98252298 & 0.01747702 \\ 0.16218110 & 0.83781890 \end{pmatrix},$$

Based on the transition probability matrix, it can be seen the probability value for holding on to low volatility is 0.98252298 and the probability of surviving at high volatility is 0.83781890. Transition probability matrix 3 regime for lending interest rate/deposit interest rate as follows.

$$P = \begin{pmatrix} 0.92741666 & 0.05977186 & 0.01281148 \\ 0.06156000 & 0.91671889 & 0.02172111 \\ 0.05040072 & 006298390 & 0.88661539 \end{pmatrix},$$

To detect a crisis, it can be seen from the lowest smoothed probability value when the crisis occurred in Indonesia (1997-1998). So that the crisis results are obtained when smoothed probability values are more than 0.89 for real interest rates on deposits and more than 0.921 for lending interest rate/deposit interest rates. The plot of the smoothed probability form real interest rates on deposits and lending interest rate/deposit interest rate can be seen in Figure 1(a) and Figure 1(b) respectively.

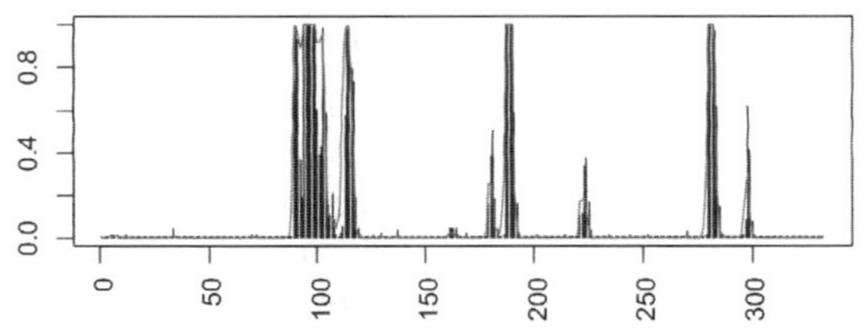

(a)

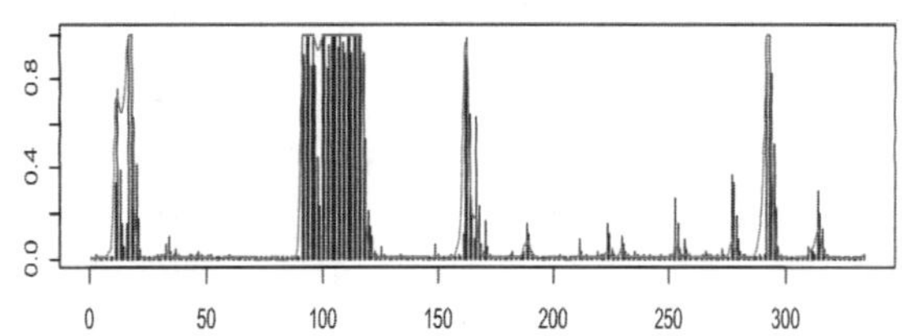

(b)

Figure 1. Smoothed probability of (a) real interest rates on deposits and (b) lending interest rate/deposit interest rate.

Figure 1(a) shows that there are 25 smoothed probability values more than 0.89 and Figure 1(b) there are 34 smoothed probability values more than 0.921. Real interest rate on deposits indicator detects crises in August 1997 - October 1998, August 1999 - October 1999, October 2005 - January 2006, July 2013 - September 2013. Whereas, for indicator lending interest rate/deposit interest rate can detect crises on June 1991 - August 1991, September 1997 - November 1999, August 2003 - September 2003, June 2014 - July 2014. Thus, it can be concluded that the MRS-ARCH(2,1) model for real interest rates on deposits and MRS-GARCH(3,1,1) model for lending interest rate/deposit interest rates can detect crises.

Based on smoothed probability values from the two indicators, the indicator of real interest rates on deposits can show signals faster than the lending interest rate/deposit interest rate indicator. This is because the actual crisis began in July 1997, as the real interest rate on deposits indicator signals a crisis starting August 1997 and the lending interest rate/deposit interest rate indicator signals a crisis starting in September 1997. Both indicators cannot capture the crisis that occurred in the month October 2008.

Furthermore, the predicted value of the smoothed probability is calculated. The calculation results predict the smoothed probability value of the next period for the real interest rate on deposits and the lending interest rate/deposit interest rate respectively shown in Table 1 and Table 2.

Table 1. Comparison of smoothed probability predictions and actual for real interest rate on deposits.

Period	Prediction of smoothed probability	Prediction crisis condition	Actual of smoothed probability	Actual crisis condition
January 2018	0.023078	stable	0.000701	stable
February 2018	0.036409	stable	0.000701	stable
March 2018	0.047345	stable	0.000553	stable
April 218	0.056316	stable	0.000554	stable
May 2018	0.063675	stable	0.000547	stable
June 2018	0.069712	stable	0.000720	stable
July 2018	0.074665	stable	0.000904	stable
August 2018	0.078728	stable	0.001699	stable
September 2018	0.082061	stable	0.002011	stable
October 2018	0.084795	stable	0.003247	stable

Table 2. Comparison of smoothed probability predictions and actual for lending interest rate/deposit interest rate.

Period	Prediction of smoothed probability	Prediction crisis condition	Actual of smoothed probability	Actual crisis condition
January 2018	0.034088	stable	0.001878	stable
February 2018	0.048773	stable	0.001040	stable
March 2018	0.061255	stable	0.000888	stable
April 218	0.071790	stable	0.001205	stable
May 2018	0.080669	stable	0.002025	stable
June 2018	0.088110	stable	0.004041	stable
July 2018	0.094321	stable	0.003915	stable
August 2018	0.099477	stable	0.003899	stable
September 2018	0.103729	stable	0.005469	stable
October 2018	0.107206	stable	0.015909	stable

Based on Table 1 and Table 2, it can be concluded that the conditions of the crisis actual and predictions are the same. This means that the model is very well used to predict the financial crisis for the next year.

Using monthly data from 1990 to 2018 can be calculated predictions of smoothed probability values for real interest rates on deposits and the lending interest rate/deposit interest rate shown in Table 3.

Table 3. Smoothed probability predictions for real interest rate on deposits and lending interest rate/deposit interest rate.

Period	Real interest rate on deposits	Prediction crisis condition	Lending interest rate/deposit interest rate	Prediction crisis condition
November 2018	0.019831	stable	0.026513	stable
December 2018	0.033450	stable	0.036354	stable
January 2019	0.044634	stable	0.045441	stable
February 2019	0.053817	stable	0.053797	stable
March 2019	0.061359	stable	0.061452	stable
April 2109	0.067551	stable	0.068441	stable
May 2019	0.072637	stable	0.074805	stable
June 2019	0.076812	stable	0.080582	stable
July 2019	0.080242	stable	0.085816	stable
August 2019	0.083056	stable	0.090547	stable
September 2019	0.085370	stable	0.094814	stable
October 2019	0.087269	stable	0.098657	stable
November 2019	0.088828	stable	0.102111	stable
December 2019	0.090108	stable	0.105212	stable

4. Conclusion

The best model for indicators of real interest rates on deposits and lending interest rate/deposit interest rates are MRS-ARCH (2,1) and MRS-GARCH (3,1,1), respectively. Based on these models the results of predictions for 2019 Indonesia have no financial crisis.

References

[1] Engle R F 1982 *Econometrics* **50** 987-1008
[2] Bollerslev T 1986 *Journal of Econometrics* **31** 307-327
[3] Hamilton J D 1989 *Econometrics* **57** 357-384
[4] Hamilton J D and Susmel R 1994 *J. of Econometrics* **64** 307-333
[5] Chang K, Cho K Y and Hong M 2010 *Journal. of Econonomic Research* **5** 249-272
[6] Sugiyanto, Etik Z and Meganisa S 2018 *Journal of Phys.: Conf. Series* **1025** 012115
[7] Sugiyanto, Etik Z and Shania P S 2018 *Journal of Phys.: Conf. Series* **1025** 012118
[8] Tsay R S 2005 *Analysis of Financial Time Series* (Canada: John Wiley and Sons) 114
[9] Gray S F 1996 *Journal of Finance Economics* **42** 27-62
[10] Kim C J and Nelson C R 1999 *State-Space Models with Regime Switching: Classical and Gibbs-Sampling Approaches with Application* (London: The MIT Press) 68
[11] Sopipan N, Sattayatham P and Premanode B 2012 *Journal of Mathematical Finance* **2** 121-131

Link quality decrement and spectrum crowded impacts to wireless local area network performances

S Suherman*, A H Rambe, N Mubarakah

Electrical Engineering Department, Universitas Sumatera Utara, Medan-Indonesia

Email: *suherman@usu.ac.id

Abstract. Wireless local area networks (WLANs) are now preferred for high speed internet connections. Compared to cable network, radio propagations are more prone to errors and interferences. The increasing internet users push more access points (APs) installations. Moreover, mobile devices tethering work at the same channels. Since the networks operate on the free license frequency bands, problems are emerged when many collocating APs work at the same area. As results, network performances may decrease. This article investigates how the radio link quality decrement as well as the crowded APs affect user performances. System modelling in Java and repeating simulation help the investigation. As results, it is proven that network performances, such as delay, throughput and packet collisions are affected by those conditions. Link quality decrement up to 1/10 causes packet transmission delay increasing 2.5 times which causes throughput drops about 30% in unsaturated condition. Meanwhile, crowded APs worsen network performances by leaving only 8% throughputs when collocating APs increase from 2 APs to 10 APs.

1. Introduction

Internet users rapidly grow anywhere in the world, where 802.11 technology is one of the popular end point access networks. This wireless local area network (WLAN) delivers internet data in a free license spectrum, where the connection depending on link quality [1], [2]. The quality varies overtime as multipath propagation, non-light of sight link and interferences exist in radio link. Link quality variations cause transmitting speed changes overtime. These changes are anticipated by using some methods such as adaptive modulations and coding. The higher number of bits per symbol results the higher bit rate and throughput [3].

However, available bandwidth decreases as number of installed access points grows. As the radio channels are limited, competitions among the APs lower network performances [4]. Various works have been carried out to overcome 802.11 dense network problems, such as by employing a non-orthogonal multiple access (NOMA) scheme [5], bandwidth partitioning [6], dynamic channel algorithm [7] and many more. Those changes may overcome the problems. Practically, users may consider access point number restriction working in one area. In order to do that, the new installation may be preceded by a performance study based on some performance parameters achieved at current number of access points [8]. Physically, link quality is measured by propagation attenuation, where the received signal level is expected to be higher than the threshold. The noises and interferences are measured in term of signal to noise ratio (S/N) or bit energy per noise (Eb/No). On medium access layer, channel quality is measured by determining the collision rate occurred which is referred to as the probability of collision. On the upper layers, parameters such as delay, throughput and packet losses are employed.

This work proposes crowded access point evaluation model based on carrier sensing multiple access/collision avoidance (CSMA/CA) algorithm by using some input parameters such as maximum and minimum bitrates, a fixed frame length, average transmission rate, and number of competing access points. The crowded access point means that number of access point is higher than the available channel. It is assumed those access point working on the same frequency, as if they work at different frequency likely will impose no major matter. The evaluated parameters are frame transmission delay, overall throughput and collision rates. The work could be used for determining maximum access point working in the same channels within the same area.

2. Research Method

In order to evaluate the performance 802.11 with co-located access points within the same frequency channel, the mathematical approximation proposed by Bianchi model is employed [9]. The collision probability is determined by the Equation 1 [9]. Since each access point is connected to some clients (i) with traffic $\emptyset_i$, the streamed traffic related to this access point is approximated by the total number of connected clients traffics ($\sum_{i=0}^{n} \emptyset_i$), where access point traffic is $\emptyset_0$. This parameter is simplified as the average transmission number, $n = i + 1$.

$$p_{collision} = 1 - (1 - \Phi)^{n-1} \quad (1)$$

The successful transmitting access point is determined randomly using Equation 1 and transmit frame size of m bytes. If the available bitrate is r, then the frame delay d, is determined by $d = {}^{r}/_{m}$. Meanwhile, the queue delay caused by the unsuccessful transmission is not counted. This is held as the counted throughput is measured at fixed time size of T. The unsuccessful transmission is counted towards the collision rate. The maximum bit rate is set 54 Mbps. The available bandwidth is calculated depending upon the signal to noise ration which is generated randomly and channel capacity which is assumed error free, is calculated by using Shannon formula [10].

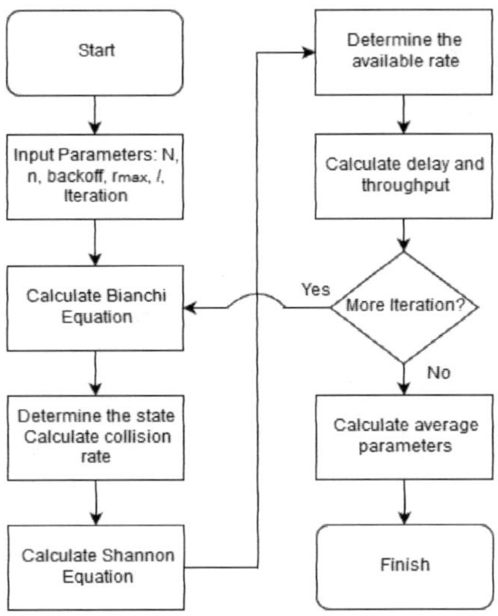

Figure 1. System model and flow

Figure 1 shows the step of model execution in order to determine the performance. Input parameters for the simulation is number of mobile transmissions, n, number of access points N, back off time, frame length and number of iterations. The system is then modelled by using Bianchi equation, collision rate is used as Shannon equation input to ensure available bandwidth, then performances were determined. The following parameters were set. The number of collocated access point N, was set from 2 to 10, the number of transmission n was set 100, maximum backoff was 1, maximum bit rate was 54 Mbps, the ethernet size l, was 2048 and simulation ran for 1000 times.

3. Results and Discussion

After running the simulation about a thousand times for many conditions and number of collocated access points, the recorded average frame delay for each channel degradation is shown in Figure 2. Channel degradation is represented in percentage which means the percent degradation to maximum bit rate of 54 Mbps. Figure 2 shows that the more significant the link quality degradation occurs, the higher the time required to send ethernet frames. On the other hand, the higher the number of collocated access points, the bigger the delay. For worst channel quality, delays vary for given number of access point. Generally, delay increases exponentially as the channel poorer.

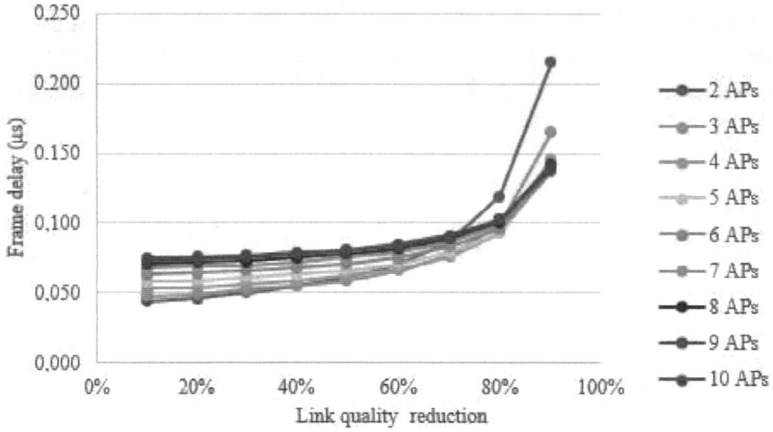

Figure 2. Frame delay performances

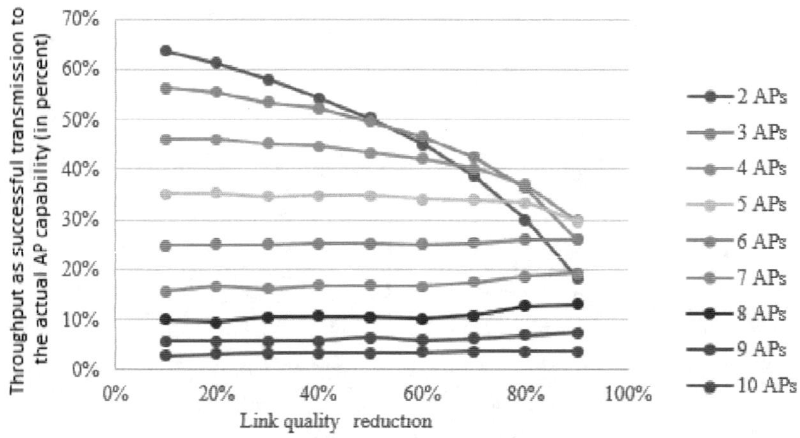

Figure 3. Throughput Performances

On the other hand, throughput decreases significantly when channel worsen for number of access point 2, 3 and 4. For higher number of wireless access point, the throughput is low and stagnant (Figure 3). For 10% of network quality is low, 28.84% of transmitted data received successfully. While decreasing link quality to 90% results throughput degradation to 28.84%. Meanwhile, impact number of access point to frame delay can be seen in Figure 4. For number of access point 2 and 3, delay varies.

But for number of access point 3 and higher, the ethernet frame delay consistently increases from 83μs to 89 μs.

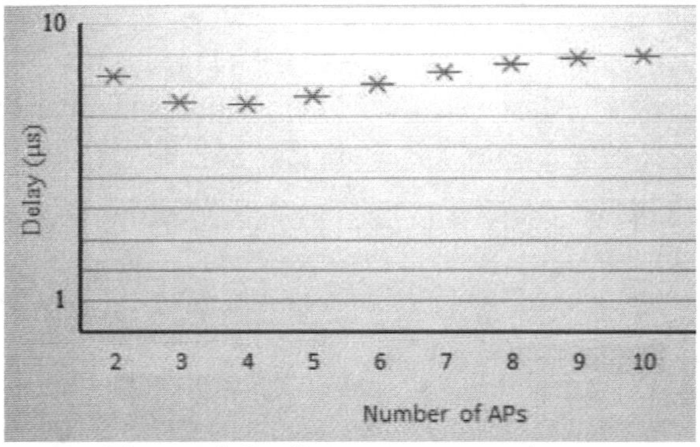

Figure 4. Performance comparisons

Throughput decreases significantly to number of transmitting access point. Throughput achieves 46.6% for number of access point 2, decreases sharply to only 3.34% (Figure 5). The throughput decrement is inversely proportional to the collision rate (Figure 6). Collision rate is about 6.76 % for two access point, decreasing tremendously to 94.25 % for 10 access points. Unfortunately, model validation by using real experiment is not finished yet. However, delay, throughput and collision characteristics are in line with the mathematic analysis presented by other researchers.

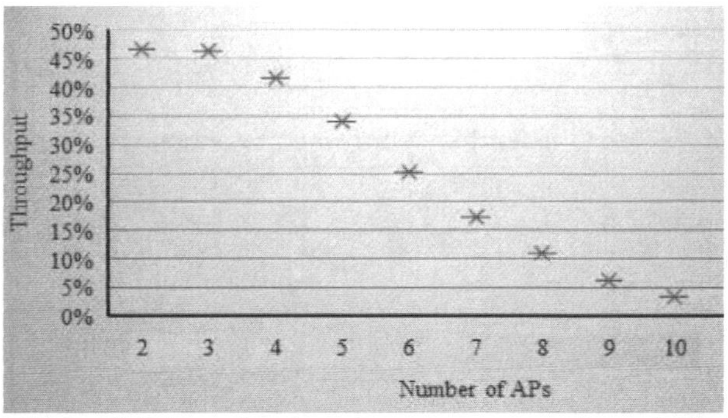

Figure 5. Throughput performance to number of APs

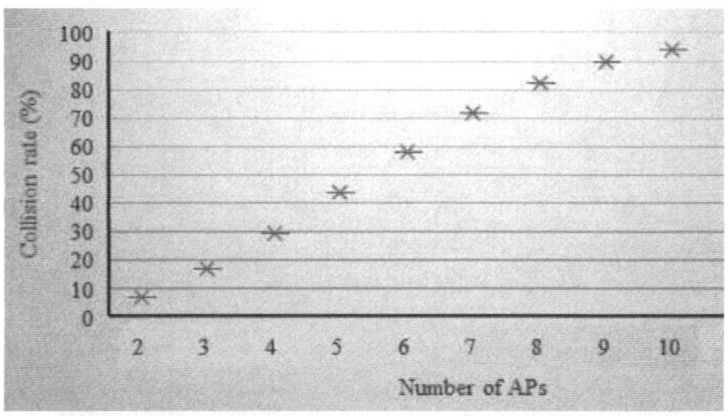

Figure 6. Collisions occurs to number of APs

4. Conclusions

This paper models link decrement and crowded access point impacts to 802.11 performance. System modelling was based on Bianchi and Shannon works with some adjustment for collocated access points. The collocated access points are assumed to work at the same frequency. Java based simulation shows that the worse the channel the higher the frame delay and the lower the system throughput. The 10% worse link leads to 2.5 times delay increment. Meanwhile, throughput dropped for more than 40% when link decreases up to 90%. The crowded APs make network performances worsen, achieving only 8% throughputs for 10 APs.

System model proposed in this paper is logically proportional to mathematic analysis. However, model should be validated to real experiment to make sure model be usable and that could be of future work, for instance, by developing MAC and PHY changeable access point. Access points that work at the same frequency as number of channels is lower than the number of activated access points should perform some techniques to avoid performance decrement, this is also of the future work.

Acknowledgement

This research has been funded by the Directorate of Research and Public Services (DRPM) Indonesia under the applied research schema, 2019.

References

[1] Soni G and Verma C 2017 *7th International Conference on Communication Systems and Network Technologies* (CSNT) 34–37.
[2] Napitupulu CJ, Sihombing P, and Suherman S 2018 *IOP Conf. Ser. Mater. Sci. Eng.* **420** 12128. 2018.
[3] Dong Q, Hayashi K, and Kaneko M 2017 *IEEE 18th International Workshop on Signal Processing Advances in Wireless Communications* (SPAWC) 1–5.
[4] Liu Y, Li X, Yu FR, Ji H, Zhang H, and Leung VCM 2017 *IEEE J. Sel. Areas Commun.* **35** 2295–2311.
[5] Ding Z, Liu Z, Choi J, Sun Q, Elkashlan M, Lin C and Poor HV 2017 *IEEE Commun. Mag.* **55** 185–191.
[6] Stefanatos S and Alexiou A 2014 *IEEE Trans. Commun.* **62** 3376–3384.
[7] Valkanis A, Iossifides A, and Chatzimisios P 2017 *Panhellenic Conference on Electronics and Telecommunications* (PACET) 1–4.
[8] Farsi A, Achir N and Boussetta K 2015 *Ann. Telecommun.* **70** 263.
[9] Bianchi G 2000 *IEEE J. Sel. Areas Commun.* **18** 535–547.
[10] Shannon CE 1948 *Bell Syst. Tech. J.* **27** 379–423.

Journal of Physics: Conference Series **1373** (2019) 012047 doi:10.1088/1742-6596/1373/1/012047

Minimizing Total Cost in Integrated Scheduling of Production and Distribution Using Ant Colony Optimization

M A Sukmarini[1,*], S Abusini[2], and C Karim[2]

[1]Master Program Mathematics, University of Brawijaya, Malang, Indonesia
[2]Department of Mathematics, University of Brawijaya, Malang, Indonesia

*Email: mitaakbar4545@gmail.com

Abstract. The integrated scheduling is a scheduling that applied in more than one problem. In this paper, we discuss about integrated scheduling of distribution and production problem in manufacturer. In the case of production, jobs are processed in several machines. In the case of distribution, one vehicle is used to distribute the product to the several customers. The objective of this paper is minimizing total cost in the production and distribution process. Since these problems are included in *NP-hard* problem, we suggest ant colony optimization method to solve both parts of the problem. We generate random data to verify the performance of the proposed method. Our result show that the performance of the method is excellent.

1. Introduction

Now days, the integrated scheduling is a topic that has been considered by researchers. This topic is more complicated than the common scheduling. The difference is the integrated scheduling can resolve more than one problem at once. Those problems can be a combination from production scheduling, maintenance, distribution, and other part of supply chain management problems. The purpose of integrated scheduling is overall optimization.

The integrated scheduling concerns on all activities of supply chain management. For example, two integrated issues discussed in [1] are machine scheduling and transportation considerations. Integrating preventive maintenance planning and production scheduling [2]. A novel genetic algorithm was proposed by [3] to solve transportation and production scheduling problem. Integrating of scheduling and packing [4]. Supply chain scheduling and coordination [5]. Batch delivery and production coordination [6]. Immune clonal selection algorithm was proposed by [7] to maximize the profit in integrated of production scheduling and maintenance planning. Three integrated issues, those are production, inventory and location-allocation decisions, discussed in [8].

In this paper, we discuss about the integrated scheduling of distribution and production issues at manufacturer. In the case of production, the job is process at several machines and the production scheduling type is job shop. Job shop scheduling is a scheduling of n jobs that is processed exactly once on m machines [9]. When the machine is started, the operation process cannot be interrupted and one machine can process one job only at the certain time. The studies that discussed about job shop, methods and techniques used for job shop scheduling [10], conventional and new solution techniques for solving job shop scheduling problem [11].

In the case of distribution, one vehicle is used to distribute the products to the customers. This problem is included on the travelling salesman problem. In a condition where a salesman goes from his own city, who wants to find the shortest path can be trough to visit all the customers in several different cities, then get back to his own city. Each of customers is visited exactly once [12]. Some of research that solved TSP are exact method for solving traveling salesman problem [13], using standard heuristics to solving traveling salesman problem of scheduling in AS/RS [14], and solving traveling salesman problem by evolutionary algorithm [15].

The objective of this paper is minimizing total cost in the distribution and production process. Since these problems are included in *NP-hard*, we suggest ant colony optimization method to solve both parts of the problem. Ant colony is used as a method to find the optimal distribution and production schedule. In the case of production, the purpose is minimizing the production cost. The minimum cost of production is obtained from the minimum number of total production time. In the case of distribution, the purpose is minimizing cost of distribution. The minimum cost of distribution is obtained from the minimum of total distance in distribution.

2. Problem characteristics and assumption

In this paper, the production type that will be optimized is job shop scheduling problem. According to [16], the characteristics of job shop scheduling problem are listed below.

1. Job shop scheduling has some job that has to be finished, represented by $J = \{J_1, J_2, \dots, J_n\}$.
2. Job shop scheduling has some machine that is used to finish every operation, represented by $M = \{M_1, M_2, \dots, M_m\}$.
3. Every job has several operations that has to be finished in the time start from ready time to due time. A job i has several operation which are represented by $O_i = \{O_{i1}, O_{i2}, \dots, O_{in}\}$.

Let a set job $J = \{Job_1, Job_2, \dots, Job_n\}$, K_i is a completion time of Job_i and total production time is symbolized by R, formulated as below.

$$R = \max\{K_1, K_2, K_3, \dots, K_n\} \tag{1}$$

In the distribution scheduling, one vehicle is used to distribute the products to the customers. This problem is included on the travelling salesman problem. Generally, the characteristic of TSP is listed below [17].

1. The tour begins and ends at the same point, that is manufacturer.
2. All the points except manufacturer, which means customers, are visited exactly once.
3. The salesman cannot be back to the manufacturer before all customers are visited.

If the customer set that must be visited is symbolized by $C_i = \{C_1, C_2, C_3, \dots, C_v\}$ and the distance from C_a to C_b is symbolized by d_{ab} is, so the total distance that can be reached at the distribution process is formulated as follows.

$$D = \sum_{C_i=1}^{C_i=v-1} d_{\pi C(i)\pi C(i+1)} + d_{\pi C(v)\pi C(1)} \tag{2}$$

R and D are the objective functions that will be optimized.

3. Ant colony optimization

We suggest ant colony optimization as a method to minimize total cost of the integrated distribution and production scheduling problem. In this paper, we use the algorithm adapted from [18].

1. Coding

   k number of ants is used to determine the optimal schedule. At every path (i,j) which selected by ant, there is additions of pheromones at that path. At the path that is not passed by ants, the pheromone levels are decreased, because there is a rate of pheromone evaporation. In this paper, the rate of pheromone evaporation is constant.

Table 1. Parameters in ant colony algorithm.

Parameters	Meaning
k	number of ants
τ_{ij}	pheromone intensity at path (i,j)
ρ	rate of pheromone evaporation
α	constant controller of the pheromone intensity
β	constant controller of visibility

In the production scheduling, $\eta_{ab} = \frac{1}{t_{ab}}$ is a value of visibility obtained from reciprocal processing time of a job a in the machine b, while in distribution scheduling, $\eta_{ab} = \frac{1}{d_{ab}}$ is the value of visibility obtained from reciprocal distance from the customer a to b.

2. Status transmission

   The rule of transmission status is showed by the following equation. This is a probability of the ant$-k$ that is lie at node i and choose to go to node j.

$$P_{ij}^{k}(t) = \begin{cases} \frac{[\tau_{ij}(t)]^{\alpha}.[\eta_{ij}]^{\beta}}{\sum_{i=1}^{n}[\tau_{ij}(t)]^{\alpha}.[\eta_{ij}(t)]^{\beta}} & ,if\ (i,j) \in S_k \\ 0 & ,if\ (i,j) \notin S_k. \end{cases} \quad (3)$$

In the production scheduling, S_k contains the operations list of every jobs must be completed at a machine. Every operation that has been done by a machine will not be included in tabu list. Meanwhile, in the distribution scheduling, S_k contains the customers list must be visited. Every customer that has been visited will be removed from the tabu list.

3. Pheromone update locally

   After every ant completes his tour, local update pheromone is carried out with this following formula.

$$\tau_{ij}(t+1) = \Delta\tau_{ij}(t) + (1-\rho).\tau_{ij}(t) \tag{4}$$

$$\Delta\tau_{ij}(t) = \begin{cases} \dfrac{[\tau_{ij}(t)]^{\alpha}.[\eta_{ij}]^{\beta}}{\sum_{i=1}^{n}[\tau_{ij}(t)]^{\alpha}.[\eta_{ij}(t)]^{\beta}} & , if\ (i,j)\ is\ the\ path\ where\ ants\ pass \\ 0 & , otherwise. \end{cases} \tag{5}$$

   The parameter of pheromone evaporation level is bounded in range $0 \leq \rho \leq 1$.

4. Pheromone update globally

   After several k ants complete their own tour, then the global pheromone update is carried out. The formulas of global update of pheromone for each problem are shown below.

$$\tau_{ij}(t+1) = \Delta\tau_{ij}(t) + (1-\rho).\tau_{ij}(t) \tag{6}$$

   In the production scheduling,

$$\Delta\tau_{ij}^{k}(t) = \begin{cases} \frac{Q}{R_k} & , if\ (i,j)\ is\ the\ path\ where\ ants\ pass \\ 0 & , otherwise. \end{cases} \tag{7}$$

   In the distribution scheduling,

$$\Delta\tau_{ij}^{k}(t) = \begin{cases} \frac{Q}{D_k} & , if\ (i,j)\ is\ the\ path\ where\ ants\ pass \\ 0 & , otherwise. \end{cases} \tag{8}$$

   Only the path included in the best tour will increase the level of pheromone.

4. Simulations

In the case of production, we use job shop scheduling problem with four jobs and four machines. In this problem, to test the performance of Ant Colony Optimization, we do the several changes of parameter value and take notes in every experiment.

The data of processing time has been generated randomly to represent job shop with four jobs and four machines. Data of processing time that we use in production part are shown in table below.

Table 2. Processing time three jobs in three machines.

	M1	**M2**	**M3**	**M4**
J1	37	58	56	40
J2	45	80	27	35
J3	64	29	46	75
J4	33	42	65	51

Journal of Physics: Conference Series **1373** (2019) 012047 doi:10.1088/1742-6596/1373/1/012047

Several experiments have been tried based on the data above. The output obtained from each experiments are shown below.

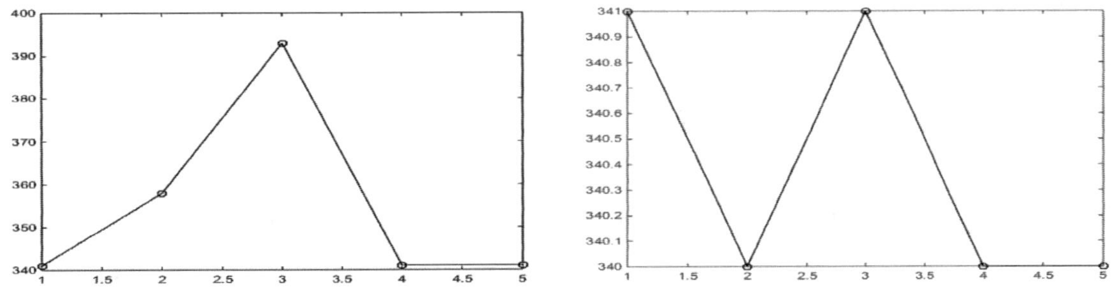

(a) (b)

Figure 1. Results of experiment with the value of parameters used are $k = 1$, $\rho = 0.1$, and the iteration is 5. (a) $\alpha = 1$ $\beta = 1$, the minimum processing time is 340 and order of job is J3-J4-J1-J2. (b) $\alpha = 0.3$, $\beta = 0.5$, the minimum processing time is 340 and order of job is J3-J4-J1-J2.

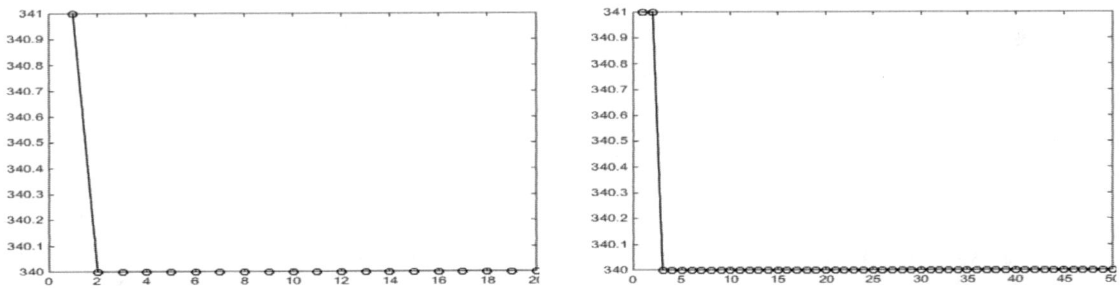

(a) (b)

Figure 2. Results of experiment with value of parameters used are $\rho = \beta = \alpha = 0.5, k = 3$. (a) use 20 iterations, the minimum processing time is 340 and order of job is J3-J4-J1-J2. (b) use 50 iterations, the minimum processing time is 340 and order of job is J3-J4-J1-J2.

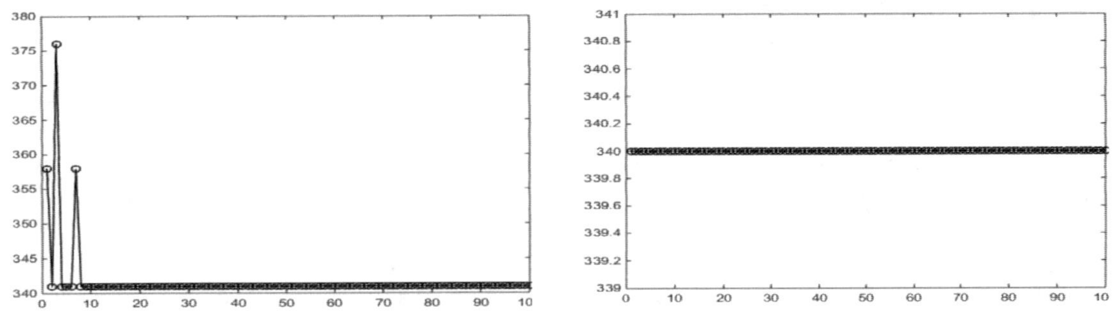

(a) (b)

Figure 3. Results of experiment with value of parameters used are $\rho = \beta = \alpha = 0.5$, and the iteration is 100 (a) the number of ant is 1, the minimum processing time is 340 and order of job is J3-J4-J1-J2. (b) the number of ant is 20, the minimum processing time is 340 and order of job is J3-J4-J1-J2.

Based on the experiments result, we obtain the minimum processing time is 340 and order of job is J3-J4-J1-J2. Let the constant cost of production process is c_1, so the minimum cost is $340(c_1)$.

We compare the total production time obtained by ant colony optimization with the manual calculation to validate the results. On job shop scheduling of four jobs on four machines, there are 24 possible order of job. All possibilities are listed below.

Table 3. 24 possible order of job

Order of job	Total processing time	Order of job	Total processing time
J1 - J2 - J3 - J4	376	J3 - J1 - J2 - J4	387
J1 - J2 - J4 - J3	408	J3 - J1 - J4 - J2	366
J1 - J3 - J2 - J4	384	J3 - J2 - J4 - J1	392
J1 - J3 - J4 - J2	358	J3 - J2 - J1 - J4	419
J1 - J4 - J3 - J2	377	J3 - J4 - J1 - J2	340
J1 - J4 - J2 - J3	377	J3 - J4 - J2 - J1	376
J2 - J1 - J3 - J4	411	J4 - J1 - J2 - J3	363
J2 - J1 - J4 - J3	430	J4 - J1 - J3 - J2	352
J2 - J3 - J4 - J1	366	J4 - J2 - J3 - J1	348
J2 - J3 - J1 - J4	384	J4 - J2 - J1 - J3	393
J2 - J4 - J1 - J3	409	J4 - J3 - J1 - J2	341
J2 - J4 - J3 - J1	398	J4 - J3 - J2 - J1	376

Total processing time from 24 order of job has been calculated with manual process. The minimum processing time from all possibilities is 340 with the optimal schedule is J3-J4-J2-J1. This comparison proves that the ant colony method works very well. Optimal result obtained from the heuristic approach is the same as exact solution. Even in the third experiment, by using more ants and iterations, the optimal solution is obtained since the first iteration. This method is really useful for solving bigger problems.

In the next discussion, distribution part, we use 40 data, one of them is represent the location of manufacturer and the others are represent the location of customers. The random data have been generated to represent the traveling salesman problem with 40 locations. Every location is represented by coordinate (x, y). All the value of x and y are shown below.

Journal of Physics: Conference Series **1373** (2019) 012047 doi:10.1088/1742-6596/1373/1/012047

TABLE 4. Location of manufacturer and customers in coordinate (x, y)

Coordinates of x	**Coordinates of y**	**Coordinates of x**	**Coordinates of y**
20	64	130	15
89	37	122	90
22	55	45	70
29	18	39	100
43	49	25	120
99	40	93	87
72	84	80	96
53	19	75	110
31	57	90	23
50	20	81	52
110	33	77	92
70	28	31	105
48	60	18	115
55	47	127	12
69	79	130	125
15	30	45	108
125	25	62	122
30	10	13	127
10	45	57	130
100	82	120	30

Several experiments have been tried bsed on the data above. With the parameters $\rho = 0.5, \alpha = 1, \beta = 1$. The output obtained from each experiments are shown below.

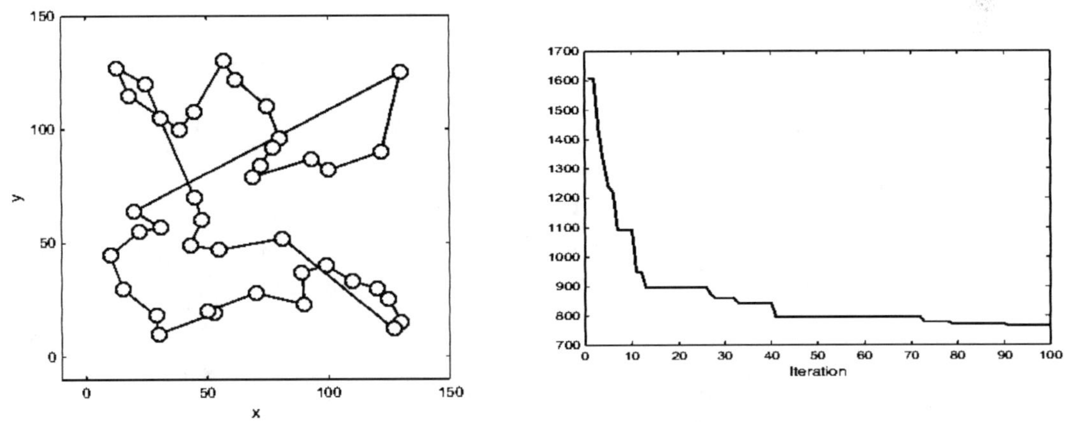

(a) (b)

Figure 4. The number of ants is 50 and iteration is 100, and the total distance obtained from this trial is 765.6428. (a) distribution route and (b) length of tour function.

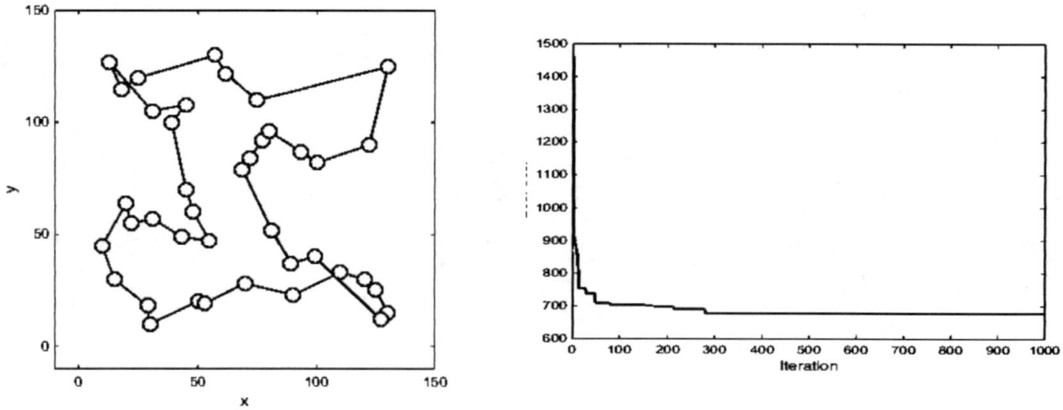

(a) (b)

Figure 5. The number of ant is 500 and iteration is 1000, and the total distance obtained from this trial is 677.8603. (a) distribution route and (b) length of tour function.

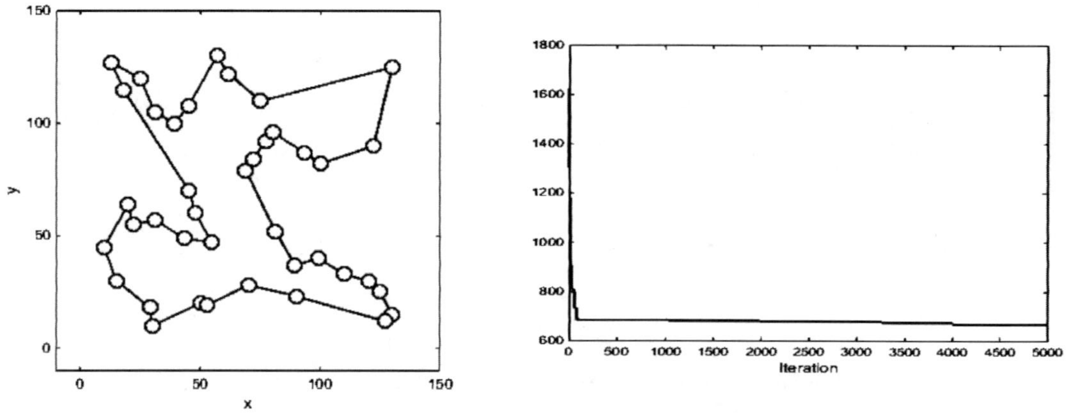

(a) (b)

Figure 6. The number of ant is 500 and iteration is 5000, and the total distance obtained from this trial is 669.0953. (a) distribution route and (b) length of tour function.

From the experiments, we know that the number of ants and the iterations used greatly affect the results. The higher number of ants and the iterations used, the better the results obtained. Based on the experiments, the minimum value of total distance of distribution route is 669.0953. Let the constant cost of distribution process is c_2, so the minimum cost is $669.0953(c_2)$.

5. Conclusion

Focus of this paper is an approach of ant colony algorithm for solving integrated scheduling problem. The problem included distribution and production scheduling. From the experiments carried out, ant colony optimization can solve the problem with great satisfaction. In the case of production, the result of the approach is the same as the exact solution, as well as in the case of distribution, ant colony optimization is able to find the optimal solution in a bigger problem.

However, the result of this study shown that solutions that obtained depend on the number of ant or iteration. Less number of ants used causes a random change of route which obtain a different output in each iterations. Meanwhile a less number of iterations allows a less variety of the obtained output, so that, it is possible that the obtained result is not optimum. While Ant Colony Optimization is applied with higher number of ant and iteration in both of problem, we obtain the great result. The experimental showed that the method we proposed is better in convergence speed and even the ability to get the best solution in a bigger problem.

References

[1] Lee C Y and Chen Z L 2001 *Journal of Scheduling* **4** 3–24
[2] Cassady C R and Kutanoglu E 2005 *IEEE Transactions on Reliability* **54** 304–9
[3] Zegordi S H, Abadi I N and Nia M A B 2007 *Computers & Industrial Engineering* **58** 373–381
[4] Chen Z L and Pundoor G 2009 *Production and Operations Management* **18** 672–692
[5] Yeung W K, Choi T M and Cheng T C E 2011 *International Journal of Production Economics* **132** 223–9
[6] Agnetis A, Aloulou M A, Fu L and Kovalyov M 2015 *European Journal of Operational Research* **241** 927–930
[7] Chen X, Xiao L, Zhang X, Xiao W and Li J 2015 *Maintenance and Reliability* **17** 70–79
[8] Alavi S, Azad N, Heydar M and Davoudpour H 2016 International Journal of Information Systems and Supply Chain Management **9** 22–42
[9] Ozguven C, Ozbakir L and Yavuz Y 2010 *Applied Mathematical Modelling* **34** 1539–48
[10] Yang Y B 1977 Methods and techniques used for job shop scheduling (University of Central Florida: STARS)
[11] Blazewicz J, Domschke W and Pesch E 1996 *European Journal of Operational Research* **93** 1–33
[12] Matai R, Singh S P and Mittal M L 2010 Traveling Salesman Problem, Theory and Applications
[13] Yatsenko V, Fast exact method for solving the travelling salesman problem Submitted to Journal of Automation and Information Science
[14] Jingfang F, Ying W, Chunli L and Rui-cheng F. 2011 Procedia Engineering **16** 601–7
[15] Cickova Z, Brezina I and Pekar J 2008 Management Information Systems **3** 17–22
[16] Zhang J, Hu X, Tan X, Zhong J H and Huang Q 2006 Transactions of the Institute of Measurement and Control **28** 93–108
[17] Dorigo M and Stutzle T 2004 Ant Colony Optimization (Cambridge: MIT Press) p 40–41, 70–73, 172–177
[18] Dorigo M, Birattari M and Stutzle T 2006 *IEEE Computational Intelligence Magazine* 28–39

Design and Development of Calculation System for Normalized Difference in Vegetation Index (NDVI) Using Landsat 8 Satellite Image

Sunardi[1,*], Abdul Fadlil[1] and Jamaludin Dwi Laspandi[2]

[1] Department of Electrical Engineering Universitas Ahmad Dahlan Yogyakarta, Indonesia

[2] Department of Informatics Technology Universitas Ahmad Dahlan Yogyakarta, Indonesian

*Email: sunardi@mti.uad.ac.id

Abstract. The satellite image is a Earth's surface image recorder that is performed without physical contact with the object. Satellite imagery can also be used as a solution to get the information of a wide area that is covered by vegetation. Normalized Difference Vegetation Index (NDVI) is an index depicting the level of a green plant with a combined tally of the Visible Red Band and the Near Infrared Band. Visible Red Band is the band of Landsat satellite image 8 whereas Near Infrared Band is a band 5 of Landsat satellite image 8. This research intends to build a system of counting of NDVI data by making use of satellite images Landsat 8 on the special region of Yogyakarta. The equation used in this study is the equation correction between Visible Red Band and Near Infrared Band from Landsat 8 satellite imagery and NDVI equation. Zoom functions and masking is also built into the system so it can be more focused on the area that wants to the canvas. The results of this research are in the form of value and a map of the NDVI that have been colors segmented in the Province of Yogyakarta.

1. Introduction

The rising temperature of the Earth's surface caused climate changing. Because of that technology that can quickly figure out and gives information regarding temperature changes are needed. Remote sensing is the science to manage and interpret an image that can be utilized in many different types of applications. Remote sensing has been recognized as a tool that is able to detect accurately land closure [1]. Remote sensing is carried out to obtain information by analyzing the data without direct interaction with the objects that will be examined [2].

The vegetation index is one of the parameters that are used to analyze the State of the vegetation of a region. Vegetation index has a range of variation of the algorithm. The decline and alternative Normalized Difference Vegetation Index (NDVI) has been proposed by various researchers to refine parameters, such as the Soil-Adjusted Vegetation Index (SAVI), the Perpendicular Vegetation Index (PVI), the Enhanced Vegetation Index (EVI), the Global Environment Monitoring Index (GEMI), and the Atmospherically Resistant Vegetation Index (ARVI). Each index is calculated by the correction factor one or some factor which becomes deficient NDVI [3]. The vegetation index is a method of transformation spectral data based images that may be used also as observation of forest and soil background effects purposes also in the analysis of vegetation [3].

Remote sensing satellite NASA spearheaded by the United States by launching the first satellite for natural resources on 23 July 1972 called the ERTS-1 (Earth Resources Technology Satellite). The second satellite was launched in 1975 named ERTS-2, the satellite carries sensors RBV (Return Beam Vidcon) and MSS (Multi-Spectral Scanner) that have a spatial resolution 80 x 80 m. After satellite launched, ERTS-1 and ERTS-2 were renamed Landsat 1 and Landsat 2 continues with the next Landsat series type 3, 4, 5, 6, 7 and later Landsat 8.

Landsat 8 is more commonly referred to as a satellite with the resume mission Landsat 7 rather than being referred to as a satellite with new specifications. Seen from characteristics Landsat 8 similar to Landsat 7, good the resolution (spatial, temporal, spectral), flying height as well as the characteristics of the sensor is brought. There are few additions that become a refinement point from Landsat 7 such as the number of more channels Landsat 8 channels than Landsat 7, the correction method, the lowest spectrum of electromagnetic waves that the sensor can capture and can see the values bit (Digital Number value ranges) of each image pixel.

Research in advance already doing research on NDVI [1][3][4][5]. But earlier studies only conduct analysis using software or existing application. In this case, the researchers will build and design a system for estimating vegetation coverage on the special region of Yogyakarta using calculation NDVI on Landsat satellite image 8. Additional features such as zoom and masking will be made on the application.

2. Methods

This research will be focused on the making of the application system of calculating the NDVI using satellite images Landsat 8 on the Yogyakarta with add mask and zoom within the application so it can ease the process of calculation.

2.1. Remote Sensing

There are a variety of definitions of remote sensing. The following definition of remote sensing, according to some experts:

- Remote sensing used to get information such as the object area by way of analyzing the data obtained using the tool without having to direct interaction on the object or area that wants to canvas [2].
- Remote sensing uses electromagnetic sensors to get images of the earth that analyzed so that useful information can be obtained [6].

Image is a continuous function of a two-dimensional image's intensity [7]. The image is obtained from the capture of the reflected signal object. The image is the output of recording devices such as cameras analog or digital. The analog or digital image is an image of a real object, while the image of the processed results of radiate heat that produces certain colors in accordance with the heat emitted is called image-based thermal [8].

The value of pixels from each gray image is at 0-255 intervals, while on the colored image there are three different values in the interval 0-255. A grayscale image is an image that only uses gray levels. Gray is the only color in RGB space with red, green and blue components which each have the same intensity value [9]. Image processing function to improve the quality of the images through some of the processes to be easy-to-read or presented [10].

The image is a picture or record object. Increasing the use of remote sensing data based on factors, such as the following [10]:

1. Get a complete overview and image information such as the object of the Earth.
2. Observations with the stereoscope can produce a three-dimensional image.
3. Produce a better and quicker image of remote area.

The characteristics of the object on the image recognition are as follows[11]:

1) Image color has a level of brightness and darkness of the object, a dark color concluded with low spectral reflection, while the high-spectral reflection symbolized by bright colors.
2) Shadows located in dark areas that hide object details.
3) The pattern is a characteristic of the mark form between human and nature.
4) The similarity of the objects to each other is called Association.
5) The form of the attribute that is clearly visible and easily recognizable.
6) The site, which is the location of the object one with other objects.
7) Size, in the form of height, breadth, distance, and volume.

The main function of image interpretation, namely tapping data taken from the image and used for certain purposes. Data capture from images presented in the form of graphs, tables, or thematic maps. The process of its performance begins by elaborate or separating different objects than for the same objects will be given withdrawal limit line. The recognized object type classified and illustrated with maps[12].

2.2. Satellite Landsat8

Landsat 8 OLI and TIRS are American earth observation satellites released on 11 February 2013. Landsat 8 is a development from Landsat 7. Initially, Landsat 8 was the Landsat Data Sustainability Mission (LDCM) which was a combination of NASA and the United States Geological Survey (USGS). Landsat 8 OLI and TIRS are equipped with 2 sensors namely sensors Operational Land Imager (OLI) and Thermal Infrared Sensor (TIRS). Landsat 8 provides global coverage of seasonal from the Mainland at any resolution. 30-meter resolution is Visible (Band 1, Band 2, Band 3, Band 4 and Band 9), NIR (Band 5) and SWIR (Band 6 and Band 7). Resolution of 100 meters, namely Thermal (Band 10 and 11 Bands), and a resolution of 15 meters is Panchromatic (Band 8). Landsat 8 can be seen in Figure 1[13].

Figure 1. Landsat 8

Table 1. Landsat7

Band	Wavelength (nm)	Name	Resolution (m)
1	0.450-0.515	Blue	30
2	0.525-0.605	Green	30
3	0.630-0.690	Red	30
4	0.775-0.900	Near-IR	30
5	1.550-1.750	SWIR-1	30
6	10.00-12.50	LWIR	60
7	2.090-2.350	SWIR-2	30
8	0.520-0.900	Pan	15

Landsat types are as follows:

1. Landsat 1 was released on 23 July 1972 used until 6 January 1978.
2. Landsat 2 was released on 22 January 1975 used until 22 January 1981.
3. Landsat 3 was released on 5 March 1978 used until 31 March 1983.
4. Landsat 4 was released on 16 July 1982 used until 1993.
5. Landsat 5 was released on 1 March 1984 are experiencing a major disruption in November 2011 and deactivated on 26 December 2012.
6. Landsat 6 was published on 5 October 1993 but the Landsat 6 failed to orbit.
7. Landsat 7 was released on 15 December 1999, suffered damage since May 2003.
8. Landsat 8 has an Onboard Operational Land Imager (OLI) sensor and a Thermal Infrared Sensor (TIRS) with 11 canals. OLI is on channels 1-9 while TIRS is on canals 10-11.

Types and comparison of the Canal and the Band on the Landsat 7 and Landsat 8 can be seen in Table 1 and Table 2.

Table 2. Landsat8

Band	Wavelength (nm)	Name	Resolution (m)
1	0.433-0.453	Coastal/Aerosal	30
2	0.450-0.15	Blue	30
3	0.525-0.600	Green	30
4	0.630-0.680	Red	30
5	0.845-0.885	Near-IR	30
6	1.560-1.660	SWIR-1	30
7	2.100-2.300	SWIR-2	30
8	0.500-0.680	Pan	15
9	1.360-1.390	Cirrus	30
10	10.30-11.30	LWIR-1	100
11	11.50-12.50	LWIR-2	100

Landsat 8 has a new specification in the Band 1, 9, 10 and 11 which served to capture the electromagnetic waves are lower compared to Landsat 7. The Band is superior in identifying characteristics of seawater at different depths. There is a one Thermal Infrared Band Satellite in the Landsat 7 Satellite, whereas in Landsat 8 has two Thermal Infrared Band. In addition to the difference in the amount of Thermal Infrared Band, a spatial resolution from both satellites was also different 60 meters on the Satellite Landsat 7 and 100 meters on the satellite Landsat 8. Satellite Landsat 8 there are two new Band is Coastal/Aerosol and Cirrus Band that doesn't exist on a Landsat 7 Satellite.

2.3. *Reflectance Correction*

The value of the pixels on the satellite image shows the exposure from the Earth's surface in the form of a Digital Number (DN) that is calibrated to a range of values. DN conversion into real exposure needs to be done for the comparative analysis of some of the images. Calculation Reflection Correction for satellite imagery can be seen in equation (1)[13].

$$\rho\lambda = \frac{M_\rho Q_{cal} + A_\rho}{sin\ (\theta_{SE})} \quad (1)$$

where:

$\rho\lambda$: TOA planetary reflectance

M_ρ: Band-specific multicative rescaling factor of metadata (REFLECTANCE_MULT_ x, where x is number band)

A_ρ: Band-specific additive rescaling factor of metadata (REFLECTANCE_ADD_BAND_x, where x is number band)

Q_{cal}: Quantized and calibrated standard product pixel values (DN)

θ_{SE}: Local sun elevation angle (SUN_ELEVATION)

Value Mρ and Aρ different in each Band of imagery acquired by each sensor. Satellite Landsat 8 has 2 sensors namely Sensors Operational Land Imager (OLI) and Thermal Infrared Sensors Sensor (TIRS). Value Mρ dan Aρ obtained from the file metadata on Landsat 8, whereas the value of the Mρ dan Aρ used in this study was obtained from the sensor OLI that is the Band 4 and the Band 5.

Q cal is the value of the DN of the band 4 and band 5 on satellite images Landsat 8. θ_SE the angle of the Sun is the value obtained from Landsat satellite imagery metadata file 8. There is only one value of the angle of the Sun on a satellite image.

2.4. Index Vegetation

The vegetation index is an index which describes the level of a greenish plant. Vegetation index usually uses a simple algebraic calculation. Vegetation indexes are designed to amplify the signal of the vegetation on the data obtained by remote sensing. Vegetation index providing the size estimates of the amount of vegetation that is green and healthy[16].

Vegetation index was used in research is the NDVI which can be seen in the equation (2). The green vegetation index is a healthy size.

$$NDVI = \frac{Band\ 5 - Band\ 4}{Band\ 5 + Band\ 4} \tag{2}$$

Where:

Band 4: Visible Red Band of image satellite Landsat 8
Band 5: Near Infrared Band 5 of image satellite Landsat 8

NDVI calculation requires two inputs i.e. the Near Infrared Band (Band 5 Landsat 8) and the Visible Red Band (Landsat 4 Band 8). The use of two inputs based on the theory that the plants are healthy to tend to give a lot of reflection on wave Near Infrared and a bit of reflection (the more absorbing). Table 2.3 is a table of the values of NDVI object against the Earth's surface[17].

Table 3. NDVI Values Against the Object Surface

Band	Wavelength (nm)
<0.1	Snow, Rocks, Badlands, Sand
0.2 – 0.5	Bushes, Grasslands, Crops Ripen.
0.6 – 0.9	Temperate Forests, Tropical Forests, Healthy Herbs

NDVI values less than 0.1 are included in the category of non-vegetation objects in the form of rocks, barren land, sand, and snow. NDVI values between 0.2 to 0.5 are included in the category of vegetation cover that is not so thick as bushes, grasslands, and aging plants. NDVI values between 0.6 to 0.9 are included in the category of tight vegetation covers such as temperate forests, tropical forests and healthy plants.

3. Proposed System

This section will discuss some of the processes that are needed to build an application system of counting of NDVI data on satellite images Landsat 8.

3.1. Data Source

Source data obtained from the official website and the source of the study of literature. In this study used satellite imagery data is a satellite image of Landsat 8 OLI and TIRS that are retrieved from the official website (https://earthexplorer.usgs.gov/). The USGS website look can be seen in Figure 2.

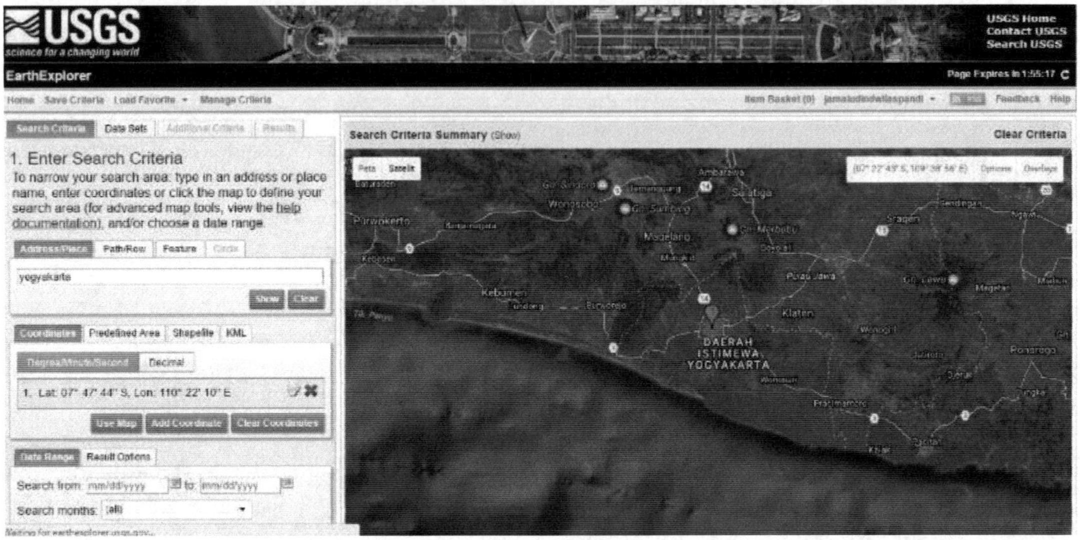

Figure 2. Search Image USGS

Fig 2 is the search function on the image data on province Yogyakarta. In the column address/place name input area is required, or by entering values for the Path/Row corresponds to the value of each region. The value Path/Row province D.I.Y is 120/65. Satellite image of the data used in this research was grouped into two with a different period of time. Satellite image data were taken during the period of 1 December 2011 to 1 December 2016 (Period A). Satellite image data within the period of 1 January 2012 to 31 December 2017 (Period B).

3.2. System Design Process

The general description of the system that describes the application flow from input to output can be seen in Figure 3.

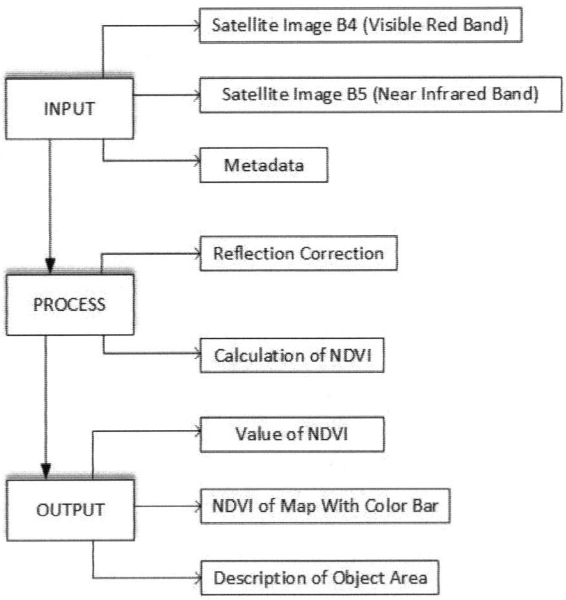

Figure 3. Application Work Process

Figure 3 is a process of the system to be built. The following of explanation application process from input to output:

1. The system that was built had 2 types of inputs, the image input, and file metadata input. The input image is a satellite image Landsat 8 Band 4 (Visible Red Band) and Band 5 (Near Infrared Band). The format of satellite images Landsat 8 is GeoTIFF. The input metadata file is input the number of variables. The value of the variable metadata that is used is the value of Multiplicative Rescaling Factor from Visible Red Band and Near Infrared Band, the value of Additive Rescaling Factor Visible Red Band and the Near Infrared Band as well as the value of the sun's elevation (height of the Sun). The metadata file is a text file that has an a.TXT format.
2. There are two processes in the system, namely the process of correction of reflectance and the process of counting of the NDVI. Reflectance correction process performed of input satellite images Landsat 8 band 4 and Band 5. NDVI calculation requires the input of Visible Red Band image and image of Near Infrared Band from Landsat 8.
3. The resulting output is the NDVI values and maps that segmented by color as well as a description of the object

3.3. Masking Image

Masking process aimed to separate the region that is going to be checked with areas that do not want to be checked. The process of masking is performed on the input image.

NDVI masking process conducted by using polygon on the image. Polygon acts as masking to cut the area that you want to use for research. Flowchart masking stages can be seen in Figure 4.

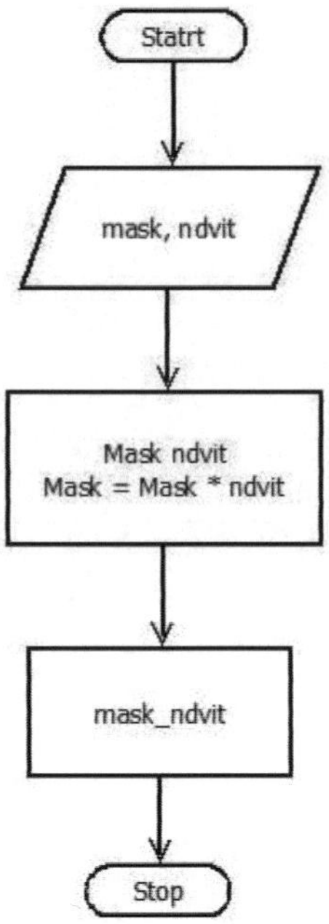

Figure 4. Flowchart Process Masking

The process of masking NDVI requires two input i.e. input image Visible Red Band (Band 4) and the Near Infrared Band (Band 5). Masking process performed on the NDVI image axes threshold in the GUI. Masking process function use implies (). The process of masking is done on a satellite image which had been modified. Image masking can be seen in Figure 5.

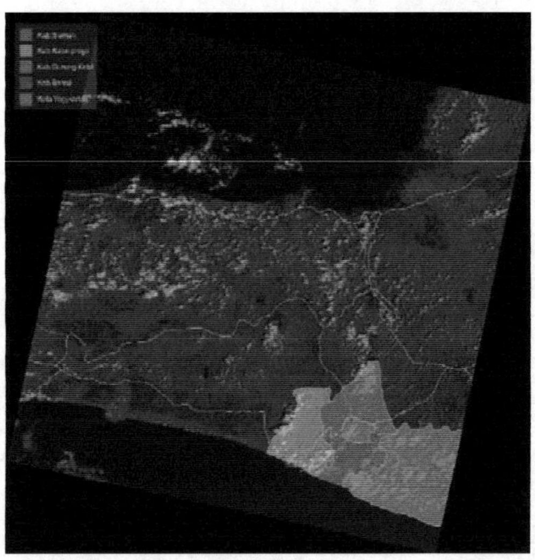

Figure 5. Image of Masking

Figure 5 is a satellite image which had been modified used in the process of masking. Image masking is given color areas and lines to the Yogyakarta which aims to ease the process of masking in accordance with the area that wants to the canvas. The colors in the image masking show each district in the Yogyakarta. There are 5 colors in image masking, color red indicates regions of Sleman, the yellow color indicates the area of Kulon Progo, purple color indicates counties Bantul, blue shows areas of the city of Yogyakarta and the green color indicates the area of Kulon Progo.

4. Result

4.1. Process Input

The application created is a GUI (Graphical User Interface). The application is made as simple as possible so that the user can operate the application easily.

The image is used in the process of counting of the GPP is a satellite image Landsat 8. Satellite images Landsat 8 TIFF format which has a size of pixel 7731X7591 and data type uint16. The result of the input image Visible Red Band and the Near Infrared Band can be seen in Figure 6.

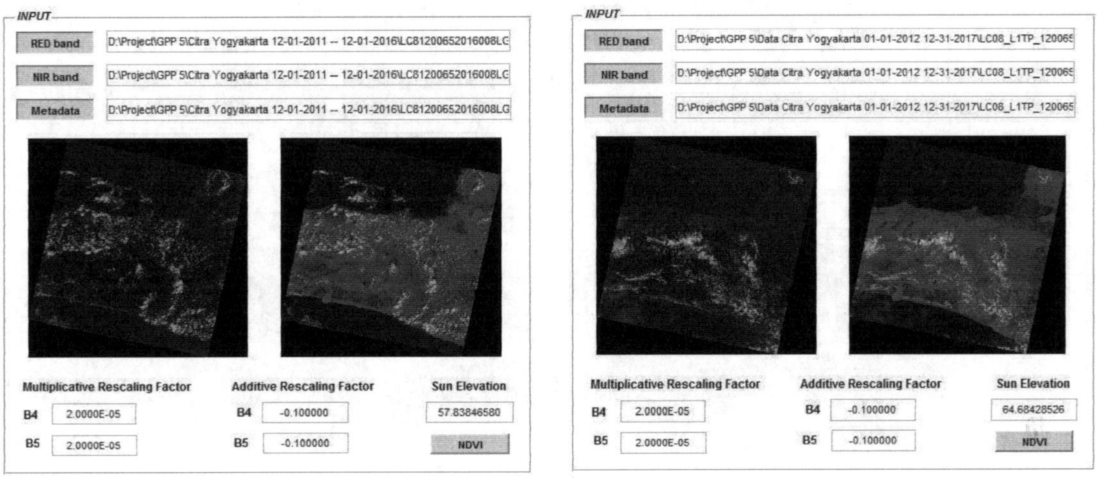

(a) (b)

Figure 6. (a) Input of Image (Period A), (b) Input of Image (Period B)

Figure 6 (a) is a satellite image input by the Date 1 December 2011 until 1 December 2016. Figure 6 (b) is the input of the satellite image of B with the date of the 1 January 2012 until 31 December 2017. The image displayed on a panel namely satellite imagery that is already subset 512x512 pixel. The left image is the input of the Visible Red Band. The left image is the result of input from the Visible Red Band while the right Image is the result of input from the image the Near Infrared Band. Multiplicative Rescaling Factor value for the image of the Visible Red Band is 2.0000-05 and image the Near Infrared Band has Multiplicative Rescaling Factor is 2.0000-05. Sun Elevation function to display the value of the angle of the Sun. The value of the variable the height of the Sun in Figure 6 (a) that is 57.83846580 whereas in Figure 6 (b) that is 64.68428526. The function used to display the variable value from the metadata file input is"set()".

The use of sections to divide the text of the paper is optional and left as a decision for the author. Where the author wishes to divide the paper into sections the formatting shown in table 2 should be used.

4.2. Process Masking

The process of masking is done manually by the user. Masking is performed on the input image and results masking in the form of image NDVI. Masking function to determine the area that wants. The display masking can be seen in Figure 7.

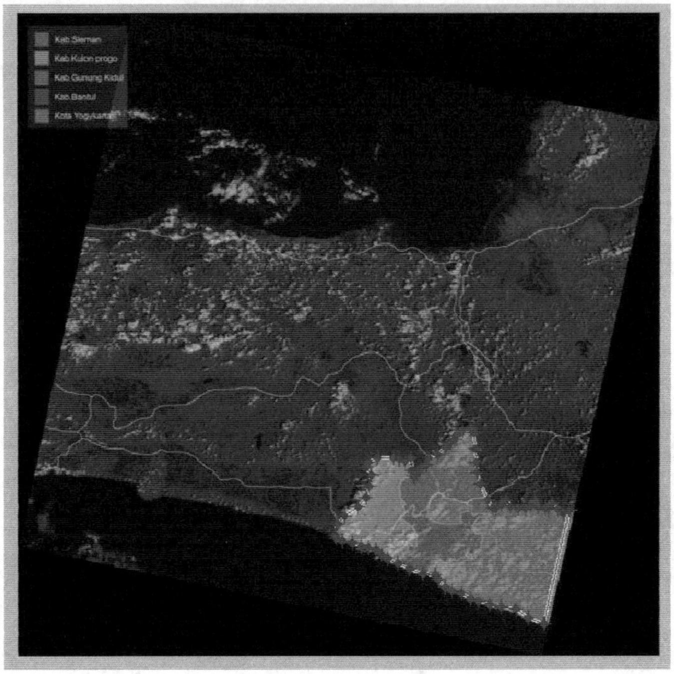

Figure 7. Process Masking

Figure 8 is the process of masking on the input image. Masking is performed on the Yogyakarta (the research areas) are marked with a red circle in Figure 8. The process of masking is only performed on an image that has been modified (Figure 5) to avoid the occurrence of errors or differences in the pixel of the image. The image of the Visible Red Band and the Near Infrared Band will be truncated automatically according to form that has been cut off earlier on image masking process. The results of the process of masking can be seen in Figure 8.

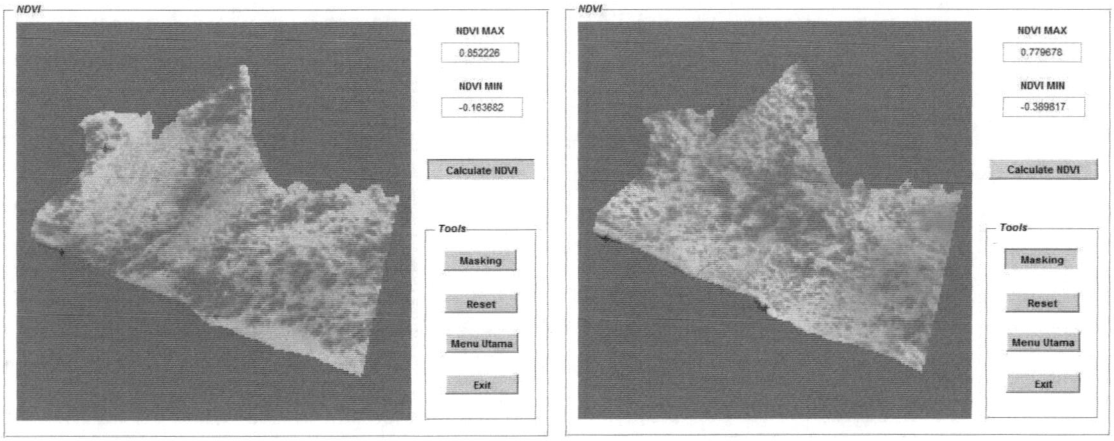

(a) (b)

Figure 8. (a) Result of Masking Process (Period A), (b) Result of Masking Process (Period B)

Figure 8 (a) is a result of the masking process on first research (period A). The results of the masking on the B can be seen in Figure 8 (b). Figure 8 (b) is the result of the process of masking on the period B.

4.3. Calculation of NDVI Output

Code program from equation 1 can be seen in Figure 9. Figure 9 is a function of the correction process reflection satellite image of the Visible Red Band and the Near Infrared Band. The variable "RED_ref" is the variable to store the results of the final calculation of the correction of reflection of the Visible Red Band. The value of the Digital image of the Visible Red Band Number multiplied by the variable "nmb4" which is the value of a Multiplicative Rescaling Factor to the Visible Red Band. then the multiplication results in total with the variable "nab4" which is the value of the Additive Rescaling Factor for the Visible Red Band and the sum divided by the results of sin from the variable "SUN_ELEVATION".

```
RED_ref =((nmb4*REDt)+(nab4))/sind(SUN_ELEVATION);
assignin('base','RED_ref',RED_ref)
NIR_ref =((nmb5*NIRt)+(nab5))/sind(SUN_ELEVATION);
assignin('base','NIR_ref',NIR_ref)
```

Figure 9. Code Program of Correction Reflektan in Matlab

The variable "NIR_ref" i.e. a variable to store the value end of the correction process reflection on the ideals of the Near Infrared Band. The variable "NIRt" Digital value is Number of input satellite imagery of Near Infrared Band. The value of Digital Number from Near Infrared Bands multiplied by the variable "nmb5" which is the value of the Multiplicative Rescaling Factor for the Near Infrared Band. The results of the multiplication combined with the variable "nab5" which is the value of the Additive Rescaling Factor for the Near Infrared Band and the result of the sum divided by the sin of the angle of the Sun value "SUN_ELEVATION".

The input image has been corrected using reflection then processed into the counting process of the NDVI vegetation cover to get value as in formula 2. Program code for the counting of the NDVI can be seen in Figure 10.

```
ndvi = (NIR_ref-RED_ref)./(NIR_ref+RED_ref);
```

Figure 10. Code Program of NDVI Calculation in Matlab

The variable "ndvi" is the variable to hold the value NDVI calculation result image. The variable "NIR_ref" is the value of the matrix image of Near Infrared Band that has corrected the previous process on reflection. Variable "RED_ref" is the value of the matrix image of the Visible Red Band that has corrected the previous process on reflection. Next image of NDVI will be shown on the GUI with the Code in Figure 11.

```
axes(handles.ndvi);
imshow(ndvi,[]);
```

Figure 11. NDVI of Print Image

Figure 11 function to display images NDVI into GUI. Axes used to accommodate the image you want to display. Images NDVI displayed using function handles so that the image can be called from the variable "ndvi" into axes on the GUI. The name of the axes to accommodate the image of NDVI which is "ndv1". Imshow () function to display the image of the NDVI.

The result of the calculation NDVI on the Yogyakarta can be seen in Figure 12.

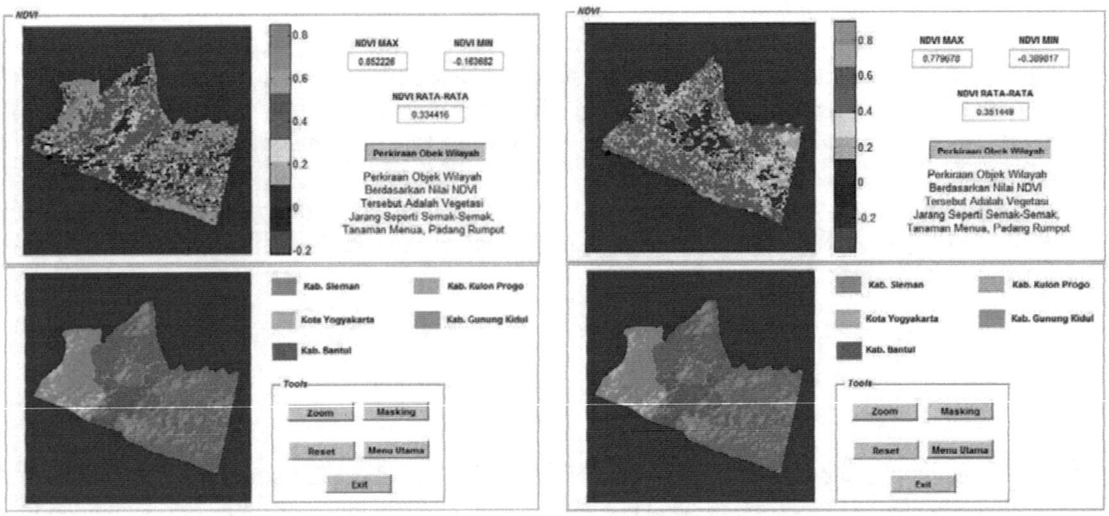

(a) (b)

Figure 12. (a) Calculation of Result NDVI (Period A), (b) Calculation of Result NDVI (Period B)

Figure 12 (a) shows the result of the NDVI calculating process from Period A (1 December 2011 until 1 December 2016). The maximum NDVI is 0.852226 values at period A is at coordinates 293/373 located in Kulon Progo Regency. The value of NDVI minimum is -0.163682 at coordinates 271/425 located in Kulon Progo Regency. The average value of NDVI is 0.334416 which included in the vegetation cover that is not so thick such as shrubs, pasture and crops ripen. Value counting of NDVI in the second study (period B) can be seen in Figure 15.

Figure 12 (b) is the result of a process of counting NDVI in period B (1 January 2012 until 31 December 2017). The maximum NDVI value at period B is 0.779678 at coordinates 343/454 in Gunung Kidul Regency. The value of NDVI minimum is -0.389817 there are at coordinates 262/420 located in Kulon Progo Regency. The average value of NDVI is 0.351449 which include the category vegetation cover that is not so thick such as shrubs, pasture and crops ripen.

5. Conclusion

Image processing of satellite imagery, correction reflectance and counting of the NDVI was successfully implemented in the application. Average NDVI values in the Yogyakarta with a period of 1 December 2011 until 1 December 2016 (Period A) is range 0.334416 and the average value of NDVI with a period of 1 January 2012 to 31 December 2017 (Period B) is 0.351449. The value of NDVI with period B is better than period A, with this it can be concluded that the NDVI value is getting higher from year to year.

An experiment using two satellite imagery with period A and period B show that there are great differences in the maximum NDVI value. The maximum NDVI values in the span of period A is 0.852226 located at Kulon Progo Regency while the maximum NDVI value at period B is located at Gunung Kidul Regency with 0.779678 NDVI values. The maximum value on the period A and B period are not so different. However, the biggest difference can be seen in its territory. The maximum NDVI value period A lies at Kulon Progo Regency while the maximum NDVI value at period B is located at Gunung Kidul Regency. It is hoped the research could help in the search for information on vegetation cover.

References

[1] Amliana D.R, Prasetyo Y, and Sukmono A 2016 *Jurnal Geodesi UNDIP* **5** 264–274
[2] Lillesand T and Kiefer R W 2015 *Remote Sensing and Image Interpretation* (USA: Wiley).
[3] Purwanto A 2018 *Proceeding of Ninth International Conference Future Energy System* (Germany) (USA: ACM DL) **18** 319–323.
[4] Dan R. V. Suyadi and Ulumudin Y. I 2013 *Ber. Biol* **12** 249–258.
[5] Arnanto A 2013 *Geomedia* **11** 155–170.
[6] Curran P. J 1985 *International Journal Remote Sensing* **6** 1765.
[7] Saifullah S, Sunardi, and Yudhana A 2017 *Jurnal Nasional Teknik Elektro* **6** 66-75.
[8] Saifullah S, Sunardi, and Yudhana A 2016 *Jurnal Teknik Informatika. dan Sistem Informasi* **2** 341–350.
[9] A. Yudhana, Sunardi, and S. Saifullah 2016 *Ilkom Jurnal Ilmiah* **8** 190–196.
[10] Anggraeni N. T and Fadlil A 2013 *Journal Sarjana Teknik Informatika* **1** 409–418.
[11] Campbell J B 2011 *Introduction to Remote Sensing* (New York: Guilford Press).
[12] Wulder M A et al 2019 *Remote Sensing of Environment* **225** 127-147
[13] United States Geological Survey 2018 Landsat Missions accessed online at https://www.usgs.gov/land-resources/nli/landsat.

Journal of Physics: Conference Series **1373** (2019) 012049 doi:10.1088/1742-6596/1373/1/012049

Identification of Pekalongan Batik Images Using Backpropagation Method

Rizky Andhika Surya [1], Abdul Fadlil [2], and Anton Yudhana [2]

[1] Department of Electrical Engineering, Universitas Ahmad Dahlan
[2]Magister of Informatics Technology, Universitas Ahmad Dahlan

*Email: fadlil@mti.uad.ac.id, eyudhana@ee.uad.ac.id, r.andhika1339@gmail.com

Abstract. Batik is an Indonesian cultural heritage that was recognized by UNESCO in 2009. One of the famous batik producing regions in Indonesia is Pekalongan City, Central Java. Pekalongan Batik has a distinctive characteristic compared to other regions, namely from the aspect of color and motif. Pekalongan batik uses bright colors and flora motifs. However, there are some batik craftsmen who still use dark colors. Identify Pekalongan's typical batik motifs become an obstacle for tourists that are visiting Indonesia. There needs to be an automatic identification system to recognize Pekalongan Batik motifs. The automatic identification system of batik images can contribute to the development of technology in the field of artificial intelligence. This research was conducted by collecting image data taken through observation, interviews and literature review. The image data obtained are four batik motifs. 5 images will be taken from each motif and will be implemented into the system using Matlab R2014a. The next process is feature extraction using the Gray Level Co-Occurrence Matrix (GLCM) method to get information in the batik image. The author then identifies image using the Backpropagation method to obtain the epoch value, learning rate, and accuracy value based on the identification process. The tested system obtained the highest accuracy of 91.2% with epoch 100 and learning rate 0.03 on Sogan batik, 89.6% accuracy with epoch 100 and learning rate 0.02 on Jlamprang batik. 87.2% accuracy was obtained from Cap Kombinasi batik and Tiga Negeri batik with epoch 100, the learning rate was 0.01 and 0.04 respectively. The results of identifying images of Pekalongan batik can be implemented in a more interactive application

1. Intoduction

Batik is a typical Indonesian clothing that has high value. Batik is spread in almost all parts of Indonesia, each region has different characteristics from each other, especially in their style and motives. [1]

At first, batik art only existed in the palace environment. This is one way to show the existence of aristocrats in the artwork produced [2]. But over time now, the art of batik has spread widely in the wider community, even the profession as a batik craftsman has become the livelihood of most people, especially women both at home and in large factories.

The term batik comes from the Javanese word, "amba", which means writing and "nitik" which means giving a point [3]. Batik formula refers to the technique of using beautiful or stamped materials and then the process of dyeing fabrics against fabric dyes. The process of describing batik motifs uses "Malam" or candles painted or superimposed on cloth. The wax is used to cover fabric fibers so that the dye of the fabric during the coloring process will not enter the pores of the fabric and produce batik motifs. Batik cloth not only stores the aesthetic value of decorations and color combinations, but more than that. Batik motifs save symbols and philosophy depending on the origin of batik cloth and based on the diversity of the community. Batik in its simplest sense is patterned images, motifs and patterns made specifically using dyeing hat techniques by writing, stamping or tasting on cloth (mori, cotton, teteron, or silk). [2]

People still don't know much about the developments and batik motifs between Indonesia and other countries because there is no computerized data collection and there is no application to analyze batik especially Pekalongan traditional batik as shown in Figure 1. A very reliable pattern recognition method is applied for the introduction of batik motifs by applying texture feature extraction features of Pekalongan batik images. This study will discuss how to design and build an image feature extraction system using the GLCM (Gray Level Co-Occurrence Matrix) method so that it can be used for the batik classification process for the data collection process. Samples of Pekalongan batik will be tested so that information about batik is known

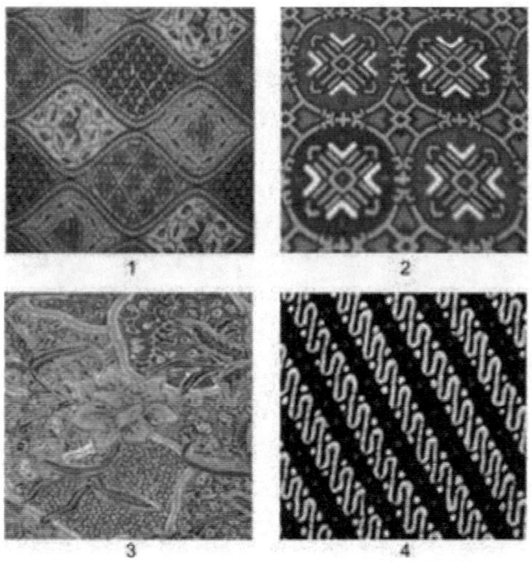

Figure 1. Example of Pekalongan Batik pattern used in this study

Figure 1 is an example of the image of Pekalongan batik that will be the object of research. There are four Pekalongan batik motifs, the first are the Sogan batik motif, the second Jlamprang motifs, the third three-country batik motifs and the last are the combination batik motif. Each number of batik motifs will be taken for the object of system training and also as a media for testing the identification system of Pekalongan batik to determine the feasibility of identification systems.

2. Theory

2.1. Image

Image is a picture, resemblance, or imitation of an object. Images can be in the form of photos, video signals (images on TV) or are digital in nature that can be directly stored in storage media. Images that can be processed by a computer are called digital images. Each image can be obtained by image

Journal of Physics: Conference Series **1373** (2019) 012049 doi:10.1088/1742-6596/1373/1/012049

acquisition, namely the process carried out to obtain digital images to imaging. Imaging is a way to change the visible image (photos, images, or sculptures) [4].

2.2. Batik

October 2 is a national batik day. UNESCO has confirmed batik as a masterpiece of oral humanity heritage and not objects to Indonesia [5]. One of the cities of batik craftsmen is Pekalongan City. Pekalongan City has a creative industry that has been able to promote the image of its city, the creative industry is the batik industry, through batik crafts [6]. Initially, batik had a limited variety of motifs and colors, and some motifs could only be worn by certain groups. [7]. But with the development of time makes each region has a diverse development.

2.3. Image Processing

Image Processing is a field of study about how an image is formed, processed, and analyzed so as to produce information that can be understood by humans. Image processing aims to:

1. improve image quality, seen from the radiometric aspect and geometric aspects. Radiometric aspects consist of contrast enhancement, image restoration, color transformation while geometric aspects consist of rotation, scale, translation, geometric transformation
2. carry out the process of withdrawing information or description of objects or the introduction of objects contained in the image.
3. To select the optimal feature image for analysis purposes
4. do data compression or reduction for the purpose of data storage, data transmission, and data processing time.

2.4. GLCM (Grey Level Co-Occurrence Matrix)

GLCM is a feature extraction method that uses texture calculations in the second order that takes into account two pixel pairs of original images, while in the first order uses statistical calculations based on the pixel value of the original image and does not pay attention to pixel density. Co-Occurrence can be interpreted as a joint event, meaning the number of events at one pixel level that is adjacent to the other pixel values based on distance (d) and orientation of an angle (Θ). Distances are represented as pixels while the orientation is represented in degrees. The orientation is formed from four angles with 45o intervals, namely 0°, 45°, 90°,and 135°, and the distance between pixels is determined by 1 pixel. These four directions are as in Figure 2.

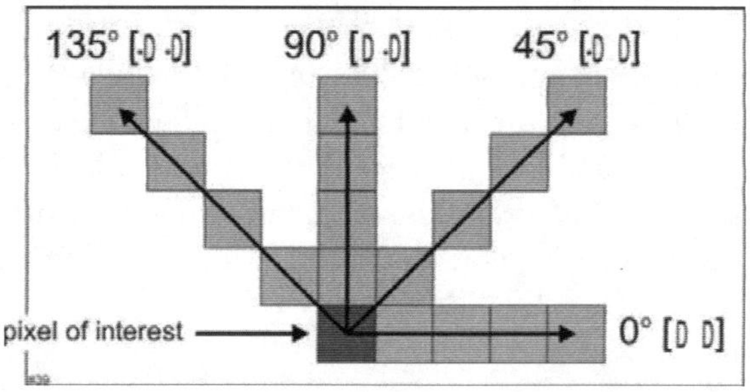

Figure 2. Four-way degree of grayness

Figure 2 describes Orientation in four angles with 45° angles, namely 0°, 45°, 90°,and 135° while the distance between pixels is usually set at 1 pixel.

Introduction Patterns can be said as the ability of humans to recognize objects based on the characteristics and knowledge they have observed from these objects. The purpose of the introduction of this pattern is to classify and describe the pattern through the knowledge of the properties or characteristics of the object. Can the pattern be said as a defined identity and can be given an identification or name. Image processing represents existing techniques for manipulating and modifying images. The procedure of image processing reflects the most important part in the construction of an identification system [8].

2.5. Artificial Neural Network (ANN)
Artificial Neural Network (ANN) is an information processing system that has characteristics resembling biological neural networks (JSB) Artificial Neural Networks are created as a generalization of mathematical models of human understanding [9]. Like the human brain, neural networks also consist of several neurons, and there is a connection between these neurons. Neurons will transform the information received through the output connections to other neurons. In neural networks this relationship is known as the weight name. This information is stored at a certain value on that weight.

In neural networks, neurons will be collected in layers called neuron layers. Usually neurons in one layer will be connected to the layer before or after except the input layer and output layer. Information provided on a neural network will be propagated from layer to layer, through from the input layer to the output layer through a hidden layer.

3. Methodology
The identification of batik images is designed as depicted in Figure 3

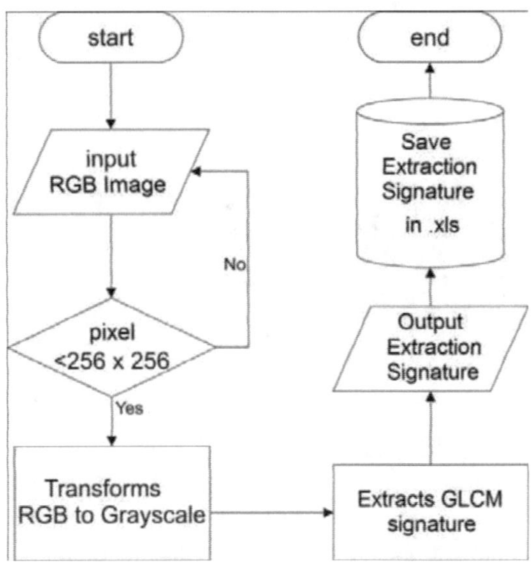

Figure 3. Flowchart system identification training

Taking batik images in *JPEG or *PNG format is 256 x 256 pixels, then preprocessing by changing into gray scale image. Feature extraction is done to find information contained in batik images regarding the value of the GLCM parameter.

After the GLM parameter values appear, Classification is processed using the Backpropagation method which will conclude the name of the batik and the accuracy of the test. Figure 4.

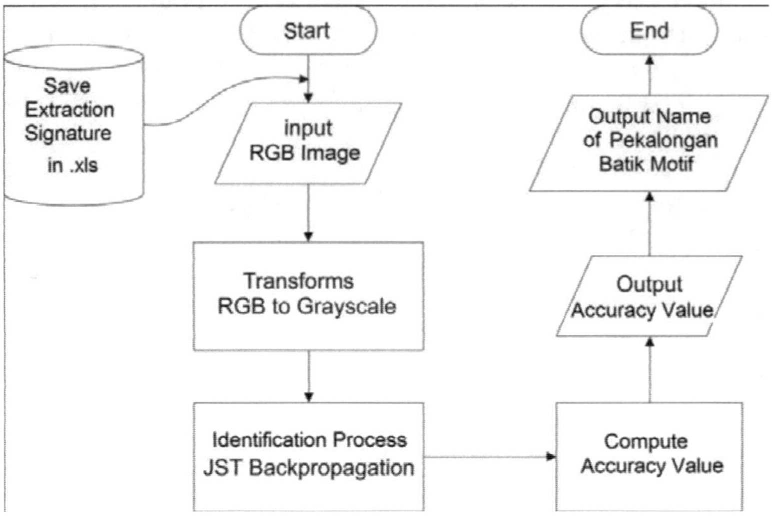

Figure 4. Flowchart system identification testing

4. Data and Database Acquisition

This study uses objects of batik imagery as a result of personal documents during research at batik craftsmen in Pekalongan City. The camera becomes a tool that serves to capture the characteristics of Pekalongan batik images. Data taken images are stored in * JPEG and * BMP files. The data obtained will be divided into two parts for the training and testing process. The data obtained in the form of images of Pekalongan batik motifs as shown in figure 5.

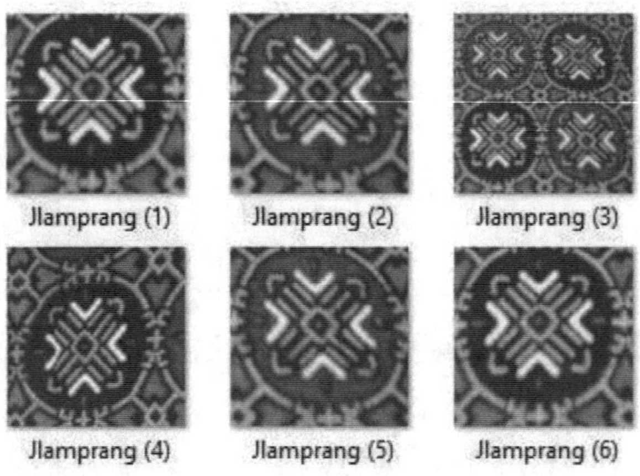

Figure 5. Sample Data Batik Pekalongan

Journal of Physics: Conference Series 1373 (2019) 012049 doi:10.1088/1742-6596/1373/1/012049

Figure 5 is one example of the Pekalongan batik motif that will be used in the training process. The name of the motif is Jlamprang batik which is a traditional batik motif that is very typical of Pekalongan City. The image is taken based on several camera points of view to test whether the system

5. Initial Process and Feature Extraction

The initial stage of the whole process of introducing Batik Pekalongan is to process it. This stage provides the components needed in the training and introduction process which includes hardware connections and processes that are carried out before entering the segmentation stage. This stage includes: image retrieval, image conversion from RGB format to grayscale format [10].

The next process is to extract features using the GLCM method to obtain information about the image of Pekalongan batik as an object of research, but the image needs to be resized to be smaller but quite representative of the truth. data [11]. The batik image data used in this study will reduce the original yellow resolution to 30 × 20 pixels. Furthermore, the image is said to be a matrix measuring 30 × 20 that is converted into a pattern in the form of a matrix measuring 600 × 1 [12].

The pattern of initial processing of batik images can be seen:

$$\psi = \frac{1}{M}\sum_{n=1}^{M}\Gamma_n \tag{1}$$

The mean image pattern Ψ and the pattern with the average reduction Φ.

$$\Phi_i = \Gamma_i - \psi \tag{2}$$

Φ i is a pattern of feature extraction that will be used as input to the classification of Artificial Neural Networks.

6. Artificial Neural Networks and End Processes

The technique of identifying Pekalongan batik uses a network of artificial conditions. The Multi Layer Perceptron consists of input layers, there is one hidden layer and the output layer, the layer process can be seen in the illustration in Figure 5. Training algorithms use backpropagation where the complete algorithm can be seen in books written in Fauset [13]. In this study the training rate was set α = 0.5.

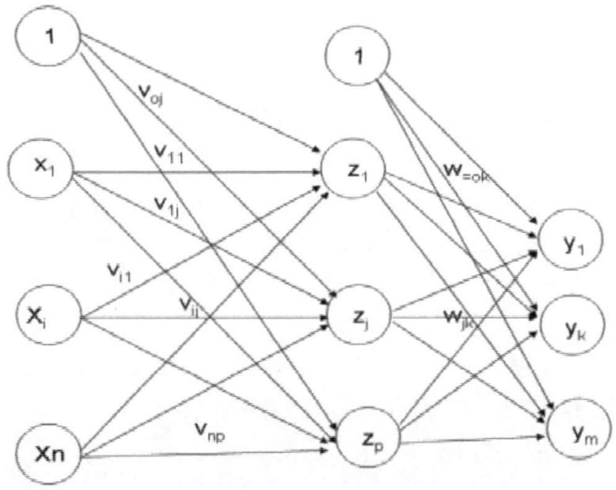

Figure 6. Architecture of Artificial Neural Network

Journal of Physics: Conference Series **1373** (2019) 012049 doi:10.1088/1742-6596/1373/1/012049

Figure 6, y k, z j and xi respectively are output values from the output layer, hidden layer, and input layer. Wjk is the weights between the output layer and the hidden layer, and Vij is the weights between the hidden layer and the input layer.

7. Result

The system for identifying the characteristics of batik images can be tested after the system has been trained first using the image of Pekalongan batik used in the system training process. The image of batik used in training is 4 batik motifs of each of 5 different images so that there are 20 training images. Testing is done by providing new image data that has not been 'known' by the system or in other words not the data that has been used in the training process. The experimental results used the backpropagation neural network classification technique as shown in Table 1:

Table 1. Results of Identification of Pekalongan batik images

Name of Motif	Original image	Epoch Value	Learning Rate Value	Accuracy Value (%)
Cap Kombinasi Tulis		100	0.01	87.2%
Jlamprang	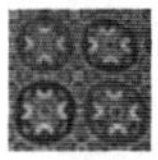	100	0.02	89.6%
Sogan		100	0.03	91.2%
Tiga Negeri		100	0.04	87.2%

Table 1 explains the results of the batik image identification system that contains the accuracy of the images of Pekalongan batik. symmetrical batik patterns and sharp colors can increase the value of the accuracy of an identification process due to its motives that can be better understood by object recognition systems.

8. Conclusion

Produce a system that can be used to identify the image of Pekalongan batik using Bacpropagation Artificial Neural Network. Sogan's motive is the result with the highest level of accuracy that reaches 91.2 %% in testing the value of age = 100 and the level of learning = 0.03. the second level of accuracy was 89.6% with age 100 and learning level 0.02 in Jlamprang batik. The last level of 87.2% was obtained from the Cap Combination batik and Tiga Negeri batik with the age of 100, the learning level was 0.01 and 0.04. High and low accuracy values can be caused by structured motives and the sharpness of batik colors will help increase the value of accuracy.

References

[1] Suliyanto, N.Weni, S. Sri 2015 *Jurnal Ekonomi Bisnis* **18** 135-144.
[2] Iskandar, Eny Kustiyah 2017 *Gema* **30** 2456-2472.
[3] Agung N 2016 *Justindo* **01** 57-63.
[4] Anton Y 2016 *Jurnal Ilmiah ILKOM* **8** 190-196.
[5] Parmono K 2013 *Jurnal Filsafat* **23** 134-146.
[6] Damayanti M, Latifah 2015 *Jurnal Pengembangan Kota* 100-111
[7] Adhi P, Singgih 2016 *Jurnal Imajinasi* **10** 51-60.
[8] Fadlil A 2007 *Seminar Nasional Aplikasi Teknologi Informasi* (Yogyakarta) G7-G10.
[9] Surya, Andika R, Fadlil A,Y Anton *Jurnal Informatika: Jurnal Pengembangan IT* **02** 23-26.
[10] Abdul F, Hidayat I, Sunardin 2008 *Seminar Nasional Aplikasi Teknologi Informasi* (Yogyakarta) J5-J8.
[11] Kumalasanti RA, Ernawati, Dwiandiyanta BW 2015 *Simposium Nasional RAPI XIV FT UMS.* E93-E100
[12] Fauset L 1994 *Fundamentals of Neural Network* USA: Prentice Hall Inc p.291-296.
[13] Fadlil A 2015 *SINERGI* (19), 181-186.

Journal of Physics: Conference Series **1373** (2019) 012050 doi:10.1088/1742-6596/1373/1/012050

Exponential Distribution Parameter Estimation with Bayesian SELF Method in Survival Analysis

Yuli Triana[1], Joko Purwadi[1,*]

[1]Mathematics Study Program, Faculty of Applied Science and Technology, Ahmad Dahlan University Yogyakarta

* Email: joko@math.uad.ac.id

Abstract. This Paper discussed the Exponensial distribution parameter estimation using Bayesian SELF method in survival analysis with $\hat{\theta}$ as SELF Bayesian estimator for the θparameter. Survival analysis corresponds to a method that relates to time starting from the start point to the occurrence of a particular event or an endpoint. The event referred in this research is the event of death. The distribution used in this research is an exponential distribution. Bayesian SELF estimation of $\hat{\theta}$ on exponential distribution for the censored data was obtained by minimizing expectations of loss function. Application of Bayesian SELF method on acute coronary syndrome patient's data, it was obtained $\hat{\theta} = 0.18583$ that indicates the patients' probability to survive is high and the probability for the patients to fail is low.

1. Introduction

Survival analysis or life-resistant analysis is a method related to the time of origin or start point until the occurrence of a special event or end point. The Data obtained in the field of health is an observation of the observed patient and the time of the failure of each individual [1]. The difference between survival analysis and other statistical analyses is the presence of censored observation concepts and uncensored observations. In analyzing the survival data, there are two models that can be used that are parametric models and nonparametric models. Some parametric models consist of the Weibull distribution, Log-Normal distribution, LOG-logistics distribution and exponential distribution. In the field of health, exponential distribution can be used to examine the survival data, where the research is derived from the survival data in acute coronary syndrome cases by estimating the parameters of the distribution. In this study used the Bayesian *SELF* method obtained by minimizing the expectation of loss function.

The purpose of this study is to determine the estimation of parameters of the exponential distribution and apply them to the case data of acute coronary syndrome sufferers. The time distribution of survival is usually described in three functions, namely survival function, probability density function and hazard function.

2. Material and Methods

Probability density function is the probability an individual dies or experiences instantaneous events in a time interval t until $t + \Delta t$ which is equivalent to $f(t)$. Survival function defined as an probbility stating that an individual can survive until the time t [1], which is defined as the following:

$$S(t) = 1 - F(t). \tag{1}$$

If T random variables are not negative at intervals $[0, \infty)$ That indicates the time of the individual experiencing events in a population, the chances that individuals experience events at intervals $(t + \Delta t)$ is defined as hazard function [1] $h(t)$:

$$h(t) = \frac{f(t)}{S(t)}. \tag{2}$$

Exponential distribution is the most commonly used distribution in the modeling of reusability and analysis of life tests. If T is the survival time of the T continuous random variable that follows the exponential distribution with θ parameters, then, probability density function of exponential distribution is the probability density function of exponential distribution [2] is:

$$f(t) = \theta e^{-\theta t}, t > 0, \theta > 0 \tag{3}$$

the cumulative distribution function for exponential distribution is:

$$F(t, \theta) = 1 - e^{-\theta t} \tag{4}$$

Bayesian is a method that views parameters as variables describing the initial information about the parameters before the observation is performed and expressed in a distribution called the prior distribution [3]. Furthermore, the posterior (posterior distribution) information is a combination of two sources of information regarding the parameters of the statistical model, i.e., likelihood from the distribution of samples and preliminary information (prior distribution). The result is expressed in a posterior distribution form which then becomes the basis in the Bayesian method.

Bayesian SELF method Defined as follow :

$$L(\theta, \hat{\theta}) = (\hat{\theta} - \theta)^2$$

Where $\hat{\theta}$ is the Bayesian SELF estimator for the θ parameter [4]. The Bayesian estimate of θ on the exponential distribution for censored data is achieved by minimizing the expectations of the loss function obtained as follows:

$$\frac{\partial \left[E\left(L(\theta, \hat{\theta})\right)\right]}{\partial \theta} = 0$$

$$\hat{\theta} = E(\theta) \tag{5}$$

Likelihood function is defined as join density function of n random variabel ([5],[6]). join density function of $f_{t_1, t_2, \cdots, t_n}(t_1, t_2, \cdots, t_n; \theta)$ That considers the function of θ. if $t_1, t_2, \cdots, t_n$ is a random sample of the density function $f(t; \theta)$, join density function is:

$$L(\theta) = \prod_{i=1}^{n} (t_i; \theta).$$

On the Bayesian approach selects the prior distribution which shows uncertainty about the unknown θ parameter being a major problem in this approach. Based on the likelihood function the prior distribution is divided into two groups [7]: 1) Based on the distribution of data pattern identification result there are prior conjugates and prior non-conjugates, 2) based on the

determination of each of the parameters of the prior distribution pattern there are prior informative and prior non-informative.
Posterior distribution is a conditional density function of θ if known as the value of T observation, on the Bayesian method, inferences are based on the posterior distribution. So the posterior distribution can be defined as [4]:

$$f(\theta|t) \propto L(t;\theta)f(\theta)$$

The symbol of $\propto$ states that the posterior distribution is propotional or comparable to the likelihood function if multiplied by the prior distribution. When it is continuous θ, the prior and posterior distribution of θ can be presented with density function ([8[[9]). Conditional density function of one random variable if known value of the second random variable is just a join density function with two random variables it is divided by a marginal density function of second random variables. Join density function of $f(\theta,t)$ and marginal density function $g(t)$ is generally unknown, Only the prior distribution and the likelihood function are usually expressed. So the posterior density function for continuous random variables can be written as [10]:

$$f(\theta|t) = \frac{f(t;\theta)f(\theta)}{\int_{-\infty}^{\infty} f(\theta)f(t;\theta)d\theta} \qquad (6)$$

The stages of research on the exponential distribution using the SELF Bayesian method is as follows:

1. Determine the survival function, and the hazard function.
2. Determine the likelihood function.
3. Formulating the prior and posterior distribution.
4. Estimate exponential distribution parameters with Bayesian SELF method.
5. Apply the estimation parameters of the Bayesian SELF method in acute coronary syndrome patient data.

3. Results and Discussion

3.1 Survival function and Hazard function of exponential distribution

Based on the equation (1), the survival function of t is:

$$\begin{aligned} S(t;\theta) &= 1 - F(t;\theta) \\ &= e^{-\theta t} \end{aligned} \qquad (7)$$

Based on the equation (2), the hazard function of t is:

$$\begin{aligned} h(t) &= \frac{f(t)}{S(t)} \\ &= \theta \end{aligned} \qquad (8)$$

3.2 Likelihood function on the Bayesian SELF method

The likelihood function is a function that is considered a function of θ, for $t_1, t_2, \cdots, t_n$ that is fixed. In the censored data of observation(t_i, δ_i) the likelihood function is:

$$L(\theta) = \prod_{i=1}^{n} [f(t_i)]^{\delta_i} [S(t_i)]^{1-\delta_i}$$

with δ_i is a censorship indicator, $\delta = 1$ if the data is not censored and $\delta = 0$ if the data is censored. The value of t_i is the survival time of the patient with $i = 1,2,3,\cdots,n$. So the likelihood function of the exponential distribution for the censored data has the following forms:

$$L(t_i;\theta,\delta_i) = \prod_{i=1}^{n}\left[\theta e^{-\theta t_i}\right]^{\delta_i}\left[e^{-\theta t_i}\right]^{1-\delta_i}$$
$$= \theta^{\sum_{i=1}^{n}\delta_i}\, e^{-\theta\sum_{i=1}^{n}t_i}$$

3.3 Forming the Prior distribution

In the Bayesian method, when a particular distribution with a parameter in it is described with θ, it is likely that the θ parameter follows a particular distribution of probabilities called the prior distribution. In this research, Gamma distribution was set as a comrade distribution for exponential distribution with the θ parameter where $0 < \theta < \infty$. The prior distribution for θ-on exponential distribution with α and β parameters, the probability density function as follows

$$f(\theta) = \frac{1}{\beta^{\alpha}\Gamma(\alpha)}\theta^{\alpha-1}e^{-\frac{1}{\beta}\theta} \qquad \alpha > 0, \beta > 0, \theta > 0$$

Determination of α and β parameters for gamma distribution (α, β) which is used as a prior distribution can be solved by matching between the mean and variance of gamma distribution with the mean and exponential distribution variance. Thus obtained $\alpha = 1$ and $\beta = \frac{1}{\theta}$. With θ is the average of failure time.

3.4 Determining Posterior Distribution

The posterior distribution can be expressed by comparison between common density functions and marginal functions. So based on equations (6) Posterior distribution can be expressed as follows:

$$f(\theta,t) = \frac{\theta^{\sum_{i=1}^{n}\delta_i}\, e^{-\theta\sum_{i=1}^{n}t_i}.\frac{1}{\beta^{\alpha}\Gamma(\alpha)}\theta^{\alpha-1}e^{-\frac{1}{\beta}\theta}}{\int_0^{\infty}\theta^{\sum_{i=1}^{n}\delta_i}\, e^{-\theta\sum_{i=1}^{n}t_i}.\frac{1}{\beta^{\alpha}\Gamma(\alpha)}\theta^{\alpha-1}e^{-\frac{1}{\beta}\theta}\,d\theta}$$
$$\propto \theta^{\sum_{i=1}^{n}\delta_i}\, e^{-\theta\sum_{i=1}^{n}t_i}\theta^{\alpha-1}e^{-\frac{1}{\beta}\theta}$$
$$\propto \theta^{\sum_{i=1}^{n}\delta_i+\alpha-1}e^{-\theta(\sum_{i=1}^{n}t_i+\frac{1}{\beta})}$$

Thus, the posterior distribution of the exponential distribution is Gamma distribution $\left(\alpha = \sum_{i=1}^{n}\delta_i + \alpha, \beta = \left(\sum_{i=1}^{n}t_i + \frac{1}{\beta}\right)^{-1}\right)$.

3.5 Estimation Parameter of Bayesian SELF Method

Because in this research using the estimation of the Square Error Loss Function (SELF) parameter, the Bayesian estimator used is the mean of the posterior distribution. So based on equations (5) obtained:

$$\hat{\theta} = E(\theta)$$
$$\hat{\theta}_{BS} = \frac{(\sum_{i=1}^{n}\delta_i + \alpha)}{\left(\sum_{i=1}^{n}t_i + \frac{1}{\beta}\right)} \qquad (9)$$

from the equation (9) can be obtained estimation of the parameters of survival function and hazard function with the Bayesian SELF of the exponential distribution method is:

$$\hat{S}_{BS}(t_i;\hat{\theta}_{BS}) = e^{-\frac{(\sum_{i=1}^{n}\delta_i+\alpha)}{\left(\sum_{i=1}^{n}t_i+\frac{1}{\beta}\right)}t_i} \quad (10)$$

$$\hat{h}_{BS}(t_i;\hat{\theta}_{BS}) = \frac{(\sum_{i=1}^{n}\delta_i+\alpha)}{\left(\sum_{i=1}^{n}t_i+\frac{1}{\beta}\right)} \quad (11)$$

3.6 Case Study In Acute Coronary Syndrome Patient Data

The data conformance test used is the Anderson-Darling test, by using the R software obtained output from the Anderson-Darling test of *p-value* = 0.08529. Because the value of *p-value* > α is 0.08529 > 0.05 then H_0 is accepted which means the data is exponential distribution.

3.7 Estimation parameters of Survival function and Hazard function

From acute coronary syndrome sufferers, it is known that $\sum_{i=1}^{n}\delta_i = 63$, $\sum_{i=1}^{n}t_i = 340$ dan $\beta = 0.227272$ So that the estimation of parameters with the Bayesian SELF method is:

$$\hat{\theta}_{BS} = \frac{(\sum_{i=1}^{n}\delta_i+\alpha)}{\left(\sum_{i=1}^{n}t_i+\frac{1}{\beta}\right)}$$

$$\hat{\theta}_{BS} = 0.18583$$

So that it can be known the Bayesian SELF estimation for the θ parameter of the exponential distribution is 0.18583 which indicates the patient's probablity to endure is high and the probability of failure are low. And can also be known estimation of Bayesian SELF for survival function and hazard function as follows:

$$\hat{S}_{BS}(t_i;\hat{\theta}_{BS}) = e^{-\frac{64}{344,400141}t_i}$$

and if from the data is taken one sample that is in the individual to 63 with a long time treated is 12 days then the estimate obtained the Survival function as follows:

$$\hat{S}_{BS}(12;\hat{\theta}_{BS}) = e^{-\frac{64}{344,400141}12}$$
$$= 0.10753$$

So if the patient's time treated is 12 days then the patient's probability to survive is 0.10753. and the hazard function estimation as follows

$$\hat{h}_{BS}(t_i;\hat{\theta}_{BS}) = \frac{(\sum_{i=1}^{n}\delta_i+\alpha)}{\left(\sum_{i=1}^{n}t_i+\frac{1}{\beta}\right)}$$

$$\hat{h}_{BS}(t_i;\hat{\theta}_{BS}) = 0.18583$$

Journal of Physics: Conference Series **1373** (2019) 012050 doi:10.1088/1742-6596/1373/1/012050

Survival Function dan Hazard Function Graph

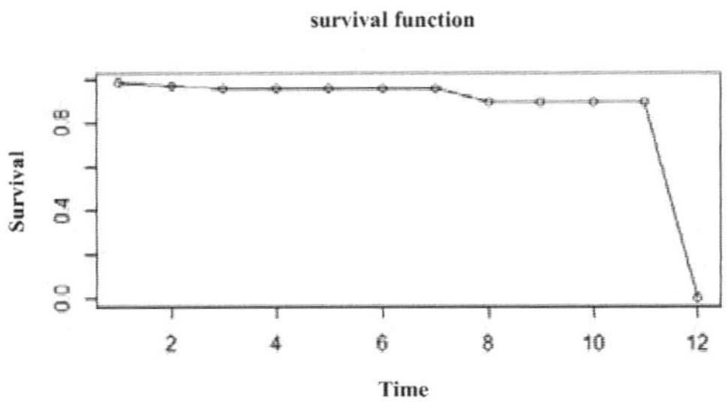

Figure 1

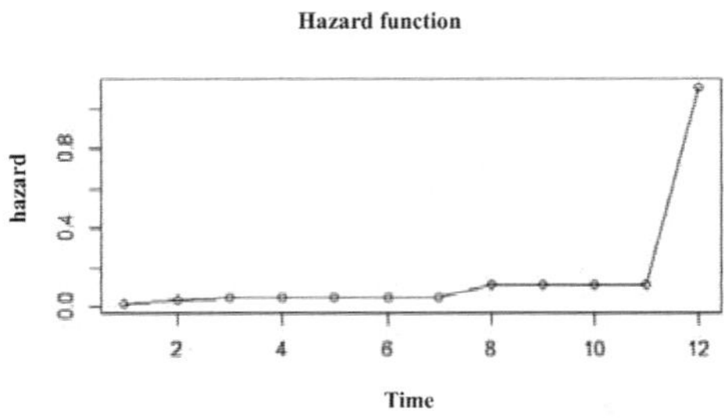

Figure 2

from Figure 1, it can be known that the graph of survival function is getting down closer to 0, which means the individual probability to survive the longer the more the decline. While from the figure 2, It can be noted that the graphic hazard is inversely proportional to the survival function graph. The graph of hazard function is getting up close to 1, That means a person's probability to experience failure or death is increasingly rising or higher.

4. Conslusion

Based on the analysis had been done, the research conclusion is as follows :

1. Estimation of exponential distribution parameters by Bayesian SELF method on survival analysis is obtained as follows:

$$\hat{\theta}_{BS} = \frac{\left(\sum_{i=1}^{n} \delta_i + \alpha\right)}{\left(\sum_{i=1}^{n} t_i + \frac{1}{\beta}\right)}$$

2. Application of estimation of exponential distribution parameters by Bayesian SELF method in survival analysis of acute coronary syndrome is $\hat{\theta}_{BS} = 0.18583$ That indicates the patient's probability to endure is high and the probability of the patient failing are low.

Reference

[1] Ibrahim J G, Ming-Hui C, Sinha, Debajyot, 2001. *Bayesian Survival Analysis.* Springer-Verlag.

[2] Ahsanullah, M. dan Hamedani G.G. 2010. *Exponentiated Distribution: Theory And Methods.* Book Series: Mathematics Research Developments New York: Nova Science Publishers, Inc.

[3] Amelia S R 2018 *Estimasi Parameter Distribusi Weibull Dua Parameter Menggunakan Metode Bayes.* Makassar: Universitas Islam Negeri Alauddin Indonesia.

[4] Luoma A 2014 *Introduction To Bayesian* Finland: University of Tampere.

[5] Bain L J and Engelhardt M 1992 *Introduction to Probability and Mathematical Statistic. California*: Duxbury of Wathfor, Inc.
[6] Rarasati I P 2012 *Estimasi Parameter Distribusi Gamma Dengan Metode Bayes* Malang: Universitas Isam Negeri Maulana Malik Ibrahim Indonesia.
[7] Box, G E P and Tiao G C 1973 *Bayesian Inference In Statistical Analysis* Addison-Wesley: Publishing Company,Inc.
[8] Collect D, 1993 *Modelling Survival Data in Medical Research* New York: Chapman and Hall.
[9] Sagita I, Satyahadewi N, Rizki S W 2018 *Bimaster* **7** 353-362
[10] Aulia R, Fajriah N, Salam N, 2011 *Epsilon* **5** 40-52

Investigation of the Number of Features and Muscles for an Effective Hand Motion Classifier Using Electromyography Signal

Triwiyanto Triwiyanto*, Bedjo Utomo, Dyah Titisari, M. Ridha Mak'ruf, Triana Rahmawati

Department of Electromedical Engineering, POLTEKKES KEMENKES Surabaya
Jl. Pucang Jajar Timur No. 10, Surabaya, Indonesia.

*Email: triwiyanto123@gmail.com

Abstract. The essential problem in the development of a prosthetic hand based on electromyography (EMG) signal is the choice of the features and number of muscles that will be used to recognize the hand motion. A minimal number of feature and channel could reduce the cost in the development of the prosthetic hand and processing time. Therefore, it is important to obtain the correct number of feature and muscles in the system. The objective of this study is to evaluate the effectiveness of using the number of feature and muscle to recognize the hand motion for amputee person using the EMG signal. In this study, the EMG dataset was obtained from five transradial amputee persons. Ten disposable electrodes (Ag/AgCl) were placed on the residual hand with equal space between the electrodes. In order to obtain the EMG features which are related to the hand motion, each of the EMG signal was extracted using six-time domain features which are root mean square (RMS), integrated EMG (IEMG), waveform length (WL), difference absolute standard deviation value (DASDV), Wilson amplitude, and myopulse rate (MYOP). Each feature was evaluated using Decision Tree, Linear Discriminant Analysis (LDA), Support Vector Machine (SVM), K- nearest neighborhood, and Ensemble machine learning (ML). The number of class to be classified in this study was eighteen motions. The effectiveness of using the number of features and muscles was evaluated by varying the number of features and muscles. Further, the accuracy, in the discrimination of the motion, was calculated and compared among the machine learnings. The results of this study show that six muscles can effectively classify the eighteen of hand motion. The ML with WL feature has the highest accuracy among the others (81.3% based on quadratic SVM). The study suggested the effective number of muscle and feature which can be used in the prosthetics hand development so that the effective prosthetic machine can be built with a low cost in the time computing and the hardware for data acquisition.

1. Introduction

Forty-one thousand people in the USA had to lose their hand due to work accident and cancer disease [1]. In this case, an amputation had to be taken, and this condition could limit their work and influence their social activity. Some prosthetics devices have been developed to substitute the hand function by controlling the devices using EMG signal or mechanical sensors such as accelerometer-gyroscope or inertial sensor. However, because of the lack of functionality, some of the users abandoned the devices. Recently, the EMG signal has been used a lot as a control signal for assistive exoskeleton devices [2] [3] [4], electronic prosthetics [5] [6] [7] and teleoperation [8] because the response is faster than the mechanical sensors [9]. However, the EMG signal has complex, non-linear and stochastics characteristics in nature [10]. Even though the EMG signal is generated from the same electrode's location, motion, and person, it will generate a different pattern for every time of measurement [10]. Previous studies had been investigated the behavior of the EMG signal in accordance with the human motion and it was found that the EMG signal (amplitude) is related to human limb (upper and lower) motion [11].

Feature extraction plays an important role to obtain the relation between human limb motion and the EMG signal. Feature extraction is classified into three domains, those are time, frequency and time-frequency domain (wavelet) [12]. The feature extraction based on frequency and time-frequency domain needs more computation time than time domain because it needs a transformation between domain. Therefore, features extraction which uses time domain is often used by previous studies. Generally, time-domain feature extraction is divided into four categories; those are based on energy (such as RMS, IEMG, MAV), complexities (WL, AAC, DASDV), frequency (ZC, MYOP, WAMP, SSC) and statistics (mean, kurtosis, maximum, minimum, skewness). Each feature has a different response for each type of motion. Therefore, an investigation of the time domain features which related the hand motion is important in order to find the best feature. In order to discriminate the type of hand motion, previous studies have investigated several EMG features to discriminate some type of hand motion for a transradial amputee. However, majority, previous studies only proposed a method to discriminate two (open and close) [7], four (open/close and supination/pronation) [13] to eight (wrist flexion, wrist extension, pronation, supination, power grip, pinch grip, open hand, and rest) [14] type of hand motions. By considering that the higher number of class then the complexities of the EMG signal will increase and also it will reduce the accuracy of the machine learning. However, a more complex hand motion is needed in order to get a prosthetics hand which is able to imitate the real hand for an amputee. A study of hand motion for complex motion has been developed to discriminate seventeen motions [15]. However, in those work, the twelve of electrode pairs has to be placed in order to discriminate the seventeen motions. In the development of the prosthetics hand, it is still a challenging study, to develop a prosthetic device with a minimum number of electrodes, in order to minimize the size of the amplifier unit and reduce the digital signal processing computation. Furthermore, machine learning is needed to encode the EMG features which are related to the hand motion. Generally, standard machine learning such as Linear Discriminant Analysis (LDA), Support Vector Machine (SVM) and K-NN were used by previous studies to discriminate the hand motion [16]. However, those machines learning used is based on a linear approach. On the other hand, the EMG signal has non-linear, random and stochastic characteristics in nature [10]. Therefore, some ML with a non-linear kernel was needed to be investigated in order to find the fittest model to recognize the hand motion. Therefore, in this study, the 22 of ML was investigated. Those ML are Fine Decision Tree, Medium Decision Tree, Coarse Decision Tree, Linear Discriminant Analysis, Quadratic Discriminant Analysis, Linear SVM, Quadratic SVM, Cubic SVM, Fine Gaussian SVM, Medium Gaussian SVM, Coarse Gaussian SVM, Fine KNN, Medium KNN, Coarse KNN, Cubic KNN, Weighed KNN, and Ensemble (such as: boosted tree, bagged tree, subspace discriminant, subspace KNN, RUS–Boosted tree).

Based on the problems that have been mentioned, consequently, the objective of this study is to investigate the performance of the twenty-two machine learnings with an approach to the quadratic, cubic, fine, medium, and coarse kernel to classify the eighteen type of hand motion based on the EMG signal from amputee person. Special aims of this study are 1) to investigate the effective number of muscles to predict the eighteen type of hand motion (including in rest condition), 2) to investigate the best EMG feature to recognize the eighteen type of hand motion. The impact of this study is toward an effective hand motion classifier so that in the future this proposed method can be used for the development of the hand prosthetics with minimum number of muscle and less computation.

2. Materials and Methods

2.1. Participant

In this study, the EMG signal (**Figure 1**) was obtained from Ninapro dataset db3 (http://ninapro.hevs.ch) which was collected from five amputee persons who have an amputation after the accident. The EMG signal was recorded using twelve electrodes (Trigno Wireless electrodes, Delsys, Incorporation, www.delsys.com). The eight electrodes were placed on the residual hand with equal spaces, two electrodes were located on the extensor carpi and flexor carpi which are the most responsible muscle in the hand motion [17]. The two other electrodes were also positioned on the biceps and triceps muscles. The preparation procedures and the placement of the electrodes have followed the SENIAM rules [18].

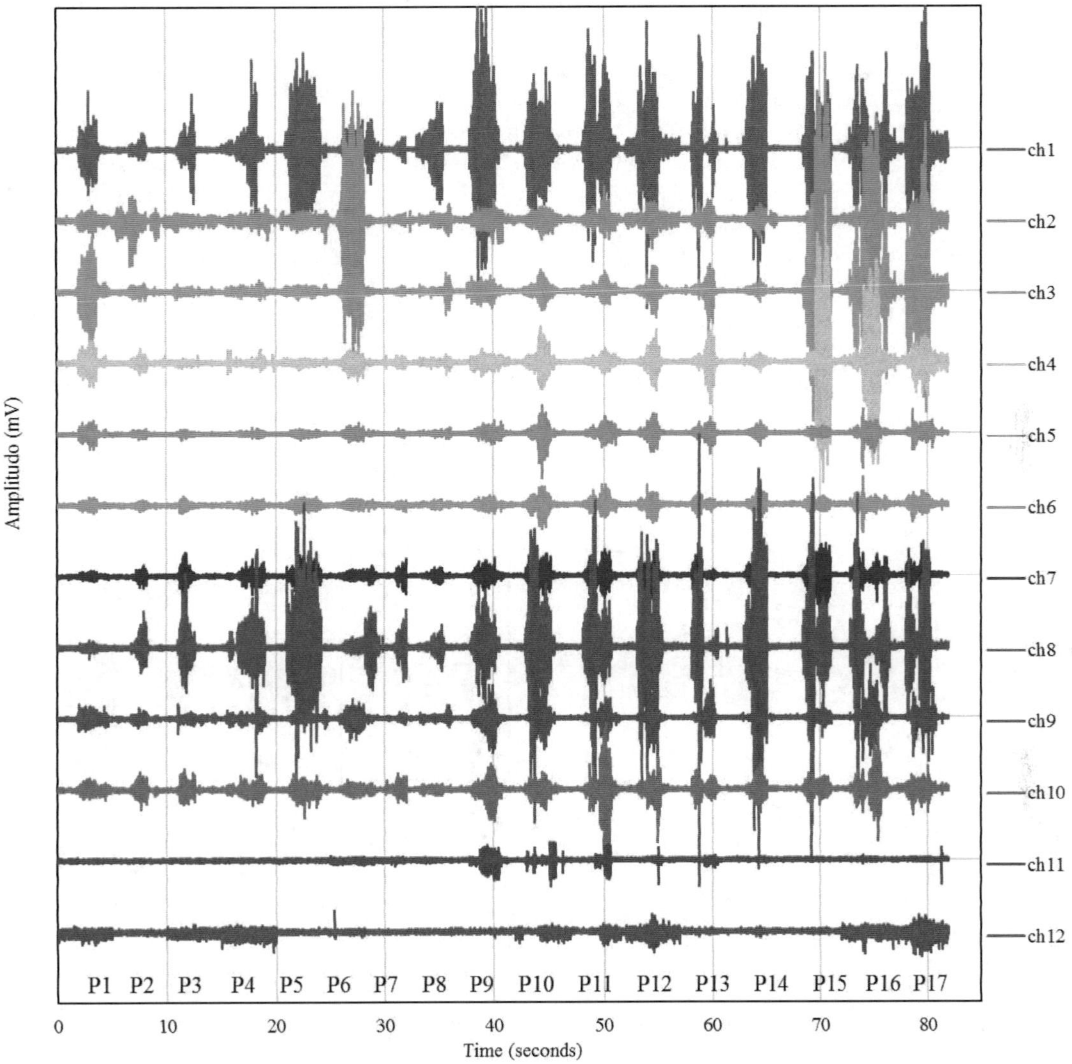

Figure 1. The representation of the EMG signal which is collected from the 12 of the channel from amputee from Ninapro DB3 database. The EMG signal is recorded while the amputees perform the 18-motion including the rest position.

2.2. Experimental Setup

In the data acquisition process, the EMG signal was recorded using a data acquisition card (NI-DAQ, PCMCIA 6024E, 12-bit resolution, National Instruments). The gain of the amplifier and frequency sampling was adjusted at 14,000 and 2,000 Hz, respectively. In the experimental stage, the amputee learned the eighteen motion by watching the hand motion on the monitor. Each motion was repeated for six-time which is consisted of (1) thumb up, (2) extension of index and middle (flexion on the others), (3) flexion of ring and little finger, (4) thumb opposing based of little finger, (5) abduction of all fingers, (6) finger flexed together in fist, (7) pointing index, (8) adduction of the extended finger, (9) wrist supination (axis middle finger), (10) wrist pronation (axis middle finger), (11) wrist supination (axis little finger), (12) wrist pronation (axis little finger), (13) wrist flexion, (14) wrist extension, (15) wrist radial deviation, (16) wrist ulnar deviation, (17) wrist extension with closed hand motion, and (18) rest. Each type of motion was assigned as P1 to P18 (including rest condition: P18). A representation of the EMG signal, from twelve channels, as shown in **Figure 1** which was generated from a subject when performed a sequential motion from motion P1 to P17. Between the motions, there was a rest condition for 3 seconds. Because of the EMG signal for channel 11 (biceps) and channel 12 (triceps) generated low and inconsistent activities to the hand motion so that in this work, the features extraction process was only performed for EMG signal from channel 1 to channel 10.

Furthermore, the EMG signal was extracted using six-time domain features which consist of root mean square (RMS), integrated EMG (IEMG), waveform length (WL), difference absolute standard deviation value (DASDV), Wilson amplitude (WAMP), and Myopulse percentage rate (MYOP) features. In order to discriminate the eighteen motions, the six-machine learning, which consists of Decision Tree, Linear Discriminant Analysis (LDA), Support Vector Machines (SVM), K-Nearest Neighborhood (KNN) and Ensemble (a combination of machine learning), were used in this work. The effectiveness of this system was evaluated by varying the number of channel and feature used in the machine learning. The experimental setup and the classification system were shown in **Figure 2**.

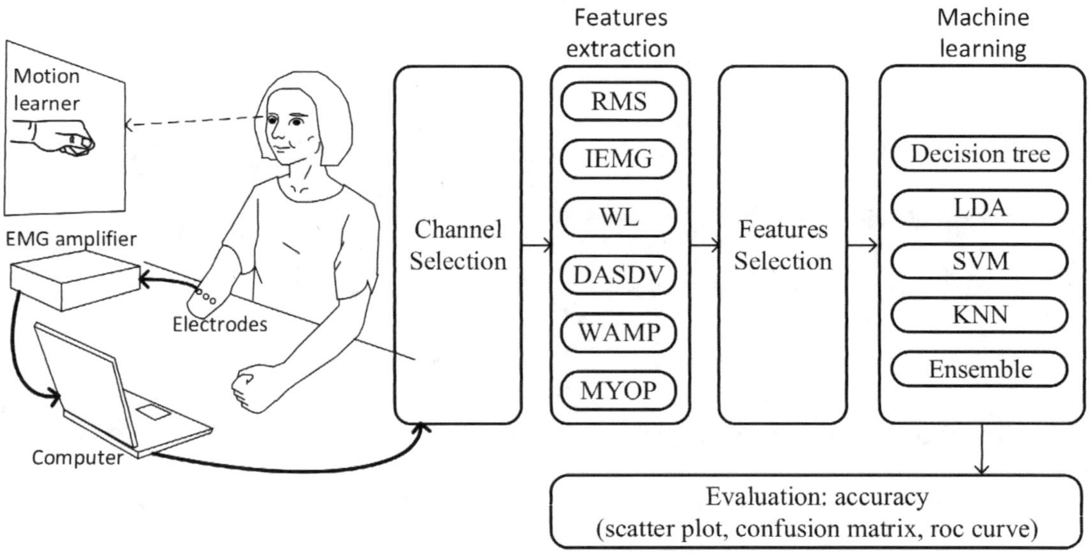

Figure 2. The experimental setup and the classification of the hand motion using machine learning.

2.3. Feature Extraction

The time domain feature was used in this study to extract the EMG signal which is related to the type of hand motion. The time domain feature was selected because the computation time is faster than other domains. A representation of the time domain feature was selected which is based on energy (RMS and IEMG), complexities (WL and DASDV), and frequency (WAMP and MYOP). Every EMG signal was extracted with a window length of 400 samples. So that, if the sampling frequency is 2,000 Hz (0.0005 seconds) then the window length in time is 0.2 seconds. In this study, the record of the EMG signal from 5 subjects consists of 5 (subjects) x 176k (raw EMG) x 12 (channel) records. Ten from twelve channels of EMG signal were extracted sequentially. Two channels (biceps and triceps) of the dataset were abandoned because those channels have a lesser response than others. The six time-domain features were written as follow (**Table 1**) [11]:

Table 1. The six-time domain features

(1) $RMS=\sqrt{\frac{1}{N}\sum_{i=1}^{N}x_i^2}$	(2) $IEMG=\sum_{i=1}^{N}\lvert x_i\rvert$
(3) $WL=\sum_{i=1}^{N-1}\lvert x_{i+1}-x_i\rvert$	(4) $DASDV=\sqrt{\frac{1}{N-1}\sum_{i=1}^{N-1}(x_{i+1}-x_i)^2}$
(5) $WAMP=\sum_{i=1}^{N-1}[f(\lvert x_i-x_{i+1}\rvert)]$ $f(x)=\begin{cases}1, if \rightarrow x\geq threshold\\ 0, otherwise\end{cases}$	(6) $MYOP=\frac{1}{N}\sum_{i=1}^{N}[f(x_i)]$ $f(x)=\begin{cases}1, if \rightarrow x\geq threshold\\ 0, otherwise\end{cases}$

where x is the EMG signal, N is the window length, and a threshold is a pre-defined constant

2.4. Machine Learning

The twenty-two classifier machines were chosen in this study to discriminate the eighteen hand motions. The type of hand motion shown on the monitor computer was indicated in **Table 2**. This motion is a standard type to be implemented on the prosthetics hand.

In this work, the 22 supervised machine learning was evaluated to classify the 18 classes of the hand motion, which are Decision Tree (DT), Linear Discriminant Analysis (LDA), Quadratic Discriminant Analysis (QDA), Support Vector Machine (SVM), K-Nearest Neighborhood (KNN), and Ensemble. However, because of the EMG features have a linear and non-linear characteristic, then we explored the machine learning with various optimization and kernel. In the Decision Tree, the maximum number of splits was varied as four (Fine), twenty (Medium), and 100 (Coarse). For Discriminant Analysis machine learning, we used the linear and quadratics function. The various kernel of the SVM was also explored such as the linear, quadratic kernel, cubic kernel, fine Gaussian kernel, medium Gaussian, coarse Gaussian. In the supervised machine learning based on KNN, we also used some variation of the number of the neighborhood (NN). In the Euclidean matrix, the NN was set to 1, 10, and 100 for fine KNN, medium KNN, and coarse KNN, respectively. For cosine, cubic, and weighted KNN, the NN was set to 10 and distance matric was based on cosine, cubic, and weight, respectively. Ensemble machine learning is a combination of some machine learning which solves the same problem. In this study, the Ensemble machine learning consists of Boosted Tree, Bagged Tree, Subspace Discriminant, Subspace KNN, and RUSBoosted Tree [19].

Before the pattern recognition stage, the EMG signal (10 channels) was extracted using the six-time domain features (RMS, IEMG, WL, DASDV, MYOP, and WAMP). thus, after the feature extraction process, each record of the dataset, we have 60 features (which consist of 6 features and 10 channels) and 1 label. The number of labels to indicate the motion is 18 motions (consist of (1) thumb up, (2) extension of index and middle (flexion on the others), (3) flexion of ring and little finger, (4) thumb opposing based of little finger, (5) abduction of all fingers, (6) finger flexed together in fist, (7) pointing index, (8) adduction of the extended finger, (9) wrist supination (axis middle finger), (10) wrist pronation (axis middle finger), (11) wrist supination (axis little finger), (12) wrist pronation (axis little finger), (13) wrist flexion, (14) wrist extension, (15) wrist radial deviation, (16) wrist ulnar deviation, (17) wrist extension with closed hand motion, and (18) rest). We performed a pre-processing dataset using Microsoft Excel (Version 2016, Microsoft Corporation, USA) before it was used as input of the classifier machines. In this study, the Classification Learner from MATLAB (Version R2017b, Math Works, Inc.,

USA) was used to perform supervised machine learning. In the Classification Learner, a number of features and PCA (principal component analysis) was selected using feature selection menu. A number of training and testing dataset was set to 70% and 30%, respectively. For validation, a 5 k-fold was used to prevent the overfitting in the training stage. The accuracy of machine learning was shown using a confusion matrix and ROC curve.

Table 2. The eighteen hand motions have to be followed by amputee on the screen monitor (rest included) [15]

No.	Hand motion	No.	Hand motion	No.	Hand motion
1.	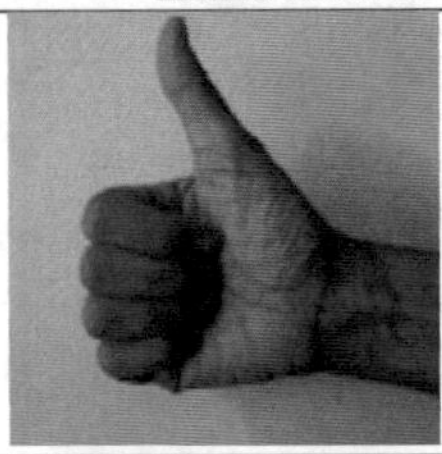	7	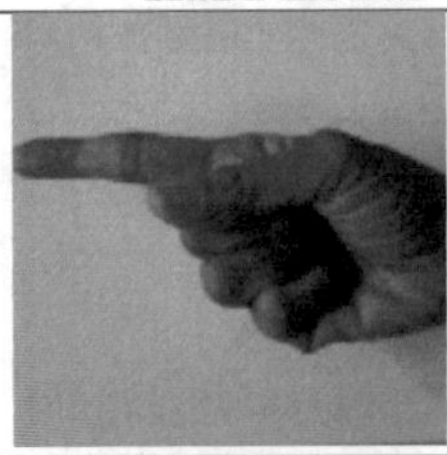	13.	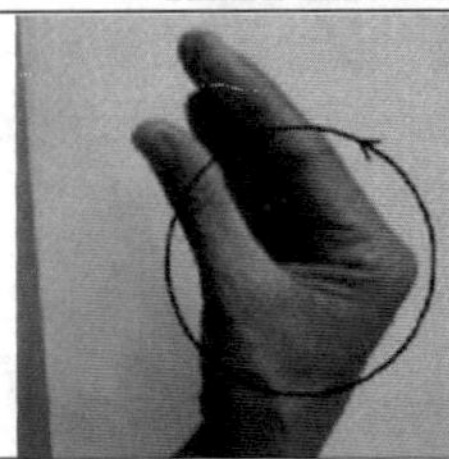
2.	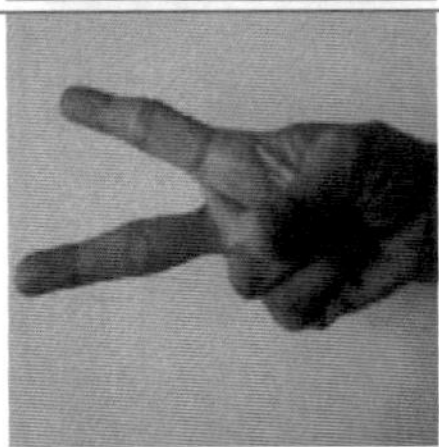	8	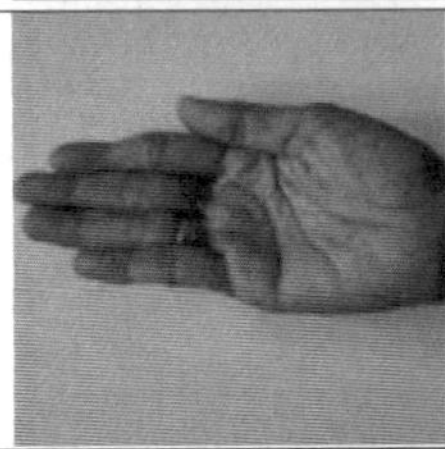	14.	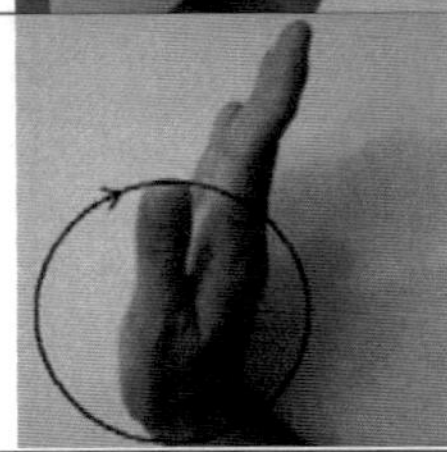
3.	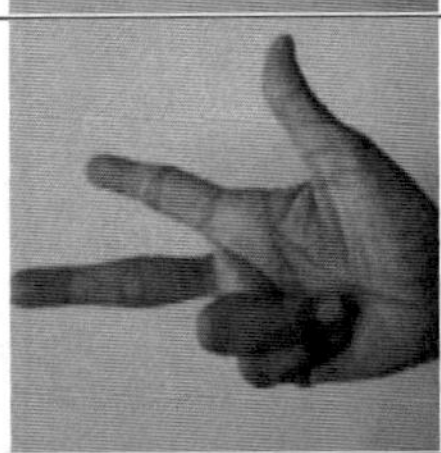	9	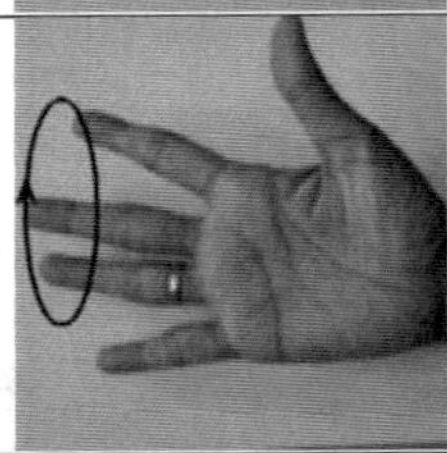	15.	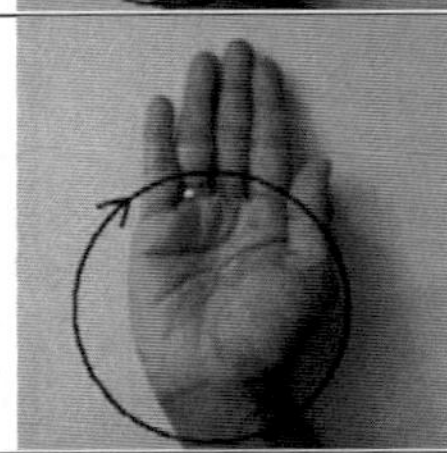
4.	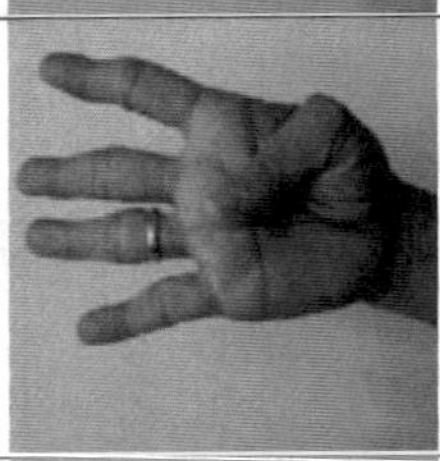	10	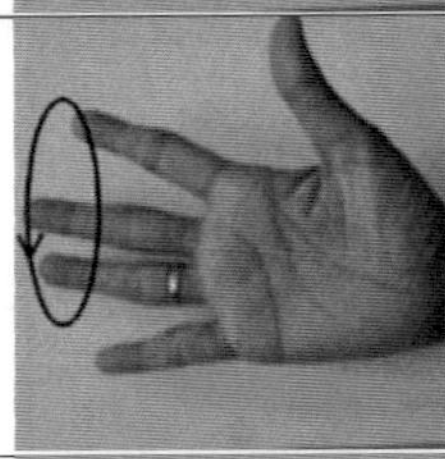	16.	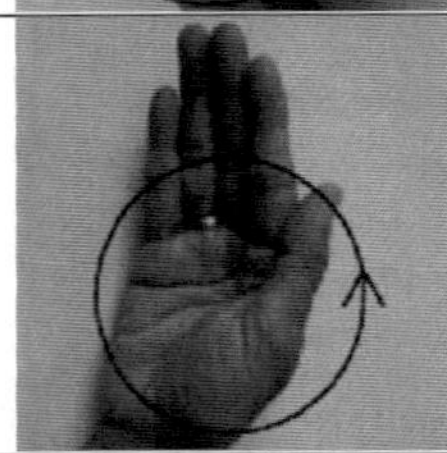
5.	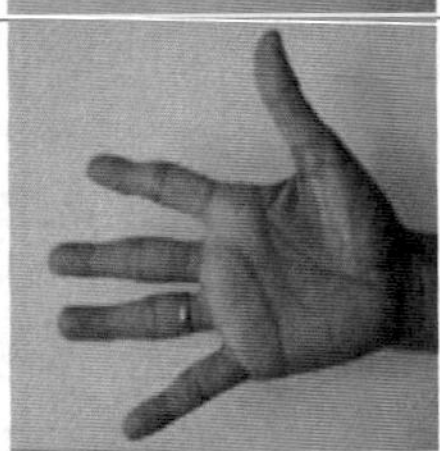	11	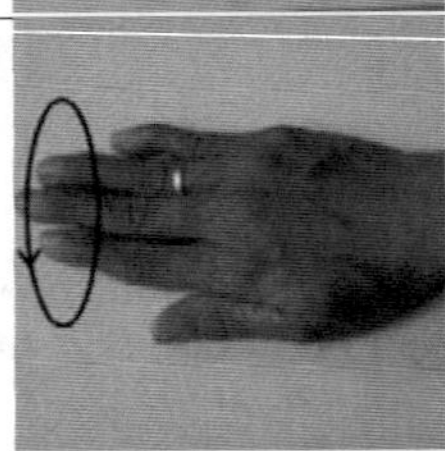	17.	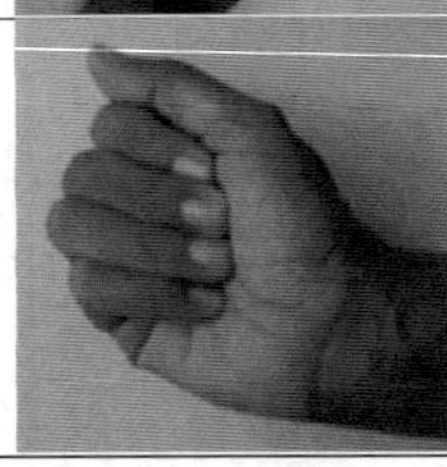
6.	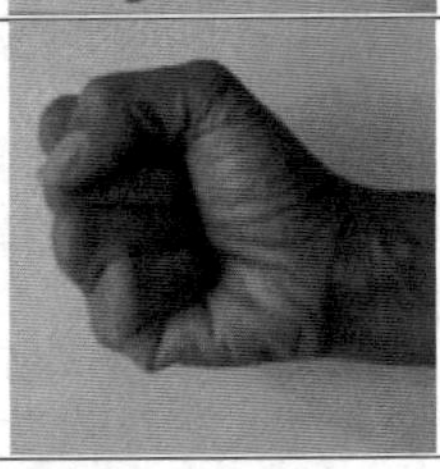	12	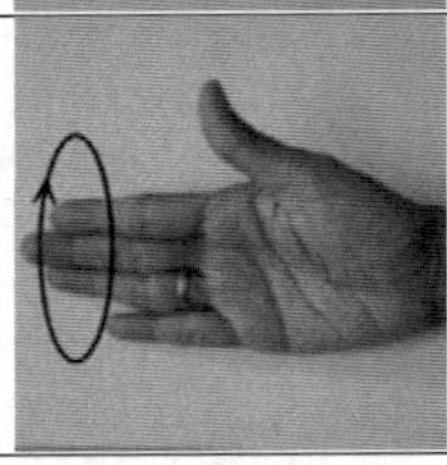	18.	**REST**

In order to know the effective number of features and number of channels to discriminate the hand motion in the classifier machines, we performed a feature selection which is able to choose the number of features (2, 4, and 6 features) and channel selection (2, 4, 6, 8, and 10 channels). The accuracy for each number of features from all the subjects was recorded and analyzed using ANOVA or T-test for the significant difference of the accuracy between a number of features and number of channels. For statistic analysis, we used Microsoft Excel 2016 Profesional (Microsoft Corporation 2016, USA.) and an add-on library from Realstats (http://www.realstats.com).

3. Result and Discussion

The EMG signal shows a different pattern for each channel and the type of hand motion as shown in **Figure 1**. When the amputee performed the motion, channel 1 to 4 shows higher activities than others.

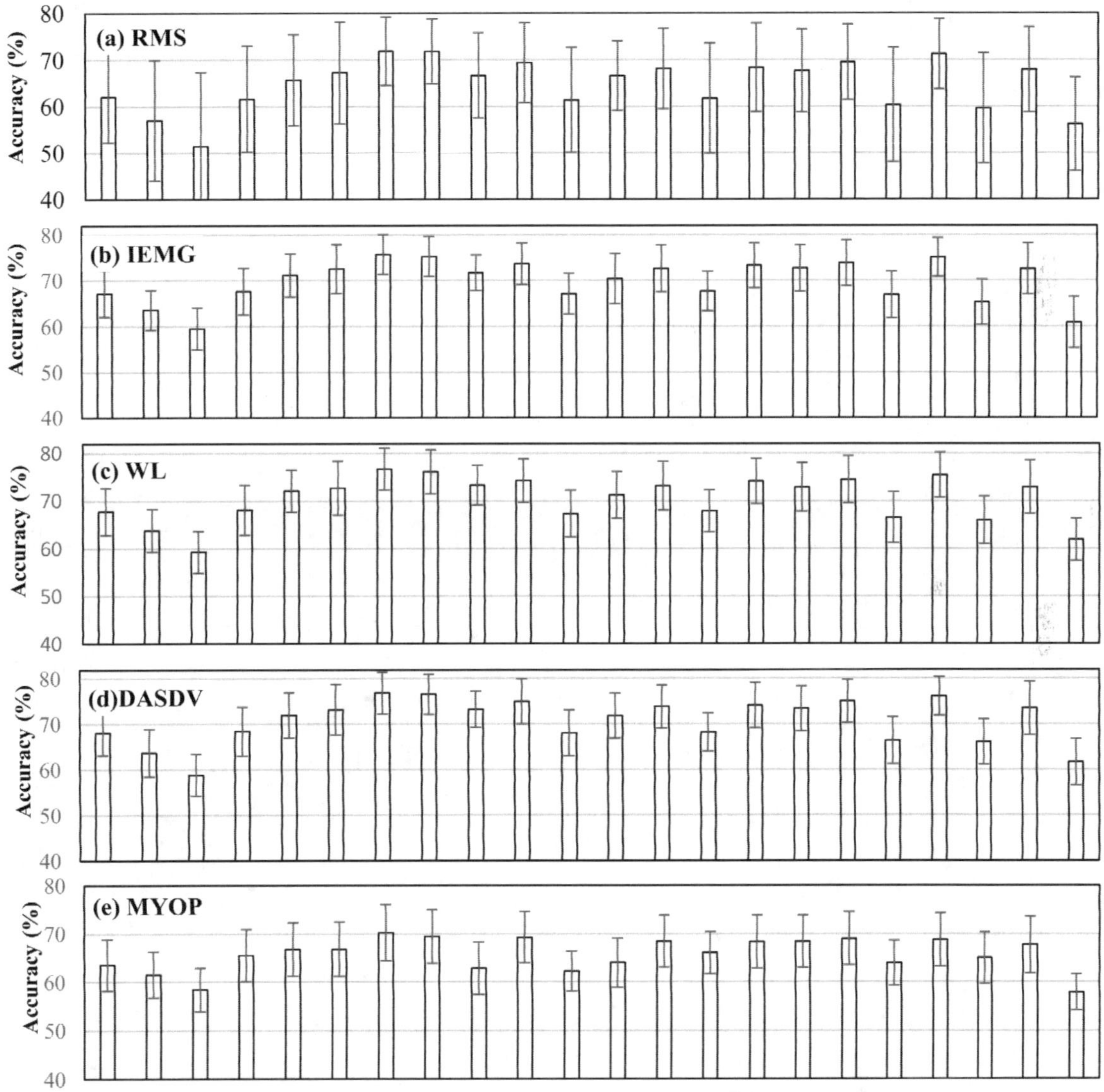

Figure 3. The accuracy of the machine learning in discrimination of 18 motion based on six-time domain feature (a) RMS, (b) iEMG, (c) WL, (d) DASDV, (e) MYOP, and (f) WAMP.

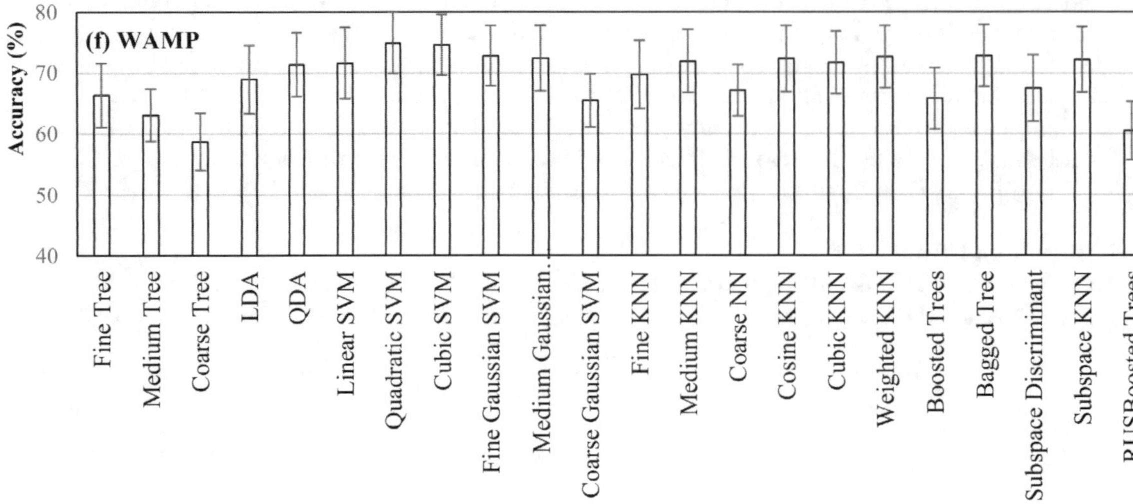

Figure 3. (Continued)

True class	P1	P10	P11	P12	P13	P14	P15	P16	P17	P2	P3	P4	P5	P6	P7	P8	P9	R
P1	79	0	0	2	0	0	0	1	1	0	0	0	0	0	0	6	3	1
P10	0	90	7	5	2	0	0	2	0	0	0	0	0	0	0	0	0	1
P11	0	0	63	11	2	2	3	4	1	0	0	5	5	3	0	2	2	1
P12	0	6	7	63	4	2	0	2	4	0	2	7	1	1	0	2	2	1
P13	2	4	3	0	59	11	5	8	1	0	0	4	0	12	0	0	0	1
P14	0	0	3	6	11	64	2	2	1	0	0	3	1	5	0	0	0	1
P15	0	0	0	0	2	13	80	2	0	0	0	0	1	6	0	0	2	1
P16	2	0	0	3	7	4	7	59	11	0	4	1	4	5	0	0	2	1
P17	2	0	0	2	5	4	2	12	73	17	0	0	0	2	0	0	1	1
P2	0	0	0	0	0	0	0	0	0	72	17	0	0	0	2	0	0	1
P3	6	0	0	0	0	0	0	1	0	21	70	4	0	0	4	4	0	1
P4	0	0	2	2	2	0	0	0	1	0	6	66	4	1	0	0	0	1
P5	0	0	2	0	4	0	0	5	1	0	0	5	72	3	0	0	0	1
P6	2	0	0	0	2	0	2	1	0	0	0	1	3	62	0	0	1	1
P7	0	0	0	0	0	0	0	0	0	2	0	0	0	0	91	0	0	1
P8	3	0	3	3	0	0	0	2	0	0	0	1	0	0	0	84	0	1
P9	6	0	5	2	0	0	0	0	5	0	0	0	0	0	0	2	85	1
R	0	0	3	2	2	0	0	1	0	5	0	0	1	3	4	0	2	85

Predicted class

Figure 4. The confusion matrices for quadratic SVM using WL feature. The accuracy is in % unit. The diagonal matrices which assigned with green color show training and testing with the same pattern. (R is rest condition)

Each EMG signal generated from channels was segmented for windows length of 400 samples and extracted using the six-time domain features. Twenty-two machine learning was explored in this study to obtain the fit model and highest performance in classifying the hand motion. The hand motion from all subjects was trained and tested with a proportion of 70 % and 30 %, respectively. Each training and testing of machine learning was performed based on the EMG features (RMS, IEMG, WL, DASDV, MYOP, and WAMP). **Figure 3** shows that machine learning is able to discriminate 18 motions (including rest condition) based on 22 machine learning as shown in Figure 3 but with different accuracy. The top of the bar chart indicated the average accuracy of the machine learning and followed by the standard deviation of the accuracy.

After the exploration (training and testing), it was proven that a quadratic SVM have the best accuracy in the classification of the eighteen hand motions. The average accuracy (from all subjects) based on the quadratic SVM for RMS, IEMG, WL, DASDV, MYOP and WAMP features were 71.9±7.31%, 75.7±4.36%, 76.68±4.42%, 76.68±4.65%, 70.22±5.81%, and 74.88±5.02%, respectively. The statistics show that the WL features have the best accuracy in classifying the 18 hand motions. Among other kernels (linear, cubic, fine, medium, and coarse Gaussian), the SVM with the quadratic kernel showed the best accuracy. This is because the EMG characteristic have a non-linear response to the hand motion and the quadratic is the fittest model in the SVM classifier. The cubic kernel was also fit to be used on the SVM model due to the characteristic of the EMG signal as approached to the cubical. On the other hand, the accuracy of the machine learning based on coarse Decision Tree showed the worst classifier in classifying the 18-hand motions. The accuracy of the coarse Decision Tree with RMS feature was 51.4±15.90%.

In this work, our finding proved that the SVM based on quadratic kernel was superior to the other machine learning. A representation of a confusion matrices which was calculated based on SVM (quadratic kernel) was shown in **Figure 4**. The accuracy on the confusion matrices was resulted based on WL feature. Each type of hand motion showed a different accuracy. In the matrices, the hand motion of P7 shows higher accuracy (91%) than the others. The lowest accuracy was founded at hand motion of P6 (62%). In the matrices, the misclassification was founded on hand motion of P2 (72%) which is influenced by hand motion of P3 (21%) and P17 (17%). This problem is reasonable because of the P2, P3 and P17 related to the same EMG activities (as shown in **Figure 1**).

The effectiveness of the number of muscles in machine learning was investigated in this study. In this investigation, we selected the SVM with the quadratic kernel as proven to be the best machine learning to classify the 18-hand motions. The highest accuracy of the SVM (quadratic kernel) for ten channels is 76.68±4.42% (mean ± standard deviation). The waveform length (WL) feature was also selected due to the higher accuracy than the others. In the investigation, we reduced the number of muscles (channels) then each number of muscles, we calculated the accuracy of the machine learning. The accuracy of the SVM (quadratic kernel) for the various number of channels is shown in **Figure 5.** It was proved that the accuracy of machine learning was influenced by the number of channels. The bar chart shows the mean value of the accuracy (from all subjects) and the error bar shows the standard deviation of the accuracy. The statistics show that there is a high relationship between the number of channels and accuracy (0.98). It means that the smaller number of channels will reduce the accuracy of machine learning. For a statistics test, all of the accuracy data from all channels (from all the subjects) were pooled and statistical ANOVA single factor was performed. The statistics show that the p-value is higher than 0.05 (p-value=0.257 and F=1.355). It means that there was no significant difference of accuracy among the number of channels. This finding shows that the machine learning was able to classify 18 hand motions (including rest condition) with 3 channels with no significant difference of accuracy if we compared to the machine learning with 9 channels (p-value>0.05). In this investigation, the minimum number of channels which are able to classify the 18-hand motions was 3 channels. When we tried the number of channels was 2, the machine learning was not able to train the model because of the number of the class was too many (18 class or hand motion). This result proved that the minimum number of channels which can be used to discriminate the 18 types of hand motion was 3 channels.

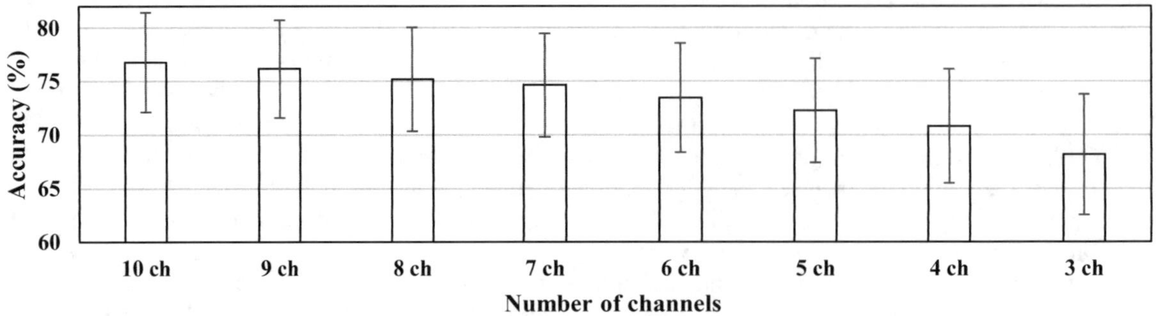

Figure 5. The accuracy of the machine learning based on a number of channels. The accuracy was calculated using quadratic SVM and WL feature.

As a comparison of this study, previous works also used this dataset (Ninapro, DB3) to investigate the hand motion from an amputee person. At the point of view of the machine learning, Atzori et. al. also founded that the SVM has better accuracy (45.19±14.75%) to discriminate the 18 hand motions than other machine learning (Random Forest and K-NN) [15] [20]. Al-timemy et. al. investigated a classifier to discriminate 6 motions (class) with 3 different forces (low, medium and high). They found the accuracy was 93%, 90.3%, and 82% for low, medium, and high force, respectively [21]. Even though the accuracy of the Al-timemy's work was better, but they used only 6 classes in their training and testing. Here, we can conclude that the number of class affects the accuracy of machine learning. The quality of the EMG signal was affected by many parameters such as 50 Hz noise from line power, noise artifact from hand motion, electrode position, sweat, and muscle fatigue. Muscle fatigue is a parameter which needs to be considered when we use the EMG signal as a control signal because the muscle fatigue could influence the amplitude and frequency of the EMG signal [22]. Therefore, this condition should be considered due to it will affect the feature extraction results and the classifier decision. Thus, in the future study, those parameters are needed to be consider in order to maintain the accuracy of machine learning.

4. Conclusion

After the investigation of 22 machine learning in classifying 18 classes (hand motion), the results showed that the SVM with the non-linear kernel (quadratic and cubic) has the best accuracy. Each feature extraction shows different accuracy, however, waveform length (WL) feature showed better accuracy than the others. We have investigated the effect of the number of channels (from 10 to 3 channels) to the accuracy of the machine learning. Up to 3 channel, the machine learning still able to recognize the18 hand motions but with different accuracy. However, the statistics show that a number of channels did not influence the accuracy significantly (p-value >0.05. In the future works, we suggested using the results of this study (which consisted of the number of channels, the type of feature extraction, and the machine learning) so that an effective prosthetics hand based on EMG signal can be developed.

References

[1] Ziegler-graham K, MacKenzie E J, Ephraim PL and Brookmeyer R 2008 *Archives of Physical Medicine and Rehabilitation* **89** 422–429.

[2] Utomo B, Triwiyanto, Suhartini, Luthfiyah S., and Mudjiono U, 2018 *International Conference on Science and Applied Science (ICSAS)* (Indonesia : Solo / AIP Published Omline) vol. 2014 p. 020105-1 - 020105-8.

[3] Leonardis D. *et al.* 2015 *IEEE Trans. Haptics* **8** 140–151.

[4] Triwiyanto, Wahyunggoro O, Nugroho HA, and Herianto, 2016 *Proc. International Annual Engineering Seminar, InAES* (Indonesia : Yogyakarta/IEEE Xplore) p 176 - 181.

[5] Segil JL, Huddle SA, and Weir RF, 2016 *IEEE Trans. Neural Syst. Rehabil. Eng.* **4320** c.25(6):1-10.
[6] Geethanjali P, 2016 *Med. Devices Evid. Res.* **9** 247–255.
[7] Engeberg ED, 2013 *Biomed. Signal Process. Control*, **8** 6–15.
[8] Kyriakopoulos KJ, Artemiadis PK, 2007 *IEEE/ASME international conference on advanced intelligent mechatronics*,(Switzerland: Zurich/IEEE Xplore) pp. 1–6.
[9] Fleischer C and Hommel G, 2008 *IEEE Trans. Robot* **24** 872–882.
[10] De Luca CJ, 1997 *J. Appl. Biomech* **13** 135–163.
[11] Triwiyanto, Wahyunggoro O, Nugroho H A, and Herianto, 2017 *Adv. Electr. Electron. Eng.* **15**, 448–458.
[12] Phinyomark A, Phukpattaranont P, and Limsakul C, 2012 *Expert Syst. Appl.* **39**7420–7431.
[13] Fougner A L, Stavdahl Ø, and Kyberd P J, 2014 *J. Neuroeng. Rehabil.* 1–13.
[14] Khushaba R N, Takruri M, Valls J, and Kodagoda S, 2014 *Neural Networks* **55** 42–58.
[15] Atzori M, Gijsberts A, Castellini C, Caputo B, Hager A-G M, Elsig S, Giatsidis G, Bassetto F, & Müller H, 2014 *Sci. Data* **1** 1–13.
[16] Khushaba R N, Member S, Al-timemy A H, Al-ani A, and Al-jumaily A, 2017 *IEEE Trans. Neural Syst. Rehabil. Eng.* **25**, 1821 - 1831.
[17] Martini, 2012 *Fundamental of Anatomy and Physiology*, 9th ed. Boston: Pearson Education.
[18] Stegeman D. and Hermens H, 2007 *Línea. Dispon. en http//www. med. ...*, pp. 108–112.
[19] Gubta V, Bhavsar A, 2017 *International Conference on Computer Analysis of Images and Patterns* CAIP 2017 p. 354–366.
[20] Atzori M, Gijsberts A, Castellini C, Caputo B, Hager A-G M, Elsig S, Giatsidis G, Bassetto F, & Müller H 2016 *J. Rehabil. Res. Dev.* **53**, 345–358.
[21] Al-timemy A H, Khushaba R N, and Bugmann G, 2015 *IEEE Trans. Neural Syst. Rehabil. Eng.*, **4320** c.
[22] Triwiyanto, Wahyunggoro O, Nugroho H A, and Herianto H, 2018 *Comput. Electr. Eng.* **71** 284–293.

Journal of Physics: Conference Series **1373** (2019) 012052 doi:10.1088/1742-6596/1373/1/012052

Adaptive Trajectory Tracking Control System of Two-Wheeled Robot

Nur Uddin

Department of Informatics, Universitas Pembangunan Jaya
Jl. Cendrawasih Raya B7/P, Ciputat, Tangerang Selatan 15413, Indonesia

E-mail: nur.uddin@upj.ac.id

Abstract. A design of two-wheeled robot (TWR) trajectory tracking control system (TTCS) using model reference adaptive control method is presented. The TTCS is to steer the TWR track a desired trajectory. The tracking system dynamic is represented by a posture error dynamic. The posture error dynamic is derived based on the robot kinematics. Assuming small heading angles the posture error dynamic is approached by a a linear system. Model reference adaptive control (MRAC) is applied in designing the TTCS based on the linear posture error dynamic. The TTCS performace is evaluated through computer simulation. The simulation results show that the designed controller is able to make the TWR track a desired trajectory.

1. Introduction

Mobile robots are robots that able to travel the whole body from one location to another location. The mobile robot is also known as the vehicle. The mobile robots have locomotion mechanisms to move in the environment. There are several types of locomotion mechanism and including walk, jump, run, slide, skate, swim, fly, and roll [1]. Roll is the most popular locomotion mechanisms due to simple and cheap in the implementation. The roll mechanism is implemented using wheel and motor. Mobile robots using the roll locomotion mechanism are known as the wheeled mobile robot.

Autonomous mobile robot is one of the most interesting research topics in the last three decade. Several types of autonomous mobile robot have been developed, for examples: unmanned aerial vehicle (UAV), unmanned ground vehicle, unmmaned water surface vehicle, and unmmaned underwater vehicle. The autonomous mobile robots have ability to move on a desired route without a human driver. Self driving car is an example of the autonomous mobile robot that is currently developed in automotive sector. One of the most essential system in autonomous mobile robot is trajectory tracking control system (TTCS). The TTCS is used to steer the robot such that the robot move from departure point to destination point through a desired trajectory. The TTCS is replacing the human driver in steering the mobile robots or vehicles.

Initial work on developing autonomous mobile robot was presented in [2]. They developed a four wheel mobile robot to be autonomous. The robot has four wheels where two wheels are active wheels and the other wheels are passive wheels. Each active wheels are driven by an electric motor. In developing a TTCS, the trajectory tracking system was formulated by the robot posture error dynamics. Posture is a description of the position and orientation of the robot. Posture error is the different of the robot posture to the reference posture. The posture error dynamics were derived based on the robot kinematics on a planar space. The derivation resulted in a non-linear dynamics system. Lyapunov-based control method was applied in design the TTCS. Experimental results showed that the designed TTCS was successfully steering the robot move autonomously form indoor to outdoor. Since then the works in [2] was the followed by several works on developing autonomous mobile robots. Difference kinds of control method have been applied in designing the TTCS for various mobile robots, for examples: backstepping [3,4], Lyapunov direct's method in [5,6], sliding mode control in [7], and adaptive control in [8].

A two-wheeled robot (TWR) is a ground mobile robot where the robot is equipped by two wheels to support the robot body. Each wheels are driven by a high torque electric motor. The use of two wheels makes the robot to be very easy in maneuver. This becomes the advantage of the TWR where the maneuverability is higher than other ground mobile robots with more wheels. However, the use of two wheels makes the TWR statically unstable. Active stabilization is required in the TWR. Several works in TWR active stabilization have been presented [9–14]. The stabilized TWR can be operated for various purposes. One of them is applying the stabilized TWR for autonomous mobile robot.

An autonomous TWR was introduced in [15]. The autonomous TWR was built by developing two control systems. The first control system is an active stabilization to stabilize the TWR. The second control system is trajectory tracking control system to steer the TWR move on a desired track. Both control sytems were designed by applying optimal control method. Experimental results show that the developed autonomous TWR was able to move autonomously on a desired straight line path from indoor to outdoor. The results show that an autonomous TWR required two control modes: TWR stabilization and trajectory tracking. Analysing the control design and experimental results show that coupling between the both control modes can be neglected for small angular velocity of heading motion. Since then research works on autonomous TWR have been presented by applying different control design methods. Partial states feedback linearization was applied in developing an autonomous TWR [16]. Adaptive control schemes were also been applied, for example: adaptive backstepping control [17], adaptive sliding mode control [18], and neural networks [19].

Investigation of the presented works on autonomous TWR show that most of the developed tracking control system works to track a reference which is a set of linear and angular velocities or certain desired trajectory but not arbitrary desired trajectory in earth-fixed coordinate system [20]. In real world application, the desired trajectory should be arbitrary. Expressing the trajectory in earth-fixed coordinate system will be an advantage in real world application. It will make the trajectory tracking control system be ready integrated with the available earth-fixed coordinate navigation system, for an example the global positioning system (GPS). A work on developing a TWR tracking control system for an arbitrary trajectory in earth-fixed coordinate system was presented in [20]. The tracking system dynamics were derived base on the robot kinematics on a fixed-frame coordinate system and robot body coordinate system. Model predictive control (MPC) was applied in design the trajectory tracking control system. A work on developing trajectory tracking system of TWR based on a fixed-frame coordinate system has been presented by applying optimal control method [21]. A simple design method through pole domination approach was also applied in designing the trajectory tracking control system based on the fixed-frame coordinate system.

This study presents a design of trajectory tracking control system based on fixed-frame coordinate system using model reference adaptive control (MRAC). The MRAC is a part of adaptive control method where a reference model is used as the reference for the closed loop system. The control system will make the closed loop system behave as the reference model. Presentation of this paper is organized as follows. The trajectory tracking system is derived based on the robot kinematics as described in Section II. It will be end up with a linear states

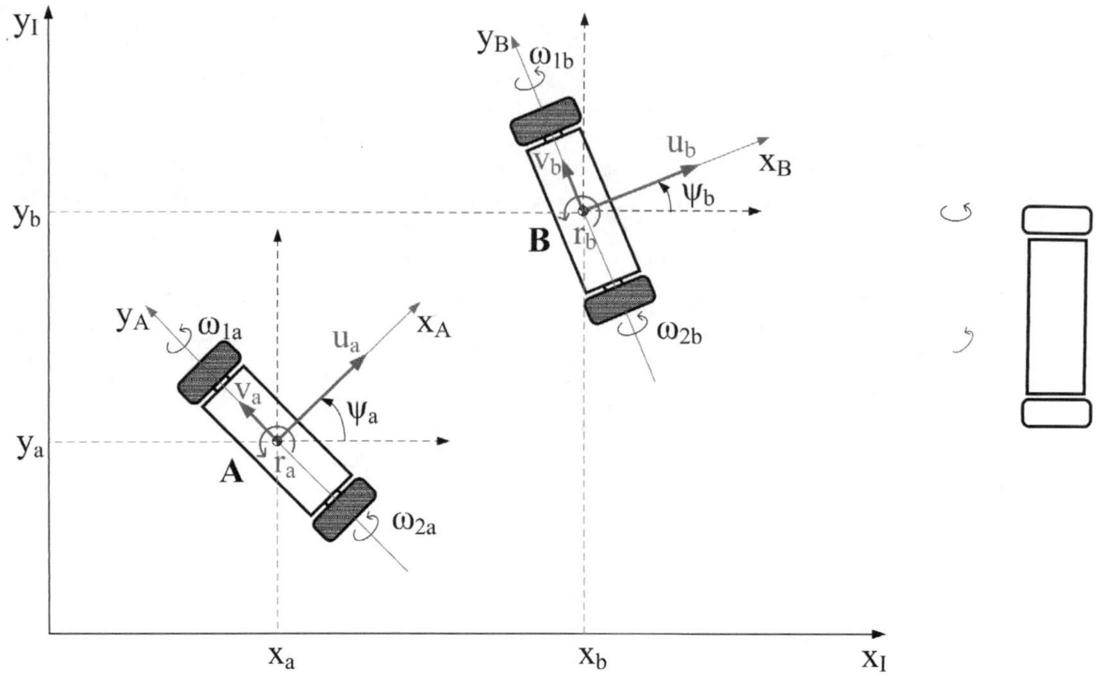

Figure 1. TWR coordinate systems.

equation of the posture error dynamics. Theory of MRAC applied in the trajectory tracking control system design is described in Section III. A study case of applying the MRAC theory in designing the trajectory tracking control of a TWR and the simulation results are presented in Section IV. Finally, conclusions of the work and the future works are presented in Section V.

2. Kinematics of Two-Wheeled Mobile Robot on Planar Space

Figure 1 shows two units of two-wheels robot (TWR), TWR A and TWR B, on a planar space. A fixed-frame coordinate system $X_I Y_I$ is used to represent position of the both TWRs. The TWR A is located at (x_a, y_a) and the TWR B located at (x_b, y_b). Both TWRs move independently on the space. The TWR A moves with linear velocity u_a and angular velocity r_a, while the TWR B moves with linear velocity u_b and angular velocity r_b. Position is not enough to represent the TWRs on the planar space. Both TWRs may have the same position in the planar space but moving direction or orientation of both TWRs may be different. Moving direction of the TWR is known as the TWR heading represented by heading angle, ψ. The TWR heading angle is varying as the TWR makes angular motion. In oder to express the angular motion, new coordinate systems call the robot body coordinate systems are defined at each TWRs. Origin of the robot body coordinate system is located at the center mass of the robot. The robot body coordinate system sticks on the robot body and moves following the robot movements. The robot body coordinate system for TWR A is $X_A X_A$ and for the TWR B is $X_B Y_B$. The x-axis of the robot body coordinate system points to the TWR forward movement and the y-axis points perpendicularly to the left side of the x-axis. Deviation in orientation of the robot body coordinate system to the fixed-frame coordinate system is representing the TWR heading heading angle. It shown in the figure, the heading angle of TWR A is represented by ψ_A which is the deviation angle of the X_A axis to the X_I axis, while heading angle of the TWR B, ψ_B, is the deviation angle of the X_B axis to the X_I axis.

The TWRs on the planar space need to be represented by position and orientation. Expression of the position and orientation is known as the posture. The position describes location of the origin of TWR body coordinate system with respect to the origin of inertial coordinate system. Position of the TWR A and the TWR B are (x_a, y_a) and (x_b, y_b), respectively. The orientation described the orientation different of the TWR body coordinate system with respect to the inertial coordinate system. Orientations of the both TWRs are denoted by ψ_a for the TWR A and ψ_b for the TWR B. Therefore, postures of the both TWR are given as follows:

$$\xi_a = \begin{bmatrix} x_a \\ y_a \\ \psi_a \end{bmatrix} \tag{1}$$

$$\xi_b = \begin{bmatrix} x_b \\ y_b \\ \psi_b \end{bmatrix} \tag{2}$$

where ξ_a and ξ_b are the postures of TWR A and TWR B, respectively.

The TWR A and TWR B move independently on the planar space. The TWR A moves with linear velocity u_a and angular velocity r_a while the TWR A heading angle is ψ_a as shown in the Figure 1. Both velocities are expressed in the TWR A body coordinate system. The TWR A movement can be expressed in the fixed-frame coordinate system through the following transformation:

$$\dot{x}_a = u_a \cos\psi_a \tag{3}$$
$$\dot{y}_a = u_a \sin\psi_a \tag{4}$$
$$\dot{\psi}_a = r_a \tag{5}$$

The TWR B with heading angle ψ_b moves with linear velocity u_b and the angular velocity r_b. Expressing the TWR B movement in the fixed-frame coordinate system is done by the following transformation:

$$\dot{x}_b = u_b \cos\psi_b \tag{6}$$
$$\dot{y}_b = u_b \sin\psi_b \tag{7}$$
$$\dot{\psi}_b = r_b \tag{8}$$

Assuming that the TWR B becomes a reference for the TWR A. The TWR A is desired to track the TWR B moving trajectory. The tracking trajectory means that the TWR A posture approaches the TWR B posture. Define a posture error as the deviation of the TWR A posture to the TWR B posture given as follows:

$$\xi_e = \xi_b - \xi_a = \begin{bmatrix} x_b - x_a \\ y_b - y_a \\ \psi_b - \psi_a \end{bmatrix} = \begin{bmatrix} x_e \\ y_e \\ \psi_e \end{bmatrix}, \tag{9}$$

where ξ_e is the posture error. For the trajectory tracking purpose, the TWR A needs to be equipped by a trajectory tracking control system (TTCS). The TTCS is to steer the TWR A for tracking the TWR B trajectory. Objective of the TTCS is to make the posture error convergence to zero. State feedback control can be applied for converging the posture error. Postures of TWR A and TWR B defined in (1) and (2) are expressed in an inertial coordinate system. Therefore, the posture error (9) is in the inertial coordinate system. The posture error is being a feedback in the TTCS. Due to TTCS is applied in the TWR A, it is necessary to

Journal of Physics: Conference Series **1373** (2019) 012052 doi:10.1088/1742-6596/1373/1/012052

express the posture error in the TWR A body coordinate system. Transforming the posture error from the inertial coordinate system into the TWR A body coordinate system is given as follows:

$$\xi_{e_A} = R_{AI}\xi_e \tag{10}$$

where ξ_e is the posture error in the inertial coordinate system, R_{AI} is the transformation matrix from the inertial coordinate system into the TWR A body coordinate system, and ξ_{e_A} is the posture error in the TWR A body coordinate system. Both R_{AI} and ξ_{e_A} are given as follows:

$$R_{AI} = \begin{bmatrix} \cos\psi_a & \sin\psi_a & 0 \\ -\sin\psi_a & \cos\psi_a & 0 \\ 0 & 0 & 1 \end{bmatrix}. \tag{11}$$

$$\xi_{e_A} = \begin{bmatrix} x_{e_A} \\ y_{e_A} \\ \psi_{e_A} \end{bmatrix}. \tag{12}$$

Substituting (9), (12) and (11) into (10) results in:

$$\begin{bmatrix} x_{e_A} \\ y_{e_A} \\ \psi_{e_A} \end{bmatrix} = \begin{bmatrix} \cos\psi_a & \sin\psi_a & 0 \\ -\sin\psi_a & \cos\psi_a & 0 \\ 0 & 0 & 1 \end{bmatrix} \begin{bmatrix} x_b - x_a \\ y_b - y_a \\ \psi_b - \psi_a \end{bmatrix}. \tag{13}$$

Due to both TWRs are moving, the TWR postures are changing every time and therefore the posture error. Dynamics of the posture error is obtained by calculate time derivative of the (10) that results in:

$$\dot{\xi}_{e_A} = \dot{R}_{AI}\xi_e + R_{AI}\dot{\xi}_e \tag{14}$$

and the complete expression is [21]:

$$\begin{bmatrix} \dot{x}_{e_A} \\ \dot{y}_{e_A} \\ \dot{\psi}_{e_A} \end{bmatrix} = \begin{bmatrix} r_a y_{e_A} + u_b \cos\psi_{e_A} - u_a \\ -r_a x_{e_A} + u_b \sin\psi_{e_A} \\ r_b - r_a \end{bmatrix}. \tag{15}$$

The posture error dynamic is used as the mathematics model in design the TTCS. The posture error dynamic is represented by a non-linear system in (15). Simplifying the TTCS design, the posture error dynamics is approached by a linear system. Assuming the heading angle error to be small such that $\psi_e \approx 0$ such that

$$\sin\psi_e \approx \psi_e \tag{16}$$

$$\cos\psi_e \approx 1 \tag{17}$$

and therefore the (15) can be approximated by:

$$\begin{bmatrix} \dot{x}_{e_A} \\ \dot{y}_{e_A} \\ \dot{\psi}_{e_A} \end{bmatrix} = \begin{bmatrix} r_a y_{e_A} + u_b - u_a \\ -r_a x_{e_A} + u_b \psi_{e_A} \\ r_b - r_a \end{bmatrix}. \tag{18}$$

The (18) is a linear system and can be represented in a state space form as follows:

$$\dot{x} = Ax + Bu \tag{19}$$

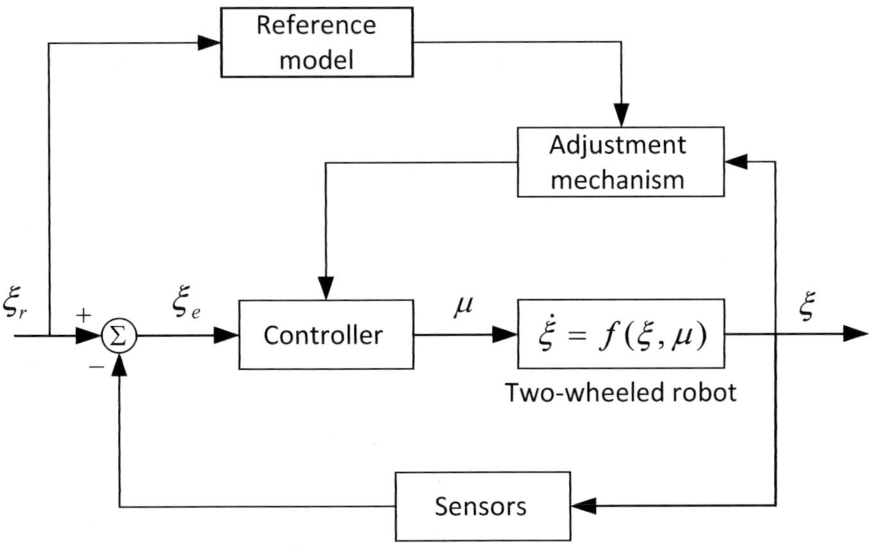

Figure 2. Block diagram of model reference adaptive control (MRAC) [22].

with

$$x = \begin{bmatrix} x_{e_A} \\ y_{e_A} \\ \psi_{e_A} \end{bmatrix}, A = \begin{bmatrix} 0 & r_a & 0 \\ -r_a & 0 & u_b \\ 0 & 0 & 0 \end{bmatrix}$$

$$B = \begin{bmatrix} 1 & 0 \\ 0 & 0 \\ 0 & 1 \end{bmatrix}, u = \begin{bmatrix} u_1 \\ u_2 \end{bmatrix} = \begin{bmatrix} u_b - u_a \\ r_b - r_a \end{bmatrix}.$$

where x is the system states, A is the system matrix, B is the input matrix, and u is the system input.

3. Adaptive Control Design

Figure 2 shows block diagram of model reference adaptive control for trajectory tracking control system. A comprehensive literature of adaptive control can be found in several text books, for example [22, 23]. Formulating adaptive control design for the trajectory control is derived as follows [23, 24].

Recall the linear trajectory tracking system (19) and given as follows:

$$\dot{x} = Ax + Bu. \tag{20}$$

Defined the following asymptotic linear system as a reference model for the (20):

$$\dot{x}_m = A_m x_m + B_m r. \tag{21}$$

where the x_m is the model reference system state, A_m is the model reference system matrix, B_m is the model reference input matrix, and r is the model reference input. Vectors and matrices in the model reference have the same size with the respectively vectors and matrices in the linear system (20). For the linear trajectory tracking system (20), define the control input u as follows:

$$u = -K_1 x + K_2 r. \tag{22}$$

where K_1 and K_2 are the control gain matrices, x is the system state of the linear trajectory tracking system (20), and r is the model reference input in (21). Applying the control input (22) into (20) results in a closed loop system as follows:

$$\dot{x} = (A - BK_1)\, x + BK_2 r. \tag{23}$$

For a simpler expression, defining $A_c = A - BK_1$ and $B_c = BK_2$ and substituting them into (23) results in:

$$\dot{x} = A_c x + B_c r. \tag{24}$$

System response of the closed-loop system (24) is desired to approach system response of the reference model (21). The following states error is defined to measure how the (24) system response approaches the (21) system response:

$$e = x - x_m. \tag{25}$$

Time derivative of the state error results in the state error dynamics and given as follows:

$$\dot{e} = A_c x - A_m x_m + (B_c - B_m)\, r. \tag{26}$$

Mathematics manipulation of (26) by adding and subtracting $A_m x$ results in:

$$\dot{e} = A_m e + (A_c - A_m)\, x + (B_c - B_m)\, r. \tag{27}$$

The expression of (27) can be simplified by defining the following two matrices: $\tilde{A} = A_c - A_m$ and $\tilde{B} = B_c - B_m$. Substituting the both matrices into (27) results in:

$$\dot{e} = A_m e + \tilde{A}x + \tilde{B}r. \tag{28}$$

Approaching of the (24) system response to the (21) system response of is indicated by the state error converging to zero. Lyapunov stability theorem is applied to make the state error e converges to zero [25]. Define the following function as a Lyapunov function candidate:

$$V = e^T P e + \mathrm{tr}\left(\tilde{A}^T \tilde{A}\right) + \mathrm{tr}\left(\tilde{B}^T \tilde{B}\right), \tag{29}$$

where P is a positive definite matrix and tr is a mathematics operator for calculating trace of a matrix. Time derivative of V is given by:

$$\dot{V} = \dot{e}^T P e + e^T P \dot{e} + 2\mathrm{tr}\left(\tilde{A}^T \dot{\tilde{A}}\right) + 2\mathrm{tr}\left(\tilde{B}^T \dot{\tilde{B}}\right) \tag{30}$$

and further calculation will result in:

$$\dot{V} = e^T \left(A_m^T P + P A_m\right) e + x^T \tilde{A}^T P e + e^T P \tilde{A} x + r^T \tilde{B}^T P e + e^T P \tilde{B} r + 2\mathrm{tr}\left(\tilde{A}^T \dot{\tilde{A}}\right) + 2\mathrm{tr}\left(\tilde{B}^T \dot{\tilde{B}}\right). \tag{31}$$

Define the following positive definite matrix Q

$$Q = -(A_m^T P + P A_m). \tag{32}$$

and substituting it into (31) results in:

$$\dot{V} = -e^T Q e + 2x^T \tilde{A}^T P e + 2r^T \tilde{B}^T P e + 2\mathrm{tr}\left(\tilde{A}^T \dot{\tilde{A}}\right) + 2\mathrm{tr}\left(\tilde{B}^T \dot{\tilde{B}}\right). \tag{33}$$

The (32) is known as the algebraic Riccati equation. The positive definite matrix P is obtained by solving the algebraic Riccati equation (32). In order to make the $\dot{V}$ in (33) negative semi-definiteness, the following condition needs to be satisfied:

$$\operatorname{tr}\left(\tilde{A}^T\dot{\tilde{A}}\right) = -x^T\tilde{A}^TPe \tag{34}$$

and

$$\operatorname{tr}\left(\tilde{B}^T\dot{\tilde{B}}\right) = -r^T\tilde{B}^TPe. \tag{35}$$

Analyzing size of the matrices and the vectors, it is known that both $x^T\tilde{A}^TPe$ and $r^T\tilde{B}^TPe$ are scalar such that:

$$\operatorname{tr}\left(x^T\tilde{A}^TPe\right) = \operatorname{tr}\left(\tilde{A}^TPex^T\right) = x^T\tilde{A}^TPe \tag{36}$$

$$\operatorname{tr}\left(r^T\tilde{B}^TPe\right) = \operatorname{tr}\left(\tilde{B}^TPer^T\right) = r^T\tilde{B}^TPe. \tag{37}$$

Substituting (36) into (34) results in:

$$\operatorname{tr}\left(\tilde{A}^T\dot{\tilde{A}}\right) = -\operatorname{tr}\left(\tilde{A}^TPex^T\right) \tag{38}$$

and a further calculation will result in

$$\dot{K}_1 = (B^TB)^{-1}B^TPex^T. \tag{39}$$

Through the same procedure, substituting (37) into (35) results in

$$\operatorname{tr}\left(\tilde{B}^T\dot{\tilde{B}}\right) = -\operatorname{tr}\left(\tilde{B}^TPer^T\right) \tag{40}$$

and a further calculation will end up with:

$$\dot{K}_2 = -(B^TB)^{-1}B^TPer^T. \tag{41}$$

Both (39) and (41) are the key to make the $\dot{V}$ negative semi-definite.Because K_1, K_2, and e are bounded, using Barbalat's lemma, the negative semi-definiteness of $\dot{V}$ is sufficient to make (28) asymptotic stable such that the error goes to zero [22].

The (39) and (41) are known as the mechanism for updating the control gain matrices. Therefore the control gain matrices in the MRAC are given as follows:

$$K_1(t) = K_1(t_0) + \int_{t_0}^{t} \dot{K}_1 dt \tag{42}$$

$$K_2(t) = K_2(t_0) + \int_{t_0}^{t} \dot{K}_2 dt \tag{43}$$

where $K_1(t_0)$ and $K_2(t_0)$ are the initial control gain matrices.

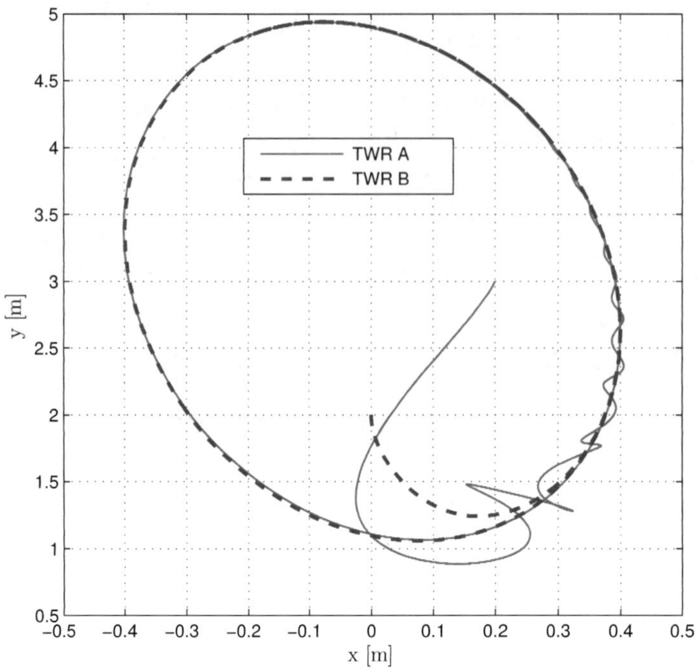

Figure 3. Trajectory the TWR A compared to the the reference trajectory (TWR B trajectory).

4. SIMULATION

A trajectory tracking control system (TTCS) of two-wheeled robot (TWR) is designed by applying the model reference adaptive control (MRAC). The TTCS is applied in a TWR which is called the TWR A. The TWR A is desired to track trajectory of another TWR that is called the TWR B. The TWR B trajectory is known as the reference trajectory for the TWR A.

Performance of the TTCS is evaluated through a computer simulation. Scenario of the simulation is described as follows. TWR A is initially at an idle position located at $(0.2, 3)$ in a fixed-frame coordinate system with heading angle $90°$. Both linear and angular velocities of the TWR A are zero. TWR B is initially located at $(0, 2)$ in the fixed-frame coordinate system and the heading angle is $0°$. The TWR B moves with linear velocity 2 m/s and angular velocity 0.5 rad/s such that the TWR B will move on an ellipsoid trajectory. The TWR A is desired to track the TWR B trajectory.

MRAC is applied to design a trajectory tracking control system (TTCS) for the TWR A. An asymptotically stable linear system with the following parameter is defined as the model reference (21) for the MRAC.

$$A_m = \begin{bmatrix} -10 & 0 & 0 \\ 0 & -2 & 0 \\ 0 & 0 & -50 \end{bmatrix}, \qquad B_m = \begin{bmatrix} 1 & 0 & 0 \\ 0 & 1 & 0 \\ 0 & 0 & 1 \end{bmatrix}$$

Control design parameter of the MRAC is represented by the matrix Q in (32). The value of matrix Q is selected as follows:

$$Q = \begin{bmatrix} 100 & 0 & 0 \\ 0 & 10 & 0 \\ 0 & 0 & 10 \end{bmatrix} \tag{44}$$

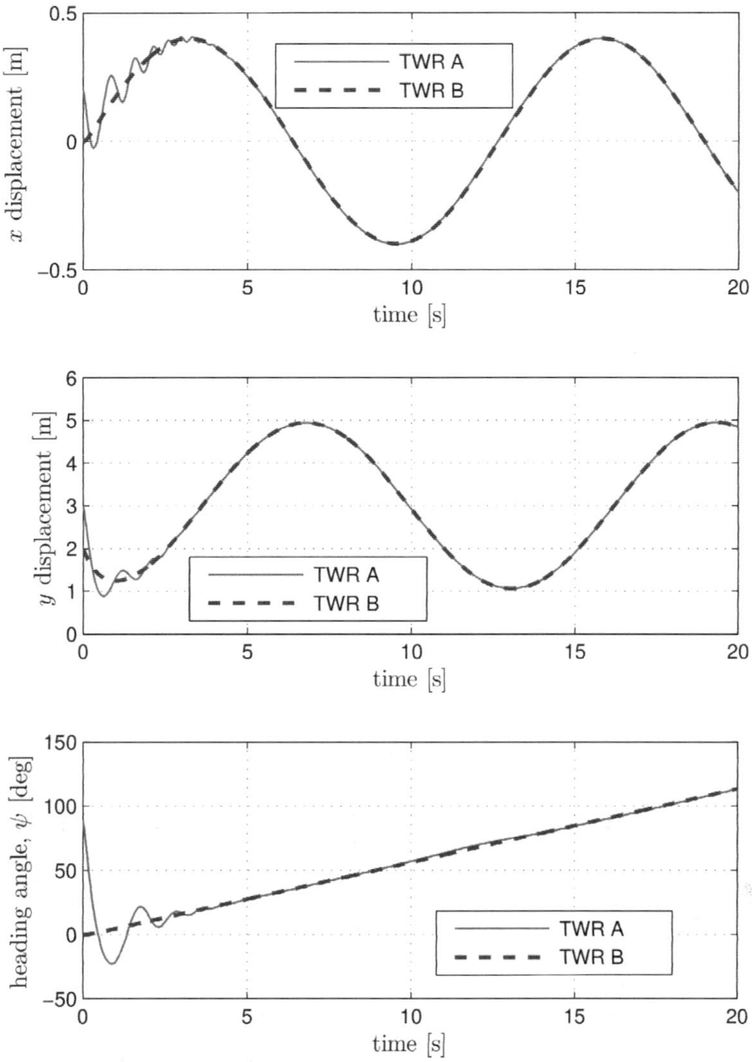

Figure 4. Posture of the TWR A compared to the the reference posture (TWR B posture).

Initial value of control gain matrices in (22) are:

$$K_1 = \begin{bmatrix} 1 & 0 & 0 \\ 0 & 1 & 0 \\ 0 & 0 & 1 \end{bmatrix} \quad \text{and} \quad K_2 = \begin{bmatrix} 1 & 0 & 0 \\ 0 & 1 & 0 \\ 0 & 0 & 1 \end{bmatrix}. \tag{45}$$

The designed TTCS is applied in the TWR A and the trajectory tracking is simulated. Simulation results are shown in the following figures. Trajectories of the TWR A and the TWR B are shown in Figure 3. Both TWRs are initially located in difference location. Figure 4 shows a comparison of the TWR A posture and the TWR B posture. The simulation results show that the designed TTCS makes the TWR A posture approches the TWR B posture as the reference posture. The simulation results show that the TWR B moves from the initial position and make an ellipsoid trajectory. The designed TTCS is able to make the TWR A track the TWR B trajectory.

5. CONCLUSION

A design of trajectory tracking control system (TTCS) of two-wheeled robot (TWR) has been presented. The TTCS was designed by applying model reference adaptive control (MRAC) method. Simulation results show that the designed TTCS was able to make the TWR A track the reference trajectory which is the TWR B trajectory. Designing TTCS using MRAC method was done by defining a model reference and a positive definite matrix Q. An asymptotically stable linear system was chosen as the model reference. The positive definite matrix Q was representing the control design parameter. The control performance is depend on the selected model reference and the matrix Q.

6. ACKNOWLEDGMENTS

The author acknowledges financial support from Universitas Pembangunan Jaya through internal research grant program on the scheme of Penelitian Unggulan Perguruan Tinggi 2018-2019.

References

[1] Siegwart R, Nourbakhsh I R, Scaramuzza D and Arkin R C. 2011 MIT press
[2] Kanayama Y, Kimura Y, Miyazaki F and Noguchi T. 1990. IEEE International Conference on 1990. Proceedings.,
[3] JIANGdagger Z P and Nijmeijer H. 1997 *Automatica* **33** 1393–1399
[4] Dhaouadi R and Hatab A. 2013 *Advances in Robotics & Automation* **2** 1–7
[5] Jiang Z P. 2002 *Automatica* **38** 301–309
[6] Mazenc F, Pettersen K and Nijmeijer H. 2002 *IEEE Transactions on Automatic Control* **47** 1759–1762
[7] Xu J, Wang M and Qiao L. 2015 *Ocean engineering* **105** 54–63
[8] Sahu B K and Subudhi B. 2014 *International Journal of Automation and Computing* **11** 299–307
[9] Uddin N. 2017 International Conference on (IEEE) pp 121–125
[10] Uddin N, Nugroho T A and Pramudito W A. 2018 Journal of Physics: Conference Series vol 1007 p 012056
[11] Uddin N, Nugroho T A and Pramudito W A. 2017 ICITISEE 2nd (IEEE) pp 229–234
[12] Wu J, Zhang W and Wang S. 2012 *Mathematical Problems in Engineering* **2012**
[13] Nomura T, Kitsuka Y, Suemitsu H and Matsuo T. 2009 ICCAS-SICE, 2009 (IEEE) pp 4687–4692
[14] Chan R P M, Stol K A and Halkyard C R. 2013 *Annual Reviews in Control* **37** 89–103
[15] Ha Y S *et al.* 1996 *Robotics and autonomous systems* **17** 65–80
[16] Pathak K, Franch J and Agrawal S K. 2005 *IEEE Transactions on robotics* **21** 505–513
[17] Cui R, Guo J and Mao Z. 2015 *Nonlinear Dynamics* **79** 501–511
[18] Yue M, Wang S and Sun J Z. 2016 *Neurocomputing* **191** 44–54
[19] Yang C, Li Z, Cui R and Xu B. 2014 *IEEE Transactions on NN and Learning Systems* **25** 2004–2016
[20] Yue M, An C and Sun J Z. 2018 *International Journal of Control, Automation and Systems* **16** 265–274
[21] Uddin N 2018 *Jurnal Infotel* **10** 90–97 ISSN 2460-0997
[22] ˚Astr¨om K J and Wittenmark B. 2013 *Adaptive control* (Courier Corporation)
[23] Ioannou P A and Sun J. 1996 *Robust adaptive control* vol 1 (PTR Prentice-Hall Upper Saddle River, NJ)
[24] Uddin N. 2018 *Proc. of the 12th SEATUC* (Yogyakarta, Indonesia)
[25] Khalil H K. 2002 *Noninear Systems* 3rd ed (Prentice-Hall, New Jersey)

Development of Door Safety Fingerprint Verification using Neural Network

Anton Yudhana[1,*], Sunardi[1] and Priyatno[2]

[1] Department of Electrical Engineering Universitas Ahmad Dahlan Yogyakarta, Indonesia
[2] Department of Informatics Technology Universitas Ahmad Dahlan Yogyakarta, Indonesia

[*]Email: eyudhana@mti.uad.ac.id

Abstract. Fingerprints can be used as natural keys. The unique fingerprint of a human being is potential for the security system. This research develops the door safety fingerprint verification using Neural Network. The component used in this research is fingerprint C3 as input processed by Arduino Uno ATMega 2560. After a fingerprint scanning process is done, the microcontroller identifies the fingerprint image. From the microcontroller forwarded the relay module process with solenoid output. Arduino can be read by a computer using the Arduino IDE. Identification during registration via C3 sensor verified using the MATLAB application with the Neural Network method. The design created produces forward and backward solenoid motion, after being given an electric current between 5 to 12 Volts. The MATLAB application will verify the fingerprint image and display the image from the results of the verification done with the user. Research has been tested 10 times with various types of fingerprints with 100% verified so that this study was declared successful.

1. Introduction

The biometric system is mostly developed for security technology because it can fulfill two functions, namely identification, and verification. Biometry has characteristics that cannot be lost, cannot be forgotten and is not easily falsified because its existence is inherent in humans, one another will not be the same so that the uniqueness is guaranteed. Among the systems of fingerprint biometry technology, hand geometry, retina (eye), sound and face systems that have been widely used are fingerprint systems.

Fingerprints are lines that are found on the skin of the fingers of the right hand or the left hand of a person. Fingerprint security systems were used in the United States by E. Henry in 1902. Henry used the fingerprinting method for his workers to identify them in overcoming multiple wages. Henry's system uses a ridge pattern (Ridge is the groove in the skin of the fingers), which is centered on the finger pattern, especially the index finger. Many studies prove that someone's fingerprints with others have unequal ridge patterns. The ridge pattern is also not the same as the offspring even though in one family. Ridge pattern is formed in the sixth and seventh weeks or when the fetus. When a 13-week-old fetus will look clear and will not change for life. Ridge patterns can change if scratched due to injury, burns, disease or other causes. [1]

Fingerprint patterns have a high level of security, this system has been widely used in offices, companies, schools, government, hospitals and so on. Besides having a high level of security, the fingerprint system is also easy to use. Safety using a biometric system is rarely found. The safety of the door using a biometric system as safety is rarely found.

Most in Indonesia, securing the door still uses a manual key, if the key is lost, it takes time to open it. When using fingerprints, one of our fingers will become a key without a forgotten effect or left somewhere. In the market, many types of equipments that use biometric systems have been sold, but most are fingerprint attendance. In fingerprint presence, the sensor with programming has become one. So the function of the fingerprint attendance tool is only used for attendance. Many companies that have produced fingerprint presence are FingerPrint.

This research makes door safety using a fingerprint sensor. The sensor used for the security of this door stands alone meaning it has not been programmed from the company. The components that will be used are a set of microprocessors contained in the Arduino Uno ATMega 2560, the input process using Fingerprint C3 and to verify the fingerprint used when the input is required for the Neural Network system which is included in the Soft Computing science category. Neural networks actually adopt the ability of the human brain that can provide stimulants or stimuli, the process carried out, and provide output. Acquisition of output comes from variations in stimulation and processes that occur in the human brain. Humans are capable of processing information is the result of the complexity of the process in the brain. For example, by children, they are able to learn to recognize friends, even though they don't know what algorithms are used. The extraordinary computing power carried out by the human brain is an advantage in the study of science.

Neural Network was first done in 1943 when Warren McCulloch and Walter Pitts had introduced the calculation of the Neural Network model. They do a combination of several simple processing units together which can provide overall development improvements in computing power. [2]

From the background above, the author will examine "making fingerprint identification machines as door protectors". Previous research [1][3]-[13] and have done research on fingerprints or micro-controller. But previous studies have not verified using the software. In this case, the researcher will build and design a system to verify fingerprints that have been scanned and stored.

2. Methods

Fingerprints have a periodic orientation and structure in the form of a composition of dark lines of rising skin (ridges) and bright lines of furrows that twist around form a different pattern. Although the lines of the hand flow are formed differently, the special properties of the fingerprints called minutiae are unique to each individual. These characteristics form a special pattern consisting of termination or branching of the groove. To check whether two fingerprints are from the same finger or not, experts detect the minutiae. Automatic Fingerprint Identification System (AFIS) The Automated Fingerprint Identification System will retrieve and compare these characteristics to determine compatibility. Dermatoglyphics or fingerprint patterns are defined as parallel images of dermal tendrils on the fingers and toes, and the palms and soles of the feet. Anatomically dermatoglyphics make rough surfaces on the palms of the fingers, soles of the feet, and toes which function in helping the process of holding or resting so as not to slip [6]. The formation of dermatoglyphics begins with the proliferation of volar epidermal basal epithelial cells around the 10th to 11th weeks of pregnancy. The cells then form folds and become epismetic.

2.1. Main Form of Fingerprint.

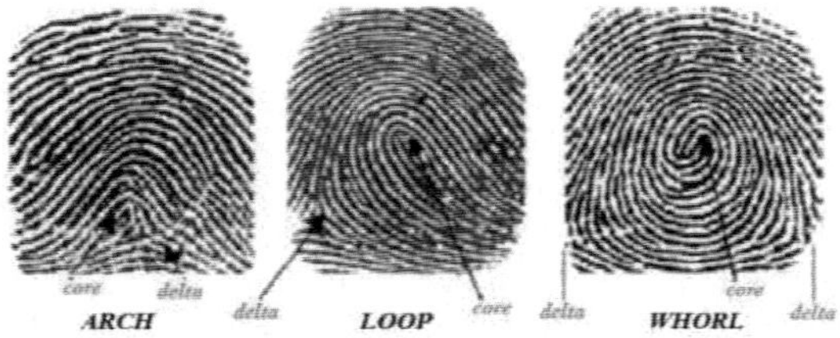

Figure 1. Fingerprint Form

There are three forms of fingerprints, namely arc (arch), hook (loop), and circle (whorl) as in Figure 1. The principal form is divided into several sub-groups, namely the shape of the arc divided into a plain arch and tented arch, the form of divided being Ulnar loop and Radial loop, while the circle form is divided into Plain whorl, Central pocket loop whorl, Double loop whorl, and Accidental whorl. The main difference between the three basic forms lies in the existence of cores and deltas in the fingerprint painting [5].

Each person's fingerprints are different from others, although there are so many fingerprints found in this world that the form of fingerprints is divided into three forms. But these three forms of fingerprints are not the same between one person and another. This is due to differences in embryo formation. As with the retina of the human eye, although many eyes are similar the retina of the eye is not the same as one person with another.

A Whorl is the main form of fingerprints, has 2 deltas and at least one circular line inside the pattern area, runs in front of both deltas. This type of whorl consists of Plain whorl, Central pocket loop whorl, Double loop whorl, and Accidental whorl. Fingerprint patterns can be seen in Figure 2.

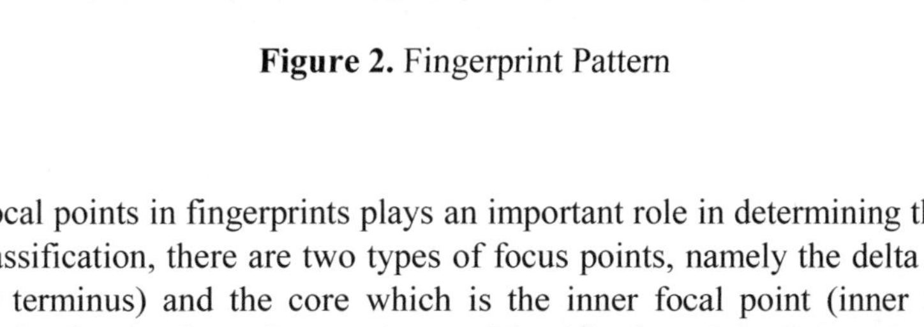

Figure 2. Fingerprint Pattern

2.2. Focus Point

The presence of focal points in fingerprints plays an important role in determining the classification of fingerprints. In classification, there are two types of focus points, namely the delta which is the outer focal point (outer terminus) and the core which is the inner focal point (inner terminus). Not all fingerprints have a focal point depending on the type/classification of the fingerprint.

A core is a midpoint that is located on the deepest loop fingerprint line and furthest from the delta as in Figure 3. It can be said that the core is the midpoint or center of the fingerprint painting. In determining the location of the core, the following provisions apply:

- The cores are placed in the loop line whose position is the deepest.
- If the innermost line does not contain an end-line or a short line that rises until the shoulder height of the core is placed on the shoulder of the cradle whose position is further from the delta position.
- If the deepest contain contains n (odd) line-end fruit that rises until the shoulder of the core is placed at the end of the center line.
- If the deepest link contains n (even) end-line fruit that rises to the shoulder of the loop core is placed at the end of the line with the middle position and is located farthest from the delta position.

The display of the midpoint of the fingerprint can be seen in Figure 3.

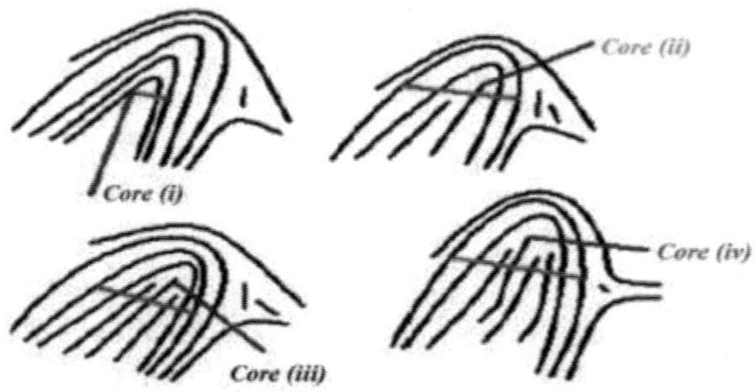

Figure 3. Middle Fingerprint

But in practice, the location of cores cannot always be determined by the rules mentioned in Figure 3. There are two cases which generally can obscure in determining the location of these cores. The first case is an additional line (appendage). The appearance of this appendage can damage the fingerprint line if the appendage appears on the fingerprint line which is located on the curved area between the shoulder of the hook line. This Appendage is considered as a stop line for hooks that are right outside. The second case is the existence of the deepest loop line (line of connection) which intersect each other (interlocking loop). In this case, the two intersecting lines are considered to be one of the hooks where the lines inside areas if the lines are rising up to the shoulder height of the loop.

Delta in the everyday sense is a cluster found in the mouth of a river of water that flows into the sea or lake always carrying mud and rock so that over time a group of islands is formed called a delta. Delta on the fingerprint is a point/line found in the center of the line type line. Delta is a focal point located in front of the center of the separation of the main lines (type lines). The main line of the painting is the two most innermost lines of a number of lines that are parallel (parallel) and separate and (inclined) to cover the principal of the painting (pattern area). The main painting is an area / white room surrounded by a line of type lines which is a place for fingerprint lines. In fact, not all fingerprints have deltas but there are also fingerprints that have more than one delta. Delta fingerprint can be seen in Figure 4.

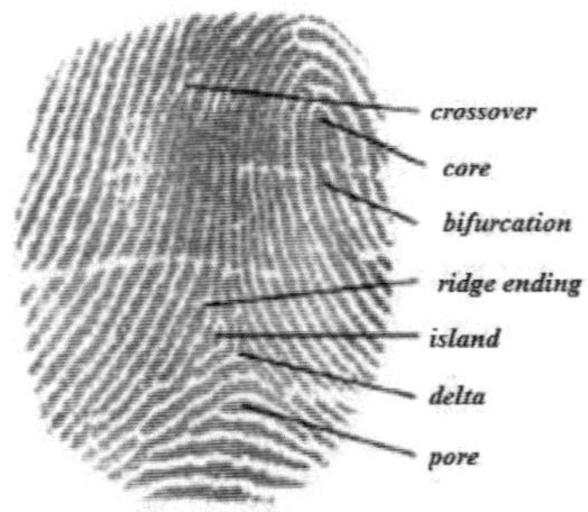

Figure 4. Delta Fingerprint

Figure 4. is a fingerprint delta image. The conditions that must be considered in determining the position of the delta are as follows:

- Delta cannot be placed on a splitting line that does not open towards the core.
- If you have to choose between dividing lines and possible deltas, then the splitting line is selected.
- If there are two or more lines that meet the delta requirements then select the one closest to the core.
- Delta must not be placed in the middle of the line between the main lines but must be placed at the end of the line that is closest to the center of the baseline.

2.3. Ridge Counting

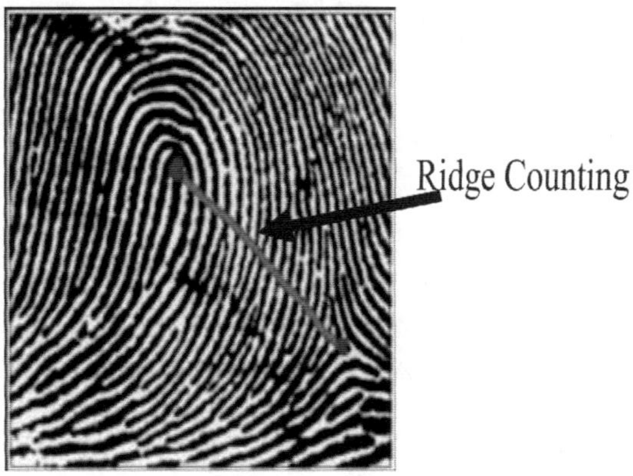

Figure 5. Ridge Counting

Ridge counting as shown in Figure 5 is a line number that touches or crosses the shadow line drawn between delta and core (delta and core are not included in the calculation of line numbers). The lines that look very fine (thin) are cracked by thick lines called insipient ridge, and this line is not counted because it usually does not always exist, however small the size of the dot (short ridge) must be treated as a line fingerprints that are counted, if they are as thick as the other lines.

Journal of Physics: Conference Series 1373 (2019) 012053 doi:10.1088/1742-6596/1373/1/012053

2.4. Fingerprint Formula

Fingerprints are unique and different forms for each person so the fingerprint formula will be different for each person. The fingerprint formulation (classification formula) is the affixing of marks on each column of the fingerprint card showing the interpretation of the principal form, the number of line numbers, the shape of the loop, and the course of the line.

3. Proposed System

The component used in this study is a series of electronic devices combined with Arduino default software and fingerprints will be verified using the MATLAB application with the Neural Network method.

3.1. Tools and materials

Tools and materials are components needed in the manufacture of door safety identification machines using the Neural Network method. In general, the fingerprint identification system to open the door automatically consists of input, process, and output.

The input part of this system is the fingerprint sensor, which uses a C3 sensor. The process part of this system is the Arduino Uno ATMega 2560 with the help of a computer. The computer is used as a bridge from the Arduino IDE application to the Arduino Uno ATMega 2560 device in order to understand the commands given by the computer. The power supply is used as a power supply or electric power so that the device remains in a standby condition. The output part is used by Relay to drive a DC motor using solenoid. Tools and materials can be seen in Table 1.

Table 1. Tools and Materials

No	Name	Information
1.	Mikrokontroler	Arduino ATMega 2560
2.	Motor DC	Selenoid
3.	RTC	Data Logger SD Card
4.	I2C	16x02 Serial LCD
5.	Fingerprint sensor	C3
6.	Current connector	Relay
7.	Resource	Adaptor 5-12 Volt

The design begins with the registration of fingerprints on the C3 sensor which is forwarded to the microcontroller. Registration is done by attaching the fingerprint twice, the first registration will be verified by the microcontroller and the second registration will be verified by the microcontroller the second time each registration and verification will be displayed by I2C. Matching the fingerprint database that has been stored in the microcontroller will be adjusted to the Arduino IDE program. If the admin or user will open the door, the fingerprint can be placed on the sensor to open the door. When the fingerprint is placed on the C3 sensor, the input is a fingerprint image that is matched with the fingerprint database if it matches the database, then the solenoid will open.

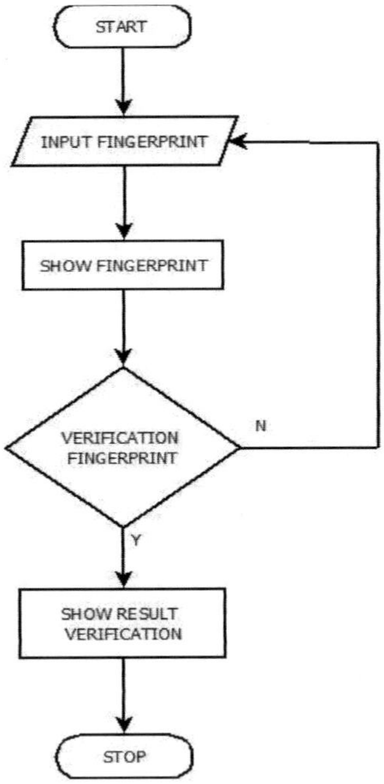

Figure 6. Flowchart Verification

Designing fingerprint verification on the MATLAB application as shown in Figure 6 is that the fingerprint that has been saved will be verified by entering the fingerprint image into the MATLAB application. The image will be displayed according to the fingerprint image selected in the database, the fingerprint image will be verified and displayed according to the selected data.

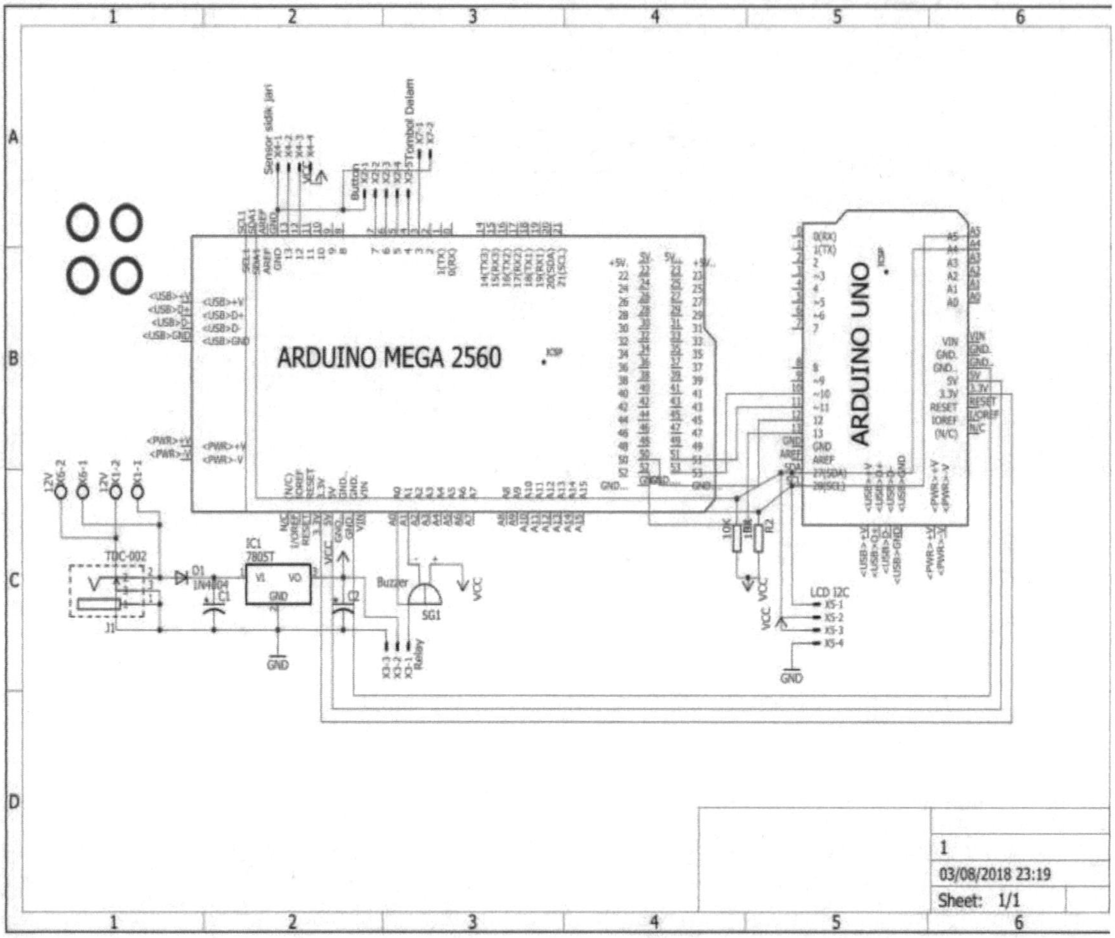

Figure 7. Tool Design

The explanation of Figure 7 is:

1. C3 sensor uses serial communication (RX and TX), RX is placed in PIN code 12 while TX is placed in PIN 13.
2. Buzzer uses A1 PIN.
3. A relay using A0 PIN.
4. Menu button uses PIN 6.
5. The + button uses a PIN 4.
6. Button - uses PIN 5.
7. Exit button uses PIN 7.
8. Open button from inside using PIN 3.

Journal of Physics: Conference Series **1373** (2019) 012053 doi:10.1088/1742-6596/1373/1/012053

9. RTC uses I2C communication.
10. SDA uses the SDA1 PIN.
11. SCL uses SCL1 PIN.
12. SD Card Module uses SPI communication.
13. MOSI is located at PIN 51.
14. MISO is located at PIN 50.
15. SCK is located at PIN 52

From the system design that has been made to ensure that fingerprints that have been registered in the identified inputs will be verified using the MATLAB application program with the Neural Network method. The verification process is done by attaching a finger to the surface of the C3 sensor which is continued with the fingerprint verification process that is received and will be compared to the fingerprint image on the database. Possibilities that occur when fingerprints have been successfully retrieved as a result of their verification is successful or failed.

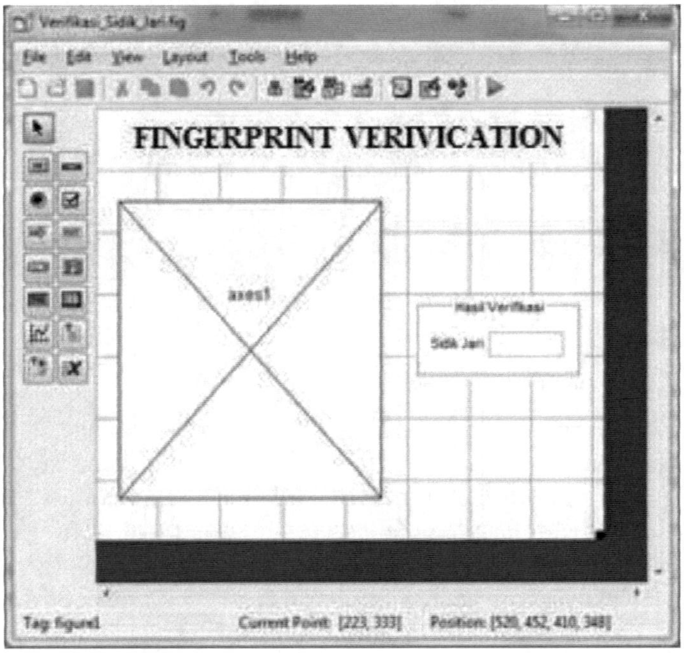

Figure 8. Design of fingerprint verification

The use of a GUI (Graphical User Interface) as shown in Figure 8 in the fingerprint verification design will look for fingerprints that have been used by the user. In this case, the MATLAB application program will search for data that has been used. Design it by pressing the Menu button then select Open. After the image appears, the Process menu is used and select Neural Network, the system will search for the fingerprint data.

Journal of Physics: Conference Series **1373** (2019) 012053 doi:10.1088/1742-6596/1373/1/012053

4. Result

4.1. Tool testing

Use of the default software from Arduino because its function is simple and user-friendly with the components used in the research. The Arduino IDE software testing is done by entering the initial display program listing as a condition for the Arduino component to be read by the computer and Arduino will be connected to the computer with the source used in the Arduino IDE. First, fill in the program listing or sketch aimed at the C3 sensor so that the C3 sensor can be read by the Arduino IDE. The second program listing is intended for I2C LCD (Inter-Integrated Circuit Liquid Crystal Display), so I2C can read the program listings used in the Arduino IDE and can show whether the fingerprint is received or not received by the Arduino IDE. The next stage is the creation of program listings on the relay so that it can be read by the Arduino IDE and can work if there is an incoming current, the relay will decide the positive and negative pins otherwise if the relay does not receive the current, the relay will connect both positive and negative pins.

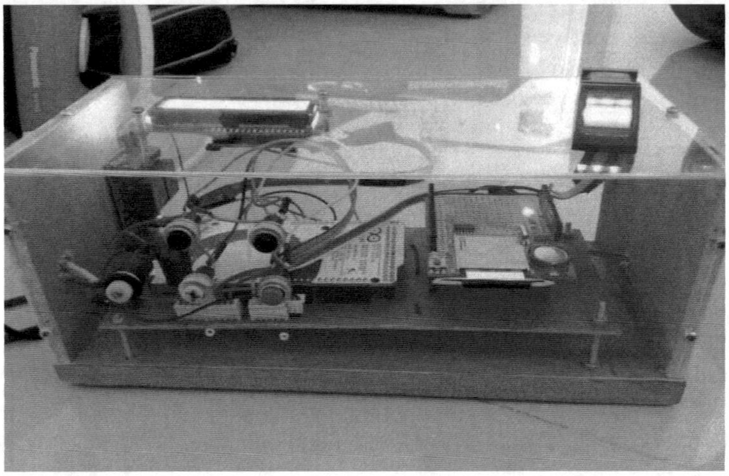

Figure 9. Connect the Arduino component

Testing of Arduino components as shown in Figure 9 becomes one of the objectives in the research. The first step taken is to connect the C3 sensor with 4 pins found in Arduino, I2C connected to the Arduino board both positive terminal and a negative terminal, the relay connected with the logger shield data is forwarded to the Arduino component. Before the sensor is used in the test first enter the current into the Arduino Uno ATMega 2560 with a power of 5-12 Volt and disconnect the USB cable from the PC because the microcontroller has stored the data provided. This system will match the fingerprint database that has been stored in the Arduino IDE program by means of fingerprints placed on the sensor to open the door. When there is a fingerprint placed on the fingerprint sensor, the input in the form of fingerprints will be matched with the database that has been stored, if the fingerprint does not match the database, then the solenoid will not open. After verification of the fingerprint in accordance with the database, the solenoid will open with a duration of 10 seconds.

4.2. Fingerprint verification using MATLAB

The Neural Network Training Process can be seen in Figure 10.

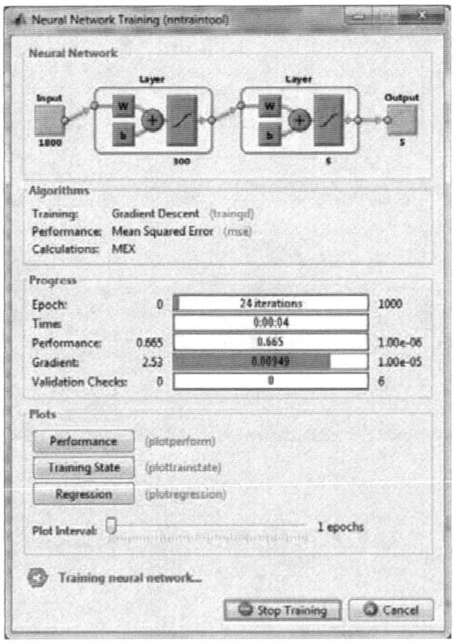

Figure 10. Training Neural Network

The training process was carried out as shown in Figure 10, indicating that there was a training process for 300 times. The next step is to build a network with 1800-300-5-5 architecture, which means it has 1800 input data where the first hidden layer contains 300 neurons and the second hidden layer contains 5 neurons. And has 5 output data. In neurons arranged in the form of layers. The simplest formation of Neural Network is a single layer. The single layer works, the input layer that comes from the source node is projected directly to the output layer of the neuron (computational node), but does not apply otherwise. This capital is a type of feedforward network, but what is meant by single layer is the output of the network. The target data is used in this training class value goal error (MSE) of 0.055494 achieved in the 24th epoch as shown in Figure 11.

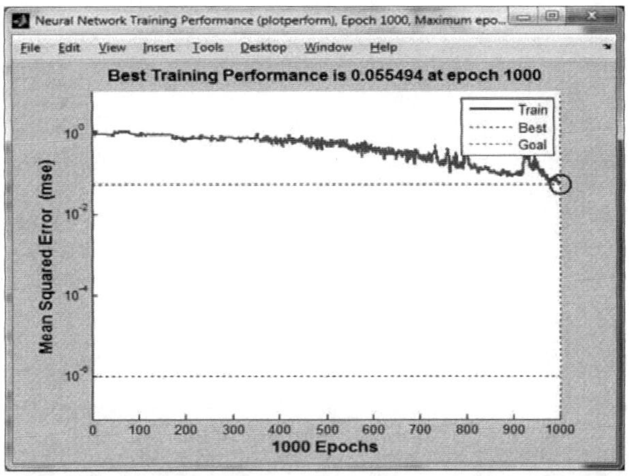

Figure 11. Plot Performance

Figure 12 shows a graph of the MSE (Mean Squared Error) value. The graph is displayed by pressing the performance button shown in Figure 10. While the appearance of the training plot can be seen in Figure 13.

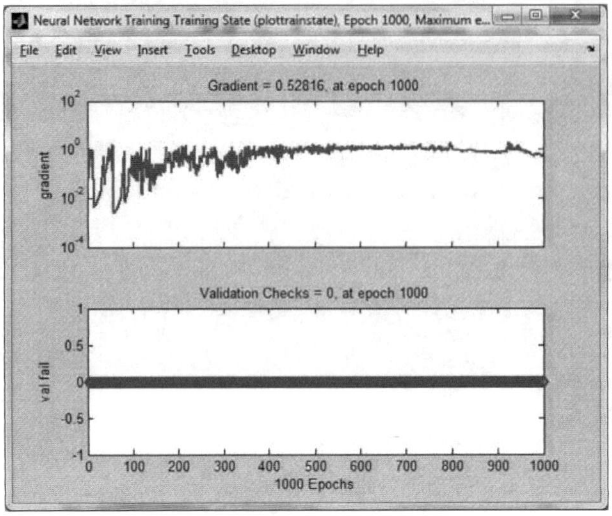

Figure 12. Training State

Figure 13 is a training plot view. The tracking plot is obtained by pressing the training state button in Figure 10. The training process with an epoch number of 1000 can produce a gradient value of 0.52816 and a check validation value of 0. While the correlation coefficient R is 0.81806 as shown in Figure 13 which shows plot regression display by pressing the Regression button in Figure 10.

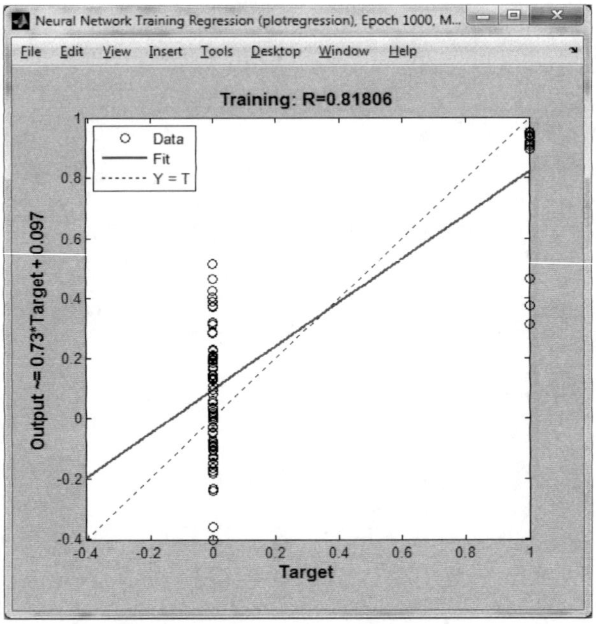

Figure 13. Plot Regression

Journal of Physics: Conference Series **1373** (2019) 012053 doi:10.1088/1742-6596/1373/1/012053

5. Conclusion

The process of making the fingerprint security door safety machine has been successfully designed and built with the following conclusions (1) The system can be applied to the Arduino component as a door safety device with a fingerprint pattern and (2) Verification made using the MATLAB application with the Neural Network method has been running well.

References

[1] Rintjap A. S, Sherwin R. U. A. 2014 *J. T. Elektro-ft :E-Journal* 1-5.

[2] Atmega M, Yuliza E, and Kalsum T.U 2015 *J. Media Infotama* **11** 1–10.

[3] Ainur A, Hastuti J, and Nugraha Z.S. 2009 *Jurnal Kesehatan dan Kedokteran Indonesia* **1** 2-11.

[4] Nugroho E 2008 *Introduction to Biometrics* [Online]. Available: https://www.scribd.com /document/375278873/pengenalan-biometrik. [Accessed: 30-May-2018].

[5] Hidayati F 2007 *AntroUnairdotNet* **4** 30-41.

[6] Anwar Y.E,Soedjarwanto N, and Repelianto A.S 2015 *Electr. J. Rekayasa Dan Teknol. Elektro* **9** 31-41

[7] Yudhana A, Sunardi, Din J, Abdullah S, and Hassan R.B 2010 *Telkomnika* **8**187–194.

[8] Kusumodestoni R.H and Zyen A.K. 2015 *Prediksi* **6** 7

[9] Muchtar H, Hidayat A. 2017 *Jurnal Teknologi* **9** 1–5.

[10] Fadlil A, Riadi I, and Aji S. 2017 *Int. J. Adv. Comput. Sci. Appl* **8** 42–50.

[11] Yudhana A, and Djalil A. 2017 Int. J. Adv. Comput. Sci. Appl **8** 509–512.

[12] Yudhana A 2018 *The Science and Information Organization* **9** 183–189.

Journal of Physics: Conference Series 1373 (2019) 012054 doi:10.1088/1742-6596/1373/1/012054

The Monitoring of Corn Sprouts Growth Using The Region Growing Methods

Anton Yudhana [1], Rusydi Umar[2], Fadlillah Mukti Ayudewi [2]

[1]Department of Electrical Engineering, Faculty of Industrial Technology, Universitas Ahmad Dahlan, Jln. Ring Road Selatan, Tamanan, Banguntapan, Yogyakarta.
[2] Magister of Information Technology, Universitas Ahmad Dahlan, Jln. Prof. Dr. Soepomo, SH, Janturan, Umbulharjo, Yogyakarta.

*Email: eyudhana@ee.uad.ac.id, rusydi@mti.uad.ac.id, fadlillah1607048015@webmail.uad.ac.id

Abstract. The aim of this research is to apply plant growth monitoring function using region growing method and Matlab application with an object of corn sprouts and parameters of length and time. The research process is done by taking pictures digitally change the color level (Red Green Blue) to grayscale, cropping the image and then calculate the growth of the plant using region growing. From the results of the study, region growing can be used to calculate the growth of plant length with parameters of length and time. The results of manual measurements show the growth of plant length by an average of 2.1 cm / day then the results of measurements with region growing show an average plant length of 2.3 cm / day, this result shows the comparison between manual and region growing measurements is 0.2 cm. The difference between manual calculation and region growing is caused by the accuracy of shooting and image cropping process.

1. INTRODUCTION

Image processing is computer science that developed in various fields such as archaeology, astronomy, biomedical, industrial and agrology. Image processing focused on how to transform image picture using a certain technique that can deliver information from inside the image easier and faster.

In general, digital image processing is image processing (two dimension image) using a computer with the aim that the image of easily interpreted and transformed into other images, by man or machine. Based on the value of the digital image pixel can be arranged into three types namely the image of colors (RGB), grayscale image, and binary image [1].

The types of digital image processing activities are grouped into two, that is improve the quality of the image, so that the image is more easily interpreted and process information in the images for the purpose of introducing an automatic object. [2]. In digital image processing, the input and output are images, but the output image has better quality than the input image [3]

Journal of Physics: Conference Series 1373 (2019) 012054 doi:10.1088/1742-6596/1373/1/012054

Indonesia is a country with a very large area and has a rapid population growth rate which of course requires abundant crops to meet the food needs of its population, so monitoring plants are needed to reduce the risk of crop failure. Monitoring of plants starts from the plants in the form of seeds until the harvest period. Monitoring plant growth must be carried out continuously and requires accuracy to produce maximum crop yields, while manual plant growth monitoring methods can be said to be less than optimal due to human physical factors, such as fatigue and inaccuracy.

Germination is the initial process of the growth of grain plants which was originally in a dormant condition which then undergoes physiological changes so that plant seeds develop into young plants called sprouts. Epigeal and hypogeal are two types of germination in which epigeal is germination that brings cotyledons out of the soil surface, one example of plants that experience epigeal germination is green beans, while hypogeal is germination which perpetuates cotyledons in the soil, one example of plants that have germination this is corn [4].

Image segmentation into homogeneous areas is one of the important problem in image processing. In image analysis, segmentation is the first step that must be done. Image processing methods, such as threshold histogram, clustering, or region growing are some of the image segmentation methods that have been used. Image processing to monitor plant growth can be done using the region growing method with parameters of length and time [5].

The segmentation process is the initial process in the separation of objects. Inaccuracies in process results are influenced by inaccurate segmentation processes. There are three types of segmentation processes, namely classification based segmentation, edge-based segmentation, and region-based segmentation [6]. Image segmentation is an image processing process where objects involving visual perception will be more easily recognized and analyzed by separating between region objects and background regions [7].

Seed Region growing segmentation is an image segmentation method that uses region-based techniques, adjacent pixels in the same area have the same visual features as gray levels, color values, or textures. The principle of Seed Region Segmentation is growing by starting the determination of seed points, from the seed area will be developed with the addition of each seed where the neighboring pixels have a characteristic similarity to the seed. Seed Region growing segmentation will try to find accurate image segmentation into regions with proportions where each component of the interconnected area meets one of the seeds [7].

The growing region is used to create a region map by grouping pixels against region-regions, where labels on image pixels are a classification of regions for those pixels [7]. This study aims to apply plant growth monitoring functions using the region growing method and the application of Matlab with corn sprout objects and parameters of length and time.

2. Literature Review

Image processing has been carried out to identify size, one of which is a study conducted by Rika Favoria Gusa entitled "Digital Image Processing to Calculate the Former Area of Tin Mining" by utilizing satellite imagery. This study used a field survey to obtain coordinate data and the size (length and width) of the former tin mining land. The initial image is a satellite image with a color image (RGB) which then changes the color level of the grayscale image. Furthermore, improvements are made to grayscale images through a contrast stretching process before the changes are made into binary images. To calculate the number of pixels in the former tin mining area, morphological images are applied, namely erosion and dilation in binary images [1].

The research was conducted by Suryani, entitled "Seed Growing Segmentation Region and Momentum Backpropagation Neural Network for Classification of White Blood Cell Types". This study discusses the growing region seed used in image preprocessing. Then feature extraction produces four numerical parameters of the image, namely the area of leukocytes, the edge of the area, the balance and the nucleus ratio. The results showed that the seed growing region can segment leukocyte images by 96.795% which can then be extracted properly [7].

The research was conducted by Syahrul Awalludin and Dessy Irmawati entitled "Image Processing for Identification of Eggs by Size". The image processing process starts with changing the color to a grayscale color level. Then the threshold method is used to change the grayscale color level to binary. The filling holes morphology is used to eliminate noise from the threshold image until it displays the egg classification decision [7].

Another study entitled "Web-Based Plant Growth Monitoring Using Machine Vision" conducted by L. Sutiarso. This study discusses the use of digital image processing to monitor plant growth in real time. Segmentation is the initial process of image processing that is used to recognize plant objects. The algorithm used is the method of excess green and color normalization. The Otsu method is used to calculate plant area. The test results show the success rate of monitoring systems using machine vision by 70% [8].

The research was conducted by W. A. Saputra entitled "Seeded Region Growing in HIS Color Space for Tuna Image Image Management" with seed and threshold parameters. The initial image is an RGB color image which is then transformed into the HSI space (hue saturation intensity). Seed region growing is used in the hue space segmentation process. RAE (relative foreground area error), MAE (misclassification error) and MHD (modified hausdroff distance) are evaluation methods used to test 30 images of tuna. The final results showed that the values of RAE, ME and MHD were 5.40%, 1.53%, and 0.41% respectively [2].

The research was conducted by F.N Riza, entitled "Region growing segmentation to detect nodules as an indication of lung cancer". This study discusses the segmentation of lung objects and nodule objects based on the morphology of the nodules themselves. Then the nodule area is calculated based on the number of pixels that form the nodule area. The test results show that segmentation can be performed on slices of lung CT images [7].

The research conducted by Diah Anggraeni entitled "Digital Image Segmentation of Fish Using Thresholding and K-Means Method". Image processing methods carried out are Thresholding and K-Means. The threshold process begins with inputting a digital image and then changing the color level to grayscale, local or global threshold selection, edge detection process using The Canny operator or Laplacian and dilation process. The K-Means process begins with inputting images, changing the level of grayscale color by the K-Means method. The results showed that the local Threshold method was able to separate fish eye objects [9].

The research conducted by A. Aprilia, discussed the Identification of the Quality of Rice Using Digital Images. Image processing is done to identify the quality of rice by using a decision tree classification. The testing process through digital image processing is done by looking at the white value, the net value, and the whole value of rice. The HSV value (Hue, Saturation, Value) is used to analyze the net value and white value of rice, while the analysis of the area of the object area is done to get the whole value of rice. The test results showed an accuracy value of 96.67% [4]. Research that has been done shows that image processing and region growing can be used to carry out classification and monitoring.

3. Methodology

The study began with planting corn sprouts for 7 days using cotton planting media instead of soil. Image processing begins with taking the image of sprouts every day for 7 consecutive days. The image obtained is in the form of a color image (RGB) which is then changed to grayscale to facilitate the subsequent image processing. The image is converted into a grayscale color level and an image cropping process is performed. The image that has gone through the cropping process is processed using data growing regions using the Matlab application. In region growing process the original image is changed to color level into gray (grayscale) level, then the image cropping process is performed. Image that has gone through the process of cropping then do data processing using the Region Growing, after that is measuring the corn spourt length by the image using image tool. The diagram flow of plant growth monitoring using region growing method are follows:

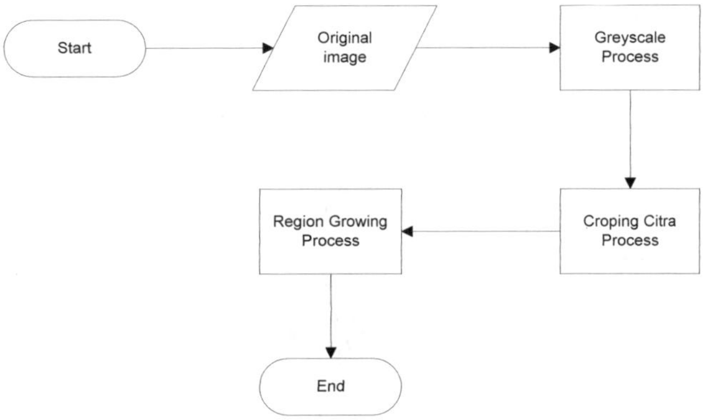

Figure 1. Flowchart monitoring growth of corn sprouts using region growing method

4. Result And Discussion

The results of the study included a change of image into a gray scale image are as follows:

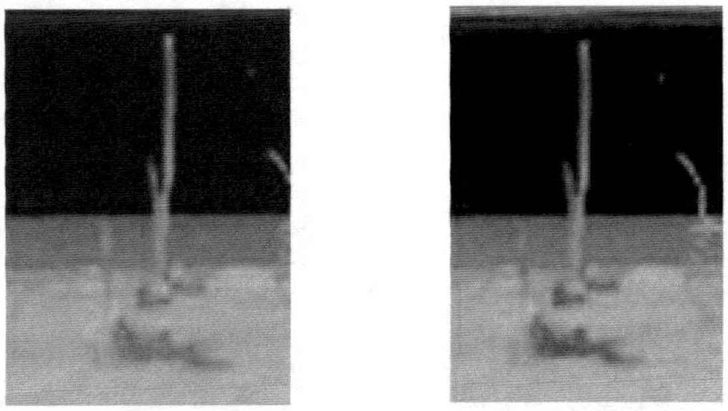

Figure 2. Color image (RGB) changes become grayscale images

The next process is cropping the image to be measured. The following is the result of image cropping that has been done:

Figure 3. Image Cropping Result

The next step after image cropping process is measuring the corn spourt length by the image using region growing method.. Figure 4 shows the results of the calculation process of corn sprout length using region growing:

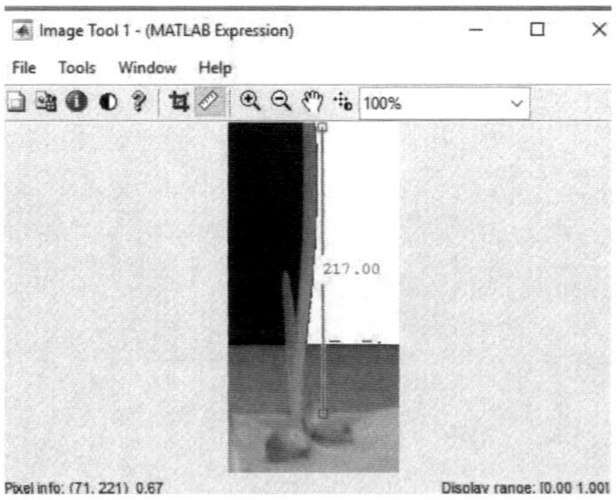

Figure 4. The results of calculations are using region growing method.

After the length growth of the sprouts is known, the next process is, data collection results of measurements of plant length to produce graph data from growth of corn sprouts length for 6 days.

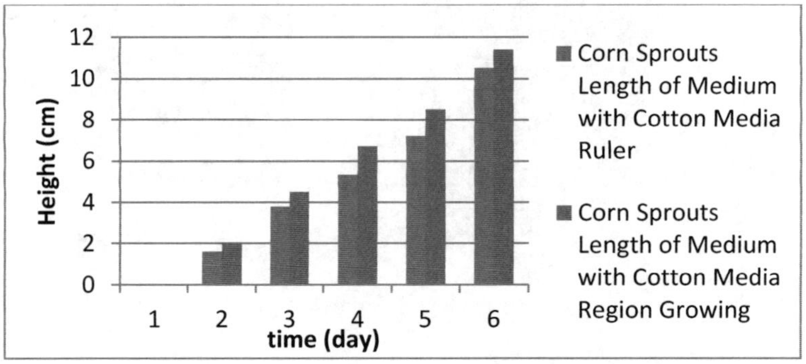

Figure 5. Corn Sprout Growth Chart for 6 Days

Figure 5 is a comparison chart of measurement of germination growth using manual measurements (ruler) which shows the growth of the average length of corn sprouts by 2.1 cm / day and the results of measurements with region growing show the average length of corn sprouts 2.3 cm / day. These results indicate a comparison between manual and region growing measurements is equal to 0.2. The difference between manual calculation and growing region is caused by the accuracy of shooting and image cropping process.

5. Conclusion

Research conducted shows that image processing using the Region growing method can be used to measure the growth of plant. The results of manual measurements show the growth of plant length by an average of 2.1 cm / day then the results of measurements with region growing show an average plant length of 2.3 cm / day, this result shows the comparison between manual and region growing measurements is 0.2 cm. The difference between manual calculation and region growing is caused by the accuracy of shooting and image cropping process.

References

[1] Gusa RF 2013 *Jurnal Nasional Teknik Elektro* **2** 7–34.
[2] Qiang S, Lin G, Ma J, Zhang H 2016 *Proc of the 2016 2nd International Conference on Control Science and Systems Engineering* (New York: IEEE).
[3] Setia A 2010 *Techno.Com* **9** 53–65.
[4] Demilly D, Ducournau S, Wagner M H 2014 *Plant Image Analysis Fundamental and Applications* (Florida: CRC Pres) p. 18.
[5] Soesanti L, Susanto A, Widodo TS and Tjokronagoro M 2010 *Forum Fam. Plan. West. Hemisph.* 33 89–96.
[6] Apriliani D and Murinto 2013 *J. Sarj. Tek. Inform.* **1** 232–240.
[7] Irshad H, Veillard A, Roux L, Racoceanu D 2013 *IEEE Reviews in Biomedical Engineering* **7** 97-114.
[8] Sutiarso L, Suyantohadi A, Kastono D and Nugroho AP **31** 359–367.
[9] Yao H, Duan Q, Li D, Wang J 2013 *Mathematical and Computer Modelling* **58** 790-798.

AUTHOR INDEX

AUTHOR INDEX